Organic Chemistry

MADE SIMPLE

Organic Chemistry

MADE SIMPLE

(Second Edition)

A.K. Srivastava

Professor and Head
Department of Chemistry
Harcourt Butler Technological Institute
Kanpur

PUBLISHING FOR ONE WORLD

NEW AGE INTERNATIONAL (P) LIMITED, PUBLISHERS

New Delhi • Bangalore • Chennai • Guwahati • Hyderabad
Jalandhar • Kolkata • Lucknow • Mumbai • Ranchi

Visit us at **www.newagepublishers.com**

Published by New Age International (P) Ltd., Publishers
First Edition : 2002
Second Edition : 2007
Reprint : 2008

ISBN 10: 81-224-2070-2

Rs. 275.00

C-07-12-2173

2 3 4 5 6 7 8 9 10

Printed in India at Mohanlal Printer, Delhi.
Typeset at Printek India, Delhi.

PUBLISHING FOR ONE WORLD

NEW AGE INTERNATIONAL (P) LIMITED, PUBLISHERS
4835/24, Ansari Road, Daryaganj, New Delhi - 110002
Visit us at **www.newagepublishers.com**

This book is dedicated
To my beloved mother ***(Late) Smt. Sharda Devi.***
She had always provided me her blessings.
I call her mother of love.

Preface to Second Edition

About the Second Edition

I am extremely happy to present 2nd edition of the book. It has been updated with new advances, for example, Chapter 12 "Polymers". The questions asked in various competitive examinations like JEE, AIEEE, UPSEAT, etc, upto 2006 have been added. A new Chapter "Molecular Rearrangement" has also been added. I am sure the revised edition will be more useful to the students.

A.K. Srivastava F. N. A. Sc
(Professor)

Preface to First Edition

This book has been written for students who are preparing for their competitive examinations (Engineering & Medical) with a view to make the fundamentals clear so that they are able to use their knowledge as and when needed. The basic approach is that a proper assimilation and application of knowledge results in good earning of marks.

The book comprises 16 chapters as per syllabi of various Engineering and Medical competitive examinations. Each chapter highlights fundamentals followed by problems of various types like objective questions, fill-in-the blanks, true or false, etc. The solution for most of the problems have been provided. I would be highly obliged to receive comments and suggestions from students and my colleagues.

I acknowledge the help rendered by my wife Dr. (Smt.) Reeta Srivastava and my sons, Er(s) Abhishek, Saurabh & Kumar Tushar. I also acknowledge blessings of my father Sri Raj Narain. In the end, I express my sincere thanks to the staff of New Age International (P) Ltd., Publishers, especially Sri K.K. Gupta, Sri Rajiv Banerji, and Sri V.R. Damodaran.

A.K. Srivastava

Contents

Chapter 1

Hybridization

1.1 INTRODUCTION

The atomic number of carbon is six. Its electronic configuration in the ground state may be written as $1s^2\ 2s^2\ 2p_x^1\ 2p_y^1\ 2p_z^0$. According to this configuration, there are two half-filled orbitals (the volume of the space, where probability of finding an electron is maximum, is termed as an orbital). Therefore, carbon should be bivalent. But, it is an established fact that carbon is tertravalent. Hence, in order to account for tetra-covalency of carbon, it is suggested that one electron jumps from $2s$ orbital to $2p_z$ orbital. The energy required for this jump is 96 kcal/mole, which is rationalized by arguing that the energy released when two additional bonds are formed would more than compensate for that required to excite the electron to a vacant $2p_z$ orbital. Thus, electronic configuration of carbon in the excited state may be written as $1s^2\ 2s^1\ 2p_x^1\ 2p_y^1\ 2p_z^1$. Although this explains well the tetravalency of carbon, but it also predicts that carbon atom forms two types of bonds, three of one type (those formed from three $2p$ orbitals) and 4th of another type (that formed from $2s$ orbitals), since s and p orbitals are different with respect to energy and shape. However, it is a well-established fact that all the four valencies of carbon are equivalent (A simple proof of this fact is that the chlorination of methane gives only one monochloromethane). Therefore, it has been suggested that all the four orbitals mix together to form new orbitals of equivalent energy. This is the basic concept of hybridization.

Hybridization may be defined as the phenomenon of mixing of atomic orbitals of *nearly equivalent* energy, involving redistribution of energy, to form new orbitals of equal energy known as hybrid orbitals. The number of hybrid orbitals is equal to the number of the orbitals hybridized. The properties of the hybrid orbitals are in between the properties of the orbitals which are hybridized.

The number of *p-orbitals*, which take part in hybridization, is called the ***hybridization index*** and is designated by 'm'. The mathematical relationship between hybridization index and the bond angle (α), formed between hybrid orbitals, is as follows:

$$\cos \alpha = -\frac{1}{m}$$

For example, for *sp* hybridization, the value of *m* is 1, therefore cos α = –1, hence α = 180°.

1.2 KINDS OF HYBRIDIZATION

On the basis of hybridization index, we can classify hybridization in three categories:

(a) **sp^3 or Tetrahedral Hybridization**: In this hybridization, one 2*s* and three 2p orbitals take part, resulting in the formation of four sp^3 hybridized orbitals. The wave-mechanical equations, showing the formation of only four sp^3 hybrid orbitals, are as follows:

$$sp_1^3 = \psi 2s + \varphi 2p_x + \varphi 2p_y + \varphi 2p_z$$

$$sp_2^3 = \psi 2s + \varphi 2p_x + \varphi 2p_y - \varphi 2p_z$$

$$sp_3^3 = \psi 2s + \varphi 2p_x - \varphi 2p_y + \varphi 2p_z$$

$$sp_4^3 = \psi 2s - \varphi 2p_x + \varphi 2p_y + \varphi 2p_z$$

where (ψ) [1]is wave function.

The mixing of atomic orbitals, which results in the formation of four sp^3 hybrid orbitals, is shown in Fig. 1.1.

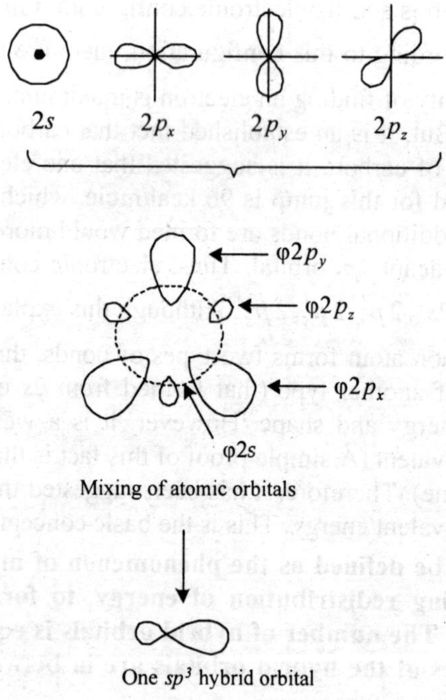

Fig. 1.1. Formation of sp^3 hybrid orbitals

Each sp^3 hybrid orbital has following characteristics:

[1] ψ (psi) is the wave function which gives probability of mixing of atomic orbitals. For details, study the Schrödinger wave equation.

(i) It is *bilobed* in shape, one lobe is large and other lobe is small. The small lobe is neglected in molecular orbital diagrams.

(ii) It has 25% s-orbital character.

(iii) Bond angle between two hybrid orbitals is 109.5 or 109°, 28″.

Thus, each sp^3 hybrid carbon atom has four sigma (σ) bonds or three sigma (σ) bonds & one –ve charge i.e.

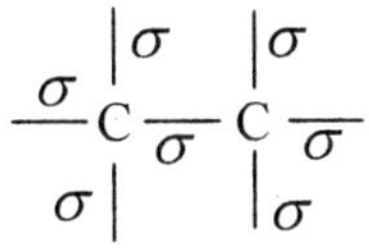

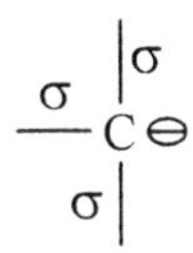

(b) sp^2 or *Trigonal Hybridization*: In this *one* 2*s* and *two* 2*p* orbitals take part in hybridization, resulting in the formation of *three* sp^2 hybrid orbitals. The wave-mechanical equations showing formation of only three sp^2 hybrid orbitals are as follows:

$$sp_1^2 = \psi 2s + \varphi 2p_x + \varphi 2p_y$$

$$sp_2^2 = \psi 2s + \varphi 2p_x - \varphi 2p_y$$

$$sp_3^2 = \psi 2s - \varphi 2p_x + \varphi 2p_y$$

Mixing of three atomic orbitals giving three sp^2 hybrid orbitals is shown in Fig. 1.2.

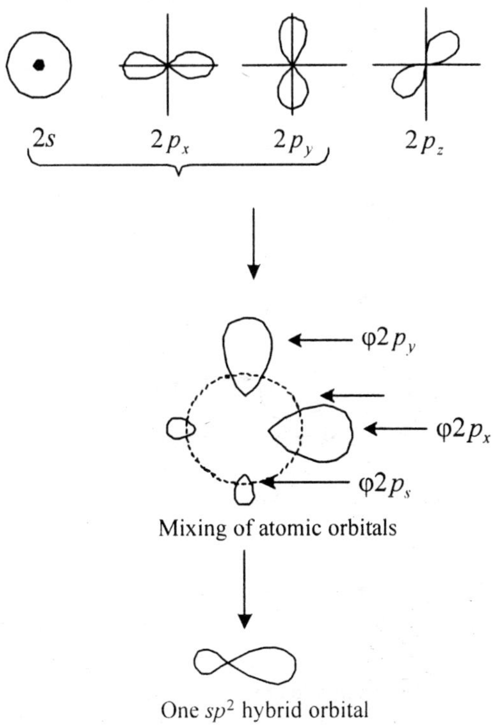

Fig. 1.2 Formation of sp^2 hybrid orbitals

Each sp^2 hybrid orbitals has following characteristics:

(i) Like sp^3 hybrid orbital, it is bilobed, one lobe is large and other is small.

(ii) Unlike sp^3 hybrid orbital, it has *33% s-orbital* character.

(iii) The bond angle between two sp^2 hybrid orbitals is 120°.

Thus, each sp^2 hybridized carbon atom has three *sigma* (σ) and one *pi* (π) bond. A double bond consists of one σ bond and one π bond.

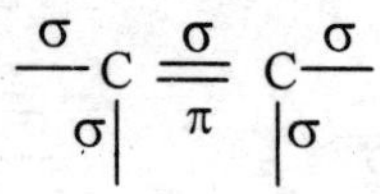

Further, a sp^2 hybridized carbon may have three sigma (σ) bonds and one +ve charge here.

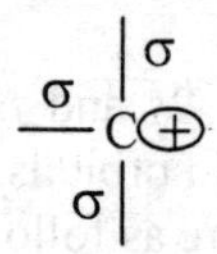

(c) *Sp or Diagonal Hybridization*: In this hybridization, one 2*s* and one 2*p* orbital take part in hybridization, resulting in the formation of two hybrid orbitals known as *sp* hybrid orbitals. The wave-mechanical equations, showing the formation of only two *sp* hybrid orbitals, are as follows:

$$sp_1 = \psi 2s + \varphi 2p_x$$
$$sp_2 = \psi 2s - \varphi 2p_x$$

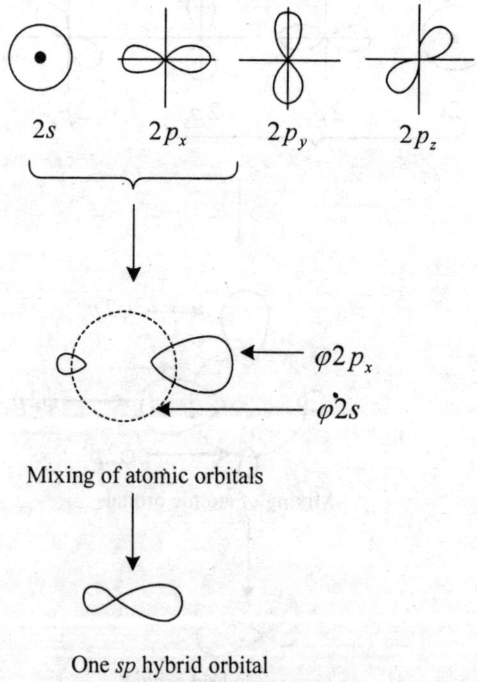

Fig. 1.3 Formation of *sp* hybrid orbital.

Each *sp* hybrid orbital has following characteristics:

(i) Its shape is same as that of sp^3 or sp^2 hybrid orbitals.

(ii) It has *50% s-orbital* character.

(iii) The bond angle between two hybrid orbitals is 180°, i.e., they are linear.

Thus, each *sp* hybrid carbon atom has only 2 sigma (σ) and 2 pi (π) bonds. A triple bond consists one σ bond and two π bonds.

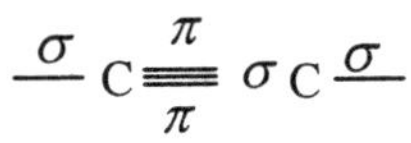

A *sp* hybridized carbon may have double bond with one positive charge or two sigma bonds and two unpaired electrons.

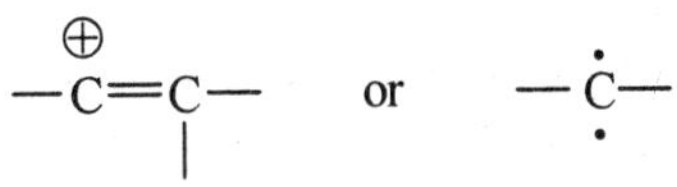

1.3 DIFFERENCES BETWEEN SIGMA (σ) AND PI (π) BONDS

The differences between sigma and pi bonds are as follows:

S. no	*Sigma bond (σ)*	*Pi bond (π)*
1.	Sigma bond is formed by co-axial or linear overlapping of atomic orbitals as shown below:  σ bond	Pi bond is formed by *lateral or parallel or sidewise* overlapping of atomic orbitals as shown below: 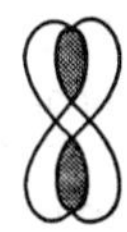 π bond
2.	Sigma bond is formed by overlapping of two hybrid orbitals or one hybrid and one pure orbital, or two pure orbitals.	It is always formed between two pure (i.e., unhybridized) orbitals.
3.	There is one electron-cloud density.	There are two electron-cloud densities, one above the other.
4.	Sigma bond is stronger because of maximum overlapping of atomic orbitals e.g., dissociation energy for C — C bond is 82 kcal/mole.	Pi bond is comparatively weaker because of less overalapping of atomic orbitals, e.g., dissociation energy for C = C bond is 142 kcal/mole and for C ≡ C. it is 198 kcal/mole.
5.	It exists independently.	It always exist along with sigma bond. i.e., it cannot exist independently.
6.	There can be only one sigma bond between two atoms.	There can be more than one pi bond between two atoms.
7.	It allows free rotation of the two carbon atoms.	It restricts free rotation of the carbon atoms.

8.	It never decolourises bromine water and Baeyer's reagent.	In alkene and alkyne it decolourises bromine water and Baeyer's reagent.
9.	It has cylindrical charge symmetry around the bond axis.	No such symmetry is present.

1.4 APPLICATIONS OF THE CONCEPT OF HYBRIDIZATION TO ORGANIC MOLECULES

1.4.1 Variation in Bond Lengths

(a) **Bond length of hydrocarbons**: The C—C bond lengths in hydrocarbons are given in Table 1.1.

As we move from left to right, we find that the bond length decreases, it is because s-orbital character of hybrid orbitals increases from 25% to 50%. *s*-orbital is small and it lies close to the nucleus; greater the percentage of *s*-orbital character, lesser is bond length.

Table 1.1 C—C Bond Length of Hydrocarbons

Hydrocarbon	—C—C— (alkane)	—C=C— (alkene)	—C≡C— (alkyne)
Bond length	1.54 Å	1.34 Å	1.20 Å
Hybridization	sp^3	sp^2	sp
% *s*-orbital character	25%	33%	50%

(b) C—H bond length of hybridization: See Table 1.2 for C—H bond length of hybridization.

Table 1.2 C—H Bond Length of Hybridization

Hydrocarbon	*Alkane*	*Alkene*	*Alkyne*
Bond length	1.11 Å in CH_4 1.10 Å in C_2H_6 and C_3H_8	1.09 Å in C_2H_4 and C_6H_6	1.06 Å in C_2H_2
% *s*-orbital character	25%	33%	50%

The reason is the same that for the C—C length variation as mentioned above.

1.4.2 Acidity of Hydrocarbons

The replacement of hydrogen atom of hydrocarbons by metals is an evidence regarding the acidity of hydrocarbons, e.g., formation of following compounds:

C_4H_9Li Butyl lithium	Ag—C≡C—Ag Silver acetylide
C_2H_5MgBr Ethyl magnesium bromide	$(C_2H_5)_4Pb$ Tetraethyl lead

The replacement of hydrogen atoms of hydrocarbon depends upn the %*s*- orbital character of hybrid orbital. Greater the % *s*-orbital character greater is the acidity of hydrocarbon. Therefore, following order of acidity is observed:

	C_2H_2	>	C_2H_4	>	CH_4	>	C_2H_6
p^{ka}	25		44		48		50

Acetylene forms acetylide because it has two hydrogen atoms which are acidic in nature due to 50% *s*-orbital character. On the contrary, dimethyl acetylene (butyne-2) fails to form acetylide because it has no such acidic hydrogen atom.

In general *ethyne and all alkynes –1 form acetylide* because they have acidic hydrogen (e.g., butyne – 1).

SUMMARY

1. Mixing of atomic orbitals of *nearly* equivalent energy to form new orbitals of equivalent energy is called *hybridization.*
2. Number of *p*-orbitals, which take part in hybridization, is called *hybridization index* (*m*):

$$\cos\alpha = -\frac{1}{m}$$

where α is the bond angle between two hybrid orbitals.

3. $$\left.\begin{array}{l}\text{\% s orbital character}\\ \text{of hybrid orbital}\end{array}\right\} \propto \left\{\begin{array}{c}\text{Acidity of hydrocarbon}\\ \dfrac{1}{\text{Bond length}}\end{array}\right.$$

PROBLEMS

1. What is hybridization? Describe sp^3, sp^2, *sp* hybridization. What is the difference between σ and π bonds? **(Roorkee, 1982)**
2. Discuss variation of bond length in hydrocarbons with C—C, C=C, C≡C. **(MNR, 1983)**
3. Explain why
   (*i*) Bond angle between two *sp* hybrid orbitals is 180°.
   (*ii*) Only three sp^2 hybrid orbitals are formed in hybridization where one 2*s* and two 2*p* orbitals take part.
4. How will you distinguish between Butyne-1 and Butyne-2? **(IIT, 1987; MNR, 1993)**
5. Find the type of hybridization of the underlined C in the following.
   (*i*) $\underline{C}H_3$ (*ii*) $CH_3\,\underline{C}H_2$— (*iii*) CH_2=$\underline{C}$H— (*iv*) CH≡$\underline{C}$—
6. Calculate the total number of σ bonds in *o*-xylene **(MNR, 1996)**
7. For each question given below, four answers are provided, out of which only one is correct. Write the correct answer giving proper reasoning:
   (*i*) 1-2 butadiene has
      (*a*) Only sp^3 hybridized carbon atom
      (*b*) sp^3, sp^2 hybridized carbon atom

(c) sp^3, sp^2 and sp hybridized carbon atom

(d) sp and sp^2 hybridized carbon atom. **(IIT, 1983)**

(ii) Ketene ($CH_2C=C=O$) has:

(a) Only sp^2 hybridized carbon atom

(b) sp^2 and sp hybridized carbon atom

(c) Only sp hybridized carbon atom

(d) Only sp^3 **(IIT, 1983)**

(iii) The C—C Bond distance is maximum in:

(a) Ethane (b) Ethylene

(c) Acetylene (d) Benzene **(IIT, 1983)**

(iv) Mycomycin is an antibiotic. Which of the following correctly describe the bond hybridization at carbon atoms: a—g?

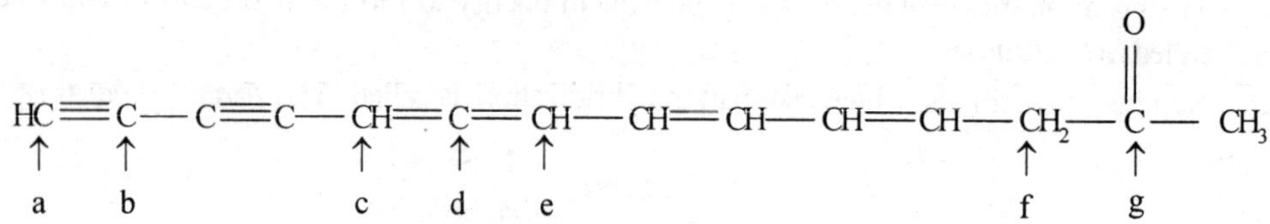

	a	b	c	d	e	f	g
(a)	sp^2	sp^2	sp	sp	sp	sp^2	sp
(b)	sp	sp	sp^2	sp	sp^2	sp^3	sp^2
(c)	sp^2	sp^2	sp^3	sp^2	sp^3	sp^3	sp^3
(d)	sp	sp	sp^2	sp^2	sp^2	sp^3	sp^2

(v) Which among the following is the correct order for increasing C—H bond length:

(a) Methane > Ethane > Ethene > Ethyne

(b) Ethyne > Ethene > Ethane > Methane

(c) Ethyne > Ethene > Methane > Ethane

(d) Ethane > Ethyne > Ethane > Methane **(GATE, 1988)**

(vi) The C—H bond distance is the longest in:

(a) C_2H_2 (b) C_2H_4 (c) $C_2H_2Br_2$

(d) C_6H_6 (e) CH_4 **(MNR, 1990)**

(vii) Which of the following would give precipitate with Ammoniacal $AgNO_3$?

(a) Butyne – 1 (b) Butyne – 2 (c) Butene – 1

(d) Both Butyne – 1 and Butyne – 2

(viii) Acidic H is present in:

(a) C_2H_6 (b) C_2H_4 (c) C_2H_2 (d) C_6H_6 **(IIT, 1985)**

(ix) The bond between the carbon atoms number 1 & 2 in the compound $N\equiv \underset{1}{C}-\underset{2}{CH}=CH_2$ involves the hybrid orbital:

(a) sp^2, sp^2 (b) sp, sp^2 (c) sp^3, sp (d) sp, sp **(IIT, 1987)**

(x) The ionic form of acetone contains:

(a) 9σ, 1π, 2 lone pairs (b) 8σ, 1π, 2 lone pairs
(c) 10 σ, 1π, 1 lone pair (d) 9σ, 1π, 1 lone pair **(IIT, 1990)**

(xi) Which of the following has maximum C—C bond length:
(a) Benzene (b) Ethane (c) Propylene (d) 1,3-Butadiene

(xii) The number of σ and π bonds in 1-butene, 3-yne:
(a) 5 + 5 (b) 7 + 3 (c) 8 + 2 (d) 6 + 4 **(IIT, 1989)**

(xiii) The compound in which carbon uses its sp^3 hybrid oribital for bond formation:
(a) HCOOH (b) NH_2CONH_2
(c) $(CH_3)_3COH$ (d) CH_3CHO **(IIT, 1989)**

(xiv) On hybridization of one *s* and one *p* orbitals we get:
(a) Two mutually-hybrid orbitals
(b) Two orbitals at 180°
(c) Hybrid orbitals directed tetrahedrally
(d) Three orbitals in plane **(IIT, 1984)**

(xv) % *s*-orbital character between two orbitals having bond angle 105° is:
(a) 20% (b) 25% (c) 20.57% (d) 33.33%

(xvi) The compound having more than one type of hybridized carbon atom is:
(a) 1, 2 Butadiene (b) 1, 3 Butadiene (c) Butane (d) Benzene

(xvii) Which of the following compounds given below have more than one kind of hybridization for carbon:
(a) $CH_3CH_2CH_2CH_3$ (b) $CH_3—CH{=}CH—CH_3$
(c) $CH_2{=}CH—CH{=}CH_2$ (d) $HC{\equiv}CH$

(xviii) In the following reaction, the hybridization state of C* atom changes from:

$$CH3—C^*ONH_2 \xrightarrow{P_2O_5} —CH_3C^*N$$

(a) sp^3 to sp^2 (b) sp^2 to sp (c) sp to sp (d) sp to sp^2

(xix) In which of the following would you expect C—C bond distance to be shortest?
(a) Benzene (b) Acetic acid (c) Acetylene (d) Diamond

(xx) In the series of ethane, ethylene, and acetylene the C—C bond energy is:
(a) Greatest in C_2H_6 (b) Greatest in C_2H_4
(c) Greatest in C_2H_2 (d) Same in all the compounds.

(xxi) The Cl—C—Cl angle in 1,1,2,2-tetrachloro ethylene and tetrachloro ethane, respectively will be about:
(a) 120° & 109.5° (b) 90° & 109.5°
(c) 109.5° & 90° (d) 109.5° & 120°. **(IIT, 1988)**

(xxii) Which of the following is is isostructural with CO_2:
(a) C_2H_2 (b) SO_2 (c) CH_4 (d) C_2H_2 **(IIT, 1988)**

(xxiii) The number and types of bonds between two carbon atoms in CaC_2 are:
(a) One σ & one π bonds (b) One σ & two π bonds
(c) One σ & a half π bonds (d) One σ bonds **(IIT, 1996)**

(xxiv) The hybridization state of carbons in allene is:

(a) sp^2, sp^2 *(b)* sp, sp *(c)* sp^2 & sp *(d)* sp^3 & sp^2 **(IIT, 1999)**

(xxv) In the compound, the $CH_2 = \overset{2}{C}H - \overset{3}{C}H_2 - CH_2 - C \equiv CH$, $C_2 - C_3$ bond is of the type

(a) $sp—sp^2$ *(b)* $sp^3—sp^3$

(c) $sp—sp^3$ *(d)* $sp^2—sp^3$ **(IIT, 1999)**

(xxvi) Identify a reagent from the following list which can easily distinguish between 1-butyne, 2-butyne **(IIT 2002)**

(a) Br_2, CCl_4 *(b)* H_2, Lindlar catalyst *(c)* Dil. H_2So_4, $HgSo_4$

(d) ammoniacal CCl_2 Cl_2 Solution

(xxvii) The nodal plane in the π bond of ethene is located in **(IIT 2002)**

(a) the molecular plane

(b) a plane parallel to the molecular plane

(c) a plane perpendicular to the molecular plane which bisects the carbon-carbon σ bond at right angle

(e) A plane perpendicular to the molecular plane which contains the carbon-carbon σ – bond.

(xxviii) Which of the following reagent can be used to distinguish butyne-1 & butyne-2 **(IIT, 2003)**

(a) Fehling solution *(b)* Grignard reagent

(c) Ammonical cuprous chloride *(d)* Tollen's reagent

8. Fill in the blanks:

(i) Nitrogen molecule has ……..bonds **(IIT, 1981)**

(ii) Benzene molecule has ………bonds **(IIT, 1982)**

(iii) In sp hybridization the two hybrid orbitals are **(IIT, 1980)**

(iv) Maximum number of σ bonds between two atom is

9. State whether the following statements are true or false. If false, then write the correct statement:

(i) The properties of hybrid orbitals are in between those of the hybridized orbitals.

(ii) All alkynes form acetylide.

(iii) Sigma bond is stronger than pi bonds.

(iv) Total number of sigma and pi bonds in $C_2(CN)_4$ is nine. **(CPMT, MP, 1986)**

(v) If the bond angle BAB in the molecule AB_2 is 180°, sp^2 hybridization would account for the shape.

10. Match the items from column A with those in column B:

	A		*B*
(i)	sp^3 hybrid orbitals	(i)	Pi bonds
(ii)	Sigma bond	(ii)	bond length
(iii)	Two electron cloud densities	(iii)	Linear overlapping
(iv)	%*s*-orbital character	(iv)	109.5°
(v)	Zero dipole moment	(v)	Acetylene

11. Thought type question:

The concept of hybridization is used to account for tetracovalency of the carbon atom. The organic compounds have tetrahedral, trigonal & diagonal hybridization. The carbon atom has two kinds of bonds, namely, Sigma & pi which are formed by linear & parallel ovarlaping of atomic orbitals respectively. The hydrocarbons are acidic in nature & alkynes with acidic H atom form acetylide.

Read the above passage carefully & answer the following questions. There may be more than one correct answer:

(i) The number of sigma bonds in sp^3, sp^2& sp hybridized carbon atoms(s) is/are:

a) 4, 3, 2 b) 2, 3, 4 c) 3, 4, 2 d) 2, 4, 3

(ii) The acidic nature of hydrocarbon is confirmed by formation of

a) Organometallic compounds b) haloalkanes

c) Aldehydes & Ketones d) Alcohols

(iii) The correct statement is:

a) Sigma bond is stronger than pi bond

b) Methane is more acidic than ethane

c) The bond dissociation energy of C-C is 82kcals mol^{-1}

d) The pKa value of acetylene is 25.

ANSWERS

4. Butyne-1 forms silver acetylide (white precipitate) whereas butyne-2 does not forms it.

5. (i) sp3 (ii) sp3 (iii) sp2 (iv) sp

6. 18; The structure of o-xylene is

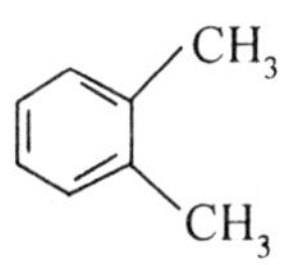

7. (i) **(c)**

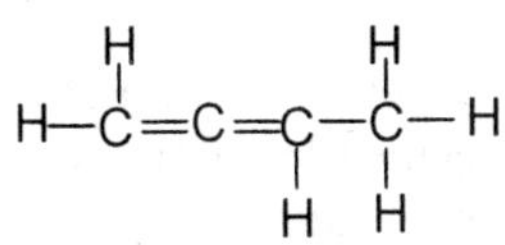

$$\begin{array}{cccc} 3\sigma & 2\sigma & 3\sigma & 4\sigma \\ 1\pi & 2\pi & 1\pi & \\ \hline sp^2 & sp & sp^2 & sp^3 \end{array}$$

(***Hint***: Write expanded structure of compound and count total number of σ and π bonds with each carbon atom. Assign hybridization state of carbon atom accordingly.)

(ii) **(b)**

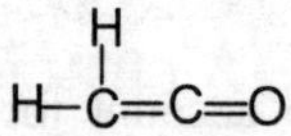

3σ 2σ
1π 2π
sp^2 sp

(iii) **(a)** See Table 1.1.

(iv) **(b)** See above problem (ii).

(v) **(a)** See Table 1.2.

(vi) **(c).**

(vii) **(a)** because it contains one acid: H atom.

(viii) **(c).**

(ix) **(b)** C(1) contains 2π & 2σ and C(2) 1π and 3σ bonds.

(x) **(a)**

(xi) **(b)** See Table 1.1

(xii) **(b)**

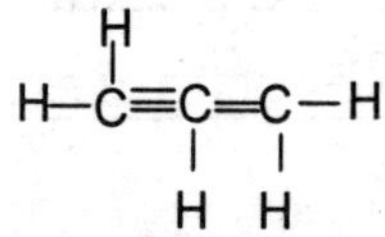

(xiii) **(c).**

(xiv) **(b)**; bond angle between two sp hybrid orbitals is 180°

(xv) **(c)**; $\cos\alpha = -\frac{1}{m}$, since $\alpha = 105°$

$$m = 3.84 \ so \ \%s = \frac{1}{4.84} = 20.57\%$$

(xvi) **(a)**, it contains sp^3, sp^2 & sp hybrid carbon atoms

Hint: $CH_2{=}C{=}CH{-}CH_3$, $H_2C{=}CH{-}C{=}CH_2$
sp^2 sp sp^2 sp^3 sp^2 sp^2 sp^2 sp^2
1,2-Butadiene 1,3-Butadiene
$CH_3{-}CH_2{-}CH_2{-}CH_3$
sp^3 sp^3 sp^3 sp^3
Butane

(xvii) **(b)**; (a) contains only hybrid carbon; (b) contains sp^3 & sp^2 hybrid C; (c) contains only sp^2; and (d) contains only sp hybrid carbon. See below:

$CH_3CH_2CH_2CH_3$
sp^3 *sp*^{3} *sp*^{3} *sp*^{3}

$CH_3-CH=CH-6CH_3$
sp^3 *sp*^{2} *sp*^{2} *sp*^{3}
Butene-2

$CH_2=CH-CH=CH_2$
sp^2 *sp* *sp* *sp*^{2}
1,3-Butadiene

$HC\equiv CH$
sp *sp*
Acetylene

(xviii) **(b)**

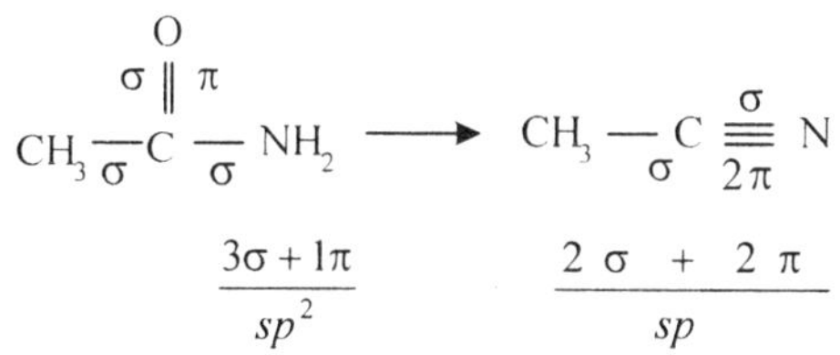

(xix) **(c)**; because acetylene contains *sp* hybrid carbon atoms; whereas benzene contains *sp*^{2}; acetic acid *sp*^{3} & *sp*^{2} and diamond only *sp*^{3}.

(xx) **(c)**; because the bond dissociation energy for C≡C is about 198 kcal/mole, C=C is 144 kcal/mole & C—C is 82 kcal/mole.

(xxi) **(a)**; the former has *sp*^{2} hybridization whereas latter has *sp*^{3} hybridization.

(xxii) **(a)**; because both CO_2 and C_2H_2 contains *sp* hybrid carbon atom.

(xxiii) One sigma and two pi bonds.

(xxiv) **(a)**.

(xxv) **(d)**.

(xxvi) **(d)**

(xxvii) **(c)**

(xxviii) **(c)**

8. (i) 2π & 1σ bonds (ii) 3π & 12σ bonds
(iii) 180° (iv) one.

9. (i) True.
(ii) False. All alkene-1 and ethyne forms acetylide because of presence of acidic H atom(s).
(iii) True.

Hint:

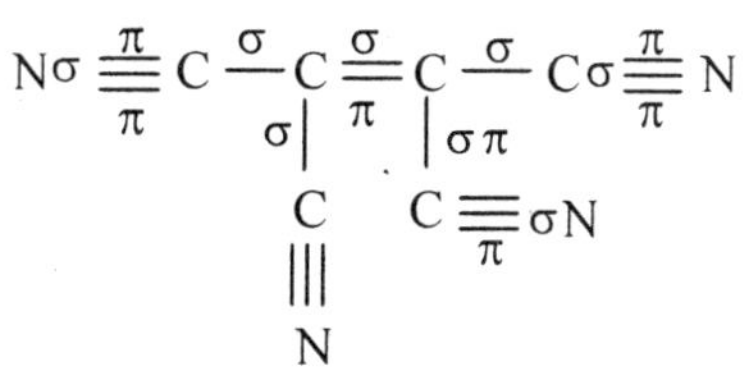

(iv) True,
(v) F alse. *sp* hybridization will account for the shape.

10.

	A		*B*
(i)	sp^3 hybrid orbitals	(iv)	109.5°
(ii)	Sigma bond	(iii)	Linear overlapping
(iii)	Two electron cloud densities	(i)	Pi bond
(iv)	%*s*-orbital character	(ii)	Bond length
(v)	Zero dipole moment	(v)	Acetylene

11 (i) **(a)**

(ii) **(a)**

(iii) **(a, b, c, d)**

Chapter 2

Fission of Covalent Bonds, Organic Reagents and Lewis Dot Formula

2.1 FISSION OF A COVALENT BOND

Let us consider a hypothetical covalent molecule A : B. Its covalent bond may be broken in two ways.

(i) *Homolysis or Homolytic Fission* (Greek: *homo-*, the same + lysis, cleavage) In homolysis, the covalent bond breaks in such a way that each resulting species get its own electron. The neutral fragments A. and B. are called free radicals

$$A : B \xrightarrow{\text{Homolytic Fission}} \dot{A} + \dot{B}$$

Free Radicals

e.g.,

$$CH_3 : Br \xrightarrow{\text{Homolytic Fission}} \dot{C}H_3 + \dot{Br}$$

Bond dissociation energy (DH°) = 293 kcal/mole

(ii) *Heterolysis or Heterolytic Fission* (Greek: hetero-, different+lysis, cleavage). This involves fission of the covalent bond in such a way that one species is deprived-off its own electron while the other species which is more electronegative gets both electrons.

$$A : B \xrightarrow{\text{Heterolytic Fission}} \underbrace{A^{+} + B^{-} \text{ or } A^{-} + B^{+}}_{\text{ions}}$$

e.g.,

$$CH_3 : Br \xrightarrow{\text{Heterolytic Fission}} \overset{\oplus}{C}H_3 + Br^{-}$$

Methyl Carbonium Ion (Since bromine is more electronegative than carbon)

It is clear that energy required to carry out heterolytic fission is always greater than that for homolytic fission. The reason seems to be the electrostatic force of attraction between the ions.

The homolytic bond dissociation energies[1] (DH°) of a variety of covalent bonds are shown in Table 2.1.

Table 2.1 Single bond homolytic dissociation energies at 25°

Bond broken	*Example*	*DH° (kj/mol^{-1})*
C—H	CH_4	435
C—C	C_2H_6	342
C—F	CH_3F	452
C—Br	CH_3Br	293
C—I	CH_3I	234
C—OH	CH_3OH	383
C—OCH_3	$C_2H_5OCH_3$	335
C—H	HC≡CH	523
O—O	$CH_3CH_2OOCH_3$	184

2.2 ORGANIC REAGENTS

Organic reagents are of two types:

A. Electrophiles (*Electro,* Electron + philic, loving) *or Lewis acids*: They are species (molecules or ions) that are short of electrons and, therefore, can accept an electron pair. In nutshell electron-aceptors are electrophiles. Electrophiles include:

(i) A positive charge ion, the proton (H^+), methyl carbonium ion

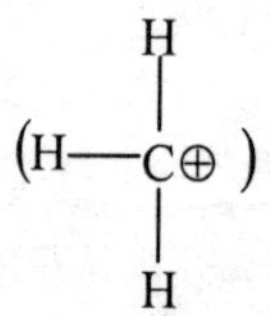

(ii) Neutral reagents with the central atom having an incomplete octet and an orbital to accept the electron pair, e.g., $AlCl_3$ and BF_3

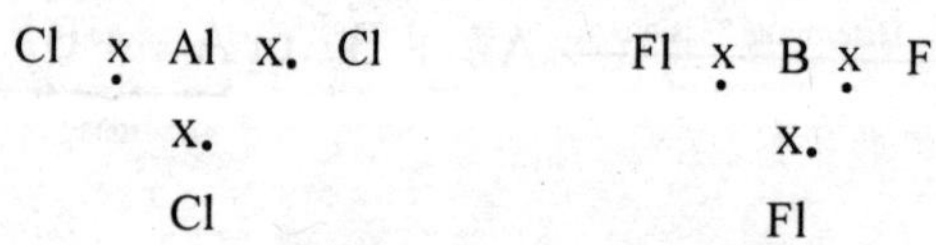

In all there are six electrons, hence there is a shortage of two electrons.

(iii) A neutral molecule which can be polarized, e.g., bromine.

(iv) Metal ion that contain vacant orbitals, e.g., silver ion (Ag^+), mercuric ion $\left(Hg_2^+\right)$.

---

1 Bond dissociation energies are usually abbreviated by the symbol (DH°).

In 1967, R.G. Pearson classified electrophiles in two categories, namely, "Hard Electrophiles" and "Soft Electrophiles". Although, there is no sharp difference between the two, both of them can be distinguished on the basis of following characteristics:

(a) Hard Electrophiles: They are characterized by small size, high positive charge and their valence electrons cannot be easily excited, e.g., H^+, Li^+, Na^+, K^+, Mg^{++}, Ca^{++}, Al^{3+}, Cr^{3+}, Fe^{3+}, BF_3, $AlCl_3$, SO_3, CO_2, $AlMe_3$, RCO, $B(OR)_3$.

(b) Soft Electrophiles: They are characterized by large size, low positive charge and their valence electrons can be easily excited, e.g., Cu^+, Hg^{++}, Pd^{++}, Br_2, CH_2 (Carbenes), I_2.

B. Nucleophiles or Lewis bases (Nucleo-, nucleus+, phile-, loving): Nucleophiles are species (molecules or ions) which are electron rich and, therefore, can donate electron pair. In nutshell *electron-donors are nucleophiles*. A nucleophile may carry:

(i) A negative charge, e.g., the hydride ion (H^-), carbonion ($-\overset{|}{\underset{|}{C}}^{\ominus}$)

(ii) π electrons, e.g., ethylene ($CH_2{=}CH_2$), benzene, acetylene ($HC{\equiv}CH$) exception: CO_2, SO_3.

(iii) A lone pair of electrons, e.g., water ($H_2\ddot{\underset{\cdot\cdot}{O}}$), alcohol ($R-\ddot{\underset{\cdot\cdot}{O}}-H$) ammonia ($\ddot{N}H_3$), amines ($R\ddot{N}H_2$)

In 1967, R.G. Pearson classified the nucleophiles in two categories:

(a) *Hard Nucleophiles*: The donor atoms are of high electronegativity and low polarizability and are difficult to be oxidized. They hold their valence electrons tightly, e.g.,

$$H_2\ddot{O}:,\ :\ddot{\underset{\cdot\cdot}{O}}H^-,\ :\ddot{\underset{\cdot\cdot}{F}}:^-,\ \ddot{N}H_3,\ R\ddot{N}H_2,\ :\ddot{\underset{\cdot\cdot}{Cl}}:^-,\ NO_3^-,\ CO_3^{--}$$

(b) *Soft Nucleophiles*: They are characterized by low electronegativity of donor atom and high polarizability. They are easy to be oxidized. The valence electrons are loosely held, e.g., I^-, RSH, RS^-, CN^-, CO, C_2H_4, C_6H_6, H^-, R^-, RMgX.

HSAB Principle: The rule states that the hard electrophiles prefer to bond to hard nucleophiles, and the soft electrophiles prefer to bond to soft nucleophiles. The rule is not a theory, but a generalization based on experimental facts. One application of the rule is observed in complexes between alkenes or aromatic compounds and metal ions. Alkenes and aromatic rings are soft nucleophiles and, therefore, prefer to complex with soft acids. Thus, Na^+, Mg^2 +, or Al^3 + complexes are rare but Ag^+, Hg^{2+} complexes are common.

Note: (1) H_2, CH_4, $(CH_3)_4N^{\oplus}$ are neither electrophiles nor nucleophiles.

2.3 LEWIS STRUCTURES (ELECTRON DOT FORMULAE)

Following rules should be observed in writing Lewis structures:

Rule 1: Element : H, C, N, O, X
Valence electrons : 1, 4, 4, 6, 7

Rule 2: Single bond (—) means 2 electrons: double bond (=) means 4 electrons, triple bond (≡) means 6 electrons.

Rule 3: Sum up the total number of valence electrons of all the atoms. If the structure is an ion, we add (in case of anion) or subtract (in case of cation) electrons equal to the charges, e.g.,

$$\begin{array}{ccc} C\cdot & H_2 & O \\ 4 & +\ 2(1)\ + & 6 \end{array}$$
$$= 12 \text{ electrons}$$

Rule.4: Write expanded structure of species (molecule or ion) keeping valency of element in mind, e.g.,

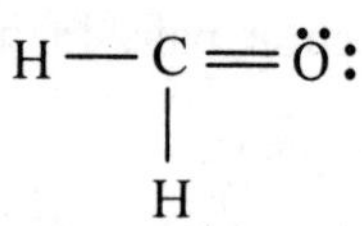 (Carbon is tetravalent, Hydrogen is univalent and Oxygen is bivalent)

Rule.5: Complete octet of each element, keeping total number of electrons in mind. In case of shortage of electrons, give preference to more electronegative element to complete octet (Please see example of NO_2).

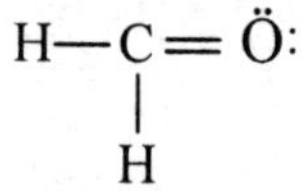

Molecules are neutral. Therefore, the sum of the formal charges on each atom making up a molecule must be zero.

Rule 6: Calculate formal charge on each atom with the help of following formula:

$$\text{Formal charge} = \text{Core charge} - \text{Electron ownership}$$
$$= \text{Number of valence electrons} - \left(\frac{1}{2}\text{ Shared electrons} + \text{unshared electrons}\right)$$

∴Formal charge for both the H atoms $= 1-(\frac{1}{2}\times 2+0) = 0$

for the carbon atom $= 4-(\frac{8}{2}+0) = 0$

for the oxygen atom $= 6-(\frac{4}{2}+4) = 0.$

Thus, there is no formal charge on any atom of formaldehyde. Therefore, the Lewis is structure of formaldehyde is as follows:

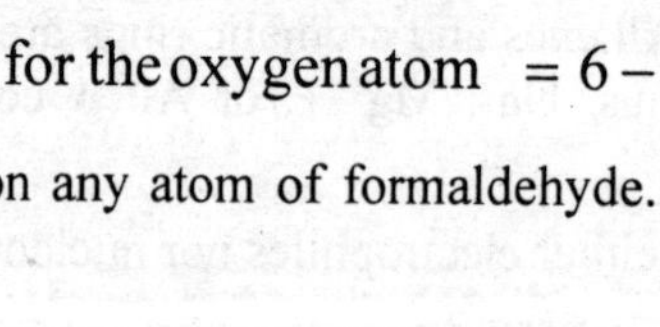

Rule 7: The arithmetic sum of all the formal charges gives the charge on the species as a whole.

Some other examples:

1. Write the Lewis structure of nitrogen peroxide (NO_2)

Rule 3: Total electrons in $NO_2 = 5 + 2\,(6) = 17$

Rule 4: O–N = 0

Rule 5: $:\ddot{O} - \dot{N} = \ddot{O}:$

Rule 6: F.C. on the left oxygen atom $= 6 - \left(\frac{2}{2} + 6\right) = -1$

F.C. on the nitrogen atom $= 5 - \left(\frac{6}{2} + 1\right) = +1$

F.C. on the right oxygen atom $= 6 - \left(\frac{4}{2} + 4\right) = 0$

Therefore, Lewis structure of NO_2 is as follows:

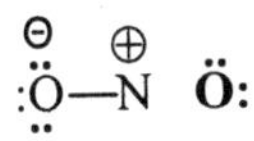

Thus, left oxygen has –1 and nitrogen has +1 formal charge.

2. Write the Lewis structure of Methylcarbonium $^{\oplus}CH_3$

Rule 3: Total electrons $C + H_3\ (4 + 3(1) - 1) = 6$

Rule 4:

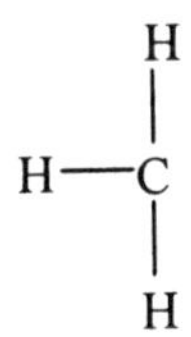

Rule 5:

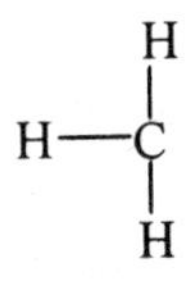

Hydrogen atoms have two electrons, but carbon has six electrons, so it has incomplete octet. We cannot complete its octet since total available electrons are six).

Formal Charge (F.C.)

Rule 6: F.C. on hydrogen atoms(s) $= 1 - \left(\frac{2}{2} + 0\right) = 0$

F.C. on carbon atoms $= 4 - \left(\frac{6}{2} + 0\right) = +1$

Therefore, Lewis structure of methyl carbonium ion is as follows:

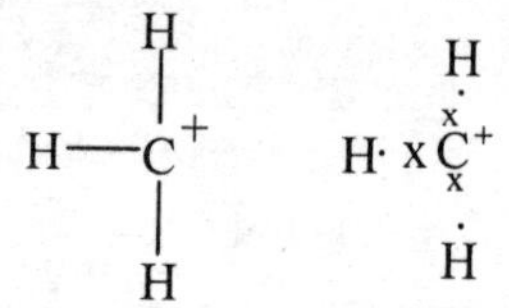

SUMMARY

1. Homolysis and heterolysis of a covalent bond produces free radicals and ions respectively.
2. As we go down a group, the bond dissociation energy decreases.
3. Electron acceptors are Lewis acid or electrophiles. They carry incomplete octet and orbital to accept the lone pair of electrons.
4. Electron donors are Lewis base or nucleophiles. They may carry negative charge or lone pair of electrons or ∏ electrons (exception CO_2 & SO_3).
5. H_2 and $(CH_3)_4 N^{\oplus}$ is neither nucleophile nor electrophile.
6. HCHO & CH_3CN acts as electrophile & nucleophile.
7. Greater the difference in electronegativity, ease of clearage of bonds is greater.

PROBLEMS

1. Amongst the four alternate answers given for the subquestions below, one or more are correct. Write down the correct answers.

(i) The compound that is not a Lewis acid is:

(a) BF_3 *(b)* $AlCl_3$ *(c)* $BeCl_2$ *(d)* $SnCl_4$ **(IIT, 1985)**

(ii) Which of the following is not a nucleophile:

(a) OH^- *(b)* NH_3 *(c)* BF_3 *(d)* HSO_3^-

(iii) In CH_3, CH_2OH, the bond that undergoes heterolytic cleavage most readily is:

(a) C—C *(b)* C—O *(c)* C—H *(d)* O—H **(IIT, 1988)**

(iv) Which of the following act neither as an electrophile nor as a nucleophile:

(a) CH_4 *(b)* H_2 *(c)* NH_3 *(d)* $AlCl_3$

(v) Which of the following acts as an electrophile as well as a nucleophile:

(a) CH_4 *(b)* NH_3 *(c)* BF_3 *(d)* CH_2O

(vi) Which of the following is not a nucleophile:

(a) O=C=O *(b)* R—C≡C– *(c)* Br_2 *(d)* SO_3

(vii) Which of the following statement is correct for NO_2 molecule:

(a) There is no formal charge on any atom

(b) Nitrogen atom carries +1 formal charge and one of the two oxygen atom carries –1 formal charge

(c) Nitrogen atom carries –1 formal charge and one of the two oxygen atom caries +1 formal charge

(d) Nitrogen atom carries +1 formal charge and both the oxygen atom carry –1 formal charge

2. State whether the following statements are true or false. If false, write the correct answer with proper reasoning. Simple negation is not an acceptable answer.

(i) Iodide ion is a better nucleophile than bromide ion. **(IIT, 1985)**

(ii) Energy required to carry out homolytic fission is more than that required for heterolytic fission.

3. Fill in the blanks with suitable formulae, symbol names:

(i) Carbonium ion is ....

(ii) It is ....to oxidize hard nucleophile.

(iii) Grignard reagent act as a nucleophile due to ....

(iv) Hard electrophile prefer to bond to ....

4. Calculate the formal charge on each bold atoms of the following molecule/ions:

(i) $\overset{\ominus}{\mathbf{C}}H_3$ (a carbon ion)

(ii) $\dot{\mathbf{C}}H_3$ (methyl radical)

(iii) $:\mathbf{C}H_2$ (a carbene)

(iv) $CH_3-\mathbf{C}\equiv C^-$ (propynide ion)

(v) 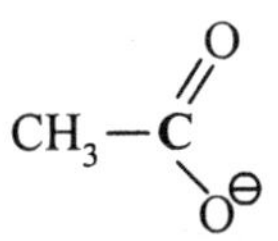

(a carboxyl ion)

(vi) 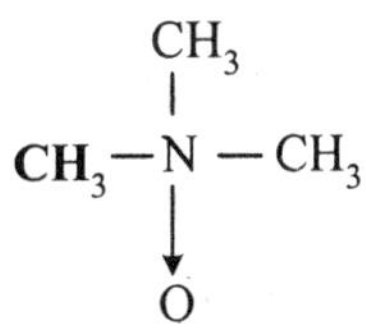

(a amine oxide)

(vii) 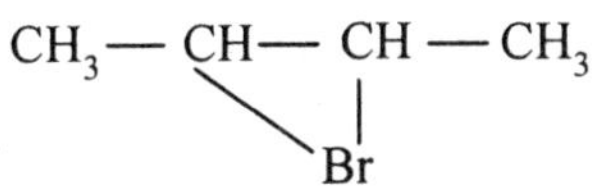

(Dimethl cyclopropyl bromide)

(viii) $\mathbf{C}H_3Cl$ (Methyl chloride)

(ix) $CH_3-C\equiv N$ (Methyl cyanide)

(x) $CH_2=\mathbf{C}=O$ (Ketene)

5. In each of the following reactions, identify each reagent as a nucleophile or electrophile:

(i) $CH_3I + C_6H_5NH_2 \rightarrow C_6H_5N + H_2CH_3 + I^-$

(ii) $CH_3OH + C_6H_5CH_2Cl \rightarrow C_6H_5CH_2OCH_3 + HCl$

(iii) $BF_3 + CH_3OCH_3 \rightarrow (CH_3)_2O^+ + BF_3^-$

(iv) $Br_2 + Br^- \rightarrow Br_3^-$

6. Match the items from column A with those in column B:

	A		B
(i)	NH_4	(i)	Soft electrophile
(ii)	BF_3	(ii)	Hard electrophile
(iii)	Carbene	(iii)	Lewis acid
(iv)	H^+	(iv)	Electrophile
(v)	Carbanion	(v)	Hard Nucleophile
(vi)	Acetylide ion	(vi)	Nucleophile
(vii)	H_2O	(vii)	Lewis base
(viii)	R.G. Pearson	(viii)	Organic chemist who classified electrophiles and Nucleophiles

ANSWERS

1. *(i)* **(c)**

(ii) **(c)**; It is because (a) and (d) carry negative charges and (b) has lone pair of electrons; which is the characteristics of nucleophiles.

(iii) **(d)**; See the bond dissociation energy given in Table 2.1.

electronegativity difference between atoms $\propto$ ease of bond fission

(iv) **(a) & (b)**; It is because in CH_4 and H_2, the atoms are neither short of electrons nor have excess electrons. NH_3 is a nucleophile and $AlCl_3$ is an electrophile.

(v) **(d)**; It is because it acts as nucleophile due to presence of negative charge on oxygen atom and as electrophile due to positive charge on carbon atom.

(vi) **(a)**, **(c)**, **(d)**; See electrophiles.

(vii) **(b)**; See Lewis structure of NO_2

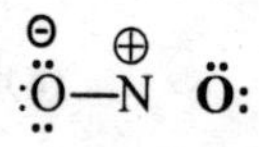

2. *(i)* True.

(ii) False. The energy needed to carry out heterolytic fission us more than that for hemolytic fission. See Table 2.1.

3. *(i)* An electrophile

(ii) Difficult

(iii) Alkyl group (R)

(iv) Hard nucleophile.

4. *(i)* –1 F.C, on C

(ii) No charge

(iii) No charge

(iv) –1 F.C. on right carbon

It has been mentioned in Table 3.2 that the inductive effect propagates through the carbon chain. However, the effect decreases as the distance between the atom or group having –IE and side chain increases (See Table 3.3).

Table 3.3 Influence of side chain and –IE on acid strength

	Acid	pKa	
–IE decreases	$CH_3CH_2CH(Cl)COOH$	2.84	Acid strength decreases
	$CH_3CH(Cl)CH_2COOH$	4.06	
	$CH_2(Cl)CH_2CH_2COOH$	4.52	

(iv) ***Influence on Basic Strength of Amines***

We have already discussed in Chapter 2 that the Lewis base are the species which donate electrons. The ammonia is a base because nitrogen atom of ammonia (NH_3) has a lone pair of electron, which it can donate. The effect of inductive effect on the basicity of amines is discussed in the following steps:

(a) When one or more hydrogen atom(s) of ammonia are replaced by the alkyl group, the electron donating power of nitrogen atom increases due to +IE (electron repelling nature of alkyl group).

Therefore, basicity of aliphatic amines increases in comparison to ammonia. However, in case of aromatic amines the electron donating power of nitrogen atom decreases due to –IE of the phenyl group; therefore the aromatic amines are less basic (see resonance) than ammonia. In a nutshell, the order of basicity is:

CH_3-NH_2	>	NH_3	>	$C_6H_5-NH_2$
methyl amine		ammonia		aniline
(aliphatic amine)				(aromatic amine)
(pKa = 10.64)		(pKa = 9.25)		(pKa = 4.62)

(b) In the gas phase, the basicities of the aliphatic amine increase in the following order:

$$(CH_3)_3N > (CH_3)_2NH > CH_3NH_2 > NH_3$$

The order can be explained on the basis of +IE of alkyl group, which stabilizes the alkyl aminium ion to a greater extent in comparison to the amine. However, in aqueous solution the order is:[1]

$$(CH_3)_2NH > CH_3NH_2 > (CH_3)_3N > NH_3$$

Therefore, the order of basicity can be easily remembered by the formula:

[1] *Pka* = - log ka; greater the *pka*, greater is the base strength.

213 $^{*}NH_3$

$\longleftarrow$

Increasing order of basicity in aqueous solution

The exact reason for this order of basicity is not known. The most probable reason is that solvation, through hydrogen bonding, stabilizes the aminium ions of secondary and primary amines more effectively than those formed from tertiary amines in aqueous solution. The tertiary amine is less basic than primary amine due to steric hindrance.

(c) The aryl amines like aniline are much weaker bases than the corresponding non-aromatic amine like cyclohexylamine:

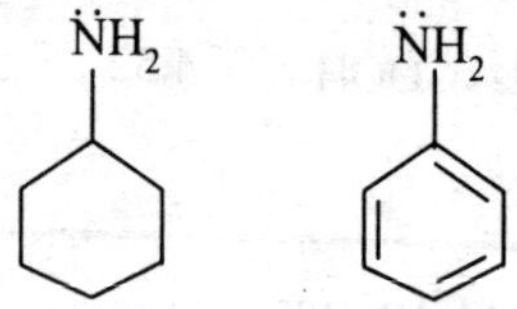

pka 10.64 4.62

This effect is explained in the following paragraphs:

+IE of cyclohexyl group which increases electron density on the nitrogen atom of the cyclohexylamine; –IE of phenyl group which decreases electron density on the nitrogen atom of the aniline. For resonance contributions to the overall hybrid of aniline; the following contributors are of significance:

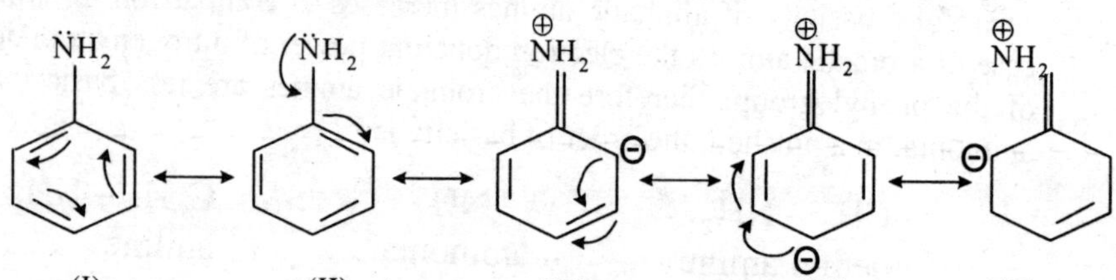

(I) (II) (III) (IV) (V)

Structures I & II are the **Kekulé structures**. Structures III-V, however, show delocalisation of the lone pair of electron on the nitrogen atom of the aniline.

(d) In the case of aryl amines, when an alkyl group is attached to the benzene ring, the order of basicity is as follows:

*2: Secondary amine; 1: primary amine; and 3: tertiary amine.

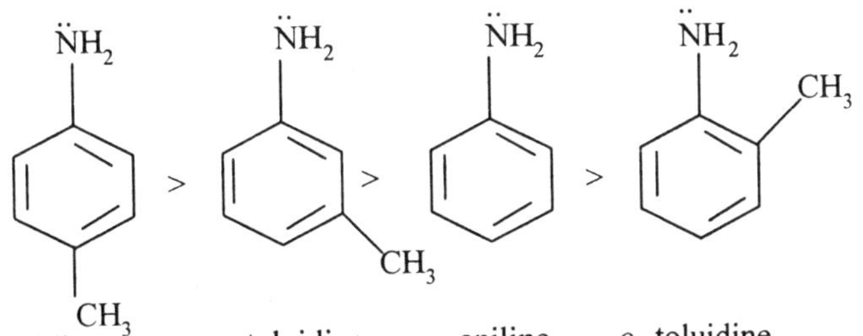

	P-toluidine	*m*-toluidine	aniline	*o*-toluidine
pka	5.00	4.67	4.62	4.38

The less basic nature of *o*-toluidine in comparison to aniline may be due to steric hindrance.

(v) In the case of aryl amines, when a methoxy group is attached to the benzene ring, the order of basicity is:

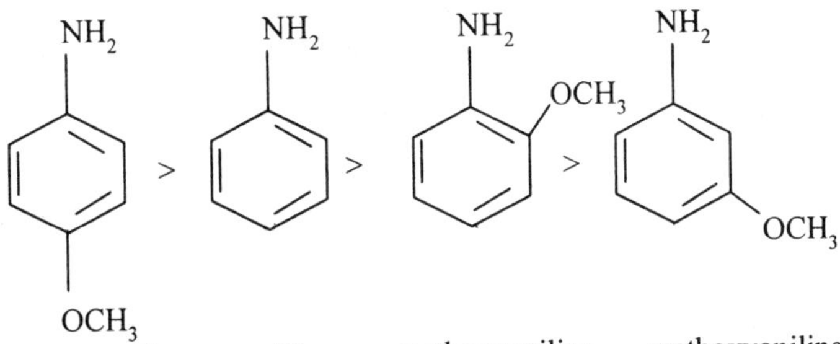

	p-methoxyaniline	aniline	*o*-methoxyaniline	*m*-methoxyaniline
pka	5.29	4.62	4.49	4.20

(vi) In the case of aryl amines, when a nitro group is attached to the benzene ring, the order of basicity is:

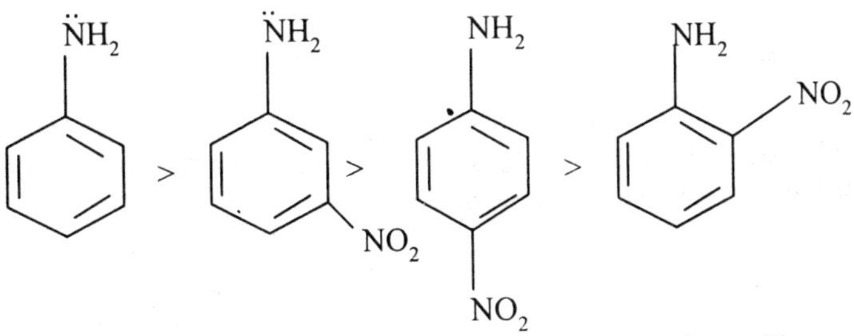

	aniline	*m*-nitroaniline	*p*-nitroaniline	*o*-nitroaniline
pka	4.62	2.45	0.98	-2.8

(vii) In the case of N-alkyl-substituted aryl amines, the order of basicity is:

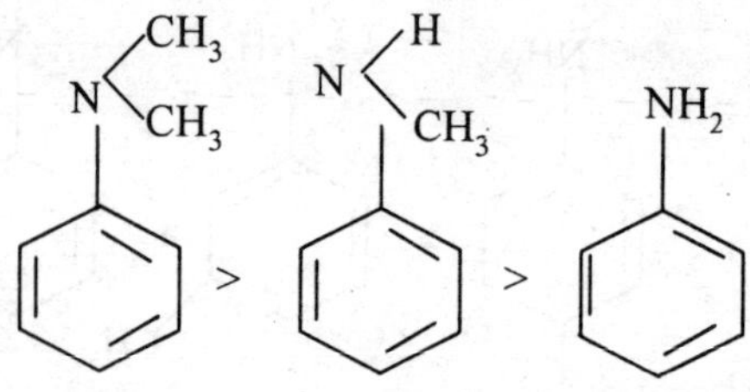

N-N-Dimethyl aniline > N-methyl aniline > aniline 4.62

3.2 ELECTROMERIC EFFECT

It is a temporary effect operating at the demand of the attacking reagent (electrophile or nucleophile). It involves complete transfer of π electrons between two atoms linked by multiple bond. The two atoms may be of same or different electronegativity. The shifting takes place towards more electronegative element in the latter case, e.g.,

(i) $>C{=}C< \xrightarrow{\text{Reagent}} >\overset{\oplus}{C}{-}\overset{\ominus}{C}<$

(ii) $>C{=}O \xrightarrow{\text{Reagent}} >\overset{\oplus}{C}{-}\overset{\ominus}{O}<$

The effect, represented by symbol  has been used to explain mechanism of a number of organic reactions, e.g., >C=C< addition reactions.

3.3 RESONANCE OR MESOMERIC EFFECTS

Structure of methane is CH_4. We can explain all the properties of methane with the help of a single structure of methane as CH_4. However, we cannot explain all the properties of benzene if we consider a single structure of benzene, i.e., . For example, C—C bond length of benzene is 1.39Å and *o*-xylene upon ozonolysis gives three products, i.e., glyoxal, methyl glyoxal and dimethyl glyoxal. However, if we consider the following two Kekulé structures for benzene, we can explain the above properties. Thus, resonance explains all properties of benzene.

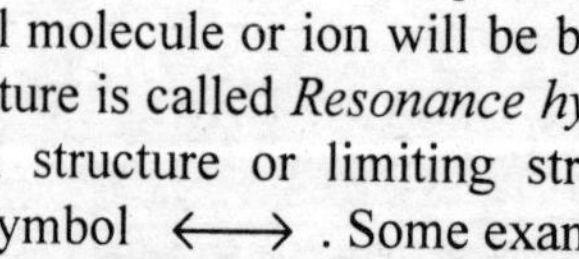

According to Ingold, when for a molecule like benzene or an ion, two or more Lewis structures that differ only in the position of the π electrons can be written, then none of these will be a correct representation for the molecule, i.e., none will be in complete accord with the physical or chemical properties of the molecule. The actual molecule or ion will be better represented by a hybrid of these structures. The real hypothetical structure is called *Resonance hybrid* and various structures are called *contributing structures* or canonical structure or limiting structures. The phenomenon is called *Resonance* and is represented by the symbol $\longleftrightarrow$. Some examples of resonance are:

Essential conditions for resonance is (i) presence of π electron and/or lone pair of electron in conjugation.

(v) –1 F.C. on one oxygen atom
(vi) +1 F.C. on N –1 F.C. on oxygen
(vii) –1 F.C. on Br
(viii) No formal charge
(ix) No. F.C. on any atom
(x) No. F.C. on any atom.

5. *(i)* CH_3I, electrophile; $C_6H_5NH_2$, nucleophile
(ii) CH_3OH, nucleophile; $C_6H_5CH_2Cl$, Electrophile
(iii) BF_3, electrophile; $CH_3O\ CH_3$ nucleophile
(iv) Br_2, electrophile; Br^- nucleophile.

6.

	A		*B*
(i)	NH_4	(iii)	Lewis acid
(ii)	BF_3	(iv)	Electrophile
(iii)	Carbene	(i)	Soft electrophile
(iv)	H^+	(ii)	Hard electrophile
(v)	Carbanion	(vi)	Nucleophile
(vi)	Acetylide ion	(vii)	Lewis base
(vii)	H_2O	(v)	Hard Nucleophile
(viii)	R.G. Pearson	(viii)	Organic chemist who classified electrophile and nucleophiles

Chapter 3

Various Effects

Various effects are as follows:

1. Inductive effect
2. Electromeric effect
3. Resonance
4. Hyperconjugation
5. Hydrogen Bonding (will be discussed later in this chapter).

3.1 INDUCTIVE EFFECT

When a covalent bond is present between the two atoms which have same electronegativity, the shared electron pair lies exactly between the two atoms. This is known as *symmetrical distribution* of electron density, e.g., $-\overset{|}{\underset{|}{C}} : \overset{|}{\underset{|}{C}}-$ However, if one of the two atoms is more electronegative, the shared electron pair shifts towards more electronegative element. This is termed as unsymmetrical distribution of electron density, e.g., $-\overset{|}{\underset{|}{C}} : Cl$. As a result of this, polarity develops in the molecule.

Inductive effect may be defined as the *induction of polarity in an otherwise covalent bond, due to incomplete shifting of the electron pair between the two atoms which have different electronegativity*,

e.g., $-\overset{|\delta^+}{\underset{|}{C}} \longrightarrow - \overset{\delta^-}{Cl}$

The characteristics of the inductive effect are as follows:

(i) It is a *permanent effect*.

(ii) It is represented by the symbol —>—.

(iii) It propagates through the carbon chain

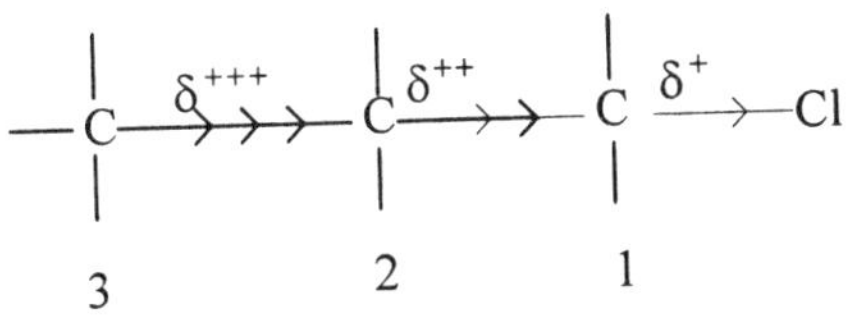

When it propagates through a solvent or space, it is known as **Field effect.**

(iv) Ingold, on the basis of the electronegativity of hydrogen atom, classified groups or atoms in two categories.

(a) Groups or atoms which are more electronegative than hydrogen atom are said to have **minus or negative Inductive Effect (–IE)**. They are electron-withdrawing in nature.

(b) Groups or atoms which are less electronegative than hydrogen atom are said to have **plus or positive Inductive Effect (+IE)**. They are electron-releasing in nature.

Following is the groups or atoms of increasing order of –IE and +IE

H— **–IE** → $-C_6H_5$, — OH, —I, —Br, — Cl, —F, —CN, $-NO_2$

Increasing Order →

H— **+IE** → $-CH_3$, $-C_2H_5$, $-C_3H_7$,

Increasing Order →

Greater the number of carbon atoms of the alkyl group, greater is the +IE.

Influence of Inductive Effect on Properties of Organic Molecules

Inductive effect has a profound effect on the properties of organic molecules. Some of them are as follows:

(i) ***Effect on Bond Length of Alkyl Halides***

An examination of the following data shows that as we move from left to right, the $-\overset{|}{\underset{|}{C}}:$ X **bond length decreases** because –IE of halogen atom **increases.**

CH_3-I	CH_3-Br	CH_3-Cl	CH_3-F
2.14 Å	1.94 Å	1.78 Å	1.38Å

→ –IE increases

(ii) ***Effect on Dipole Moment of Alkyl Halide***

Dipole moment of a molecule is a method to express the polarity of the molecule, and is equal to the product of charge and distance (see following data).

CH_3 — I	CH_3 — Br	CH_3 — Cl	CH_3 — F
1.64 D	1.79 D	1.87 D	1.81 D

It is clear that the value of dipole moment increase from CH_3—I to CH_3—CI because minus inductive effect of halogen increases, which causes more charge separation between carbon atom and halogen atom. Although the charge separation is more in CH_3—F, the shorter C—F distance results in a smaller (electron charge distance) product. **Therefore, dipole moment of CH_3—F is less than that of CH_3—Cl.**

(iii) ***Influence on Acid Strength of Carboxylic Acids***

Examine Table 3.1, which shows influence of –IE on acid strength. It is clear that as we go downwards in the Table, –IE **increases, acid strength also increases** (pKa value decreases).

Table 3.2 shows influence of +IE on acid strength. It is clear from Table 3.2 that as we go downwards in the table, +IE increases, while acid strength decreases.

Table 3.1 Influence of –IE on acid strength

	Acid	pKa	
	CH_3COOH	4.67	
	$C_6H_5CH_2COOH$	4.31	
	$HOCH_2COOH$	3.83	
–IE increases	ICH_2COOH	3.12	Acid strength increases
	$BrCH_2COOH$	2.86	
	$ClCH_2COOH$	2.86	
	FCH_2COOH	2.66	
	$Cl_2CHCOOH$	1.29	
	$Cl_3C{\cdot}COOH$	0.65	
	$F_3C{\cdot}COOH$	0.00 (zero)	

Table 3.2 Influence of +IE on acid strength

pKa = –log Ka, Ka = *dissociation constant of acid*			
	Acid	pKa	
	HCOOH	3.77	
	CH_3COOH	4.76	
+ IE increases	$(CH_3)_2CH$—COOH	4.86	
	CH_3CH_2COOH	4.88	Acid strength decreases
	$CH_3(CH_2)_nCOOH$ (n = 2 to 7)	4.82–4.95	
	$(CH_3)_3C$. COOH	5.05	

Another important factor explaining 1s and stability of phenoxide ion. powerful electron-withdrawing effect o

(b)	
Cresols	pKa
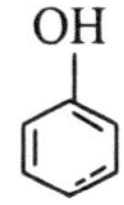 phenol	9.9
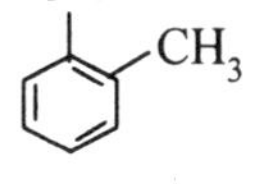 *o*-cresol	10.28
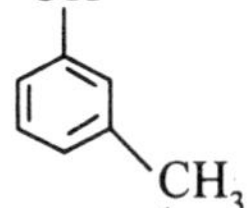 *m*-cresol	10.08
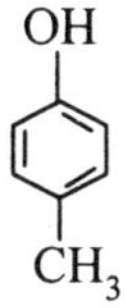 *p*-cresol	10.19

(viii) *Acidity of phenols:* Phenol dissolves in
$NaHCO_3$. It is attributed to low acidit
resonance. In fact, phenol itself is
resonance-stabilized anion, namely, ph

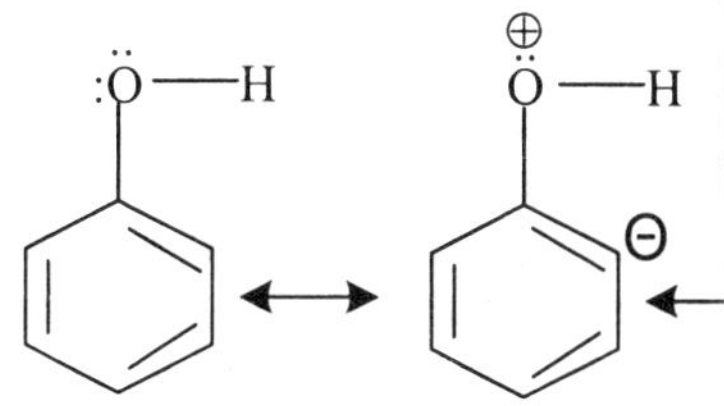

Contributing resonance st

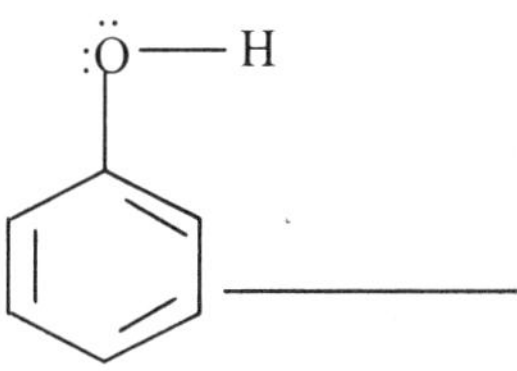

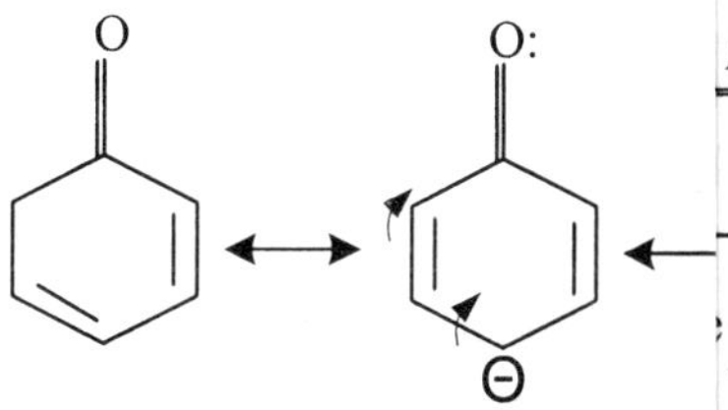

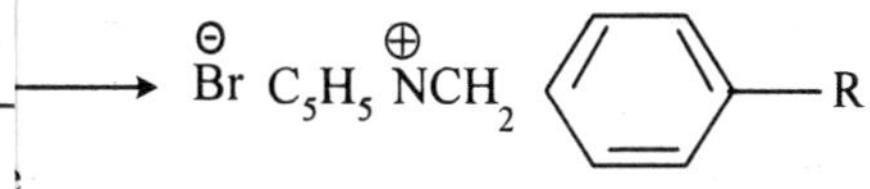

Resonating stru) at 20° C $\times 10^4$

2.02

1.81

1.63

If a group having π electrons is attache
substituted phenoxide anion. Therefor
contrary if an alkyl group is attached t
cresols (methyl-substituted phenols).
nitrophenols > phenol > cresol [*For det*

3.4 HYPERCONJUGATION OR NO-E NATHAN EFFECT

tional mechanism of electron release by an alkyl ethyl group. They suggested that the electron pairs lized than those of C—C bond and are able to unshared electron pairs.

In 1935, Baker and Nathan observed that the incr
benzyl bromide with pyridine was roughly opp
release.

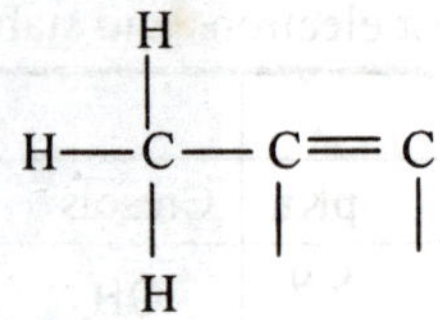

or

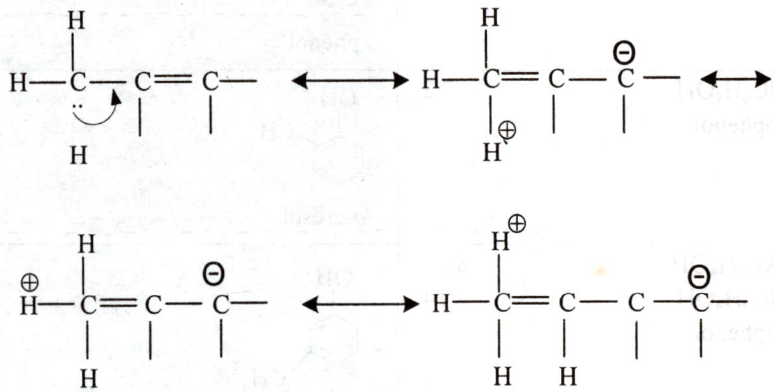

As you have read, the π bonds are essential for the resonance. So, hyperconjugation may be defined as extension of the concept of resonance to sigma electrons. In other words, the delocalisation of sigma electrons is called Hyperconjugation.

or

Hyperconjugation = Sigma electrons+ Resonance

Kinds of Hyperconjugation

According to Muller and Mulliken (1958), hyperconjugation can be classified into two categories:

(i) *Sacrificial Hyperconjugation:* In this hyperconjugation, the ground state of the neutral molecules, the canonical forms involve not only the no bond resonance but also a charge separation., not posed by the main form.

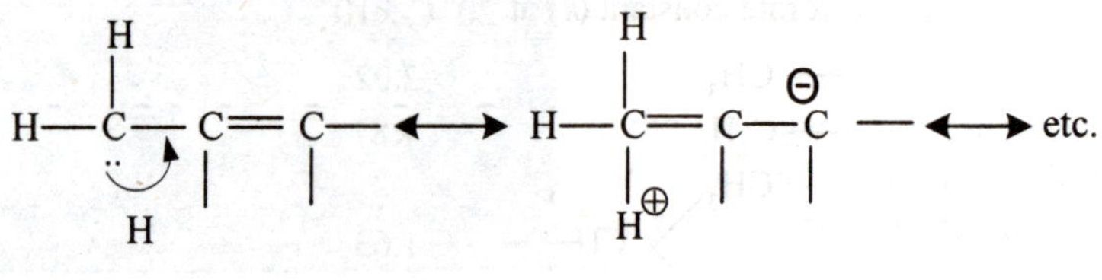

(ii) *Isovalent Hyperconjugation:* In the free radicals and carbonium ions, the canonical forms do not display more charge separation than the main form.

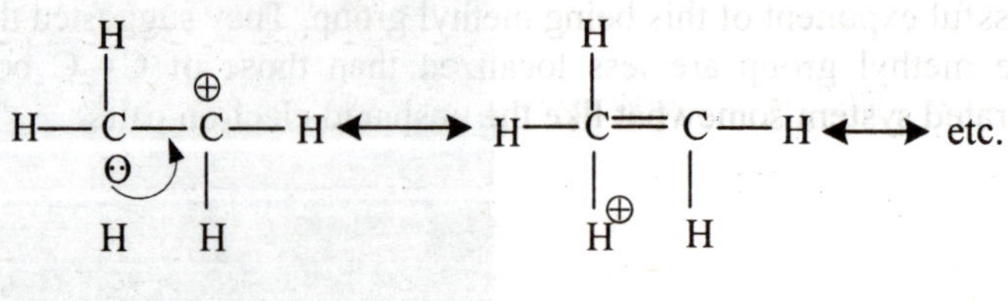

(i) Carbonate ion (CO_3^{-2})

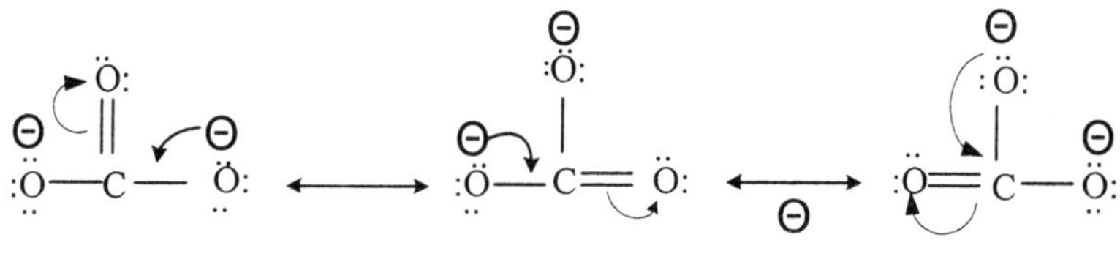

(ii) Aniline $(C_6H_5NH_2)$

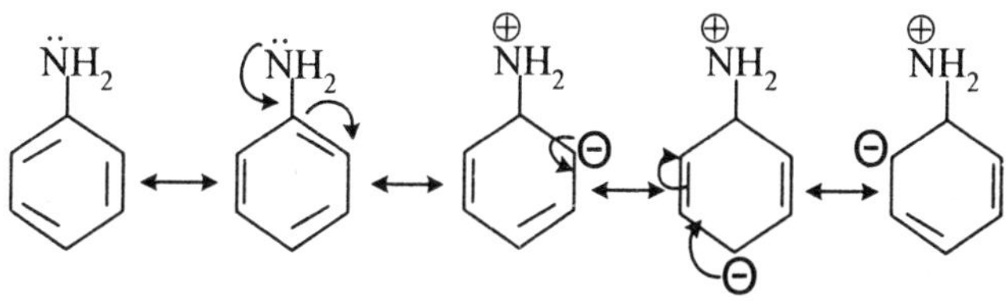

(iii) Allyl carbonium ion

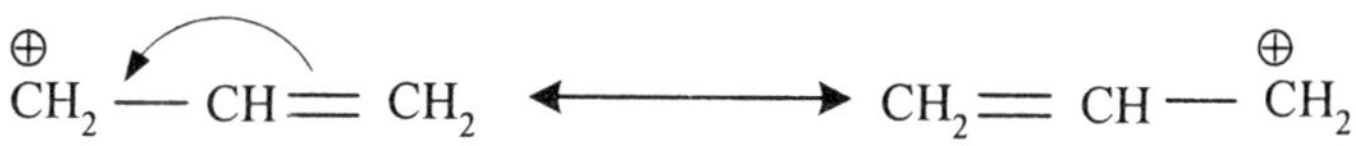

Rules of Resonance

The description of a molecule in terms of resonance is governed by a set of rules which are as follows:

1. [1]Total number of paired and unpaired electrons should be same in all the contributing structures; only their position should be different. For example,

(a) 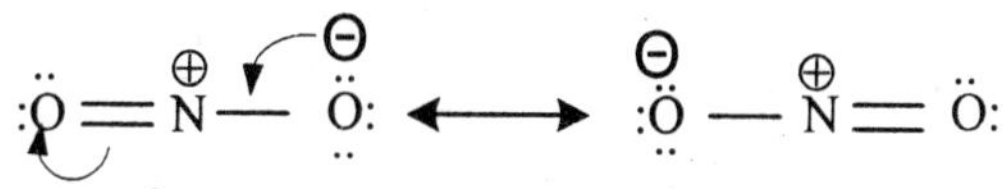

I II

Both structures contain five-paired and one unpaired electrons.

(b) 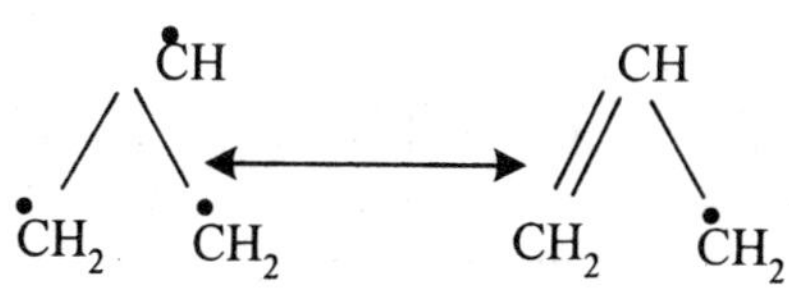

This is not a proper resonance structure for the allyl radical o.

2. The position of atomic nuclei in all the contributing structures must be the same. Resonance differs from Tautomerism in this respect. It is further explained like this:

[1] e.g., in cricket total players are 11, only their position is changed.

(i) 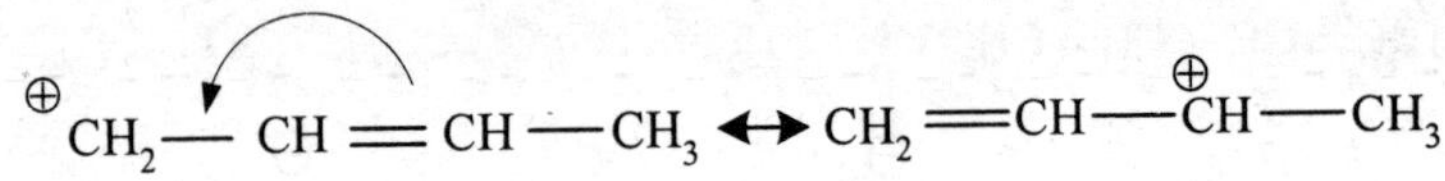

are the proper resonance structures but $\overset{\oplus}{C}H_2-CH_2-CH=CH_2$ is not a proper resonance structure because one H atom has moved.

(ii) 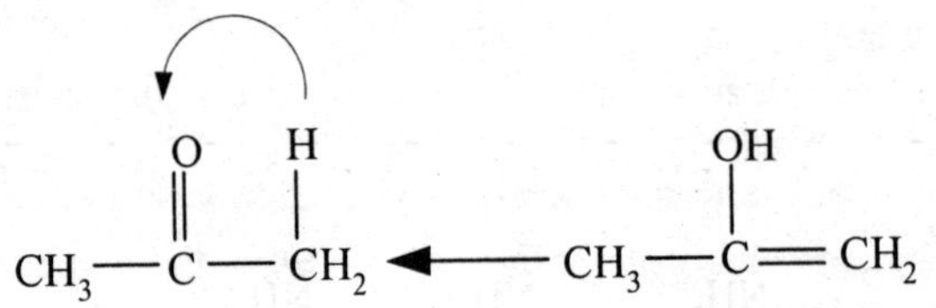

Keto form Enolic form

are examples of Tautomerism where movement of atomic nuclei is allowed.

3. The energy content of all the contributing structures must be nearly the same.
4. All the structures must be proper Lewis structures. We should not write structures in which carbon is pentavalent, e.g.,

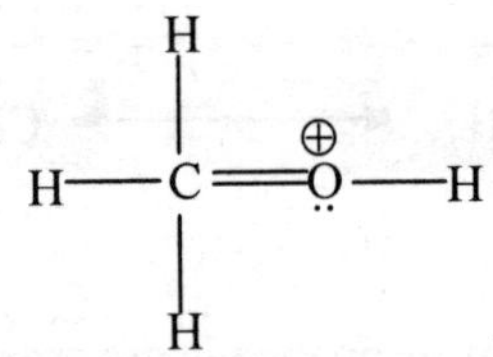

This is not a proper resonance structure for methanol.

5. Resonance structures exists only on paper.
6. All atoms that are a part of the delocalized system must lie in a plane or be nearly planar, e.g., 2, 3-di-tertbutyl–1, 3-butadiene does not show resonance

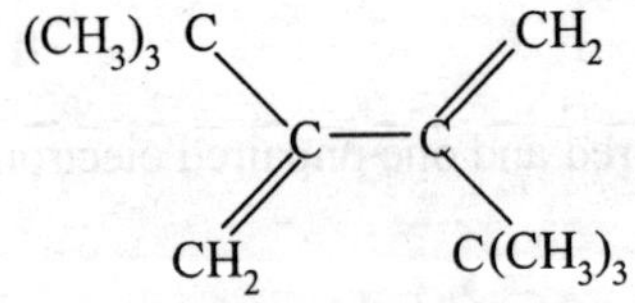

7. All the contributing structures do not have equal contribution towards the real structure of molecule. Their contribution is a direct function of their stability. The stability of a contributing structure depends upon a number of factors which are as follows:
   (a) Other things being similar, greater the number of the covalent bonds, greater would be the stability of the structure. For example, structure (I) of 1,3-butadiene is more stable than structure (II)

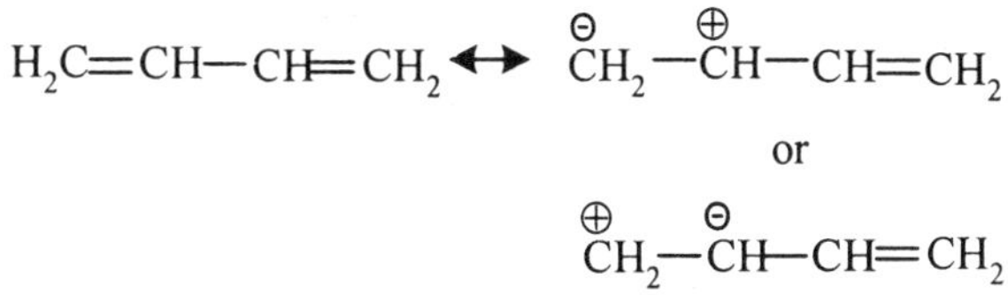

(11 covalent bonds) (10 covalent bonds)

(I) (II)

(b) Other things being similar, structures having no formal charge are more stable than structures having formal charge. For example, structure (I) of 1,3-butadiene, which has no formal charge, is more stable than structure (II).

(c) Other things being similar, structures in which negative formal charge is present on a more electronegative element is more stable than structures in which negative formal charge is present on a less electronegative element. For example, structure (I) of aldol anion is more stable than structure (II).

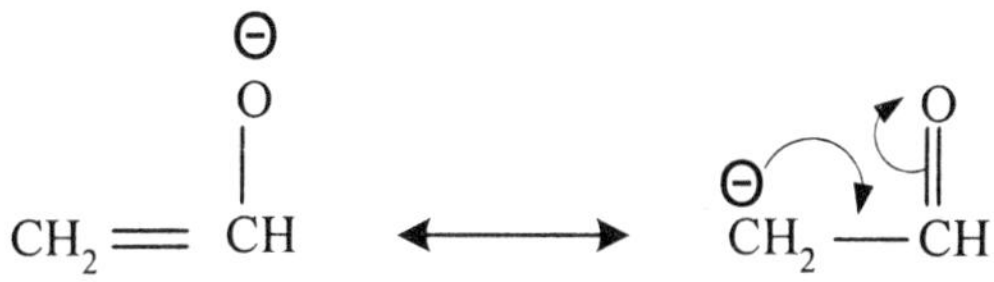

(I) Negative formal charge on more electronegative oxygen atom

(II) Negative formal charge on less electronegative carbon atom

(d) Other things being similar, structures in which positive formal charge is present on a less electronegative element are more stable than the structures in which positive formal charge is present on a more electronegative element. For example, structure (II) of pyridine is more stable than (III).

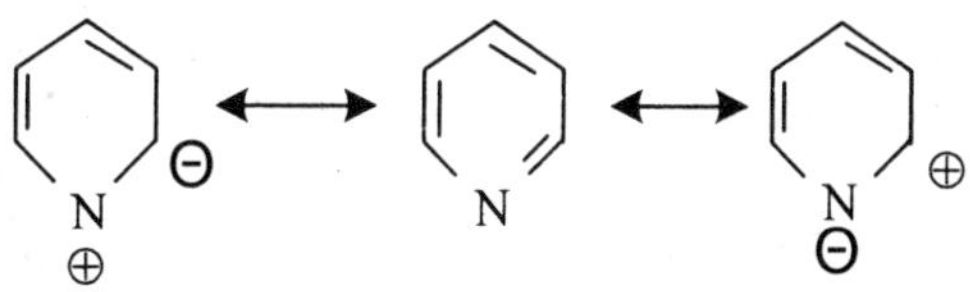

(III) (I) (II)

The formal charge present on more electronegative nitrogen atom

The formal charge present on less electronegative carbon atom

(e) Other things being similar, structures, in which there is maximum distance between like charges or minimum distance between unlike charges, are more stable. For example, structure (I) is more stable than structure (I)

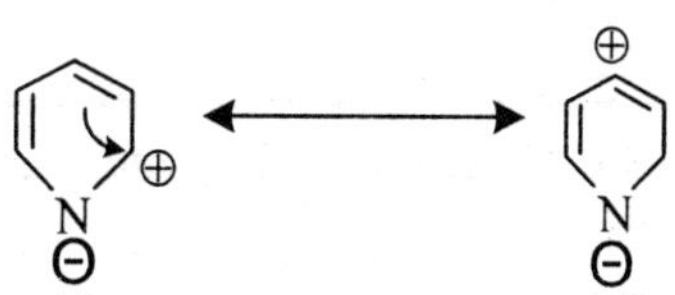

(I) (II)

8. *Resonance energy*: The heat of hydrogenation of cyclohexene is –28.8 kcal/mole, which is very much similar to that of any similarly-substituted alkene.

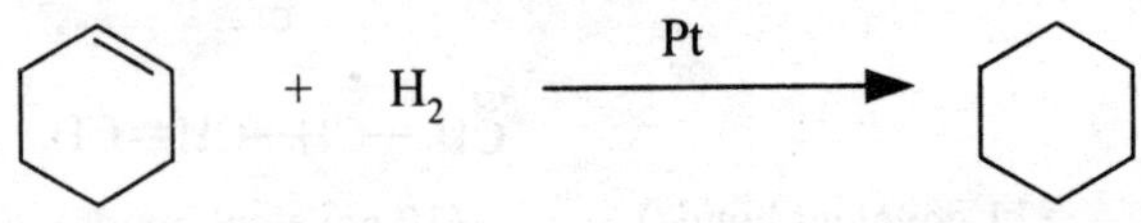

Cyclohexene Cyclohexane

If we extend this kind of thinking, and if benzene is 1,3,5-cyclohexatriene, it could be well predicted that benzene would liberate approximately 85.8 kcal/mole (3 × (–28.8) kcal/mole) when it is subjected to hydrogenation. However, the experimental data shows that the value is –49.8 kcal/mole.

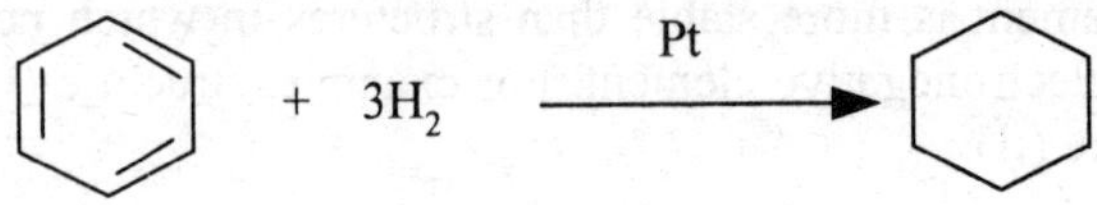

Heat of hydrogenation

Calculated 3 × (–28.8)	=	– 86.4 kcal/mole
Observed	=	– 49.8 kcal/mole
Difference	=	36.6 kcal/mole

Thus, the energy content of a resonance hybrid of benzene is less than that of the most-stable contributing structure by 36.6 kcal/mole. This is the *resonance* energy for benzene.

Thus, the energy content of a resonance hybrid is always less than that of the most stable contributing structure by an amount known as resonance energy. It has been observed that greater the number of π bonds, greater would be the number of contributing structures and hence greater would be its stability. See Table 3.4.

Table 3.4 π bonds in some compounds and their resonance energies.

S. no.	*Name of compound*	*Structure*	*No. of bonds*	*Resonance energy (kcal/mole)*
1.	Benzene	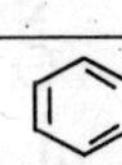	3	36.6
2.	Styrene	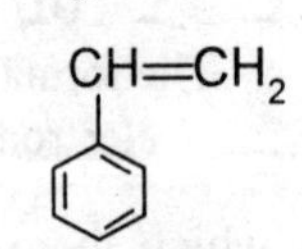	4	51.2
3.	Naphthalene	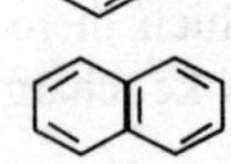	5	63.1
4.	Diphenyl	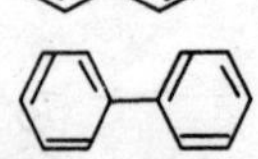	6	71.0

It is clear that as we go down the table, the boiling/melting point of the compounds increases, i.e., the stability increases. Therefore,

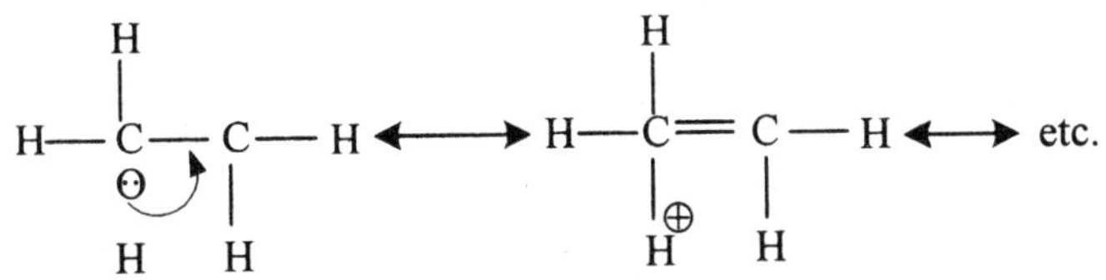

Application of Hyperconjugation

(i) Hyperconjugation explains stability of alkene

We know that:

Group	No. of hyperconjugating structures
—CH_3 (methyl)	3
—C_2H_5 (ethyl)	
—C_3H_7(propyl)	2
or —C_4H_9 and so on	
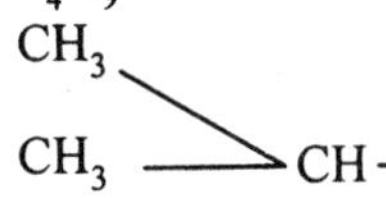 — Isopropyl group	1

Now replace H atoms of ethylene with the group and note the number of hyperconjugating structures. Greater the number, greater is the stability of alkene. Since stability of alkene is related to its heat of hydrogenation. We can write:

$$\text{Stability of alkenes} \propto \left\{ \begin{array}{l} \text{(i) No.of hyperconjugating structures.} \\ \text{(ii)} \quad \dfrac{1}{\text{Heat of hydrogenation}} \end{array} \right\}$$

For example,

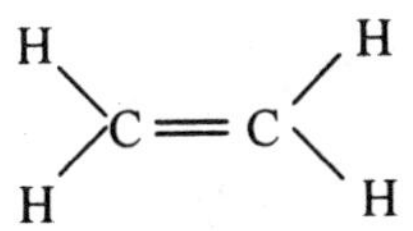

(a) No hyperconjugating structure

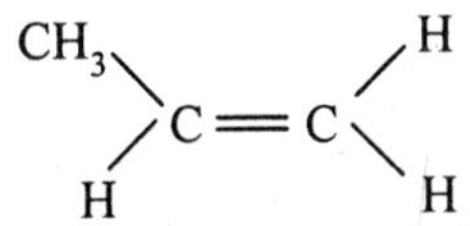

(b) 3-hyperconjugating structures

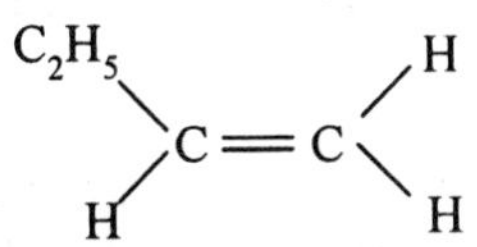

(c) 2-hyperconjugating structures

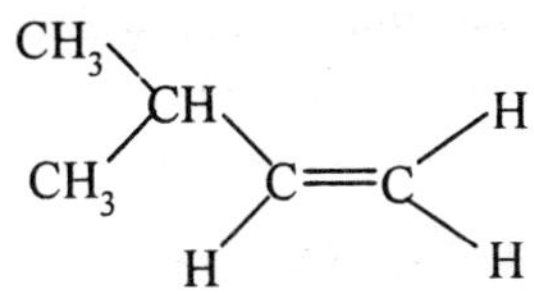

(d) 1-hyperconjugating structures

So, the order of stability (b) > (c) > (d) > (a). In general, the order of stability is

$$R_2C = CR_2 > R_2C = CHR > R_2C = CH_2,\ RCH = CHR > RCH = CH_2 > CH_2 = CH_2$$

(ii) *Heat of combustion:* *n*-butene and isobutene both have same number of C—C, C=C, C—H bonds; yet the heat of combustion is 646.1 kcal/mole and 649.8 kcal/mole, respectively. It is because *n*–butene $CH_3CH_2CH{=}CH_2$ has two hyperconjugating structures whereas isobutene 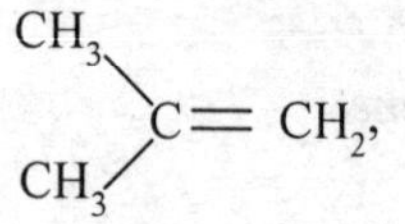 has six hyperconjugating structures.

(iii) *C—C bond length in propylene:* In propylene, all the C—C bond length values are 1·488 Å, in comparison with C—C bond length of 1.54 Å in alkane and C—C bond length of 1.34 Å in alkene. It is because hyperconjugation gives double-bond character to C—C bond and single-bond character to C=C.

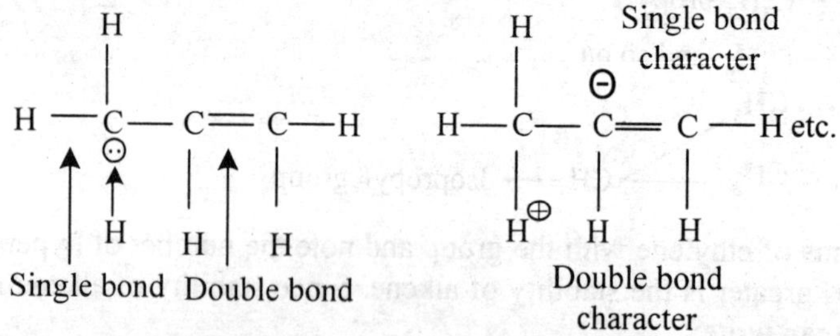

(iv) *o, p-directing nature of alkyl benzene:* *o, p*-directing nature of alkyl benzene like toluene can be explained on the basis of hyperconjugation.

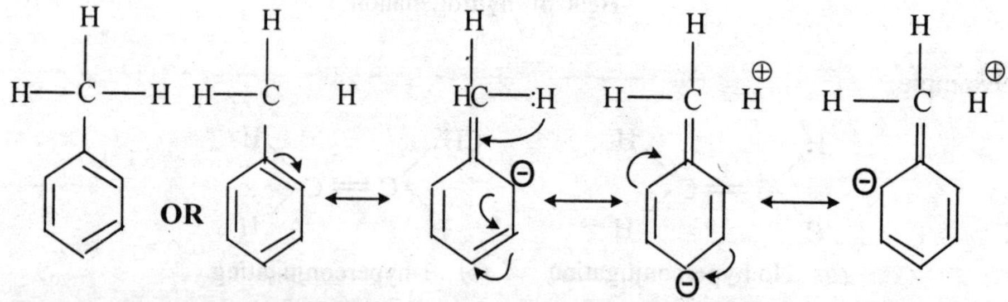

(v) *Dipole moment of benzene:* Hyperconjugation also explains the fact that the dipole moment of benzene is zero but that of toluene is 0.4 D.

(vi) *Low ionization potential for propylene:* Hyperconjugation explains low ionization potential value for propylene (9.76 eV) in comparison to that of ethylene (10.62 cV). The former is more stable due to hyperconjugation.

3.5 HYDROGEN BONDING

Hydrogen bonding may be defined as an association of the molecules through a hydrogen atom attached to a small, highly electronegative element. In fact, it is a bond between a functional group A–H and an atom or group of atoms B in the same or different molecule, i.e.,

A—H .... B

Hydrogen bond

where, A is oxygen, nitrogen or fluorine and B is oxygen, nitrògen or fluorine

The oxygen may be singly or doubly bonded and the nitrogen singly, doubly, or triply bonded. Some examples are as follows:

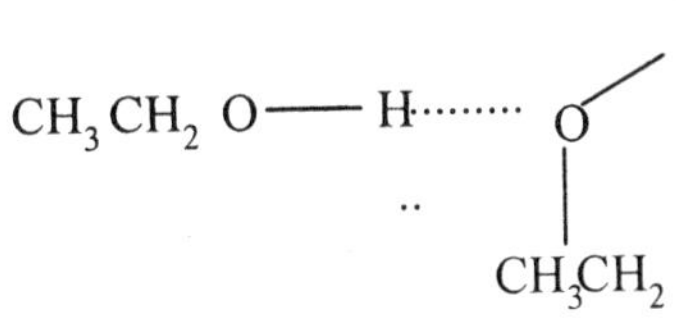

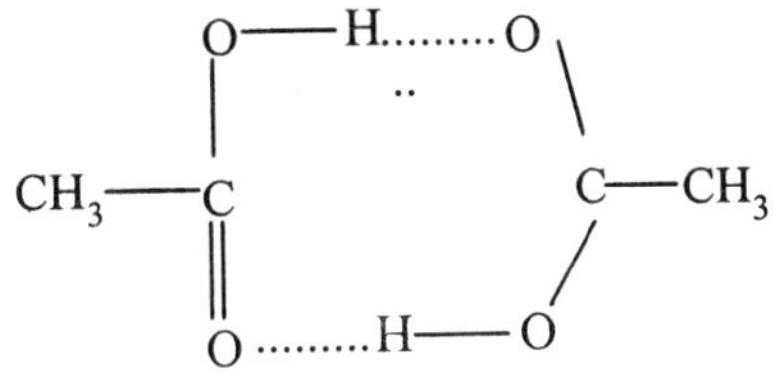

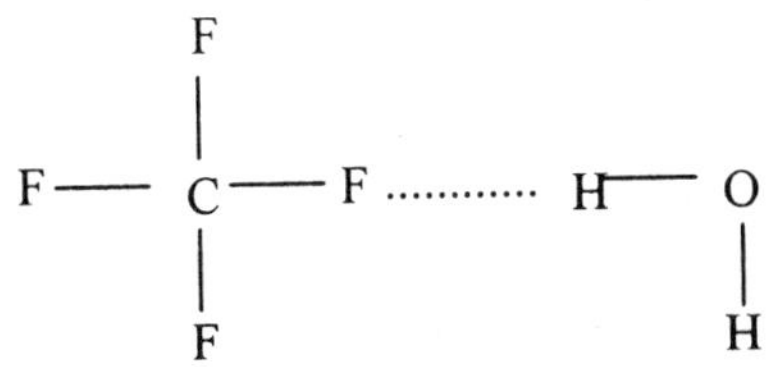

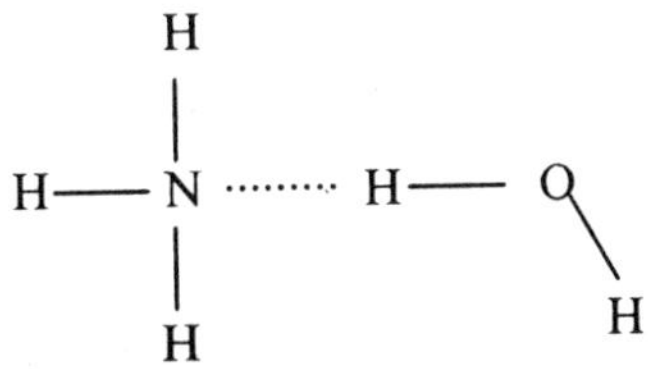

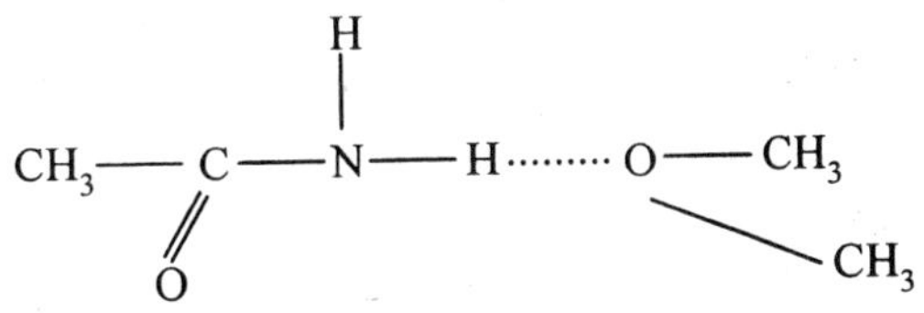

Hydrogen bonds exist in all the three states, i.e., solid (ice), liquid (water), and gaseous (acetic acid exist as a dimer in gaseous state). The strength of hydrogen bond is as follows:

F—H ------ F	50	kcal/mole
O—H ----- O	6–8	kcal/mole in carboxylic acid
	3–6	kcal/mole in alcohols
N—H ----- N	3–6	kcal/mole in amines

In general, greater the electronegativity difference between H and B, greater would be hydrogen bond strength. The hydrogen is on the straight line formed by A and B within about 15°. In most of the cases, hydrogen bonding is possible where six-membered ring can be formed, while five membered rings are less common. The distance of hydrogen from A and B generally not equal, e.g., the O—H distance is 0.97 Å in ice while the H....O distance is 2.75 Å.

Hydrogen bonding can be detected in a number of ways like dipole-moment measurements, solubility behaviour, depression in freezing point, heats of mixing and by infra-red spectroscopy (hydrogen bonding always moves the bond towards lower frequencies). Hydrogen bonding changes with temperature and concentration (in case of intermolecular hydrogen bonding).

Kinds of Hydrogen bonding and their effects on the properties of compounds are elaborated in the following paragraphs.

(i) *Intermolecular hydrogen bonding:* This hydrogen bonding occurs between a number of molecules of the same compound or of difference compounds. The intermolecular hydrogen bonding between the molecules of the same compound results in contraction of the volume. For example, if 30 ml methanol is added to 20 ml water, the total volume of the solution would be slightly less than 50 ml because the intermolecular hydrogen bonding of the molecules of methanol with those of water is stronger than that of the molecules of methanol with each other, and molecules of water with each other. In general, the number of hydrogen bonds per molecule is the sum of the number of hydrogen atoms attached to more electronegative element and the number of lone pairs of electrons present on electronegative element. For example, in water there are 4 intermolecular hydrogen bonds per molecule:

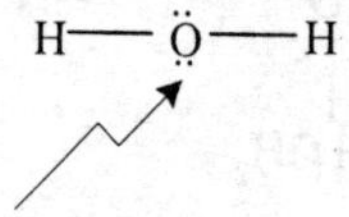

(a) Boiling point of alcohols is higher than the isomeric compounds either due to intermolecular hydrogen bonding, e.g., boiling point of ethanol & dimethyl ether are 78.3°C & 34.5°C, respectively.

(b) If intermolecular hydrogen bonding is possible between two compounds, one of them would be soluble or miscible in other, e.g., alcohols are miscible in water but ethers are not.

(c) Water is liquid due to intermolecular hydrogen bonding whereas hydrogen sulphide (H_2S) is a gas.

(ii) *Intramolecular hydrogen bonding*: Hydrogen boding within the same molecule is called intramolecular hydrogen bonding, e.g.,

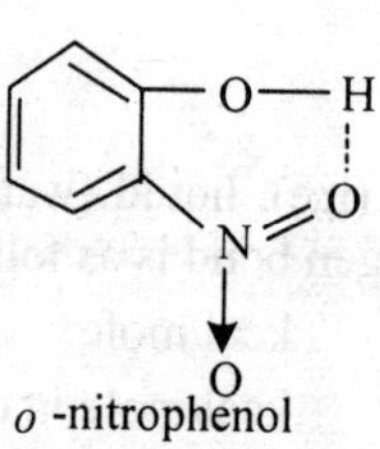

o -nitrophenol

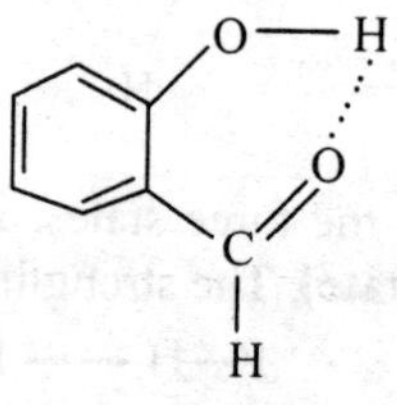

Salicylic aldehyde

Effect of intramolecular hydrogen bonding on a few properties of organic compounds is given below.

(a) Intramolecular hydrogen bonding lowers the boiling point.

(b) Intramolecular hydrogen bonding makes a substance steam volatile, e.g., *o*-nitrophenol is steam volatile but *p*-nitrophenol is not steam volatile.

(c) If intramolecular hydrogen bonding is present between more than two –OH groups, the compound is likely to be sweet in taste; e.g., Glucose and Fructose are sweet.

SUMMARY

1. –IE ∝

(i) $\dfrac{1}{\text{bond length of alkyl halide}}$

(ii) Dipole moment (exception $CH_3Cl > CH_3F$) of alkyl halide

(iii) Acid strength

(vii) $\frac{1}{\text{Base strength}}$

2. $+IE \propto$

(i) Base strength (ii) $\frac{1}{\text{Acid strength}}$

3. Basicity of amines: Aliphatic amines > NH_3 > Aryl amines.

4. Resonance energy = Energy content of most stable contributing structure – Energy content of resonance hybrid.

5. Stability of a contributing structure $\propto$

(i) Number of covalent bonds

(ii) $\frac{1}{\text{Formal charge}}$

(iii) –ve formal charge on more electronegative element

(iv) +ve formal charge on less electronegative element

(v) Maximum distance between like charges

(vi) Minimum distance between unlike charges

(vii) Resonance energy

6. Resonance increases acidity of phenols but decreases basicity of amines.

7. Number of hyperconjugating structures $\propto$

(i) Stability of alkenes

(ii) $\frac{1}{\text{Heat of hydrogenation of alkene}}$

8. Intermolecular hydrogen bonding $\propto$

(i) Boiling point

(ii) Solubility or miscibility

9. Intramolecular hydrogen bonding $\propto$

(i) Steam volatility

(ii) $\frac{1}{\text{Boiling point}}$

PROBLEMS

1. Amongst the four answers to the questions given below one or more may be correct. Tick the correct answer(s):

(i) Which of the following is the strongest acid:

(a) HCOOH (b) Cl_3CCOOH

(c) FCH_2COOH (d) CH_3COOH

(ii) Which of the following is the strongest acid:

(a) $CH_3CH_2CH_2\ COOH$

(b) $CH_3CH_2CH\ COOH$ (with Cl on the CH)

(c) $CH_3CH\ CH_2COOH$ (with Cl on the CH)

(d) $CH_2CH_2CH_2\ COOH$ (with Cl on the first CH_2) **(CP.M.T., 1995)**

(iii) Which of the following is most basic

(a) 

(b) 

(c) 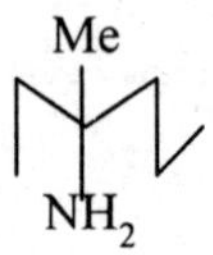

(d) 

(iv) Which of the following is most basic

(a) 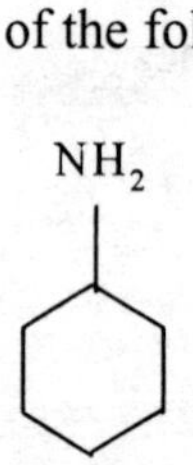

(b) 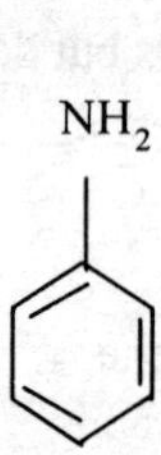

(c) 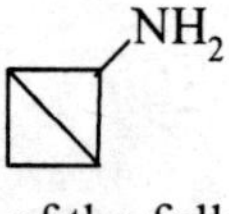

(d) 

(v) Which of the following is most basic.

(a) FNH_2 (b) $ClNH_2$ (c) $BrNH_2$ (d) NH_3

(vi) Polarization of electrons in acrolein may be written as

(a) $\overset{\delta^-}{CH_2}{=}CH{-}\overset{\delta^+}{CH}{=}O$

(b) $\overset{\delta^-}{CH_2}{=}CH{-}CH{=}\overset{\delta^+}{O}$

(c) $\overset{\delta^-}{CH_2}{=}\overset{\delta^+}{CH}{-}CH{=}O$

(d) $\overset{\delta^-}{CH_2}=CH-CH=\overset{\delta^+}{O}$ **(IIT, 1988)**

(vii) Which of the following is most basic

(a) benzylamine (b) aniline

(c) acetanilide (d) *p*-nitroaniline **(IIT, 1990)**

(viii) Phenol is less acidic than

(a) Acetic acid (b) *p*-methoxy phenol

(b) *p* -nitrophenol (d) ethanol

(ix) Resonance structures of a molecule should have

(a) identical arrangement of atoms

(b) nearly the same energy content

(c) same number of paired electrons

(d) identical bonding **(IIT, 1984)**

(x) Aryl halides are less reactive towards nucleophilic substitution reaction as compared to alkyl halides due to

(a) formation of a less stable carbonium ion

(b) resonance stabilization

(c) longer carbon-halogen bond

(d) inductive effect **(IIT, 1990)**

(xi) *o,p*-directing nature of the methyl group can be explained on the basis of:

(a) resonance

(b) inductive effect

(c) electromeric effect

(d) no-bond resonance

(xii) Which of the following is most stable:

(a) 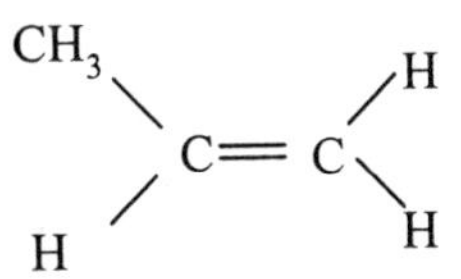

(b) 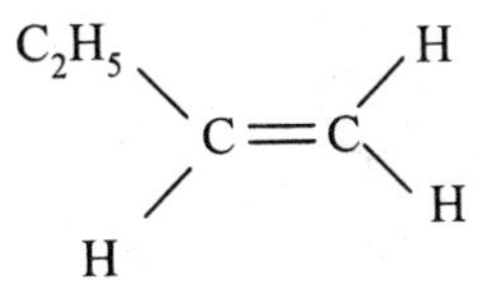

(c) 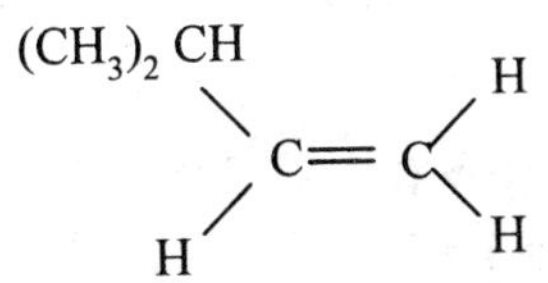

(d) 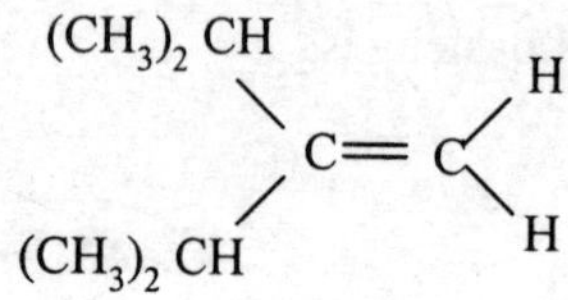

(xiii) Examine the following two structures for the anilinium ion and choose the correct statement from the ones given below:

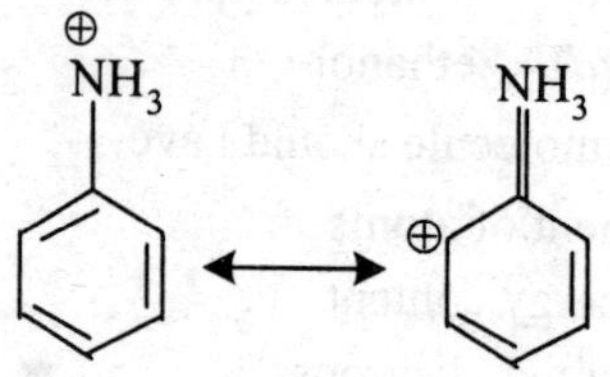

(I) (II)

(a) (II) is not an acceptable canonical structure because carbonium ions are less stable than ammonium ions
(b) (II) is not an acceptable canonical structure because it is non-aromatic
(c) (II) is not an acceptable canonical structure because the nitrogen has 10 valence electrons
(d) (II) is an acceptable canonical structure **(IIT, 1993)**

(xiv) Which of the following has the smallest heat of hydrogenation per mole
(a) butene-1 (b) trans-2-butene
(c) *cis*-2-butene (d) 1,3-butadiene **(IIT, 1993)**

(xv) If 50 ml methanol is mixed with 50 ml water, the volume of resulting solution is:
(a) exactly 100 ml
(b) slightly more than 100 ml
(c) slightly less than 100 mol
(d) 50 ml

(xvi) In the following compounds:

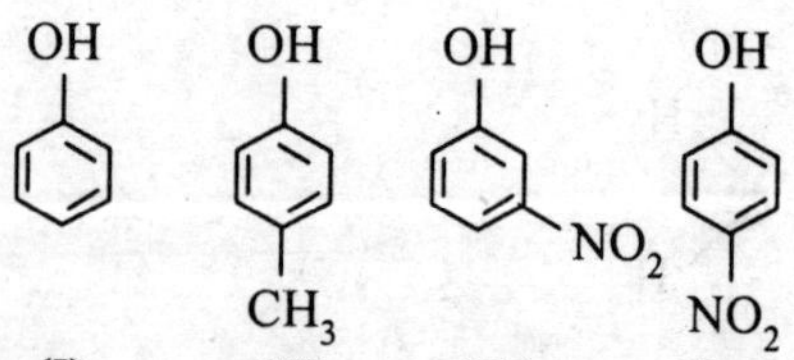

(I) (II) (III) (IV)

The order of acidities is:
(a) III > IV > I > II (b) I > IV > III > II
(c) II > I > III > IV (d) IV > III > I > II **(IIT, 1996)**

(xvii) In the following compounds, the order of basicity is:

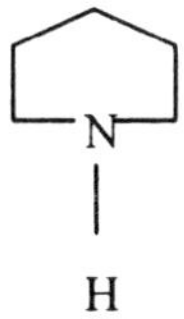
(I)

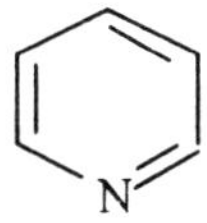
(II)

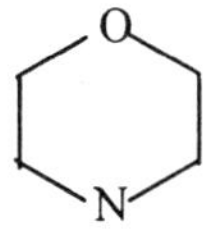
(III)

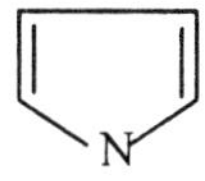
(IV)

(a) IV > I > III > II *(b)* III > I > IV > II

(c) II > I > III > IV *(d)* I > III > II > IV **(IIT, 1997)**

(xviii) The most unlikely representation of resonance structures of *p*-nitrophenoxide ion is:

(a)

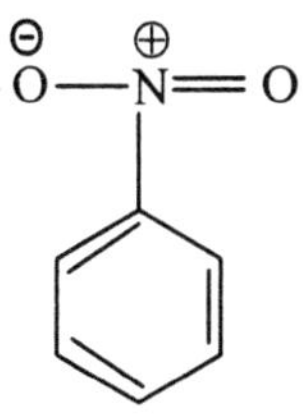

(b)

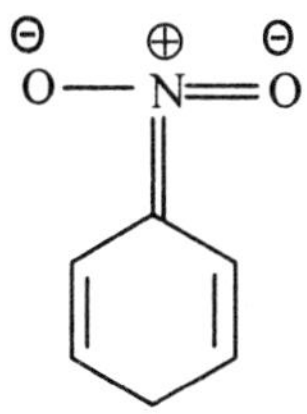

(c)

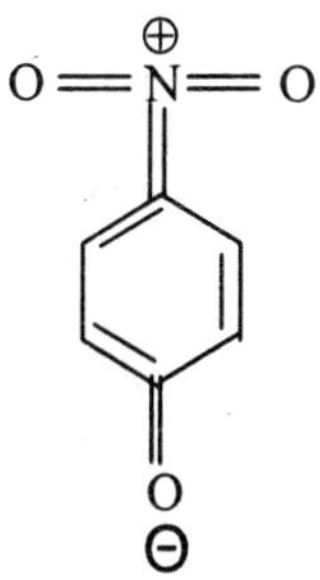

(d)

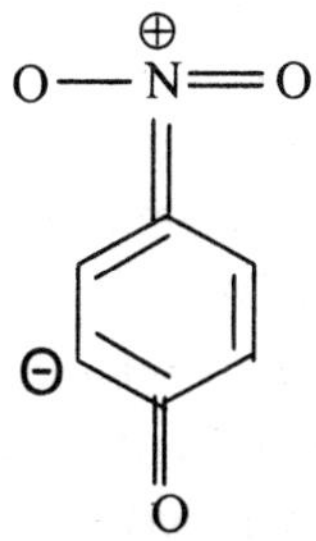

(xix) Which of the following acids has the smallest dissociation constant?

(a) $CH_3CHF\ COOH$ *(b)* FCH_2CH_2COOH

(c) $BrCH_2CH_2COOH$ *(d)* $CH_3CH\ Br\ COOH$

(xx) Identify the correct order of boiling points of the following compounds **(IIT, 2002)**

$CH_3CH_2\ CH_2\ CH_2OH$, $CH_3CH_2CH_2CHO$, $CH_3CH_2CH_2COOH$

1 2 3

(a) 1 > 2 > 3 *(b)* 3 > 1 > 2 *(c)* 1 > 3> 2 *(d)* 3 > 2 > 1

(xxi) Which of the following is the strongest base **[CBSE (Engg) 2004]**

(a) 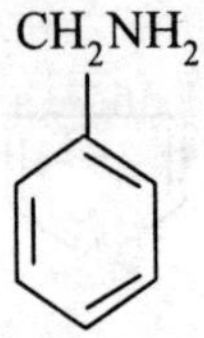 *(b)* 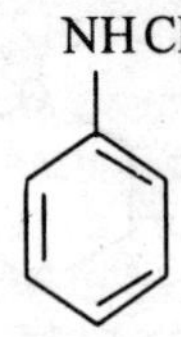

(c) 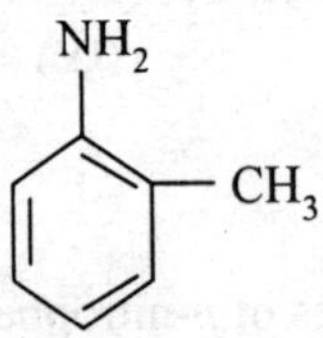 *(d)* 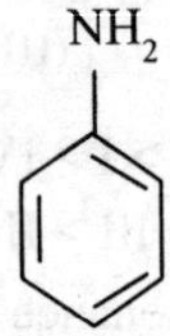

(xxii) The correct order of increasing acid strength of the compounds. **(AIEEE 2006)**

(1) CH_3COOH (2) $MeOCH_2COOH$

(3) CF_3COOH (4) $Me_2CHCOOH$

is:

(a) 1 < 4 < 3 < 2 *(b)* 2 < 4 < 1 < 3 *(c)* 4 < 1 < 3 < 2 *(d)* 4 < 1 < 2 < 3

2. Fill in the blanks:

(i) Phenol is acidic because of resonance stabilization of its conjugate base, namely............. **(IIT, 1990)**

(ii) pKa value of trifluoroacetic acid................

(iii) Most stable alkene is............................

(iv) Strongest aliphatic carboxylic acid is called .... **(GATE, 1992)**

(v) The kind of delocalization involving sigma bond orbitals is called .... **(IIT, 1994)**

3. State whether TRUE or FALSE:

(i) Iodide ion is a better nucleophile than bromide ion.

(ii) Dipole moment of CH_3F is more than that of CH_3Cl. **(IIT, 1993)**

(iii) Acetate ion is more basic than the methoxide ion.

(iv) $\overset{\ominus}{C}H_2-CHO$ is more stable than $CH_2=CH-O^{\ominus}$

4. The heat of sublimation for hypothetical form of ice, in which there are no hydrogen bonds, was 2.6 kcal/mole. If the observed heat of sublimation is 12.2 kcal/mole, what is energy of hydrogen bond in ice.

5. The heat of fusion of ice is 1.4 kcal/mole and that of a hypothetical form of ice, in which there are no hydrogen bonds is 0.30 kcal/mole. Assuming the energy of hydrogen bond is 4.8 kcal/mole, determine the percentage of hydrogen-bonds in ice which were broken during fusion.

6. Explain why

(i) Formic acid is stronger than acetic acid. **(MNR, 1992)**

(ii) Cyclohexylamine is a stronger base than aniline.

(iii) Carbon-oxygen bond lengths in Na_2CO_3 are equal. **(MNR, 1992)**

(iv) C—C bond lengths in propene are equal **(MNR, 1992)**

(v) Carbon oxygen bond lengths in formic acid are 1.23 Å and 1.36 Å and both the carbon oxygen bonds in sodium formate have the some value, i.e., 1.27 Å **(MNR, 1988)**

(vi) In acylium ion, the structure $R—C{\equiv}O^+$ is more stable than $R—\overset{+}{C}{=}O$. **(IIT, 1994)**

(vii) Aryl halides are less reactive than alkyl halides towards nucleophiles.

(viii) CH_3CONH_2 is much weaker base than $CH_3\,NH_2$. **(SSJM KU; Kanpur, 1996)**

(ix) Glucose is sweet in taste and soluble in water.

7. Match the items from column A with those in B:

	A		*B*
(i)	—>—	(i)	Tautomerism
(ii)	↷	(ii)	Electrometric effect
(iii)	$\rightleftharpoons$	(iii)	Inductive effect
(iv)	$\longleftrightarrow$	(iv)	Resonance
(v)	$CH_3—F$	(v)	pKa = Zero
(vi)	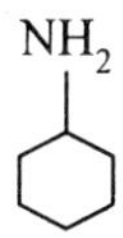	(vi)	Dipole moment 1.81 d
(vii)	CF_3COOH	(vii)	pKa value is 5

8. What effect should the following resonance structures of vinyl chlorides have on its dipole moment?

$$CH_2{=}CH—Cl \longleftrightarrow \overset{\oplus}{C}H_2—CH{=}\overset{\ominus}{Cl}$$ **(IIT, 1987)**

9. For nitromethane molecule, write, structure(s) showing significant resonance stabilization.

(IIT, 1986)

10. Write three different resonating structures of 1,3- butadiene and methyl cyanide and arrange in decreasing order of stability.

ANSWERS

1.

(i) **(b)**, because of –IE of three chlorine atoms.

(ii) **(b)**, because –IE of chlorine atom on carbon no. 2 is more prominent (see propagation of inductive effect).

(iii) **(b)**, It is a secondary amine. (a) and (d) are primary amines and (c) is a tertiary amine.

(iv) **(a)** Due to +IE of cyclohexyl group (c) and (d) contained four and three carbon atcms. In case of (b) –IE of phenyl group decreases basicity.

(v) **(d)** because (a), and (b) and (c) are less basic due to –IE group of halogen atoms.

(vi) **(d)** See rule No. 7 (d) and (e) (–ve formal charge on electronegative element and +ve formal charge on less electronegative element).

(vii) **(a)**.

(viii) **(a)** and **(c)** see Rules due to greater resonance in (c).

(ix) **(a), (b) (c)** see Rules.

(x) **(b)** and **(d)** see Rules

(xi) **(d)**.

(xii) **(a)** It is because is has three hyperconjugating structures, (b) and (d) have two and (c) has one hyperconjugating structures.

(xiii) **(c),** (a) is wrong since carbonium ion is more stable than ammonium ion, (b) is acceptable canonical structure with reference to aromatic nature, (d) is wrong since valency of nitrogen is not acceptable.

(xiv) **(d)** It is because $H_2C{=}CH{-}HC{=}CH_2$ has more resonating structures and hence is most stable, hence smallest heat of hydrogenation.

(xv) **(c)**, because intermolecular hydrogen bonding between different molecules is stronger than that between the same molecules.

(xvi) **(d)**.

(xvii) **(d)**.

(xviii) **(c)**.

(xix) **(d)**

(xx) **(b)**, because of H-bond strength.

(xxi) **(a)**

(xxii) **(d)**

2.

(i) Phenoxide anion *(ii)* Zero *(iii)*

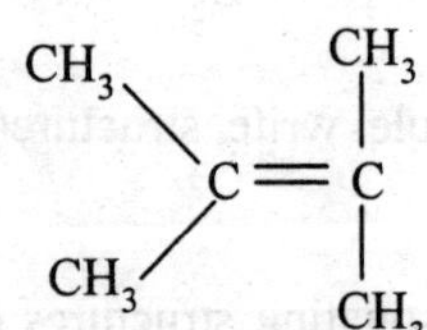

2, 3-Dimethyl butene -2 Methylene

(iv) CF_3COOH (v) Hyperconjugation

3.

(i) *True.*

(ii) *False.* `Although charge separation in CH_3F is more due to greater –IE of F atom in comparison to that of Cl atom, the charge X distance product is less.

(iii) *False*, because acetic acid is a strong acid in comparison to methanol, so its conjugate base is weak.

(iv) *False*. It is because 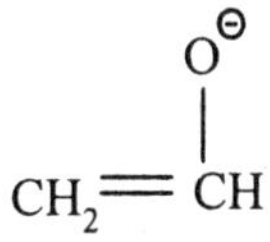 is more stable due to negative formal charge on more electronegative element oxygen.

4. The difference, due to hydrogen bonding, between the observed heat of sublimation of ice and the value calculated for a hypothetical form of ice is 12.2 – 2.6 + 9.6 kcal/mole/1. In ice each water molecule participates in four hydrogen bonds, but each bond is shared by two water molecules. So the net number of hydrogen bonds broken per molecule, when a mole of ice sublimes, is two. Therefore, energy per hydrogen bond is $\frac{9.6}{2} = 4.8$ kcal/mole.

5. Heat of fusion of ice and a hypothetical form in which there are no hydrogen bonds is 1.4 – 0.30 = 1.1 kcal/mole. Energy needed to break two moles of hydrogen bonds in ice is 2 × 4.8 = 9.6 kcal/mole. Hence, the fraction of hydrogen bonds broken per H_2O molecule in fusion is $\frac{1.1}{9.6} = 0.11$ or 11%.

6.

(i) Acetic acid is a weak acid compared to formic acid due to +IE of methyl group.

(ii) Cyclohexylamine is a stronger base due to +IE of cyclohexyl group. Aniline is less basic due to –IE of phenyl group and the resonance.

(iii) It is because of the resonance

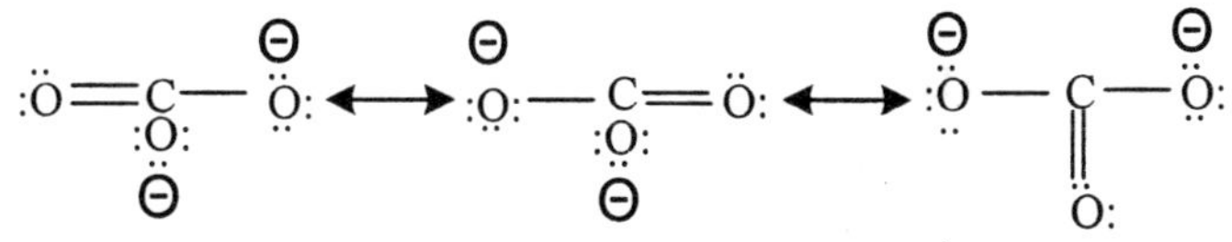

(iv) It is due to hyperconjugation.

(v) It is because of resonance

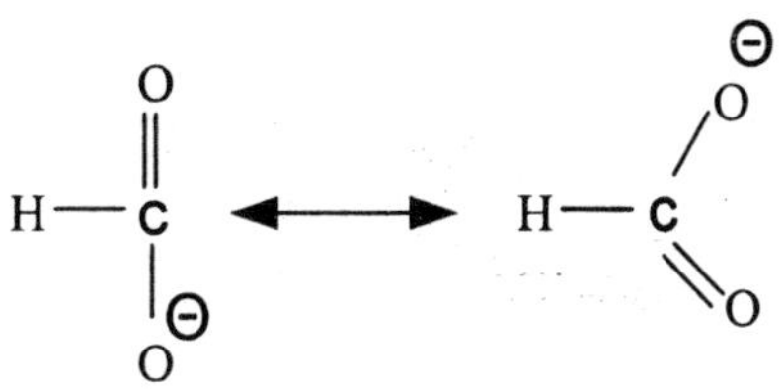

(vi) In $R-C\equiv\overset{\oplus}{O}$, the octet of every atom is complete whereas in $R-\overset{\oplus}{C}=O$ the carbon atom has only 6 electrons.

(vii) It is due to the resonance.

(viii) It is due to resonance in acetamide.

(ix) It is due to intramolecular hydrogen bonding.

7.

	A		*B*
(i)	→	(iii)	Inductive effect
(ii)	⌒ (curved arrow)	(ii)	Electrometric effect
(iii)	⇌	(i)	Tautomerism
(iv)	↔	(iv)	Resonance
(v)	CH_3—F	(v)	1.81 D
(vi)	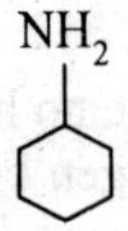	(vi)	pKa= 5.00
(vii)	CF_3COOH	(vii)	pKa = zero

8. Dipole moment will increase as the distance increases due to resonance.

9. 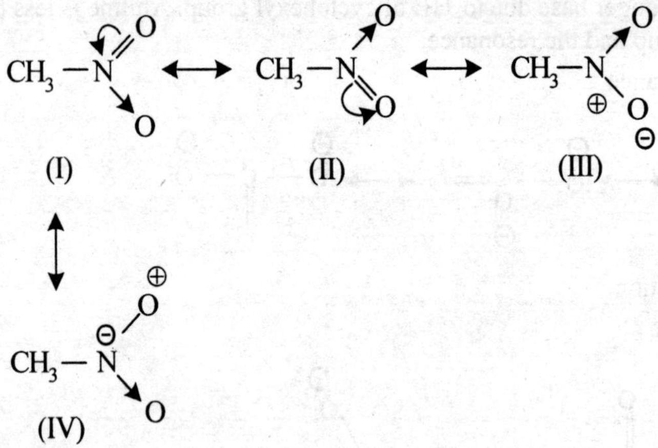

10. *(i)*

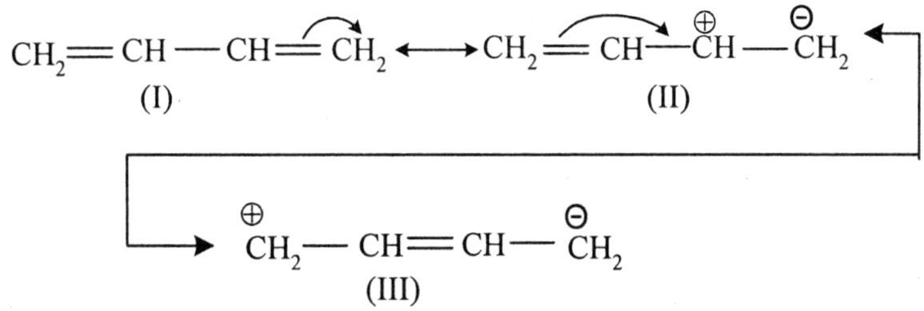

Order of stability: I > II > III

Resonating structures of 1,3,-Butadiene

(ii)

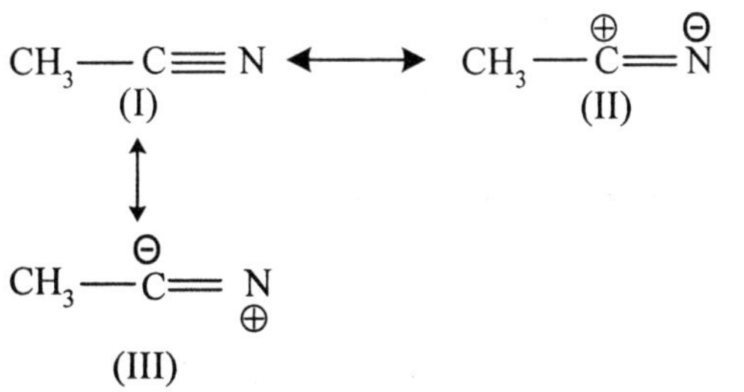

Order of stability: I > II > III

Resonating structures of methyl cyanide

Chapter 4

Reaction Intermediates

4.1 INTRODUCTION

Let us consider the possible paths given below, by which a reactant 'A' is converted into the product 'D'.

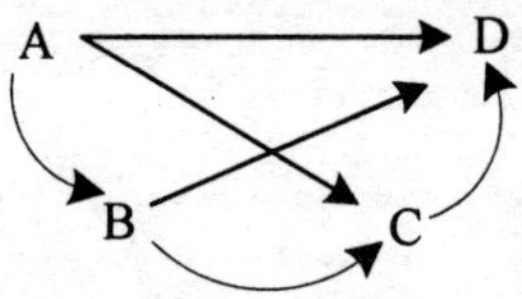

The path by which a reactant is converted into the product is called, 'Mechanism of Reaction and various species that are formed during the course of the reaction but are not isolated, are known as "reaction intermediates'. For example, if mechanism follows (A→B→C→D) path, then B and C are intermediates, if it follows path (A→C→D) then C is the intermediate and if follows path (A→B→D) the B is the intermediate.

The reaction intermediates formed in an organic reaction are as follows:

1. Carbonium ions for carbocations
2. Carbanions
3. Free radicals
4. Carbenes
5. Nitrenes

4.2 CARBONIUM IONS

(*Carb, Carbon + onium, positive charge*) Carbonium ion is the reaction intermediate in which carbon atom carries +ve charge. It may be represented as: $-\overset{|}{\underset{|}{C}}{}^{+}$ and the carbon atom is sp^2 hybridized. The structure may be written as given below and shape is trigonal planar:

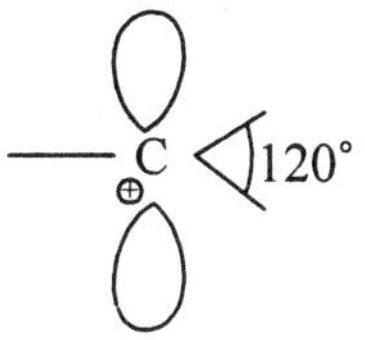

Carbonium ion is electron deficient, hence acts as an electrophile in an organic reaction.

(a) Preparation

Carbonium ions can be prepared by following methods.

(i) By heterolytic fission of a covalent bond in which carbon is attached to more electronegative element.

$$-\overset{|}{\underset{|}{C}}-X \xrightarrow{\text{Heterolytic fission}} -\overset{|}{\underset{|}{C}}{}^{\oplus} + X^{-}$$

(ii) By protonation of alkene or alcohols.

(a) 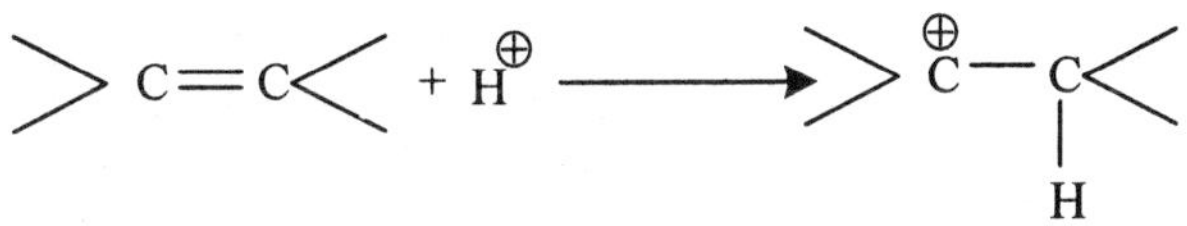

(b) $$RCH_2\ddot{O}H + \overset{+}{H} \longrightarrow \underset{\text{Protonated alcohol}}{RCH_2-\overset{\oplus}{\ddot{O}}-H_2} \xrightarrow{-H_2O} R-\overset{\oplus}{C}H_2$$

(iii) By thermal decomposition of diazo compound.

(b) Kinds of Carbocations or Carbonium Ions and Their Stability

The carbocations are of two types:

(i) Non-classical Carbonium Ion

In these carbonium ions, positive charge is delocalized over a number of carbon atoms. It is stabilized due to resonance, e.g.,

(i) $\overset{\oplus}{C}H_2-CH=CH_2 \longleftrightarrow CH_2=CH-\overset{\oplus}{C}H_2$

Allyl carbonium ion

(ii)

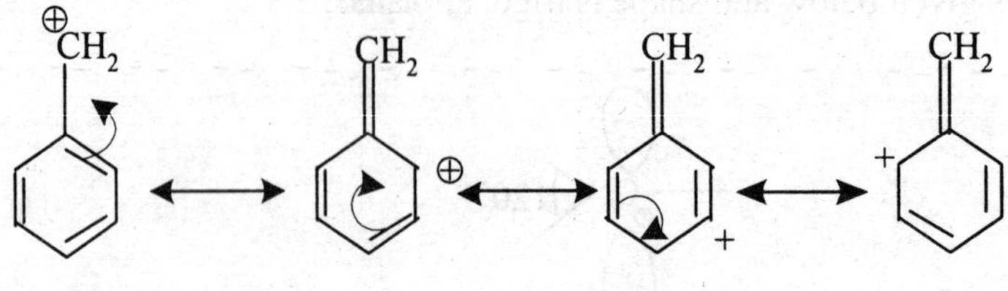

Benzyl carbonium ion

(ii) Classical Carbocation

In these carbonium ions, the positive charge is localized on single carbon atom, e.g., $\overset{+}{C}H_3$. The classical carbonium ions have been classified in there categories depending upon the nature of carbon atom bearing positive charge: (a) Primary carbonium ion, (b) Secondary carbonium ion, and (c) Tertiary carbonium ion.

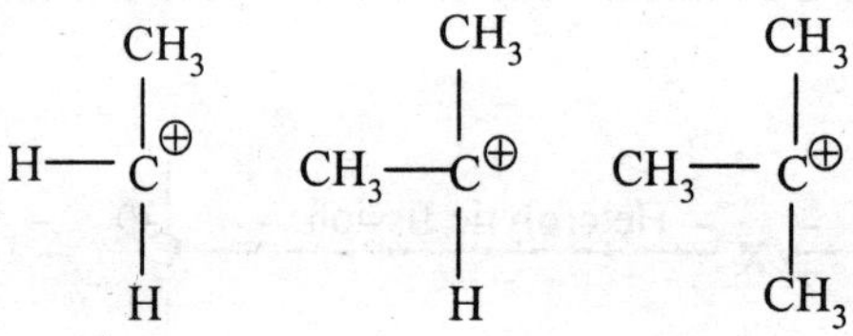

Primary < Secondary < Tertiary

The stability of classical carbonium ions can be explained on the basis of number of hyperconjugating structures. The four hyperconjugating structures for primary carbocation are as follows:

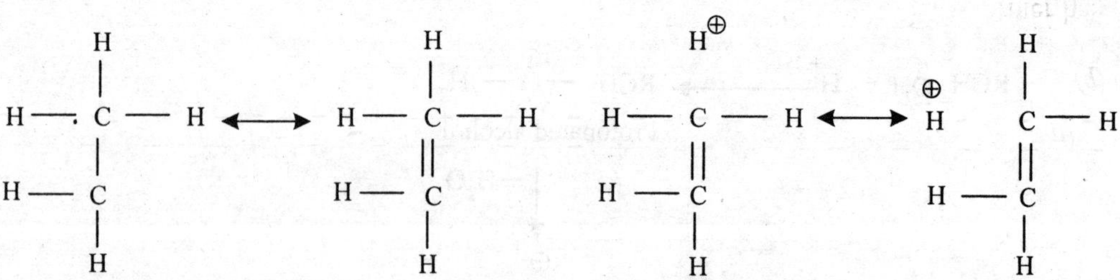

I II III IV

Similarly, the number of hyperconjugating structures for secondary and tertiary carbocation is 7 and 10, respectively.

The order of stability of classical carbonium ion can also be explained on the basis of +ve Inductive effect of alkyl group. It is clear that lesser the positive formal charge on the carbon atom of the carbocation due to +IE of alkyl group, greater is the stability of classical carbonium ion. Briefly, it can be shown as below:

$$\left.\begin{matrix}\text{Stability of}\\ \text{classical carbocation}\end{matrix}\right\} \propto \begin{cases}\text{Number of alkyl groups}\\ +\text{IE}\\ \dfrac{1}{\text{Electrophilic nature of carbon atom of the carbocation}}\end{cases}$$

In case hyperconjugation and +IE oppose each other, then hyperconjugation dominates e.g., $Me_3 C$ is more stable than $(C_2H_5)_3{}^+C$

(c) Reaction Involving Carbonium Ions

(i) Electrophilic Addition Reaction: Examples are as follows:

(A) 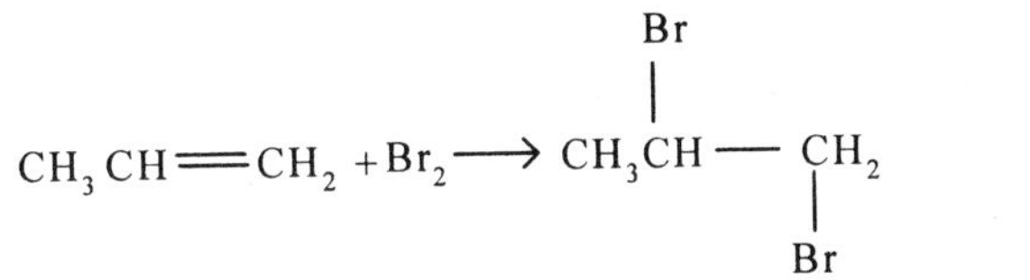

1,2-Dibromopropane

The steps are as follows:

(i) $Br - Br \xrightarrow{\text{Due to } \pi \text{ electrons}} \overset{\delta+}{Br} - \overset{\delta-}{Br}$

(ii), *(iii)*, *(iv)* 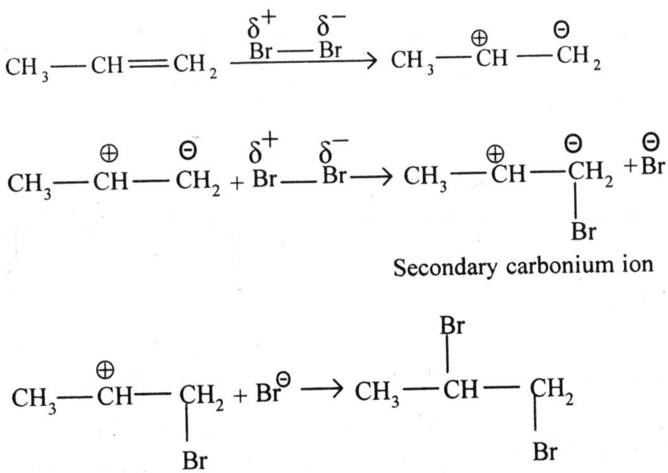

Secondary carbonium ion

(B) 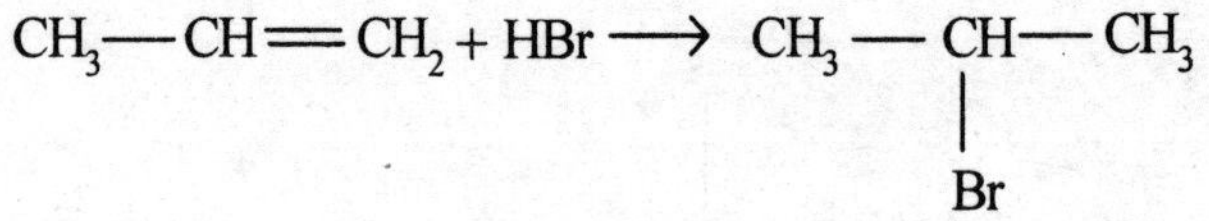

Propene 2-Bromopropane

(ii) *Unimolecular Nucleophilic Substitution Reactions* ($S_N{}^1$): Examples are as follows:

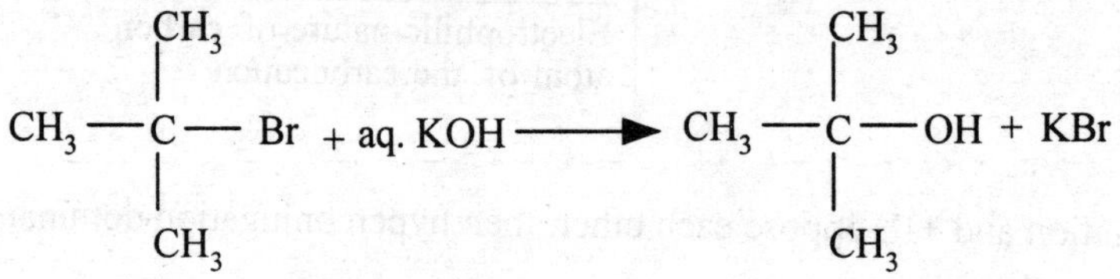

Tertiary butyl bromide

(i) 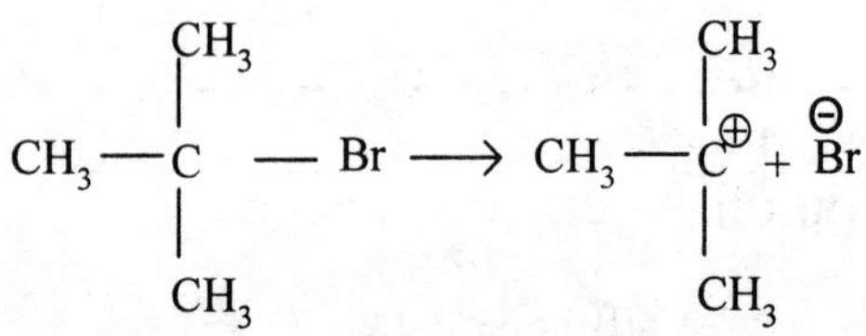

(ii) 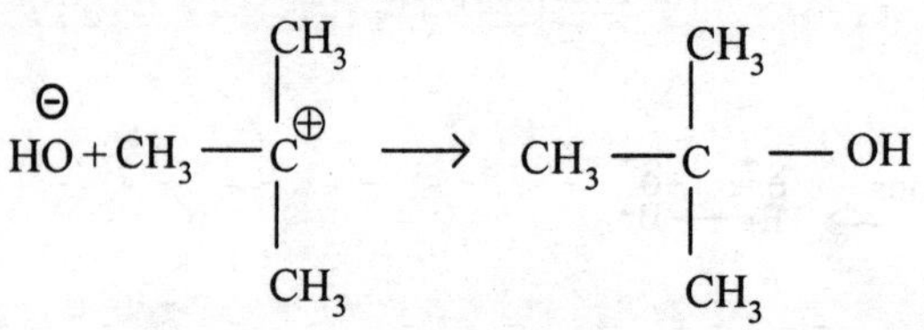

Tertiary butyl bromide

(iii) *Unimolecular Elimination Reactions (E_1)*: Examples are as follows:

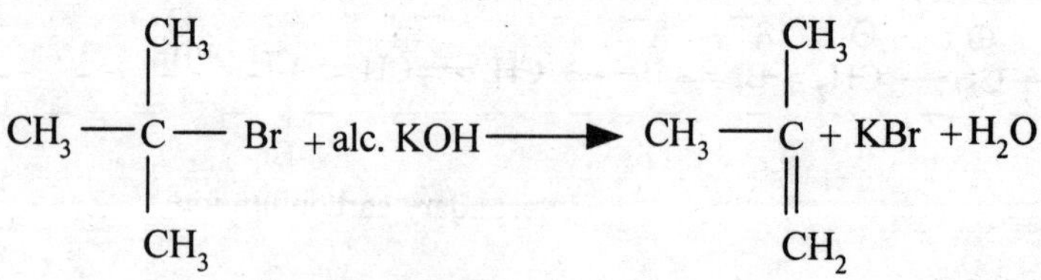

Tertiary butyl bromide 2-Methylpropene

The various steps are as follows:

(i)

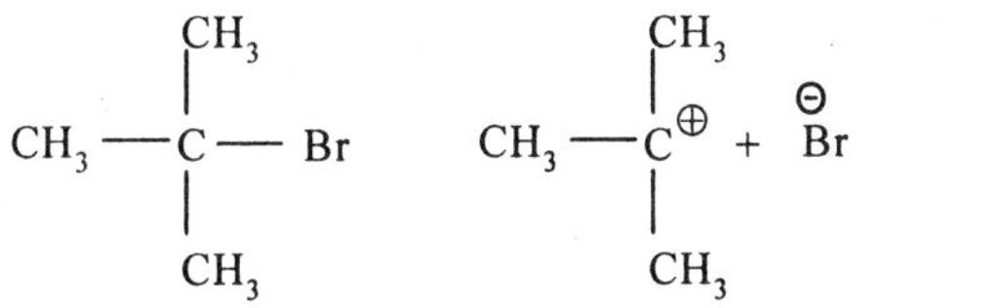

Tertiary carbonium ion

(ii)

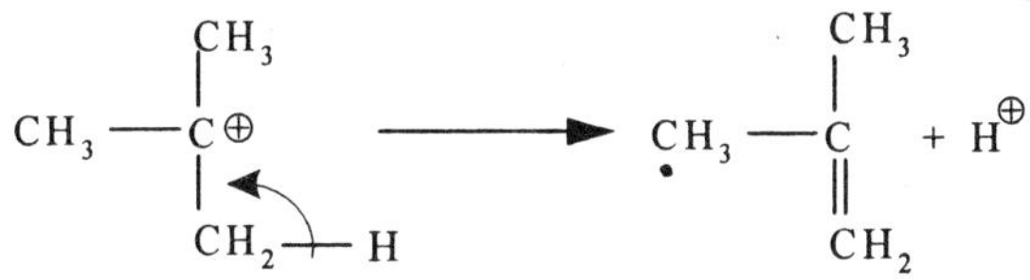

(iv) *Electrophilic Substitution Reactions*: Examples are as follows:

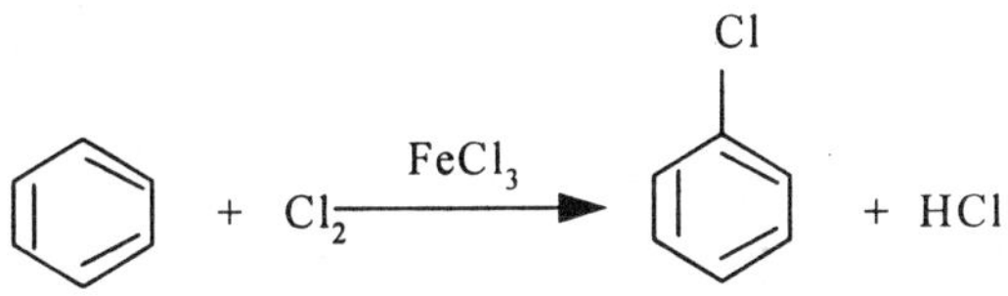

The mechanism consists of following steps:

(a) $Cl_2 + FeCl_3 \longrightarrow \overset{\oplus}{Cl} + \overset{\ominus}{FeCl_4}$

(b)

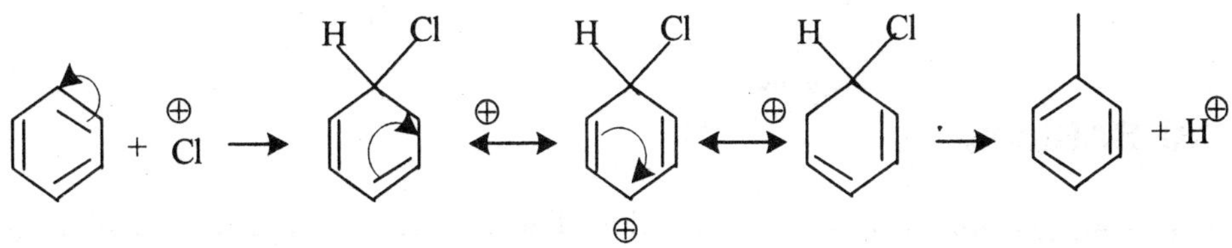

(c) $\overset{\oplus}{H} + \overset{\ominus}{FeCl_4} \longrightarrow HCl + FeCl_3$

(v) *Rearrangements like Pinacol-Pinacolone rearrangement*: Examples are as follows:

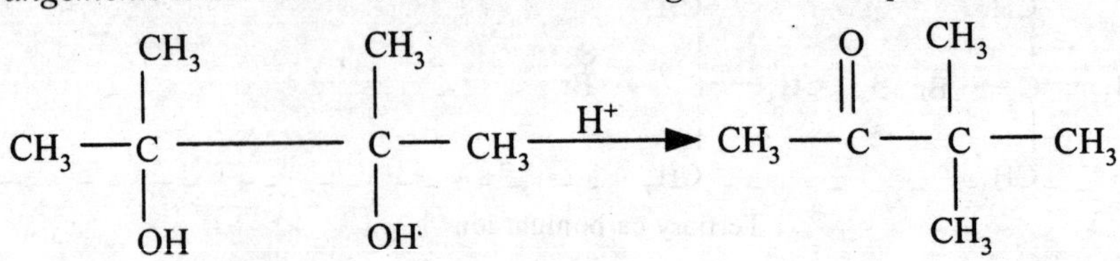

Pinacol Pinacolone

The mechanism consists of following steps:

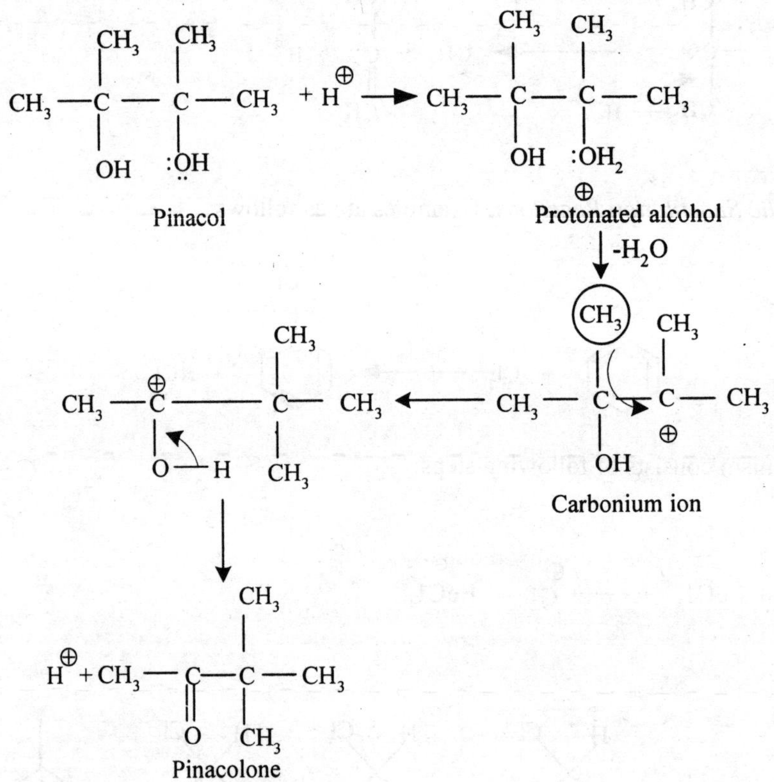

4.3 CARBANIONS

Carbanion is the reaction intermediate in which carbon atom carries negative charge, i.e., $-\overset{|}{\underset{|}{C}}{}^{\ominus}$.
The carbon atom is sp^3 hybridized, hence tetrahedral may be represented as below:

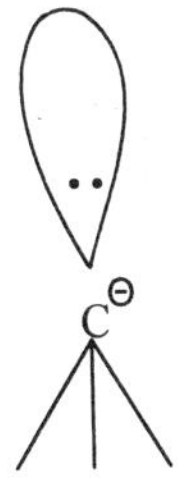

(a) Preparation

(i) By heterolytic fission of a covalent bond in which carbon is attached to a less electronegative element.

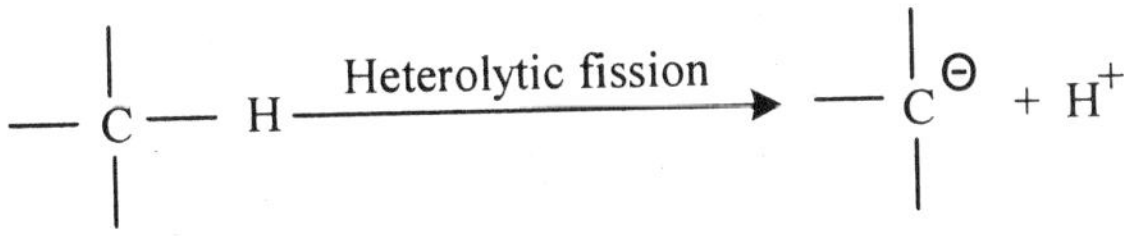

(ii) By addition of an anion to an alkene:

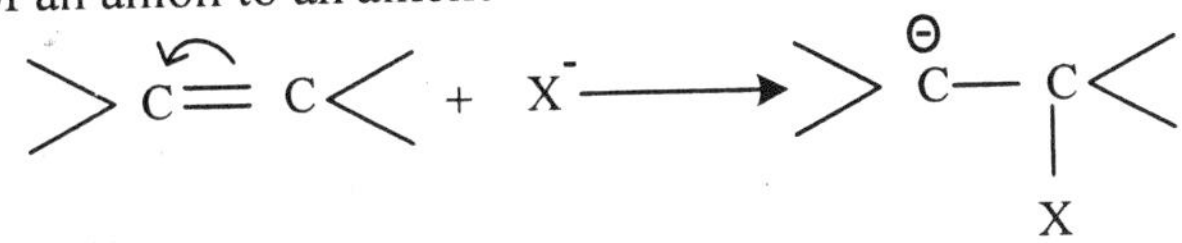

(b) Kinds of Carbanions and Their Stability

The carbanions are of two types:

(i) Non-classical carbanions, and

(ii) Classical carbanions.

(i) *Non-classical carbanions:* In these carbanions, negative charge is delocalized over a number of carbon atoms. It is stabilized due to resonance, e.g.,

$$\overset{\ominus}{C}H_2-CH=CH_2 \longleftrightarrow CH_2=CH-\overset{\ominus}{C}H_2$$

Allyl carbanion

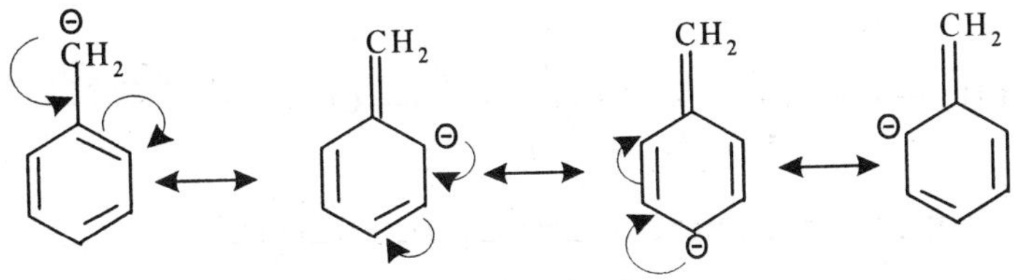

Benzyl carbanion

(ii) *Classical carbanions*: In these carbanions, negative charge is localized on a carbon atom. These have been further classified in three categories:

(a) Primary carbanion,

(b) Secondary carbanion, and

(c) Tertiary carbanion.

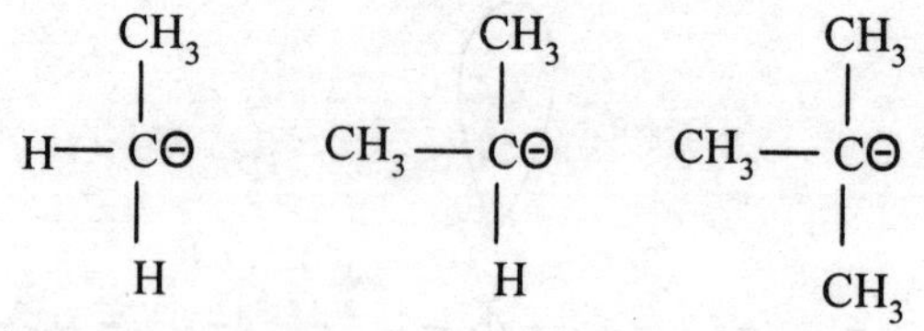

Primary > Secondary > Tertiary

The order stability of carbanions can be explained on the basis of +inductive effect of alkyl group. It is clear that greater the negative formal charge on carbon atom of the carbanion due to +IE of alkyl group, lesser is the stability of carbanion (*See* Chapter 3: Rules for Resonance).

Briefly, it can be shown as below:

$$\left.\begin{matrix}\text{Stability of}\\ \text{classical carbanion}\end{matrix}\right\} \propto \left\{\begin{matrix}\dfrac{1}{\text{Number of alkyl groups}}\\ \dfrac{1}{+\text{IE}}\\ \dfrac{1}{\text{Nucleophilic nature of carbon atom of the carbocation}}\end{matrix}\right.$$

(c) Reaction Involving Carbanion

The examples are as follows:

(a) Aldol condensation:

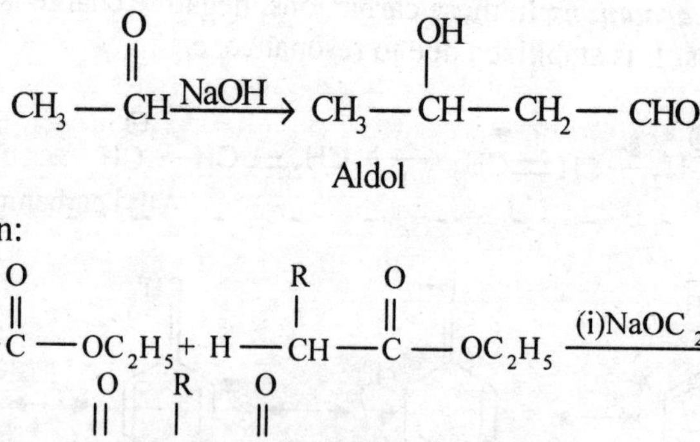

Aldol

(b) Claisen condensation:

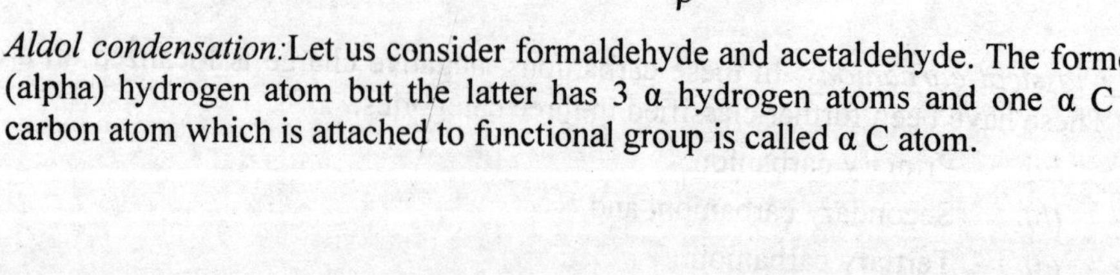

β Ketoester

(a) *Aldol condensation:* Let us consider formaldehyde and acetaldehyde. The former has no α (alpha) hydrogen atom but the latter has 3 α hydrogen atoms and one α C atom. The carbon atom which is attached to functional group is called α C atom.

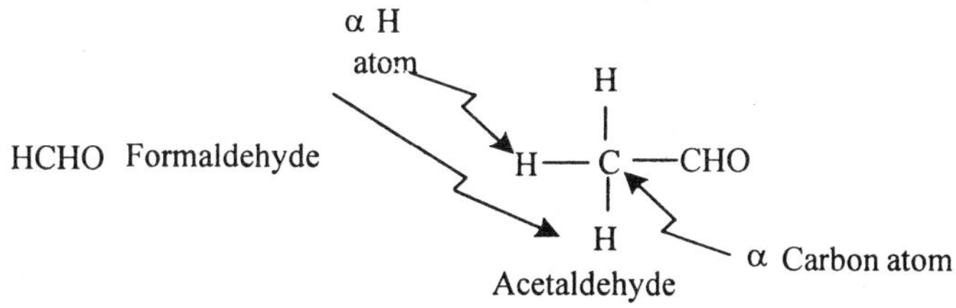

Aldehydes, which have an α hydrogen atom, undergo Aldol condensation forming β-hydroxyl aldehyde (commonly called, Aldol) when treated with an alkali. For example,

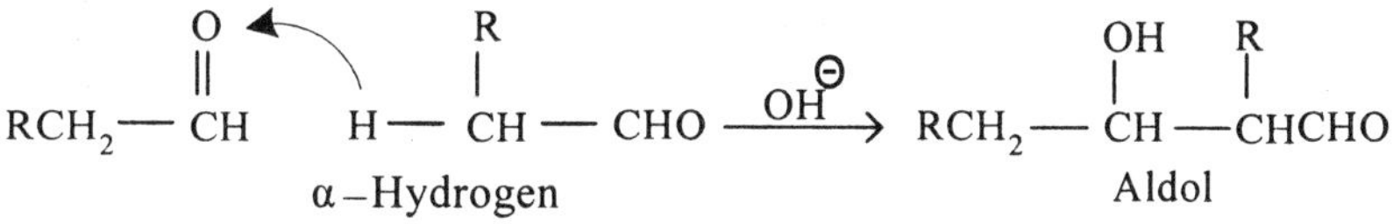

where R = H, alkyl group

(i)

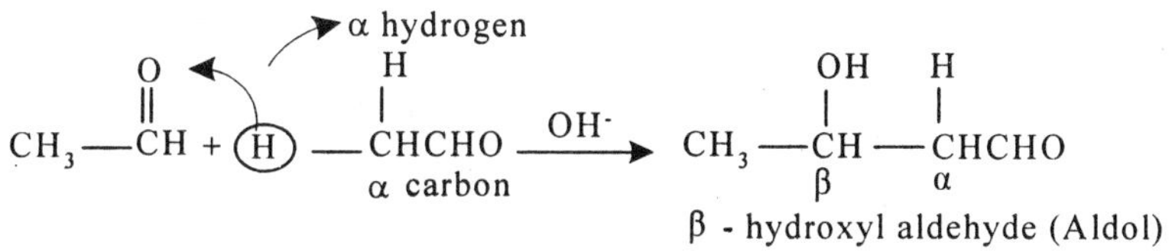

(ii)

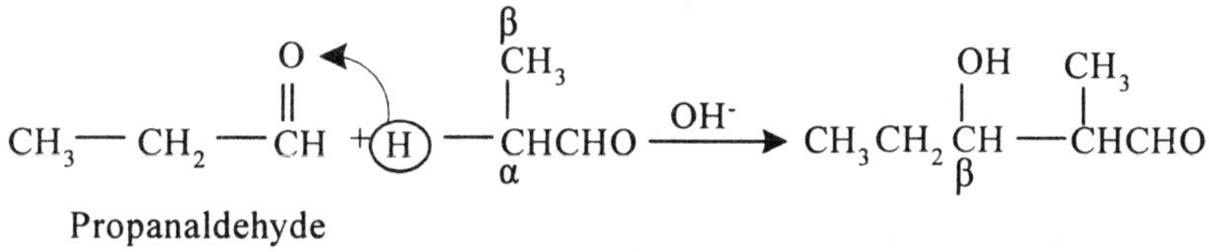

Some ketones also undergo Aldol condensation but here we will consider only the case of aldehydes.

Mechanism of Aldol Condensation

The mechanism involves carbanion as the reactions intermediate. The various steps involved are as follows:

(i)

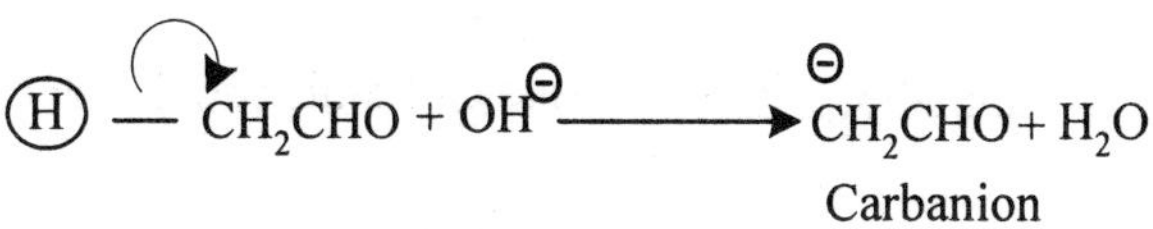

(ii)

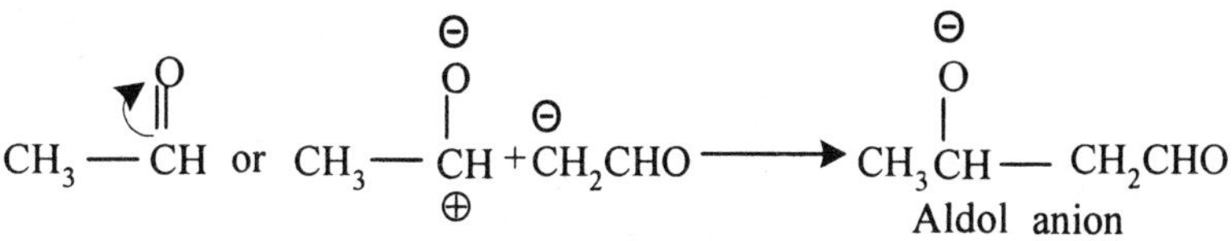

(iii)

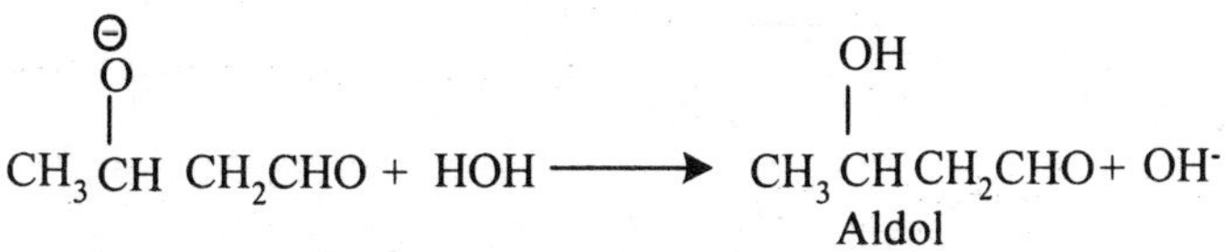

Claisen Condensation

The mechanism consists of the following steps:

(*i*)

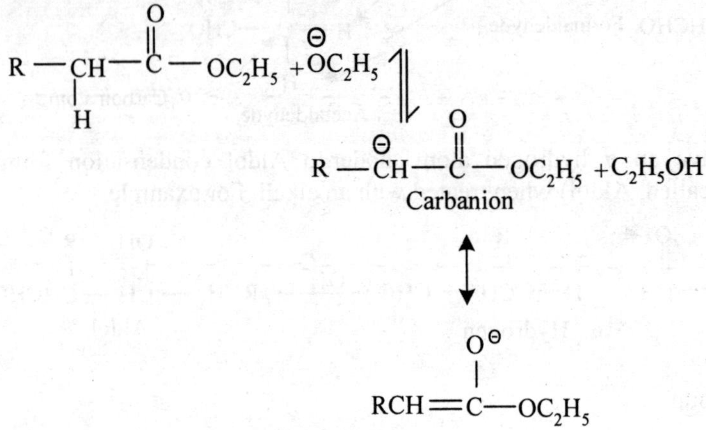

(*ii*)

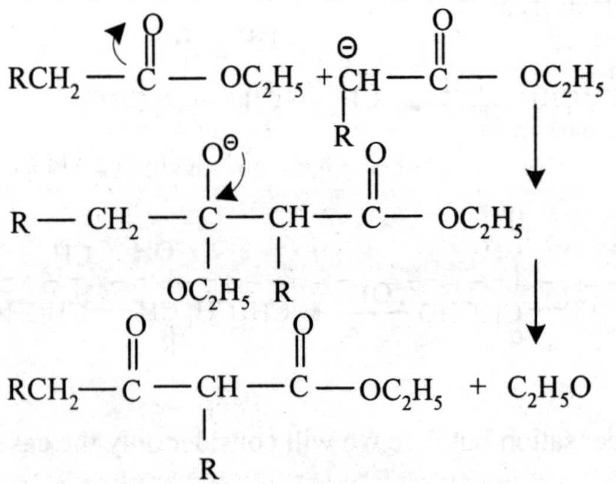

4.4 FREE RADICALS

Free radical is a reaction intermediate which may be defined as an atom or a group of atoms carrying unpaired electron, e.g., H•, Cl•, —C• and HO_2, ȮH. Like carbonium ions, the carbon atom of a free radical is sp^2 hybridized, i.e.,

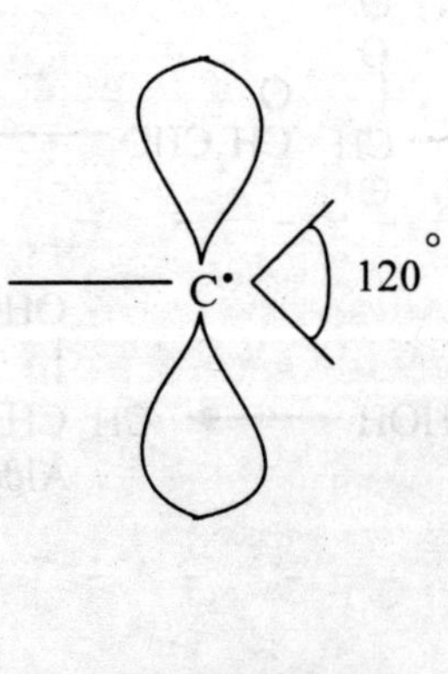

The first stable organic radical was triphenylmethyl radical 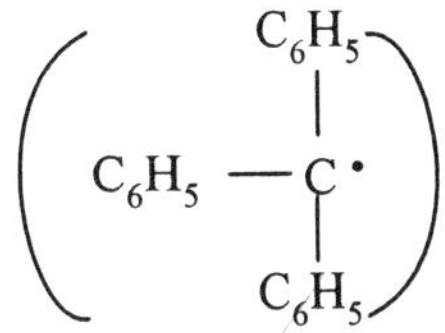 and it was isolated by Professor Gomberg.

(a) Preparation

Free radicals can be generated by a number of methods, some of which are as follows:

(i) By homolytic fission of covalent bond, e.g.,

$$CH_3-Br \xrightarrow{\text{Homolytic fission}} \dot{C}H_3 + \dot{B}r$$

(ii) By thermal decomposition of peroxides, azo compounds

$$\left(\begin{array}{ccc} -N{=}N- & R-N{=}N-R & R-N{=}NX \\ \text{Azo group} & \text{Azo compound} & \text{Diazo compound} \end{array}\right)$$

(a) 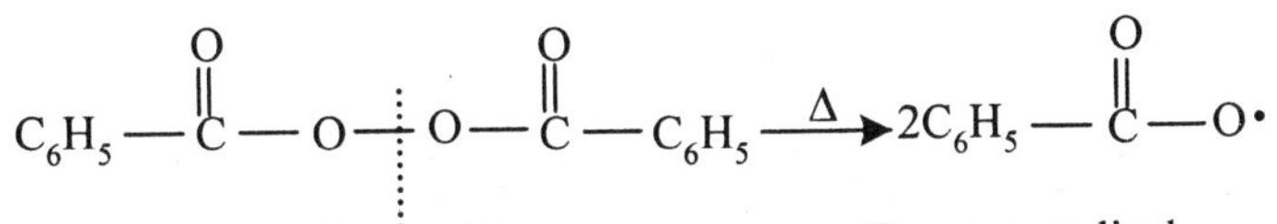

Benzoyl peroxide Benzoxy radical

(b) 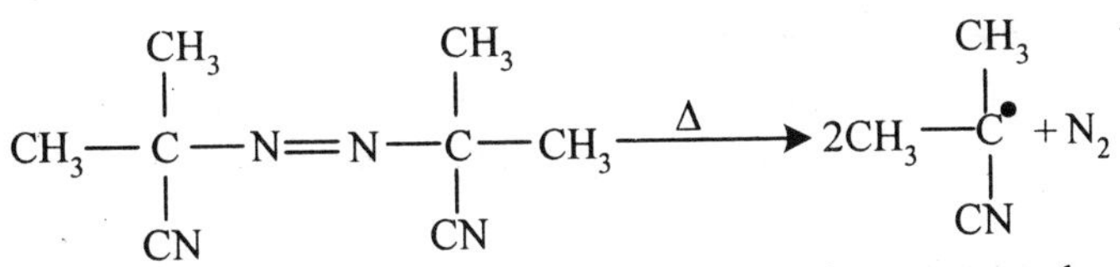

αα′-azobis isobutyronitrile 2-cyanopropyl radical

(iii) By photodecomposition of molecules. This involves decomposition of molecule using sunlight.

$$CH_4 + Cl_2 \xrightarrow{\upsilon} CH_3Cl + HCl \searrow CCl_4 + HCl$$

(a) $$Cl_2 \xrightarrow{h\upsilon} Cl^{\bullet} + Cl^{\bullet}$$

(b) $$CH_4 + \dot{C}l \longrightarrow \dot{C}H_3 + HCl$$

(c) $$\dot{C}H_3 + Cl_2 \longrightarrow CH_3Cl + Cl^{\bullet}$$

(b) Kinds of Free Radicals and Their Stability

Free radicals have been classified in two categories:

(i) Non-classical free radicals, and

(ii) Classical free radicals.

(i) *Non-classical free radicals:* In these free radicals the unpaired electron is delocalized over a number of carbon atom, e.g.,

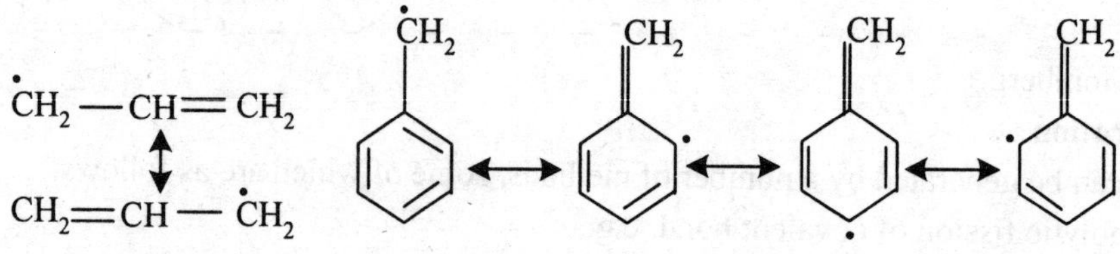

Allyl radical Benzyl radical

(ii) *Classical free radicals*: The unpaired electron is localized on single carbon atom. These have been further classified on the basis of nature of carbon atom that has the unpaired electron:

(a) Primary free radical.

(b) Secondary free radical, and

(c) Tertiary free radical.

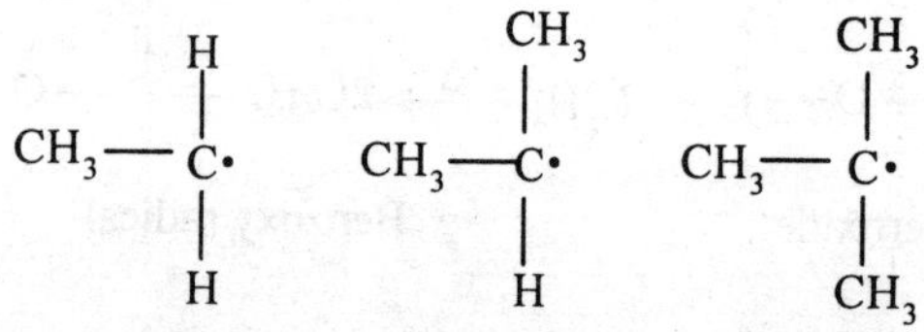

Primary < Secondary < Tertiary

The order of stability can be explained on the basis of hyperconjugation. The number of hyperconjugating structures, like carbonium ion, for primary, secondary and tertiary radicals are 4, 7 and 10, respectively.

(c) Characteristics of Free Radicals

(i) Free radicals are paramagnetic, and

(ii) Free radicals can be detected by radical scavenger like hydroquinone and electron spin resonance spectroscopy.

(d) Reactions Involving Free Radicals

(i) Free radical polymerization

$$\underset{\text{ethylene}}{nCH_2{=}CH_2} \longrightarrow \underset{\text{Polyethylene}}{\left[CH_2 - CH_2\right]_n}$$

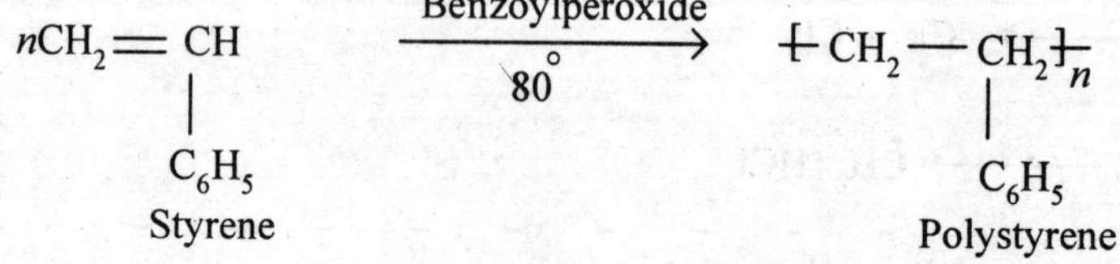

Styrene Polystyrene

(ii) *Halogenation of alkanes:*

$$CH_4 + Cl_2 \xrightarrow{\text{Sunlight}} CH_3Cl \xrightarrow{Cl_2\ \&\ \text{Sunlight}} CH_2Cl_2 + HCl \xrightarrow{Cl_2\ \&\ \text{Sunlight}} CHCl_3 + HCl \xrightarrow{Cl_2\ \&\ \text{Sunlight}} CCl_4 + HCl$$

(iii) *Peroxide effect or Kharasch Effect or Anti-Markovnikov addition:* Addition of HBr to an unsymmetrical alkene (different number of hydrogen atoms are attached to carbon atoms linked by a double bond, e.g., $CH_3–CH{=}CH_2$) in the presence of a peroxide or sunlight is called peroxide or Anti-Markovnikov addition.

$$CH_3—CH{=}CH_2 + HBr \xrightarrow{\text{Peroxide}} \underset{\substack{\text{1-Bromopropane} \\ \text{(88\% yield)}}}{CH_3CH_2CH_2Br}$$

Mechanism of the reaction involves following steps:

Initiation

(i) $$\underset{\text{Benzoyl peroxide}}{C_6H_5—\overset{\Large O}{\overset{\|}{C}}—O \vdots O—\overset{\Large O}{\overset{\|}{C}}—C_6H_5} \longrightarrow \underset{\text{Benzoxy radical}}{2C_6H_5—\overset{\Large O}{\overset{\|}{C}}—\dot{O}} + 87\ \text{kcals/mole}$$

(ii) $$C_6H_5—\overset{O}{\overset{\|}{C}}—\dot{O}—H—Br \longrightarrow C_6H_5—\overset{O}{\overset{\|}{C}}—OH + Br^{\bullet}$$

Propagation

(iii) $$CH_3CH{=}CH_2 + \dot{B}r \longrightarrow \underset{\text{Secondary radical}}{\dot{C}H_3\ CH—CH_2—Br}$$

(iv) $$CH_3—\dot{C}H—\underset{\displaystyle Br}{\underset{|}{CH_2}} + H \not{\cdot} Br \longrightarrow CH_3CH_2—\underset{\displaystyle Br}{\underset{|}{CH_2}} + \dot{B}r$$

Termination

(v) $$\dot{B}r + \dot{B}r \longrightarrow Br_2$$

(vi) $$CH_3\dot{C}HCH_2Br + Br \longrightarrow CH_3-CH(Br)-CH_2Br$$

Peroxide effect is not observed with HF, HCl, and HI. It is because the bond dissociation energy of HF (about 129 kcals/mole) and HCl (about 101 kcals/mole) is high, and therefore step (ii) does not take place. In case of HI, however, the bond dissociation energy is less (about 71 kcals/mole) but Iodine radical (i) formed in step (ii) fails to add to alkene and therefore step (iii) does not take place.

4.5 CARBENES

Carbenes are reaction intermediates in which carbon atom carries two unpaired electrons. For example,

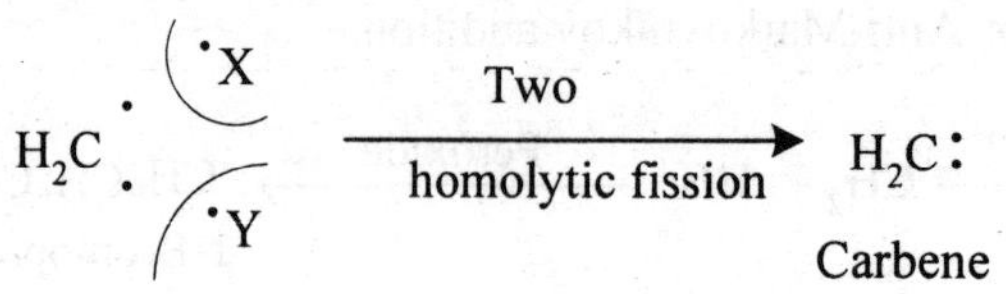

It is to be noted that the two unpaired electrons of carbene are derived from two different bonds. Thus carbene is electron-deficient species having only **six** electrons in the outermost shell. These are considered as diradicals and ultimately break into free radicals.

(a) Preparation

(i) $$\underset{\text{diazo methane}}{CH_2N_2} \xrightarrow[\text{decomposition by heat}]{\text{Photochemical}} :CH_2 + N_2$$

(ii) By the action of base on suitable polyhalogen compounds:

$$CHCl_3 + OH^- \longrightarrow :\overset{\ominus}{C}Cl_3 + H_2O$$

$$:\overset{\ominus}{C}Cl_3 \longrightarrow :CCl_2 + Cl^{\ominus}$$

(b) Kinds of Carbenes

Carbenes have been classified as follows:

(i) *Single carbene*: Both electrons are in one orbital with antiparallel spins and therefore it has not magnetic moment, i.e., it is diamagnetic

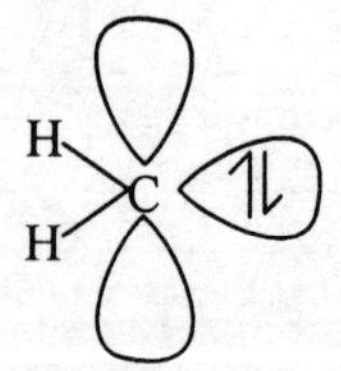

(ii) *Triplet carbene*: Two electrons are present in two different orbital with parallel spins therefore it has magnetic moment, i.e., it is paramagnetic.

$:CH_2 > :CF_2 > :CCl_2 > :CBr_2$

Reactivities of carbenes are as follows:

(c) Reactions Involving Carbenes

(i) Carbylamine reaction

$$RNH_2 + CHCl_3 + 3KOH \rightarrow RNC + 3KCl + 3H_2O$$

(: CCl_2 dichlorocarbene is reaction intermediate)

(ii) Reimer–Tiemann Reaction

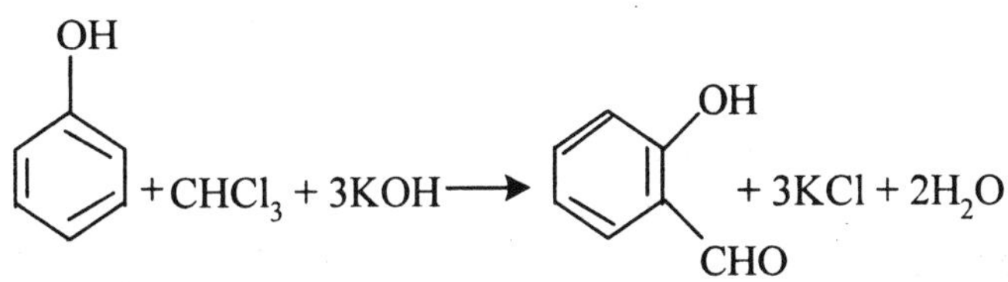

Salicylicaldehyde

(: CCl_2 dichlorocarbene is reaction intermediate)

PROBLEMS

1. Amongst the four alternate answers given for the subquestions below, one or more are correct. Write down the correct answer(s).

(i) The compound which gives the most stable carbonium ion on dehydration is:

(a) $CH_3—CH(CH_3)—CH_2—OH$ *(b)* $CH_3—C(CH_3)_2—OH$

(c) $CH_3CH_2CH_2CH_2OH$ *(d)* $CH_3\,CH(OH)—CH_2—CH_3$ **(IIT, 1989)**

(ii) In the reaction of phenol with $CHCl_3$ and aq. NaOH at 70°C, the electrophile attacking the ring is:

(a) $CHCl_3$ *(b)* $CHCl_2$ *(c)* CCl_2 *(d)* $COCl_2$

(iii) The most stable carbonium ion is

(a) Methyl carbonium ion

(b) Primary carbonium ion

(c) Secondary carbonium ion

(d) Tertiary carbonium ion **(MNR, 1988)**

(iv) Which of the following would undergo dehydration fast?

(a) 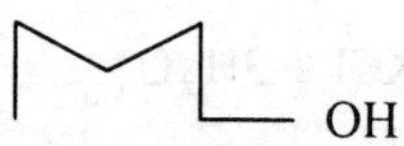

(b) 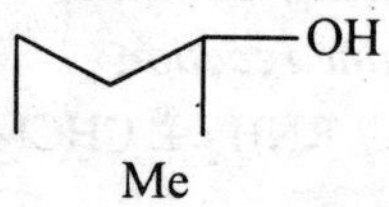

(c) 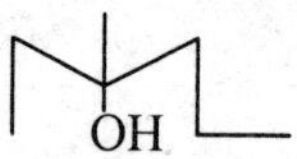

(d) 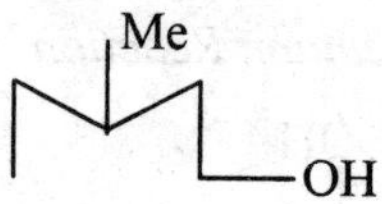

(v) Which of the following is most stable?

(a) $CH_3\overset{\oplus}{C}H_2$ *(b)* $(CH_3)_2\overset{\oplus}{C}H$

(c) $(CH_3)_3\overset{\oplus}{C}$ *(d)* $C_6H_5\overset{\oplus}{C}H_2$ **(DPMT, 1982)**

(vi) When acetaldehyde reacts with propionaldehyde in the presence of a base, the number of products is:

(a) 1 *(b)* 2 *(c)* 3 *(d)* 4

(vii) When propionaldehyde is treated with aq. NaOH, the product is

(a) $CH_3CH_2\underset{|}{\overset{OH}{C}}H—\underset{|}{\overset{CH_3}{C}}H— CHO$

(b) $CH_3\ CH_2\overset{OH}{\overset{|}{C}}H—CH_2CH_2\ CHO$

(c) $CH_3CH_2COOCH_2CH_2CH_3$

(d) $CH_3CH_2COCH_2CH_2CHO$ **(MNR, 1982; IIT, 1992)**

(viii) Which of the following would undergo Aldol condensation fast?

(a) CH_3CHO *(b)* CH_3CH_2CHO

(c) $CH_3—\overset{CH_3}{\overset{|}{C}}H\ CHO$ *(d)* $CH_3\ CH_2\overset{CH_3}{\overset{|}{C}}H\ CHO$

(ix) Addition of HI on a double bond of propene yields isopropyl iodide and not *n*-propyl iodide as the major product. This is because the addition proceeds through:

(a) A more stable carbonium ion

(b) A more stable carbanion

(c) A more stable free radical

(d) None of the above, being a concerted reaction **(CPMT, 1989)**

(x) The most stable free radical among the following is

(a) CH_3CH_2 *(b)* CH_3CHCH_3

(c) $CH_3CHC_6H_5$ *(d)* $C_6H_5CH_2CH_2$ **(GATE, 1993)**

(xi) Which of the following species is/are involved in Reimer-Tiemann Reaction?

(a) 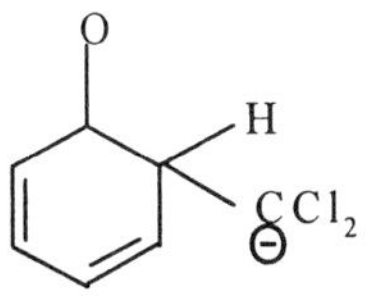

(b) 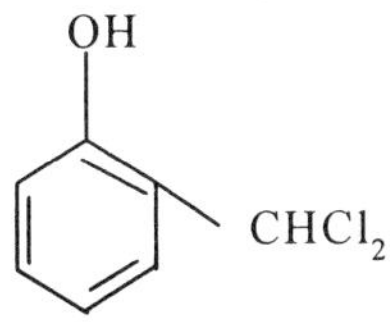

(c) 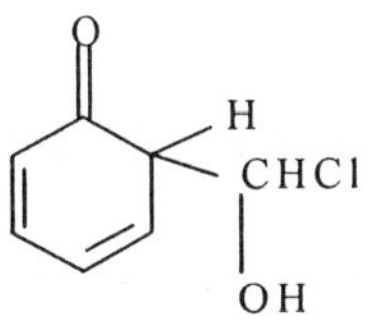

(d) 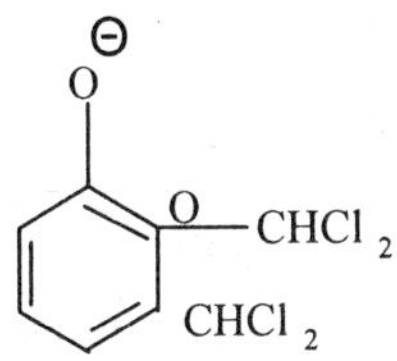

(xii) The order of reactivity of the following alcohols is:

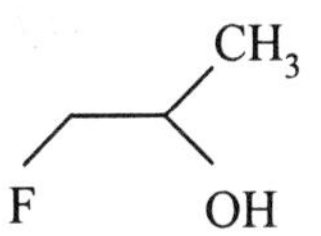

I

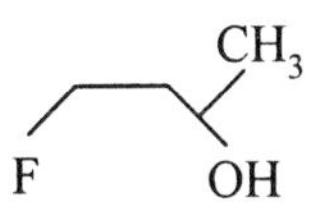

II

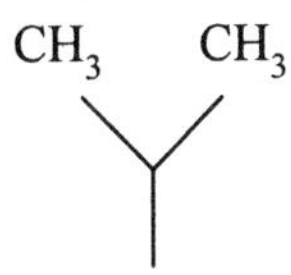

III

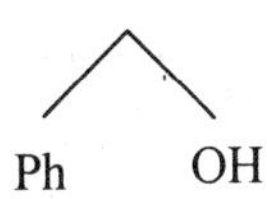

IV

(a) I > II > III > IV *(b)* I > III > II > IV

(c) IV > III > II > I *(d)* IV > III > I > II

(xiii) In the presence of peroxide, hydrogen chloride & hydrogen iodide do not give anti-Markovnikov addition to alkene because

(a) Both are highly ionic;

(b) One is oxidizing and the other is reducing;

(c) One of the steps is endothermic in both the cases;

(d) All the steps are exothermic in both the cases. **(IIT, 2002)**

(xiv) Consider the following reaction:

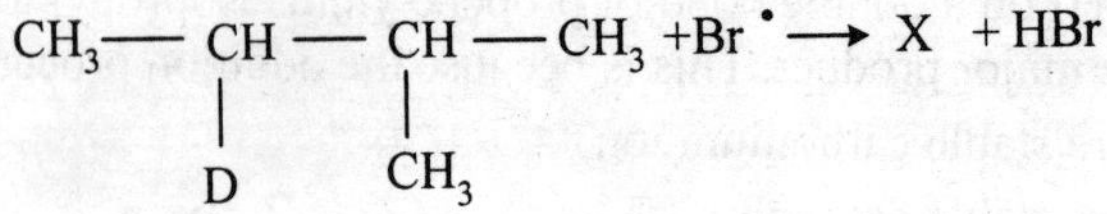

Identify the structure of the major product `X'

(a) 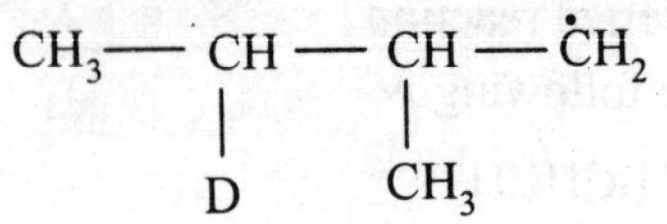
(b) 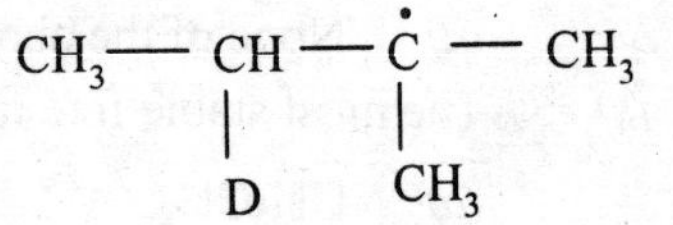

(c) 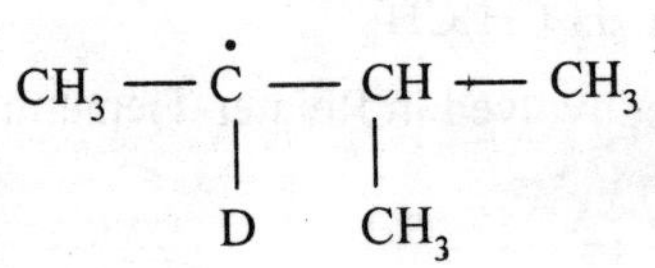
(d) 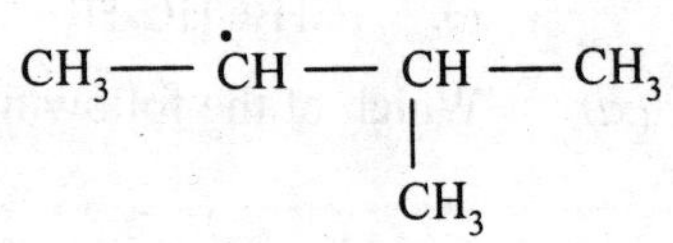

2. Fill in the blanks:

(i) The shape of $\overset{\oplus}{C}H_3$ is..................... **(IIT, 1990)**

(ii) The shape of $\overset{\ominus}{C}H_3$ is..................... **(SSJM Kanpur, 1996)**

3. State whether the following statement is True or False:
The formation of CH_3Cl, by the gas phase ultraviolet irradiation of a mixture of CH_4 and Cl_2 involves free-radical intermediate. **(MNR, 1991)**

4. 3,3-Dimethyl-butanol-2 loses a molecule of water in the presence of conc. H_2SO_4 to give tetramethylethylene as a major product. Suggest a suitable mechanism. **(IIT, 1996)**

5. Complete the equation

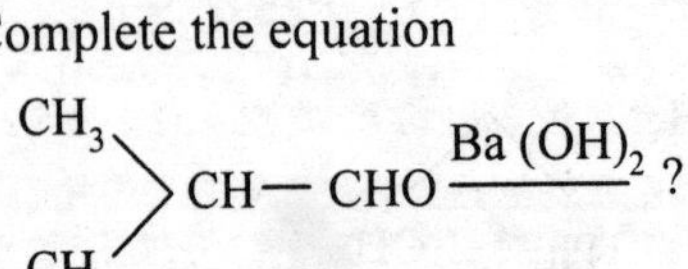

ANSWERS

1. (i) **(b).** Remove —OH from the carbon atom to which it is attached and put +ve charge atom to get carbonium ion.

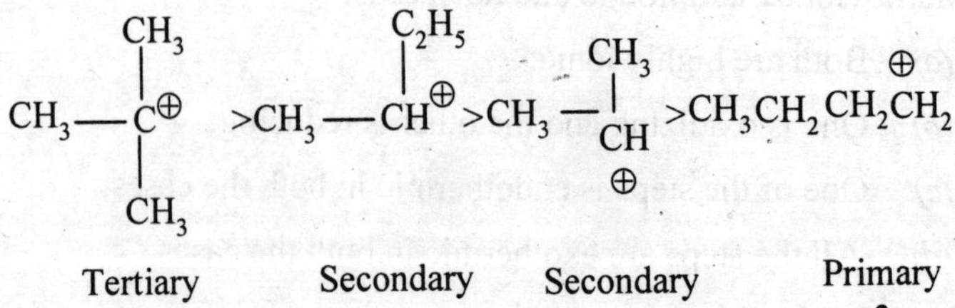

Tertiary	Secondary	Secondary	Primary
10	6	7	3

No. of Hyperconjugating structures

(ii) (**c**) See page (69)

(iii) (**d**) See page (55)

(iv) (**c**) Examine the stability of carbonium ion formed as reaction intermediate

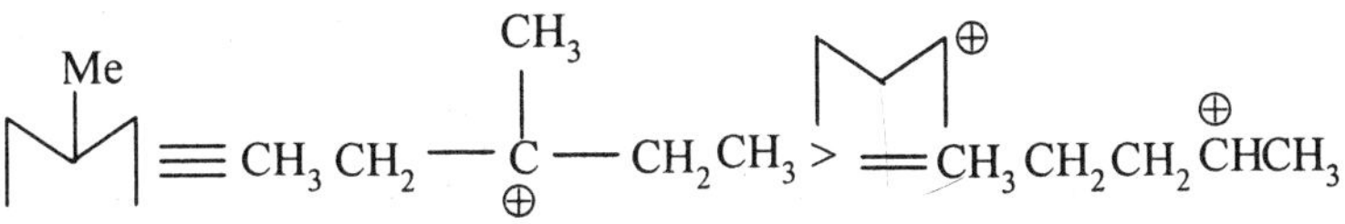

Tertiary Secondary

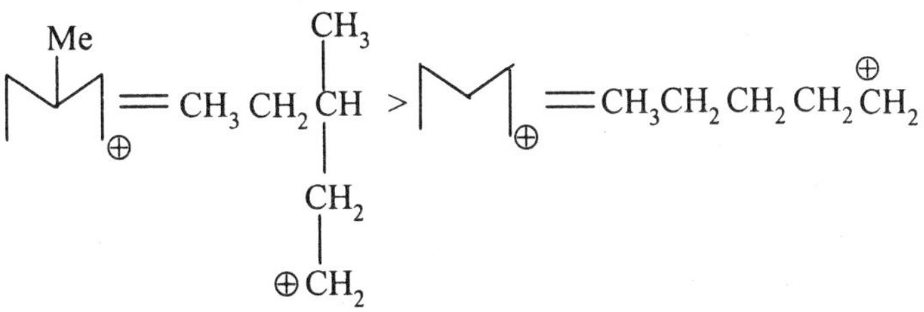

(v) (**d**) See page (55)

$$C_6H_5\overset{\oplus}{C}H_2 > (CH_3)_3\overset{\oplus}{C} > (CH_3)_2\overset{\oplus}{C}H > CH_3\overset{\oplus}{C}H_2$$

Non-classical Benzyl carbonium ion, Tertiary, Secondary, Primary

(vi) (**d**) Four products are obtained: See below: It is called **crossed aldol condensation** reaction

$$CH_3CHO + CH_3CHO \xrightarrow{OH^{\ominus}} CH_3\underset{}{CH}(OH)CH_2CHO$$

First Product

$$C_2H_5CHO + C_2H_5CHO \xrightarrow{OH^{\ominus}} CH_3CH_2CH(OH)—CH(CH_3)CHO$$

Second Product

$$CH_3CH_2\overset{O}{\overset{\|}{C}}H + \overset{\ominus}{C}H_2CHO \xrightarrow{OH^{\ominus}} CH_3CH_2CH(OH)—CH_2CHO$$

Third Product

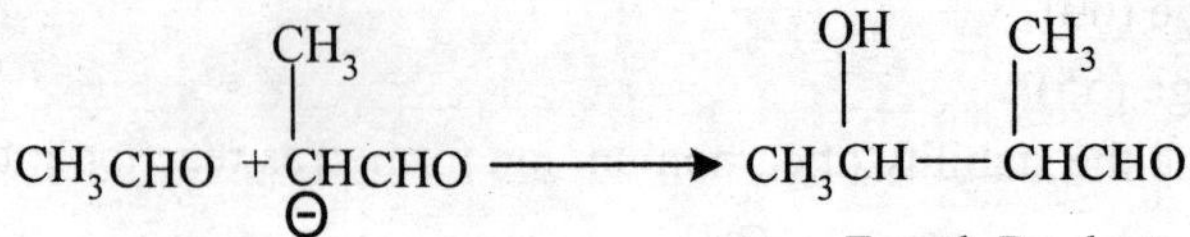

Fourth Product

(vii) **(a)**Answer (b) is possible when you consider β H atom for formation of —OH group. See —OH is on γ carbon atom.

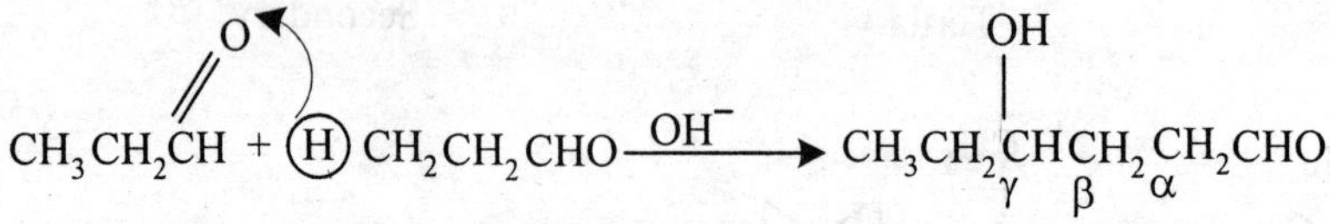

(viii) **(a)** See the stability of carbanion:

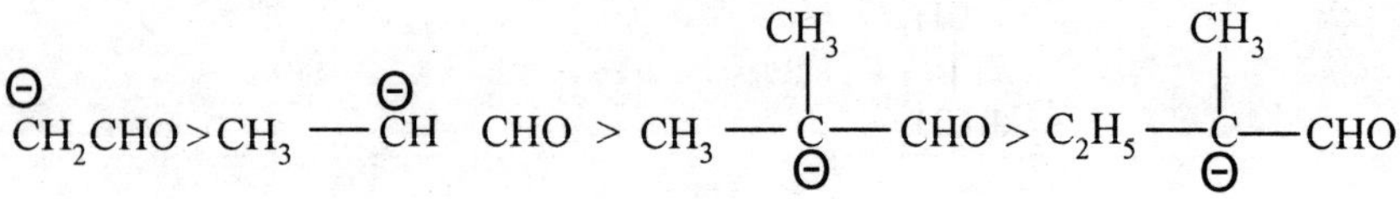

(ix) **(a)** See mechanism on page (57).

(x) **(c)** $C_6H_5\ CH_2\ CH_2$ and $CH_3\ CH_2$ are primary radicals

CH_3—$CHCH_3$ is a secondary radical

C_6H_5—CH—CH_3 is a non-classical radical.

(xi) **(a)** and **(d)**

(xii) **(c)**

(xiii) **(c)**

(xiv) **(b),** because it is tertiary radical.

2. *(i)* Trigonal planar due to sp^2 hybridization of carbon atom.

(ii) Tetrahedral

3. True:

4. 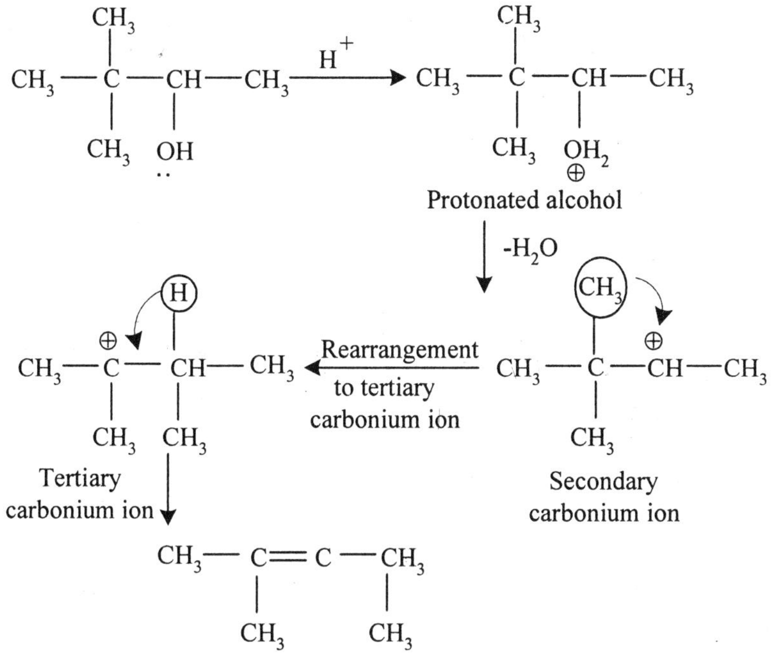

Note: Secondary carbonium ion changes to more stable tertiary carbonium ion.

5. $$\begin{matrix} CH_3 \\ CH_3 \end{matrix}\!\!> CH-CHO \xrightarrow{Ba(OH)_2} \left(\begin{matrix} CH_3 \\ CH_3 \end{matrix}\!\!> CH-COO\right)_2 Ba + \begin{matrix} CH_3 \\ CH_3 \end{matrix}\!\!> CH-CH_2OH$$

Chapter 5

Organic Reactions and Their Mechanism

5.1 INTRODUCTION

The following chart shows various types of organic reactions:

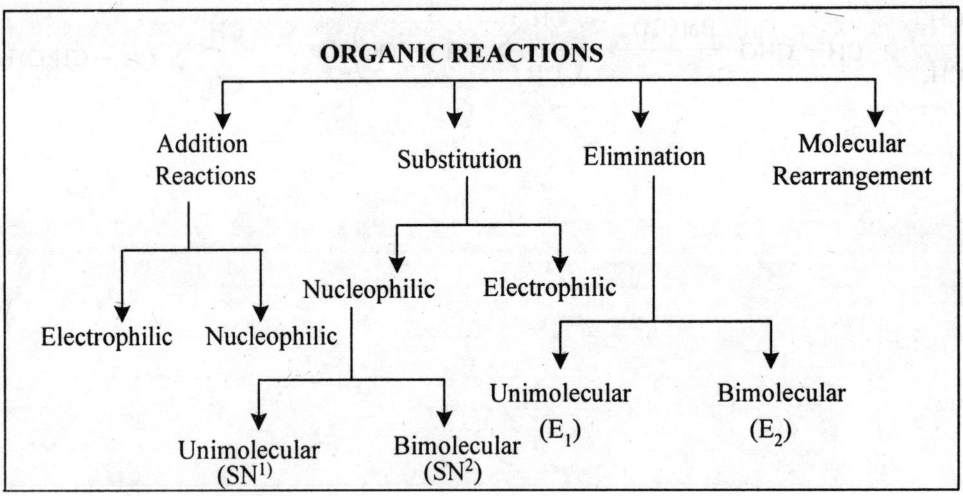

5.2 ADDITION REACTIONS

Any organic reaction in which one π bond and one σ bond is converted into 2σ bonds is called an addition reaction. The heat evolved in making 2σ bonds exceeds that needed to break one σ and one π bond; so addition reactions are usually exothermic.

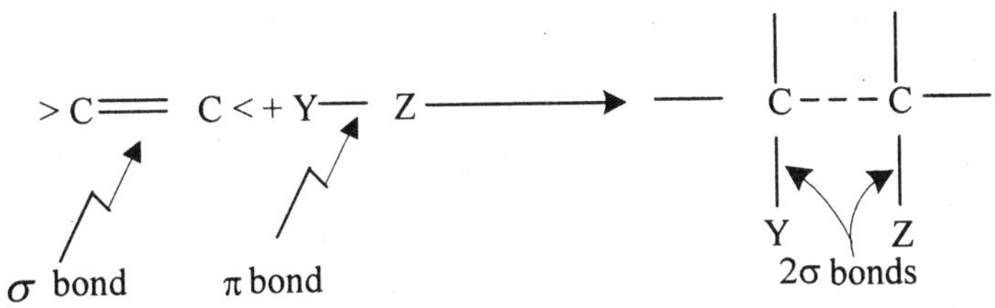

If this change is brought about by an electrophile, it is called electrophilic addition reaction and it involves Carbonium ion as reaction intermediate. On the contrary, in case of nucleophilic addition reactions, the change is brought about by a nucleophile and it involves an anion as a reaction intermediate.

Examples of Electrophilic Addition Reactions

(a) ***Addition of halogens to alkenes***

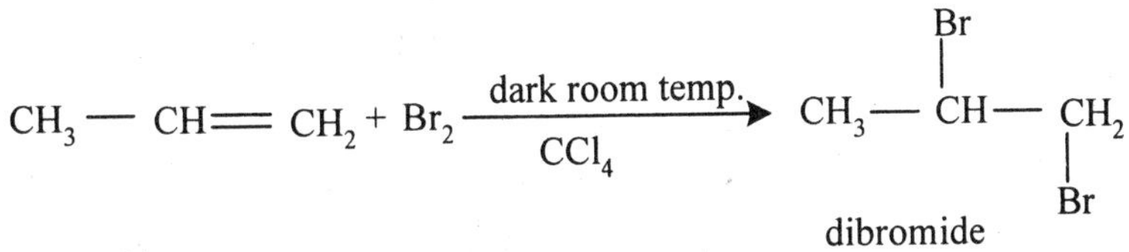

Various steps involved in the reactions:

(i)

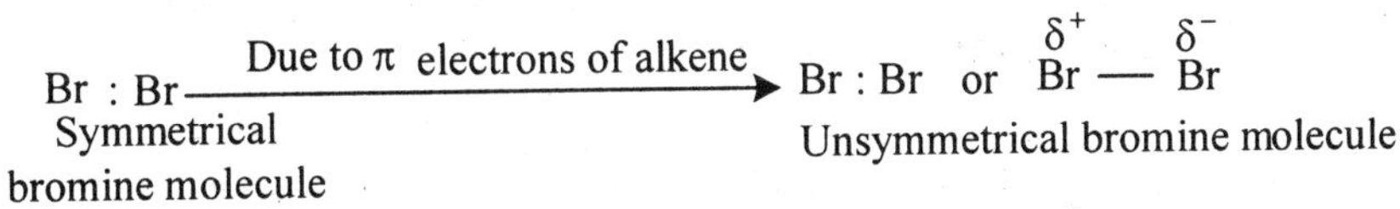

(ii)

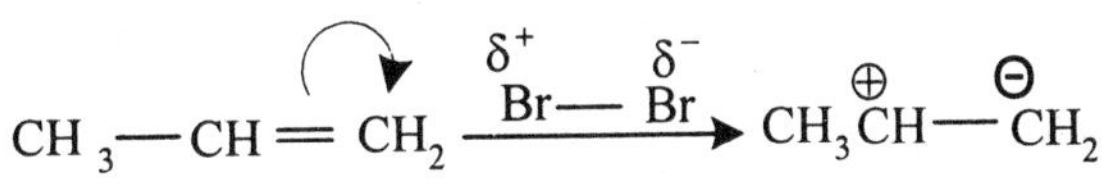

The shifting of π electrons towards right side is because of the electron-repelling nature of $-CH_3$ group and it leads to the formation of a secondary carbonium ion, which is more stable.

(iii)

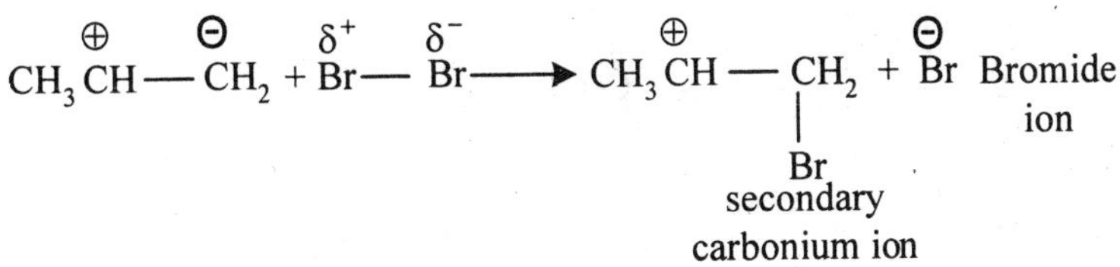

(iv)

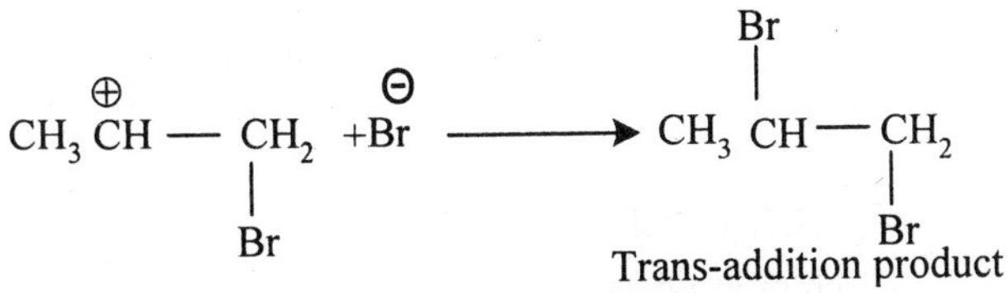

(b) ***Addition of Hydrogen Halides to Alkenes (Markovnikov's Rule)***

Markovnikov's Rule states that when an unsymmetrical molecule adds to an unsymmetrical alkene, the negative part of the addendum attaches to the carbon atom bearing less number of hydrogen atoms.

Br: Br	H: Cl	$CH_2=CH_2$
Symmetrical addendum (electronegativity of both the atoms is same)	Unsymmetrical addendum (electronegativity of both the atoms is different)	Symmetrical alkene (same number of hydrogen atoms is attached to the carbon atoms linked by double bond)

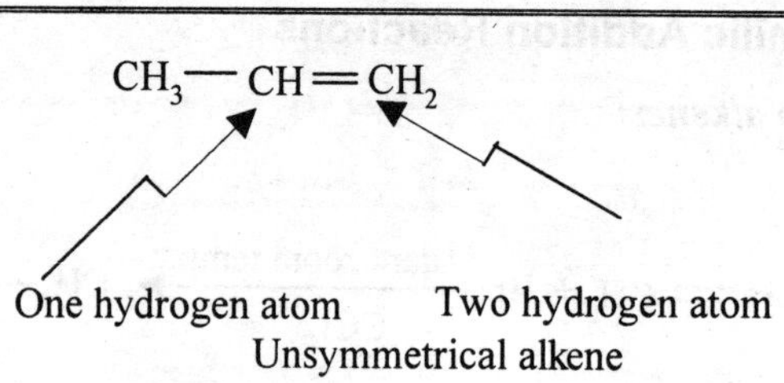

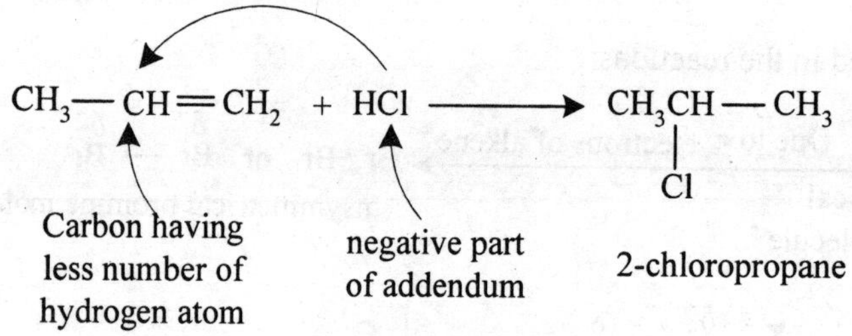

Reaction mechanism consists of the following steps:

(i) $\overset{\delta^+}{H}-\overset{\delta^-}{Cl} \xrightarrow{\text{Due to } \pi = \text{electrons of alkene}} \overset{\oplus}{H}-\overset{\ominus}{Cl}$

(ii) $CH_3-CH=CH_2 \xrightarrow{\overset{\oplus}{H}-\overset{\ominus}{Cl}} CH_3-\overset{\oplus}{CH}=\overset{\ominus}{CH_2}$

(iii) $CH_3\overset{\oplus}{CH}-\overset{\ominus}{CH_2} + \overset{\oplus}{H}-\overset{\ominus}{Cl} \xrightarrow{\text{Slow}} CH_3-\overset{\oplus}{CH}-CH_3 + Cl^{\ominus}$

Secondary carbonium ion

(iv) $CH_3\overset{\oplus}{CH}CH_3 + Cl^- \longrightarrow CH_3\underset{}{\overset{Cl}{\overset{|}{C}}}HCH_3$

2-chloropropane

In step (iii), the alkene accepts a proton from the hydrogen halide and forms a carbonium ion. This step is highly endothermic and has high energy of activation. Consequently it is a slow step. In step (iv), the highly reactive carbonium ion stabilizes itself by combining with a halide ion. This exothermic step has a very low energy of activation and takes place very rapidly.

Theoretical explanation of Markovnikov's Rule: Step (iii) of the reaction mentioned above could conceivably lead to two different carbocations:

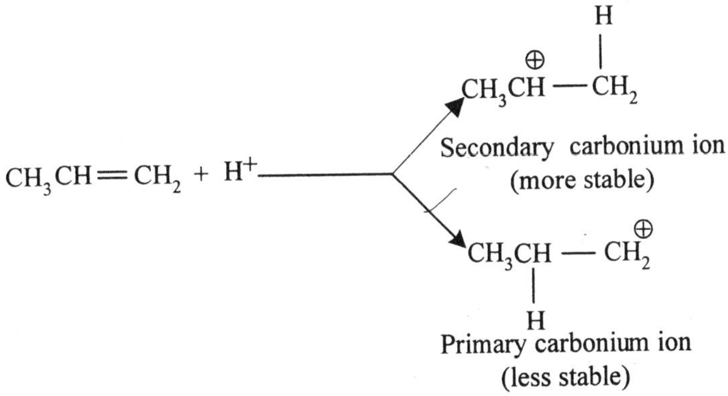

The secondary carbonium ion is more stable, and it is the greater stability of the secondary carbonium ion that accounts for the correct prediction of the overall addition by Markovnikov's rule. Therefore, we can now give the following revised statement of Markovnikov's rule:

"In the ionic addition of an unsymmetrical reagent to a double bond, the positive portion of the adding reagent attaches itself to carbon atom of the double bond so as to yield the more stable carbocation as an intermediate."

Another interesting example of Markovnikov's rule is

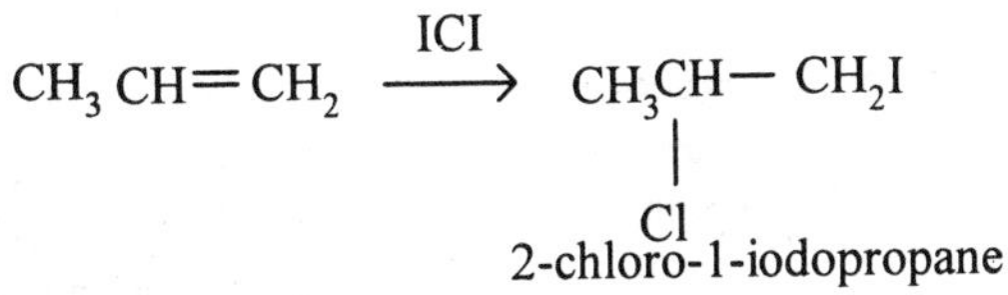

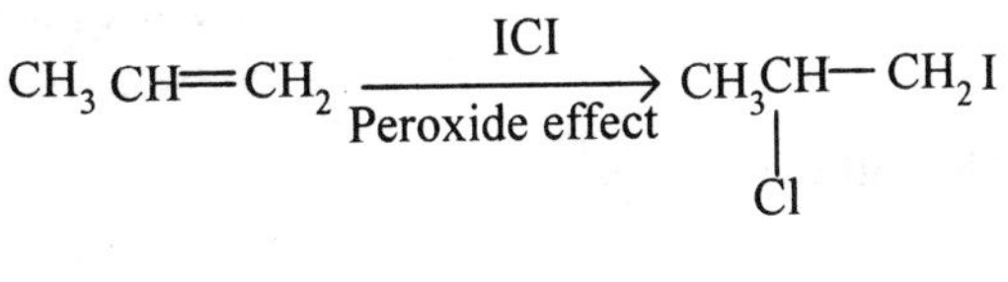

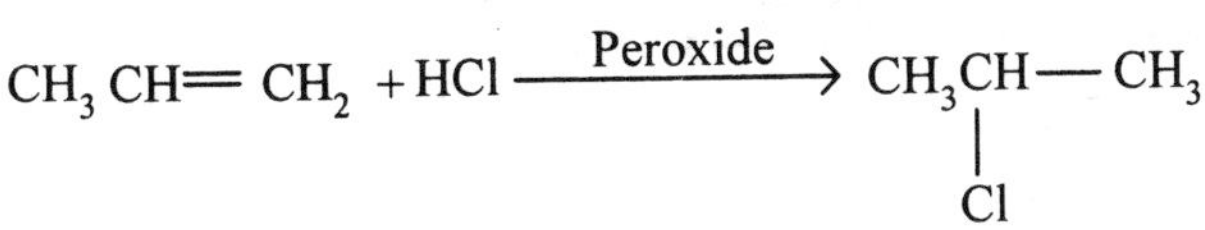

Note: In all above cases, the addition is according to Markovnikov's rule.

The Markovnikov addition of hydrogen halides to alkenes is said to be *Regioselective* (when a reaction that can potentially yield two or more constitutional isomers actually produces only one or a predominance of one, the reaction is said to be regioselective).

Examples of Nucleophilic Addition Reactions

(a) *Addition of Hydrogen Cyanide to Carbonyl Group*

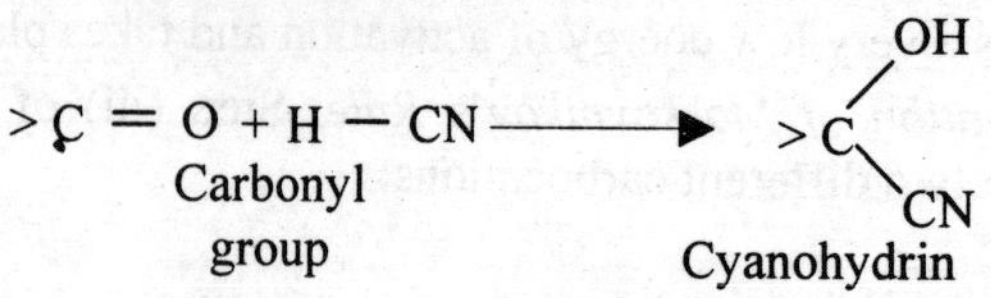

According to Arthur Lapworth (1903) the mechanism consists of the following steps:

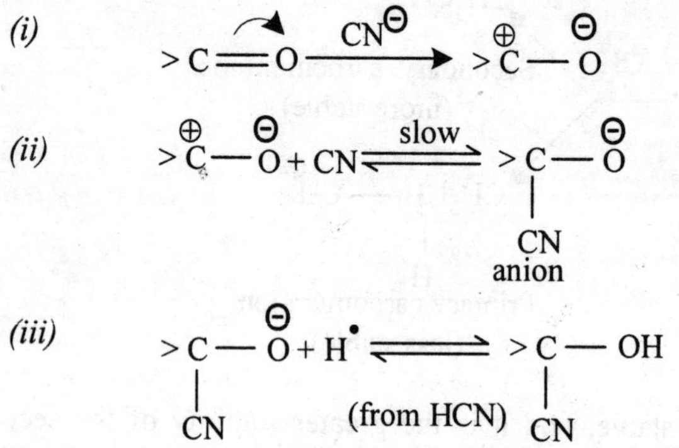

The reaction occurs very slowly since HCN is a poor nucleophile. However, the addition of potassium cyanide or any base that can give CN^- from the weak acid HCN increases the rate very fast. The cyanide ion, being a stronger nucleophile, is able to attack the carbonyl carbon atom more rapidly than hydrogen cyanide itself. Bases stronger than cyanide ion catalyze the reaction by converting HCN ($Ka = 10^{-9}$) to cyanide ion in an acid-base reaction.

(b) *Addition of Sodium Bisulphite to Carbonyl Group*

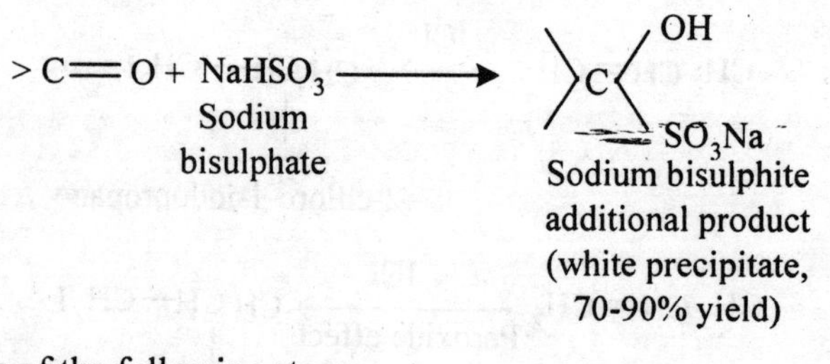

The mechanism consists of the following steps:

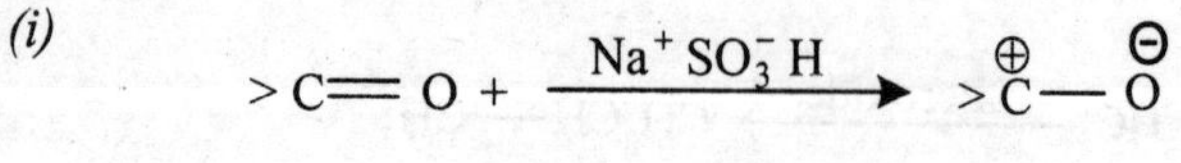

(ii)

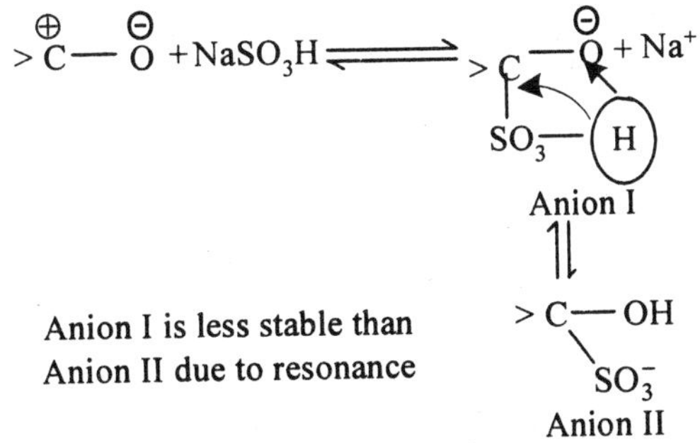

(iii)

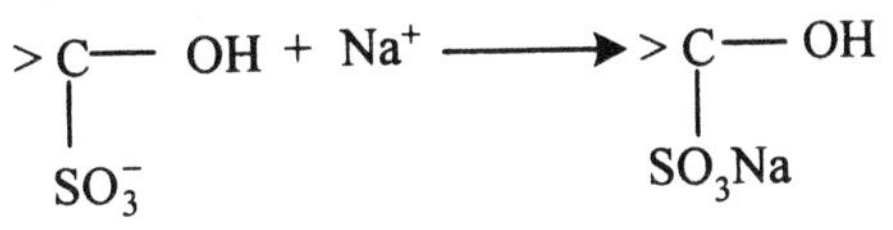

Higher ketones generally do not give bisulphate addition products in appreciable amounts because the addition is very sensitive to steric hindrance. However, yield can be improved by using excess $NaHSO_3$. This reaction is used to separate aldehydes and ketones from other organic compounds like alcohols.

5.3 NUCLEOPHILIC SUBSTITUTION REACTIONS

Organic reaction in which a group or an atom is replaced by a nucleophile along with transfer of unshared electrons pair from nucleophile to reaction center and from reaction center to the leaving group is known as a nucleophile substitution reaction. In general, it may be represented as:

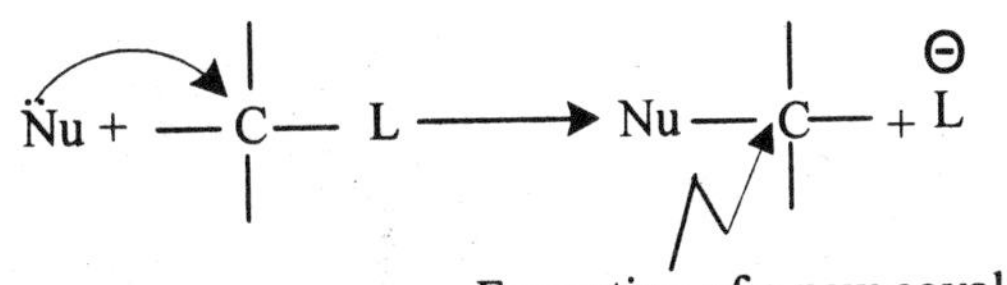

where, Nu = Nucleophile; L = Leaving group

Following are some of the examples:

(i) $H\ddot{O}^{\ominus} + CH_3-Cl \longrightarrow CH_3-OH + Cl^{\ominus}$

(ii) $CH_3O^{\ominus} + CH_3CH_2-Br \longrightarrow CH_3CH_2-OCH_3 + Br^{\ominus}$

(iii) $I^{\ominus} + CH_3CH_2CH_2-Cl \longrightarrow CH_3CH_2CH_2-I + Cl^{\ominus}$

Two changes take place in the reactions: formation of a new covalent bond and breaking of the old covalent bond. Therefore, nucleophilic substitution reactions, on the basis of kinetic studies, have been classified in two categories.

(A) Unimolecular nucleophilic substitution reactions (SN^1)

(B) Biomolecular nucleophilic substitution reactions (SN^2).

(A) *Bimolecular nucleophilic Substitution reaction* (SN^2) Kinetic studies show that the rate of this type of reaction depends upon the concentration of both the species, i.e., substrate and nucleophile taking part in the reaction, hence the word 'bimolecular' is used. For example:

$$CH_3CH_2\,Br + \overset{\ominus}{O}H \longrightarrow CH_3CH_2\,OH + :\ddot{\underset{..}{Br}}:^{\ominus}$$

So we can write rate of reaction $(-dx/dt) \propto$ [Substrate] $[\ddot{N}u]$.

The mechanism involves transition state, i.e., making of a new covalent bond and breaking of the old covalent bond occurs simultaneously. The reaction proceeds with *inversion of configuration.* It is stereochemistry of the reaction, i.e., if substrate is dextrorotatory, the product is laevorotatory and vice versa because the nucleophile attacks from the side remote (opposite side and in the same plane) to the leaving group.

Example: Hydrolysis of primary alkyl halide

$$CH_3CH_2\text{—}Br + aq.\ KOH \longrightarrow CH_3CH_2\,OH + KBr$$

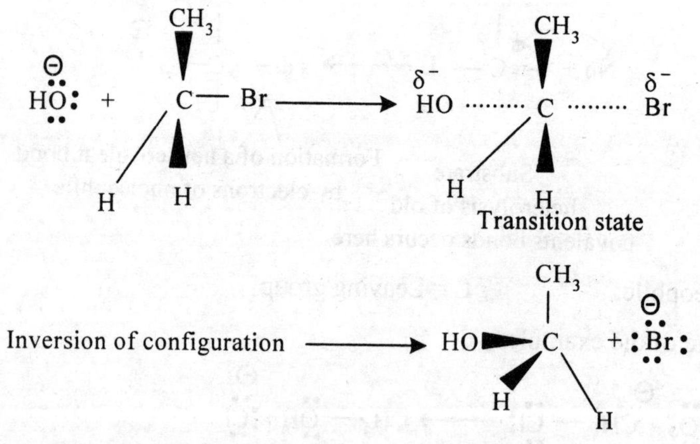

The rate of SN^2 increases in polar aprotic solvents, that is, the solvents whose molecules do not have a hydrogen atom that is attached to an atom of highly electronegative elements like oxygen, e.g.,

Dimethyl sulfoxide $(CH_3\overset{O}{\overset{\|}{S}}CH_3)$, N, N-dimethyl formamide HCO—N $(CH_3)_2$.

(B) *Unimolecular nucleophilic substitution reaction* (SN^1): Kinetic studies reveal that the rate of this type of reaction depends only upon the concentration of the substrate and is independent of that of the nucleophile, i.e., rate of reaction ($-dx/dt$) $\propto$ [Substrate] $[Nu]^{\circ}$ hence the word 'unimolecular' used.

The old covalent bond is broken first so that the carbonium ion is formed as reaction intermediate. This is followed by addition of the nucleophile. Since carbonium ion is flat, planar species, the nucleophile can attack either from the front side or from the side remote to the leaving group, with equal probability. In fact, the nucleophile does so, and therefore, the reaction proceeds with racemization (this is the stereo-chemistry of the reaction), i.e., the substrate may be dextro or laevorotatory but the product is always a mixture of the two forms (racemic mixture).

Example: Hydrolysis of tertiary alkyl halide

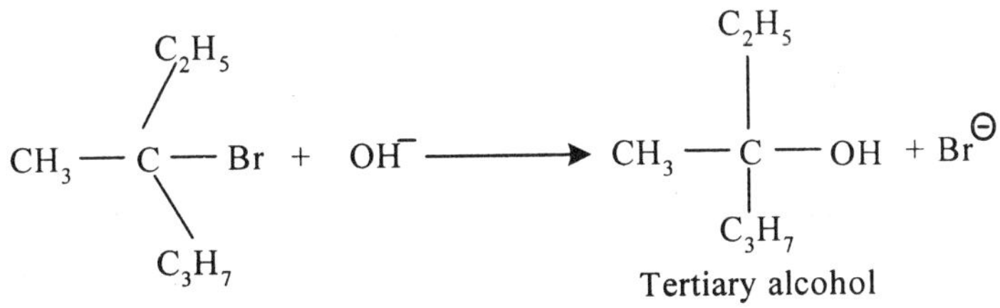

(i)

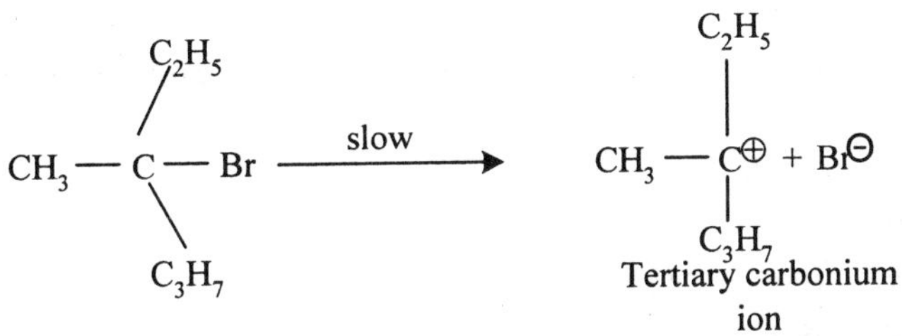

(ii)

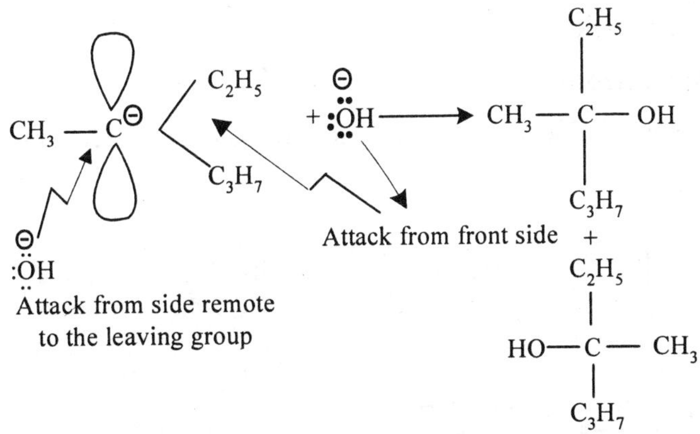

The rate of SN^1 reaction increases in the polar protic solvents (the solvents in which a hydrogen atom is attached to an atom of high electronegative element like oxygen, e.g., Alcohol, Water).

5.4 ELECTROPHILIC SUBSTITUTION REACTIONS

Benzene contains six delocalized π electrons. Therefore, it acts as a nucleophile and undergoes electrophilic substitution reactions, i.e., one or more hydrogen atom(s) of benzene can be replaced by electrophile(s), i.e.,

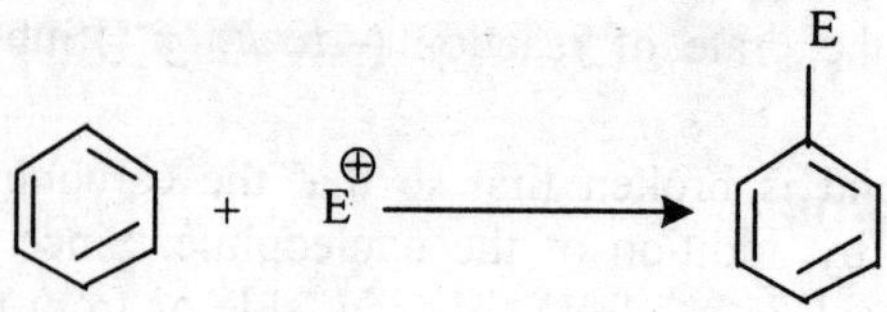

The general mechanism of the electrophilic substitution reaction is as follows:

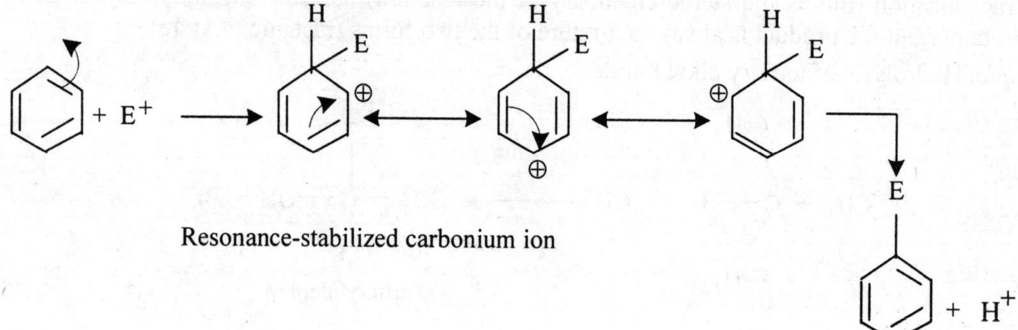

Resonance-stabilized carbonium ion

Electrophilic substitution reactions have been used to introduce following substituents in the benzene ring:

(A) A halo substitutent, —X
(B) A nitro substituent, —NO_2
(C) A sulphonic acid group, —SO_3H
(D) An alkyl group, —R; and
(E) An acyl group, COR

(A) Halogenation of Benzene

Benzene reacts with bromine or chlorine in the presence of Lewis acid like $FeCl_3$, $FeBr_3$, $AlCl_3$.

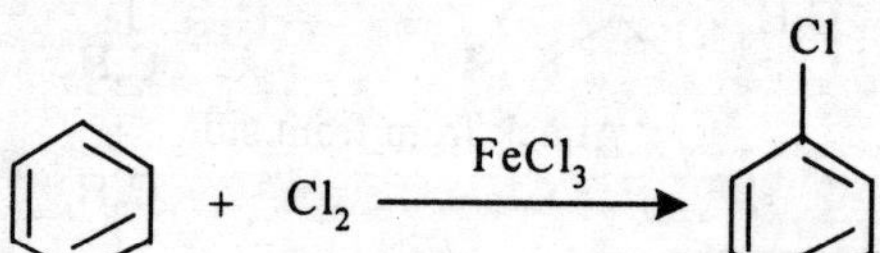

The mechanism is as follows:

(i) $Cl—Cl + FeCl_3 \longrightarrow FeCl_4^{\ominus} + Cl^{\oplus}$

(ii) $E^+ = Cl^+$, hence

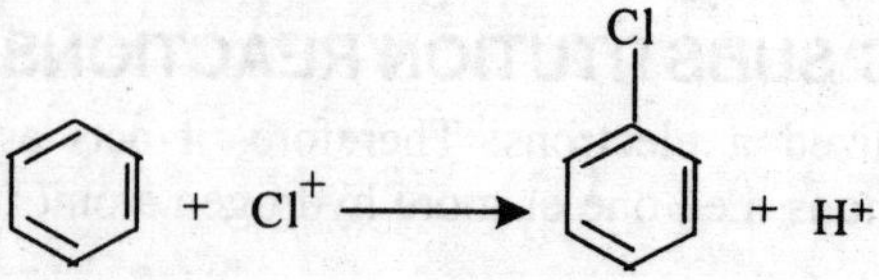

(iii) $FeCl_4^{\ominus} + H^+ \longrightarrow HCl + FeCl_3$

Fluorine reacts rapidly and it is difficult to limit the reaction to monofluorination. Iodination needs oxidizing agent like HNO_3.

(B) Nitration of Benzene

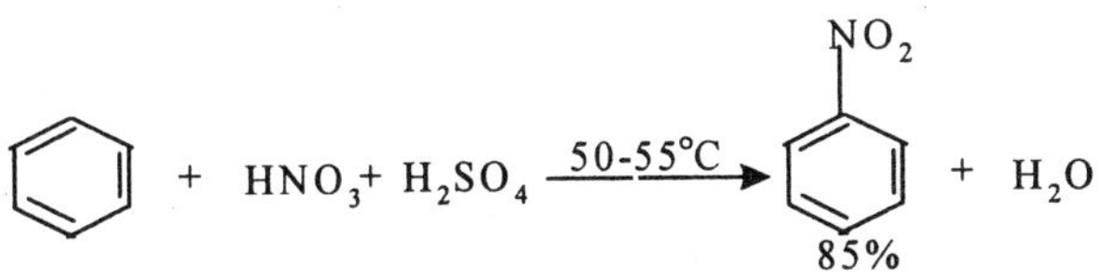

(i) $HO{-}NO_2 + 2H^{\oplus}HSO_4 \longrightarrow \overset{\oplus}{N}O_2 + H_3O^+ + 2HSO_4^-$

Nitronium ion

(ii) $E^+ = \overset{\oplus}{N}O_2$, hence

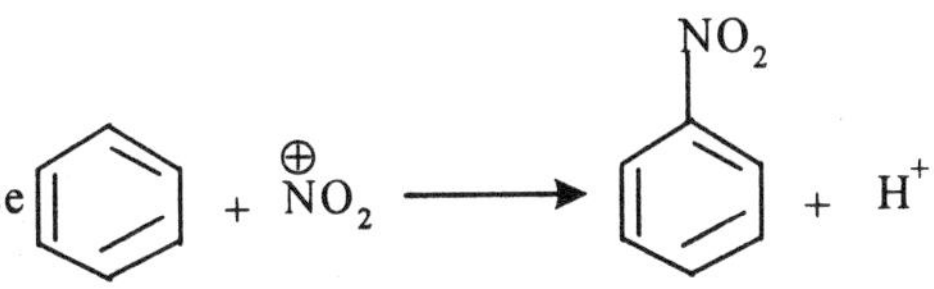

(iii) $H^+ \ HSO_4^- \longrightarrow H_2SO_4$

(C) Sulphonation of Benzene

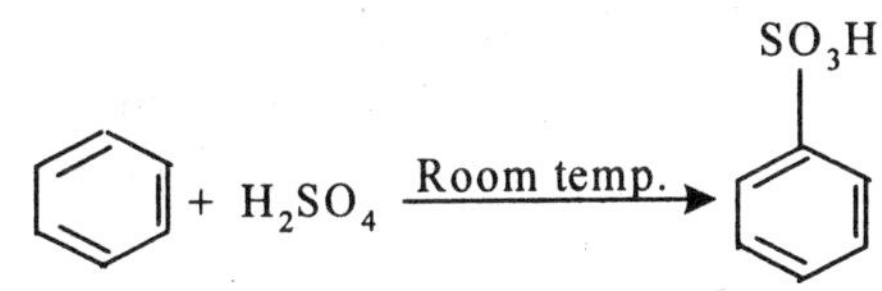

Benzene Fuming

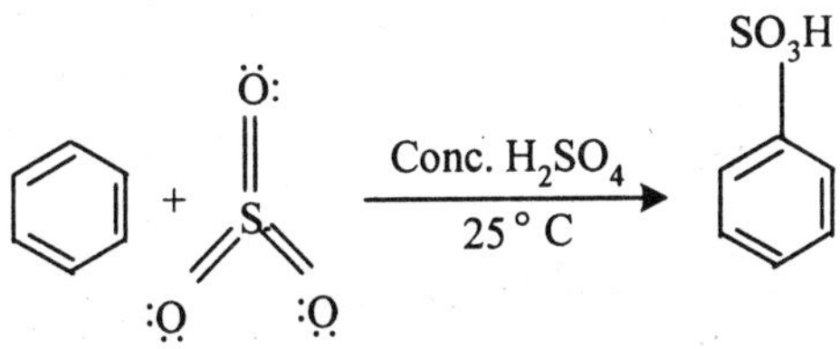

(i) $2H_2SO_4 \rightleftharpoons SO_3 + H_3O^+ + HSO_4^-$

(ii)

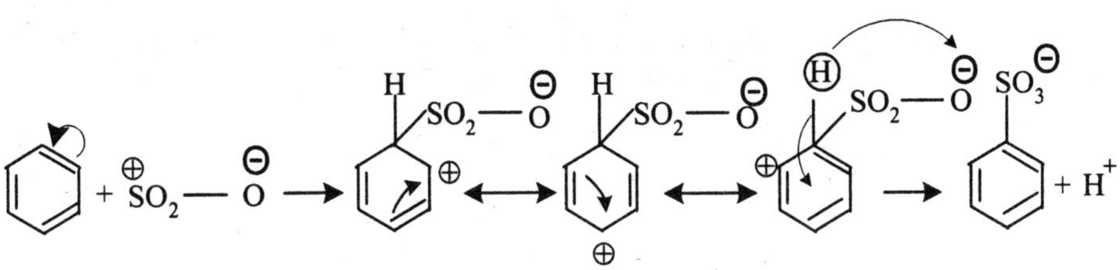

(iii) 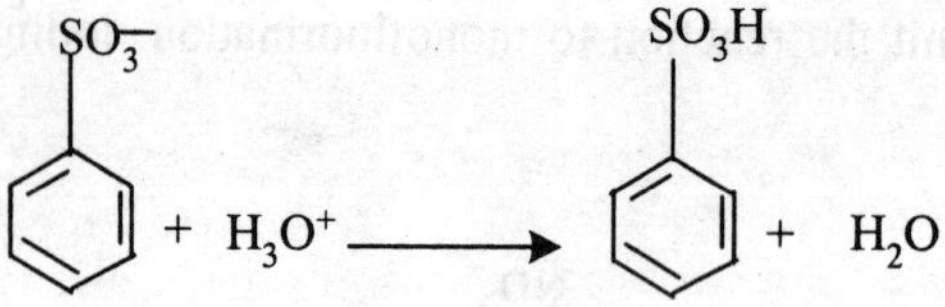

(iv) $H^+ + HSO_4^- \longrightarrow H_2SO_4$

(D) Friedel-Crafts Alkylation Reaction

It is a method for converting benzene into higher homologues. The reaction consists of the treatment of benzene with alkylating agent like alkyl halide, in the presence of a Lewis acid like anhydrous $AlCl_3$, $FeCl_3$, $ZnCl_2$, BF_3. Few examples are as follows:

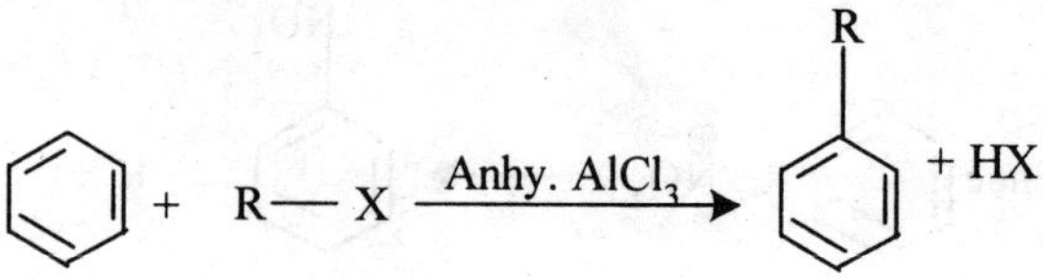

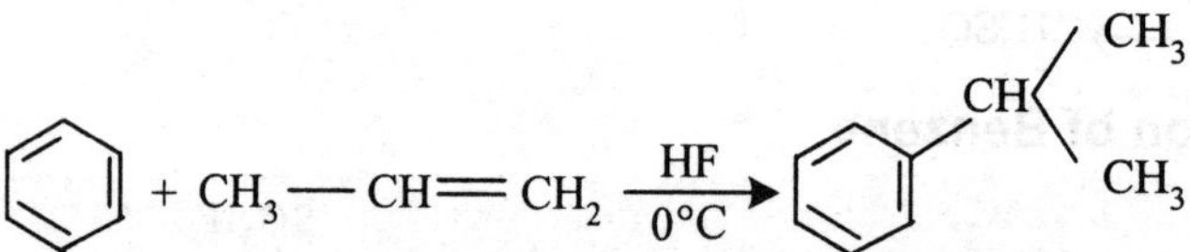

(84%)
Isopropyl benzene

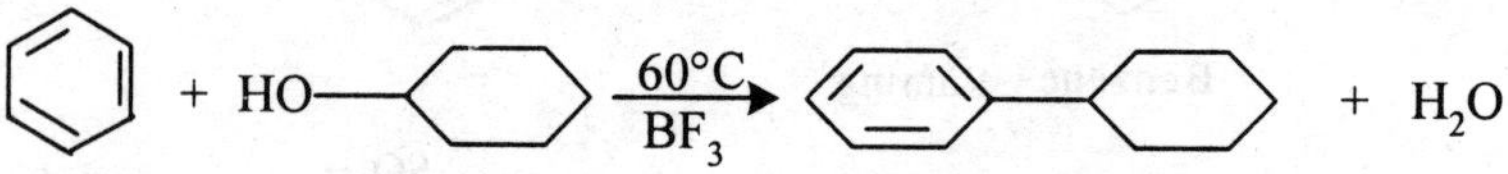

Cyclohexanol (56%)
Cyclohexyl benzene

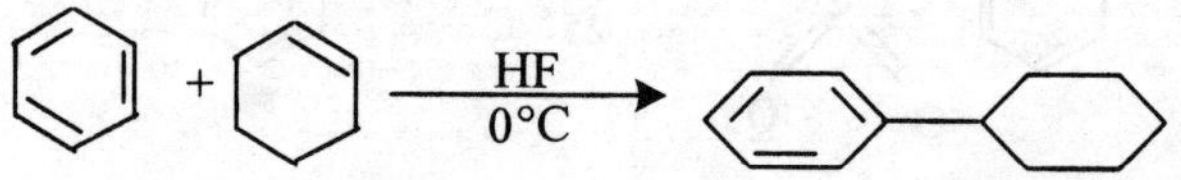

Cyclohexene (62%)
Cyclohexyl benzene

Mechanism involves carbonium ions as reaction intermediates. If possible, the primary carbonium ion rearranges to secondary or tertiary carbonium ion. The various steps are as follows:

(i) $R - Cl + AlCl_3 \longrightarrow R^+ + AlCl_4^-$

(ii) $E^+ = R^+$; hence 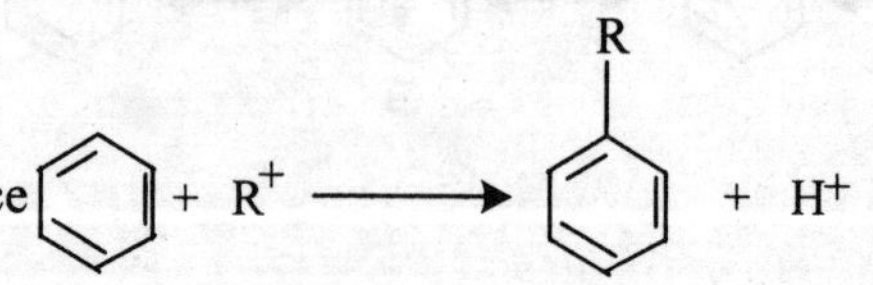

(iii) $H^+ + AlCl_4^- \longrightarrow HCl + AlCl_3$

Limitations of the Friedel-Craft Alkylation

(a) When the carbonium ion formed from an alkyl halide, alcohol, or alkene can undergo rearrangement to a more stable carbonium ion, it normally does so and the major product obtained from the reaction is usually the one from the more stable carbonium ion.

Let us consider following examples where carbonium ion undergoes rearrangement:

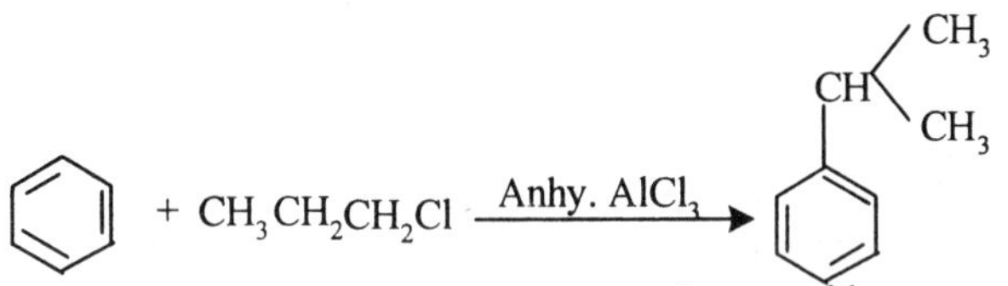

Isopropyl benzene

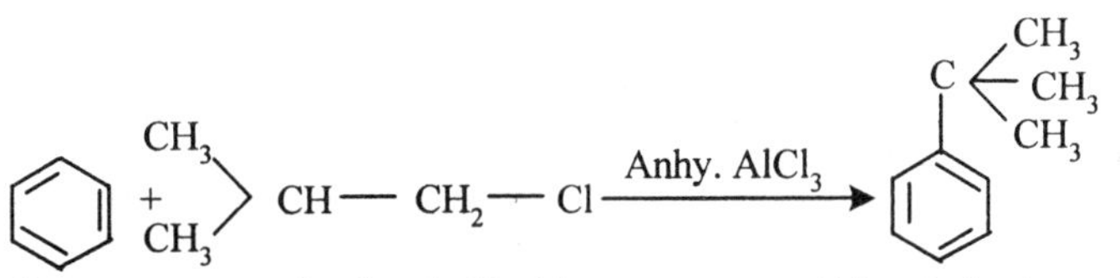

Benzene Iso-butyl chloride Trimethyl toluene

The formation of primary carbonium ion and its rearrangement is as follows:

(i)

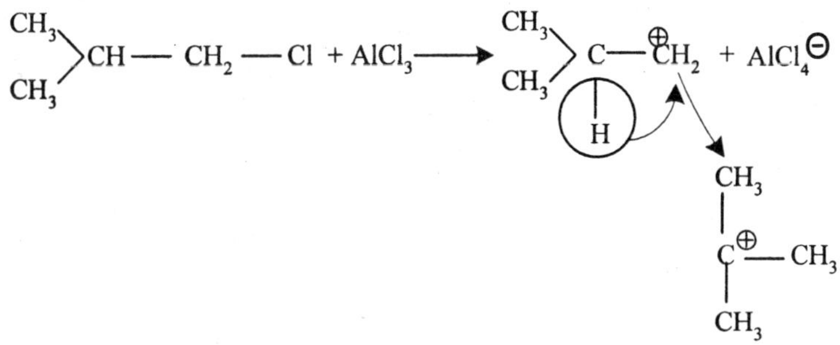

(ii) $E^+ = (CH_3)_2\overset{\oplus}{C}-CH_3$

(b) It does not take place when powerful electron-withdrawing groups like NR_3, $-NO_2$, $-SO_3H$, $-\overset{O}{\overset{\|}{C}}-H$, $-\overset{O}{\overset{\|}{C}}-R$, $-C-OOH$ are present on the aromatic ring or when the ring bears an $-NH_2$ or $-NHR$ group.

(c) Any meta-directing group makes an aromatic ring too electron-deficient to undergo a Friedel-Crafts reaction, e.g., the $-NH_2$, $-NHR$ and $-NR_2$ groups are changed into powerful electron withdrawing groups by the Lewis acids needed to catalyze Friedel-Crafts reaction:

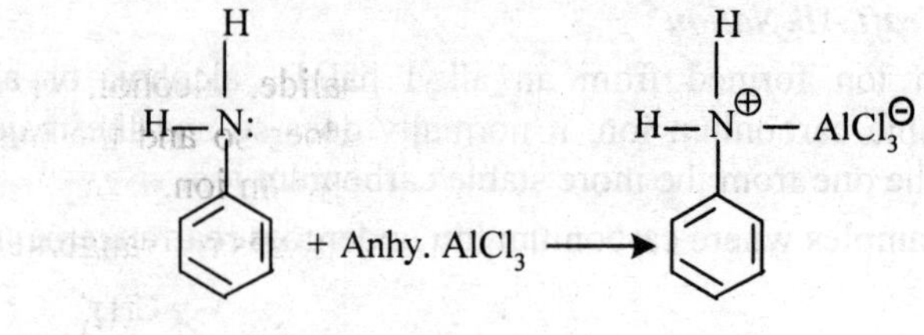

Further —$\ddot{N}H_2$ forms a complex with $AlCl_3$.

(d) Aryl and Vinylic halides cannot be used as the halide component, since they do not form carbonium ion due to resonance stabilization

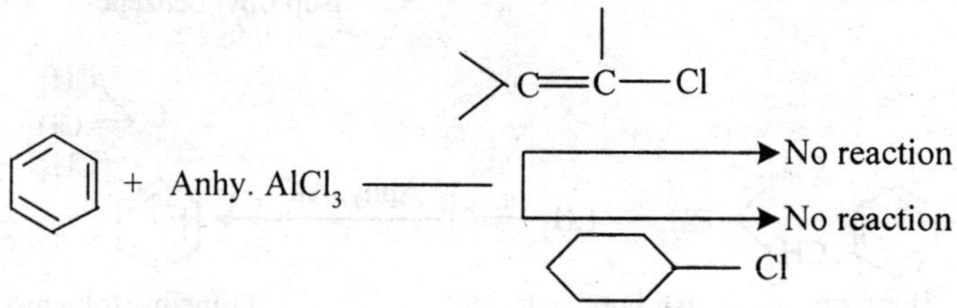

(e) Sometimes polyalkylation occurs, e.g.,

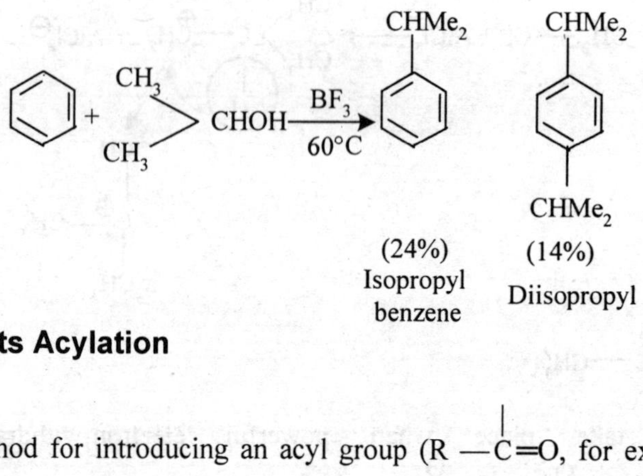

(24%) Isopropyl benzene (14%) Diisopropyl

(E) Friedel-Crafts Acylation

It is an important method for introducing an acyl group (R —C=O, for example CH_3 –C=O an acetyl group, C_6H_5 —C=O (a benzoyl group) into an aromatic ring.

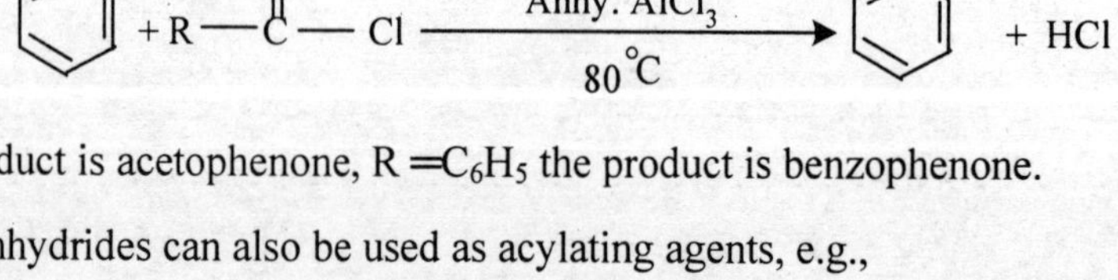

if R =CH_3 the product is acetophenone, R =C_6H_5 the product is benzophenone.

Carboxylic acid anhydrides can also be used as acylating agents, e.g.,

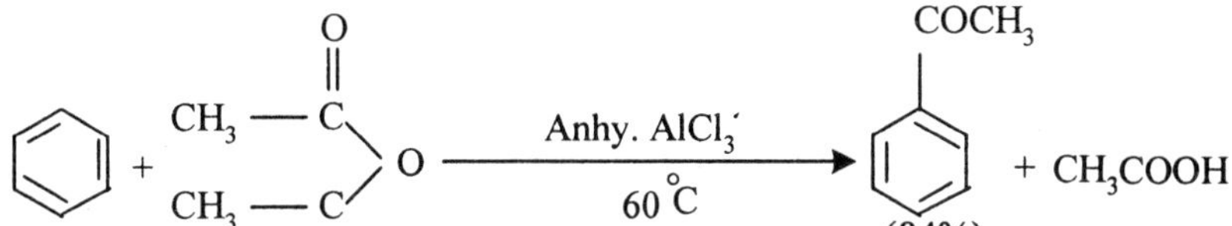

(84%)
Acetophenone

Mechanism:

(i)

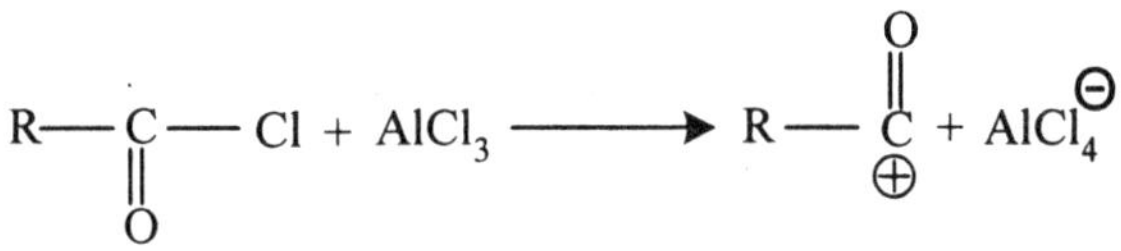

(ii)

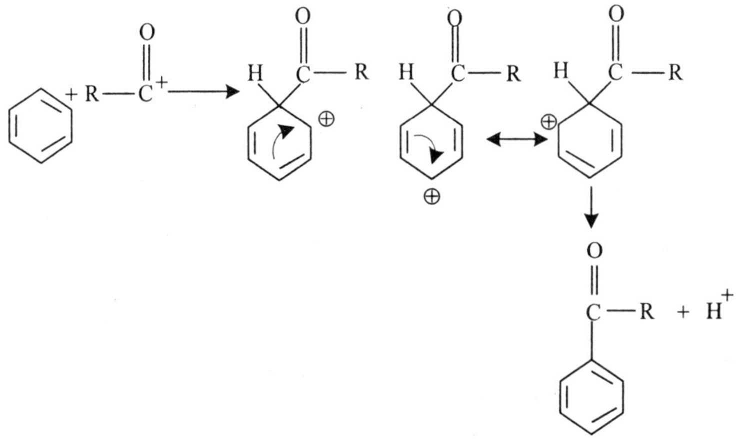

(iii) $H^+ + AlCl_4^- \longrightarrow AlCl_3 + HCl$

Rearrangement of the carbon chain does not occur in Friedel-Crafts acylation, because the acylium ion is more stable than most other carbocations. It is, therefore, used in synthesis of unbranched alkyl benzenes, e.g., synthesis of propyl benzene from benzene.

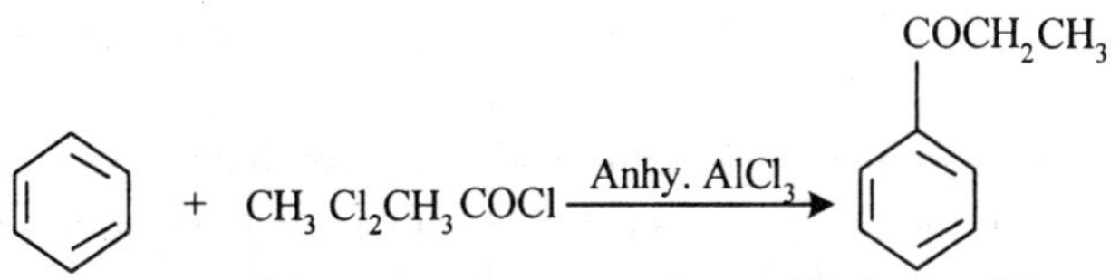

Ethyl phenyl ketone

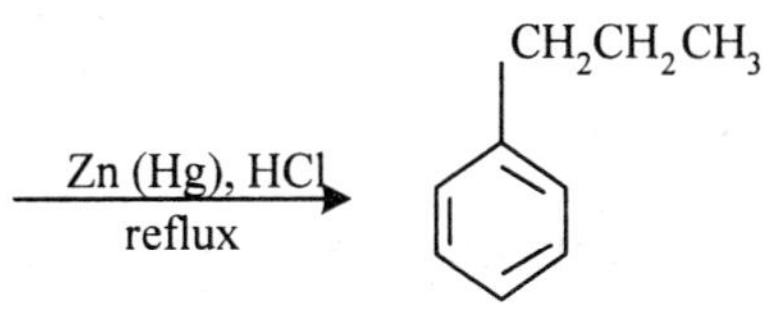

Propyl benzene

5.5 ELIMINATION REACTIONS

Elimination reactions may be defined as the organic reactions in which two atoms or groups are removed simultaneously, one in the form of an electrophile (generally H^+) and other in the form of a neucleophile (generally halide ion), thereby leading to the introduction of a multiple bond.

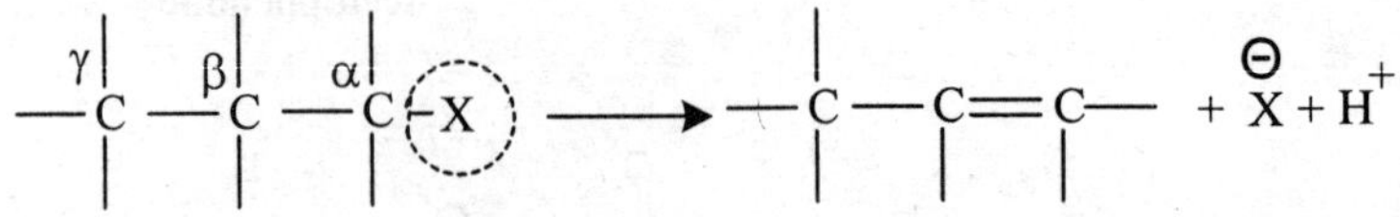

If both the groups and atoms are removed from adjacent atoms, it is known as β-*elimination reaction.* B-elimination reactions have been classified in two categories on the basis of the kinetic studies: (A) Unimolecular elimination reaction (E^1), and (b) Bimolecular elimination reaction (E^2). First, we will consider bimolecular elimination reactions.

Bimolecular Elimination Reaction (E^2) The rate of reaction depends upon the concentration of both species, hence it is called *bimolecular*. The mechanism involves transition state (*Example*: Dehydrohalogenation of primary and secondary alkyl halides).

$$CH_3CH_2Br + \text{alc. KOH} \longrightarrow CH_2{=}CH_2 + KBr + H_2O$$

Mechanism:

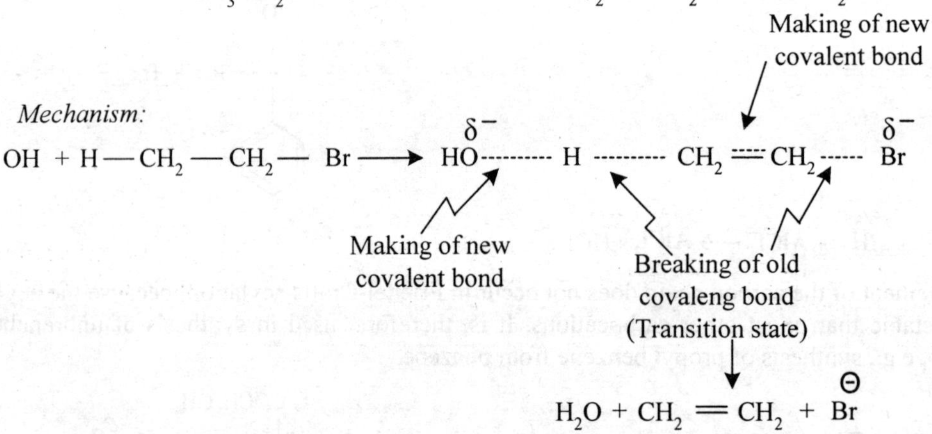

Unimolecular Elimination Reaction (E^1) Kinetic studies reveal that the rate of the reaction depends only upon the concentration of the substrate, hence it is called *unimolecular*. The mechanism involves carbonium ion as the reaction intermediate. The examples are as follows:

(a) *Dehydrohalogenation of tertiary alkyl halide*

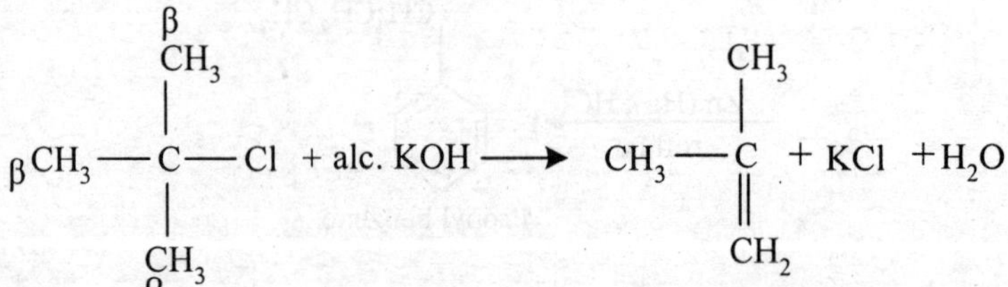

2-methylpropene

Mechanism:

(i)

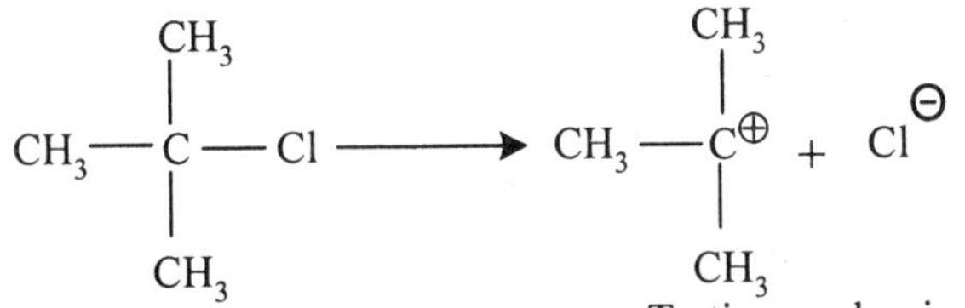

Tertiary carbonium ion

(ii)

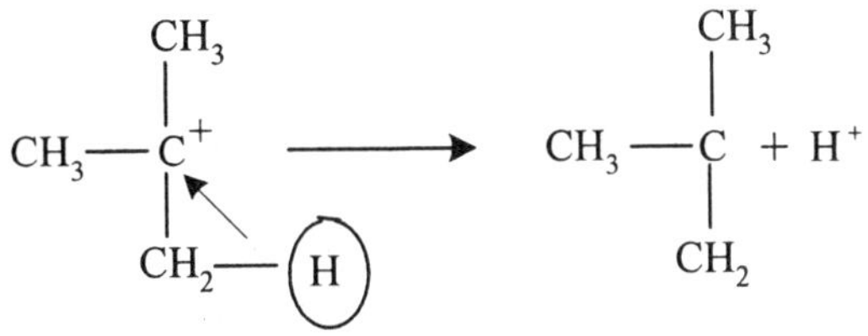

(b) *Dehydration of Alcohol*

$RCH_2CH_2OH + H^+ \longrightarrow RCH{=}CH_2 + H_2O$

Mechanism:

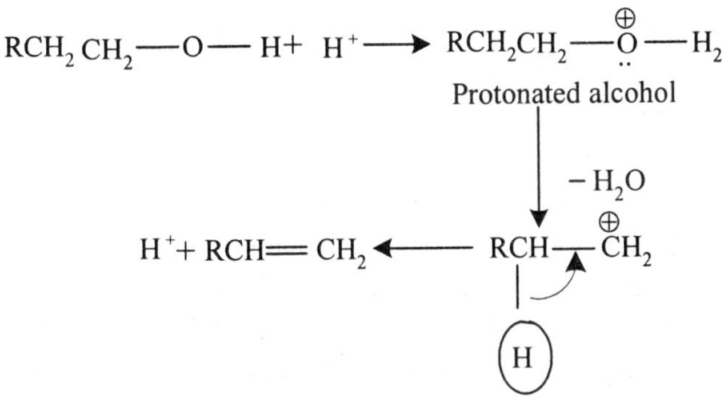

Ease of dehydration of alcohols:
Primary alcohol < Secondary alcohol < Tertiary alcohol

(c) *Saytzeff Rule:* It states that in the elimination reactions of neutral organic compounds, hydrogen is eliminated from that β carbon atom which carries less number of hydrogen atoms. The course of reaction has been explained on the basis of the stability of the alkene formed, e.g.,

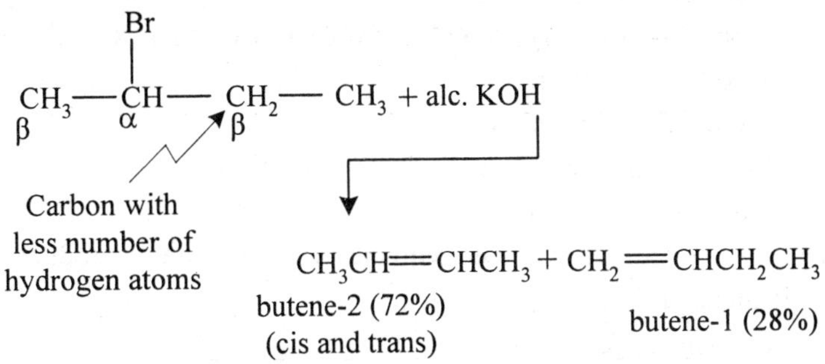

(d) *Hofmann Rule:* It states that in the elimination reaction of charged organic compounds like tertralkyl ammonium salt or trialkyl sulphonium salt, hydrogen atom is removed from that β carbon atom which carries maximum hydrogen atoms, e.g.,

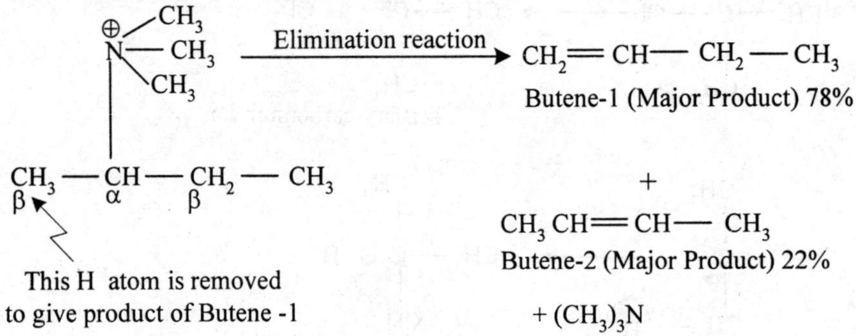

PROBLEMS

1. Amongst the four alternate given for the subquestions below one or more are correct. Write down all the correct answers.

(i) The formation of cyanohydrin from a ketone is an example of

(a) Electrophilic addition
(b) Nucleophilic addition
(c) Nucleophilic substitution
(d) Electrophilic substitution **(IIT, 1990)**

(ii) Consider the following facts about SN^2 reaction:

(a) Rate of reaction is independent of the [Nu]
(b) Nucleophile attacks on the carbon from the side opposite to the leaving group
(c) The bond making and bond breaking takes place in only one step.

Which of the above statements are correct?

(a) (x) and (y)
(b) (x), (y), (z)
(c) (x) and (z)
(d) (y) and (z)

(iii) The reaction $>C{=}O + NH_2OH \rightarrow >C{=}NOH + H_2O$ is an example of:

(a) An addition reaction
(b) An elimination reaction
(c) Addition followed by an elimination reaction
(d) A substitution reaction

(iv) When 2-bromopentane is treated with alc. KOH, the major product formed is:

(a) Pentene -1
(b) *cis* – pentene-2
(c) *trans*-pentene-2
(d) Mixture of *cis* and *trans*-pentene-2 **(GATE, 1991)**

(v) Addition of aq. HCN is possible with

(a) $C_6H_5CH{=}CH_2$

(b) $C_6H_5COCH_3$

(c) $C_6H_5COOC_2H_5$

(d) $C_6H_5CONH_2$ **(GATE, 1991)**

(vi) SN^1 characteristics would be exhibited in hydrolysis of

(a) Chlorobenzene

(b) 1-chloronaphthalene

(c) Propenyl chloride

(d) 2-chloro-1-propene **(GATE, 1992)**

(vii) The chlorination of methane to give CCl_4 is an example of

(a) An addition reaction

(b) An elimination reaction

(c) A reduction reaction

(d) A chain reaction

(viii) The reaction $CH_2{=}CH_2 + Br_2 \longrightarrow CH_2(Br){-}CH_2(Br)$ is an example of

(a) A nucleophilic addition

(b) An electrophilic addition

(c) An elimination

(d) A nucleophilic substitution

(ix) Consider the following reaction:

$$CH_3{-}C{\equiv}CH \xrightarrow{X} CH_3\underset{\substack{|\\Br}}{C}{=}CH_2 \xrightarrow{Y} CH_3\underset{\substack{|\\Br}}{CH}{-}CH_2Br$$

The reagents required at 'X' and 'Y' are:

(a) HBr and then HBr

(b) HBr + peroxide and then HBr

(c) HBr and then HBr + peroxide

(d) HBr + peroxide in both cases

(x) 1-chlorobutane on reaction with alc. KOH gives

(a) 1-butene

(b) 1-butanol

(c) 2-butene

(d) 2-butanol **(IIT, 1991)**

(xi) The intermediate during the addition of HCl to propene in the presence of peroxide is:

(a) $CH_3CH\ CH_2Cl$ *(b)* $CH_3-\overset{\oplus}{C}H-CH_3$

(c) $CH_3CH_2CH_2$ (d) $CH_3CH_2-\overset{\oplus}{C}H_2$

(xii) The reaction of propene with HOCl proceeds via the addition of:

(a) $H^{\oplus}$ in the first step

(b) $Cl^{\oplus}$ in the first step

(c) $OH^{\ominus}$ in the first step

(d) $Cl^{\oplus}$ & $OH^{\ominus}$ in a single step **(IIT, 2001)**

(xiii) An SN^2 reaction at an asymmetric carbon of a compound always gives

(a) An enantiomer of the substrate

(b) A product with opposite optical rotation

(c) A mixture of diastereomers

(d) A single stereoisomer. **(IIT, 2001)**

(xiv) $CH_3Br + \overset{\ominus}{Nu} \longrightarrow CH_3-Nu + \overset{\ominus}{Br}.$

The decreasing order of the rate of the above reaction with nucleophiles (Nu) A to D is:

$\overset{\ominus}{Nu}$ = (A) $Ph\overset{\ominus}{O}$, (B) $Ac\overset{\ominus}{O}$, (C) $H\overset{\ominus}{O}$, (D) $CH_3\overset{\ominus}{O}$

(a) B > D > C > A (b) D > C > A > B

(c) D > C > B > A (d) A > B > C > D **(AIEEE 2006)**

(xv) The structure of the major product formed in the following reaction,

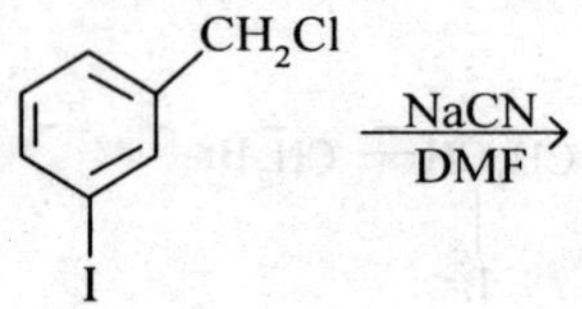

is:

(a) 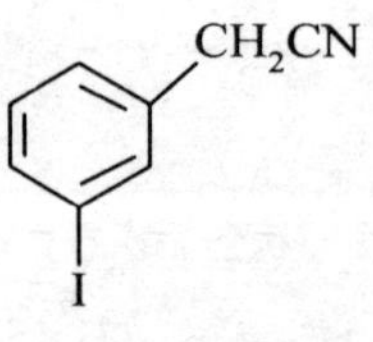

(b) 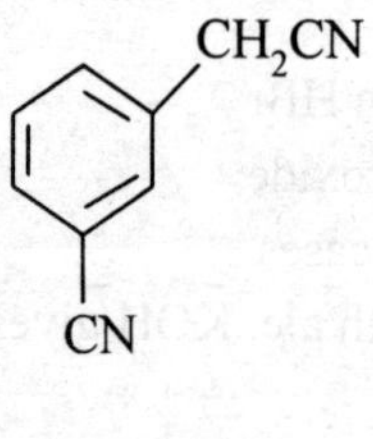

(c) 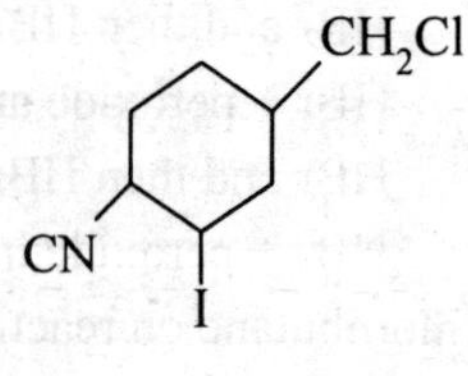

(d) 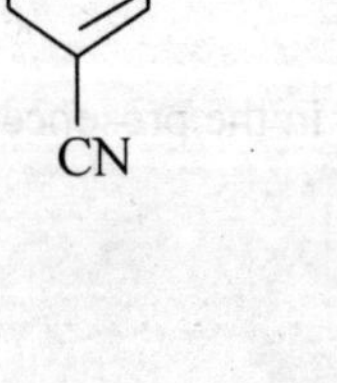

(AIEEE 2000)

(xvi) Reaction of trans -2Phenyl –1 – bromocyclopentane on the reaction with alcohol KOH produces:

(a) 3-Phenylcyclopentene *(b)* 4-Phenyclcyclopentene

(c) 2-Phenylcyclopentene *(d)* 1-Phenyl cyclopentene **(AIEEE 2006)**

(xvii) The alkene formed as a major product in the below eliminations reactions

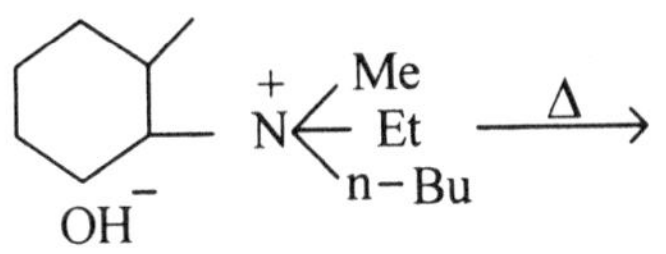

(a) 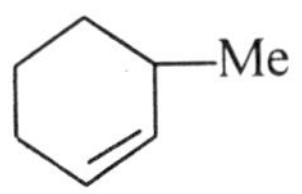

(b) 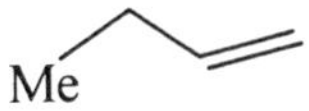

(c) $CH_2 = CH_2$ *(d)* 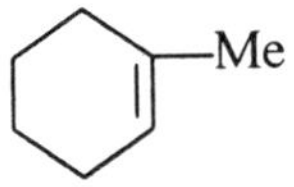

(AIEEE 2006)

2. Fill in the blanks:

(i) 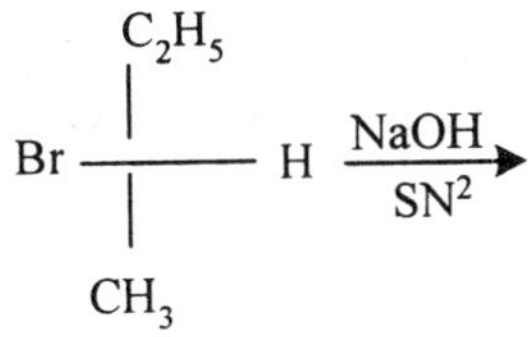

 ---------- **(IIT, 1994)**

3. State whether True or False:

(i) During SN^1 reaction, the leaving group leaves the molecule before the incoming group is attached to the molecule. **(IIT, 1990)**

(ii) SN^2 reaction proceeds with racemization.

4. Discuss the mechanism of the following reaction: **(IIT, 1981)**

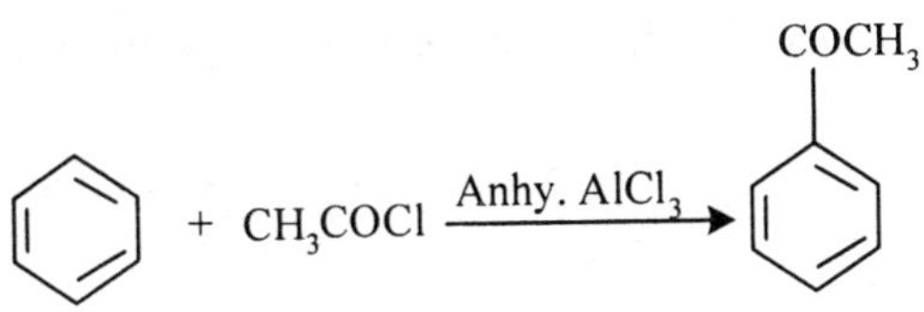

5. Predict the structure of the product in the following reaction: **(IIT, 1996)**

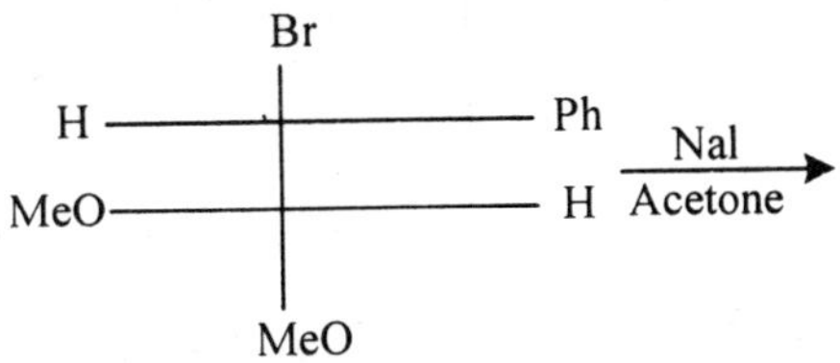

 ----

6. Classify following solvents into polar aprotic and polar protic solvents:

(i) Formic acid *(ii)* Acetone *(iii)* Sulfur dioxide *(iv)* Formamide

ANSWERS

1. *(i)* **(b)**
   (ii) **(d)**
   (iii) **(c)**
   (iv) **(c)**, see Saytzeff Rule.
   (v) **(b)** because rest of the organic molecules do not contain carbonyl group.
   (vi) **(c)**.
   (vii) **(d)**
   (viii) **(b)**
   (ix) **(c)** Markovinkov's Rule and then peroxide effect.
   (x) **(a)** It is the elimination reaction since alc. KOH is used.
   (xi) **(b)**
   (xii) **(b)**
   (xiii) **(b)**
   (xiv) **(b)**
   (xv) **(a)**
   (xvi) **(d)**
   (xvii) **(a)**
2. It is a secondary alkyl halide hence undergoes SN^2 reaction. Therefore, the product due to inversion of configuration is

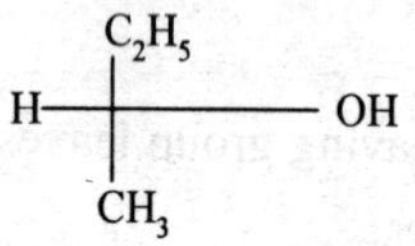

3. *(i)* *True*
   (ii) *False.* SN^2 reaction proceeds with inversion of configuration.
4. See page (88)
5. See SN^2 reactions:

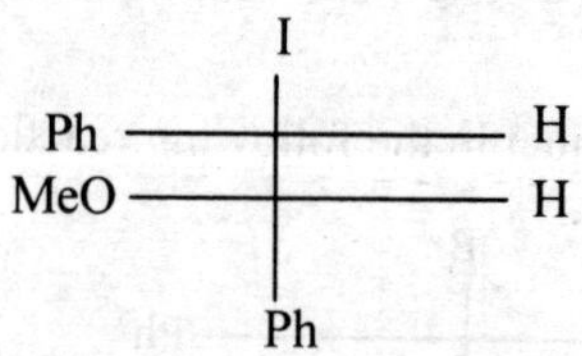

6. *(i)* Polar protic solvent
   (ii) Polar aprotic solvent
   (iii) Polar aprotic solvent
   (iv) Polar aprotic solvent

Chapter 6

Organometallic Compounds (Grignard Reagent)

6.1 INTRODUCTION

The compounds in which a metal is directly attached to carbon are known as organometallic compounds, i.e., 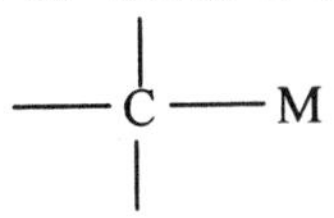

Some examples of these compounds are:

$(C_2H_5)_4$ Pb	$(C_2H_5)_2$ ZN	CH_3MgBr	etc.
Tetraethyl lead	Diethyl zinc	Methylmagnesium bromide	

The nature of the carbon-metal bonds varies widely, ranging from bonds that are essentially ionic to those that are primarily covalent, e.g.,

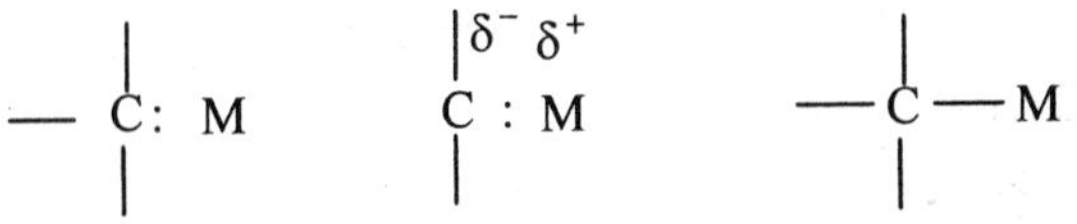

Primarily ionic (M = Na or K) (M = Mg or Li) Primarily covalent (M = Pb. Sn, Hg or TI)

The reactivity of organometallic compounds increases with the present ionic character of the carbon-metal bond.

6.2 GRIGNARD REAGENT

Organomagnesium halides are known as Grignard Reagents. They have a general formula RMgX, where R is univalent radical like alkyl or aryl radical and X is generally bromine or iodine. The order of reactivity of halides with magnesium is RI > RBr > RCl. The reactivity also decreases with size of the alkyl radical.

Preparation: Grignard reagent is prepared by addition of alkyl halide to Magnesium.

$$RX + Mg \xrightarrow{\text{ether}} RMgX$$

Magnesium is rubbed with sandpaper to remove the layer of magnesium oxide. Alkyl halide and ether is treated with a dehydrating agent like P_2O_5, anhy. $CaCl_2$, etc., before use.

(b)Synthetic Applications of Grignard Reagent

Grignard reagent acts as a nucleophile (stronger base) and has been used to prepare almost all classes of organic compounds. Some applications are as follows:

1. *Synthesis of Alkanes*: Alkanes can be prepared by reaction of Grignard reagent with a compound containing **active hydrogen atom**. A compound is said to contain an active hydrogen atom when it has a hydrogen atom attached to a more electronegative element from which it can be easily detached. Such compounds are: H_2O, ROH, NH_3 and RNH_2.

The reactions are as follows:

(a) $$\boxed{R}\boxed{MgX + HO}\boxed{H} \longrightarrow RH + MgXOX$$

e.g., $$CH_3MgI + HOH \longrightarrow CH_4 + Mg\langle^{I}_{OH}$$

(b) $$R\ \boxed{MgX + R'O}\ H \longrightarrow RH + MgXOR'$$

e.g., $$CH_3\ \boxed{MgI + C_2H_5O}\ H \longrightarrow CH_4 + MgIOC_2H_5$$

1 mole, 1 active H atom → 1 mole

Note that 1 mole of an alcohol containing one active H atom per mole yields 1 mole of alkane. Therefore, we can write:

Moles of alcohol × active H atoms per mole = moles of alkane

moles of alkane = volume (litre) of alkane at STP/22.4

(a) $RMgX + HNH_2 \rightarrow RH + MgXNH_2$

e.g., $CH_3\ MgI + HNH_2 \rightarrow CH_4 + MgINH_2$

Ammonia

(b) $RMgX + R'NH_2 \rightarrow RH + MgXNHR'$

Primary amine

$CH_3MgI + C_2H_5NH_2 \rightarrow CH_4 + MgINHC_2H_5$

Alkanes can also be prepared by the reaction of Grignard reagent with **terminal hydrogen atoms of alkynes –1.**

$$RMgX + HC{\equiv}CR' \rightarrow RH + MgXC{\equiv}CR'$$

$$CH_3MgBr + HC{\equiv}CH \rightarrow CH_4 + HC{\equiv}CMgBr$$

2. *Synthesis of Alkenes:*

$$RMgX + BrCH_2—CH{=}CH_2 \longrightarrow RCH_2—CH{=}CH_2 + MgXBr$$

allyl bromide

3. *Synthesis of Alkynes:*

$$RMgX + BrCH_2—C{\equiv}CH \longrightarrow RCH_2—C{\equiv}CH + MgXBr$$

3-Bromopropyne

4. *Synthesis of Alkyl Halides:*

$$RMgX + X'_2 \rightarrow RX' + MgXX'$$

halogen

5. *Synthesis of Primary Amines:*

$$RMgX + ClNH_2 \longrightarrow RNH_2 + MgXCl$$

Chloroamine

6. *Synthesis of Alkyl Cyanides:*

$$RMgX + CNCl \rightarrow RCN + MgXCl$$

Cyanogen chloride

7. *Synthesis of Primary Alcohols:*

(i)

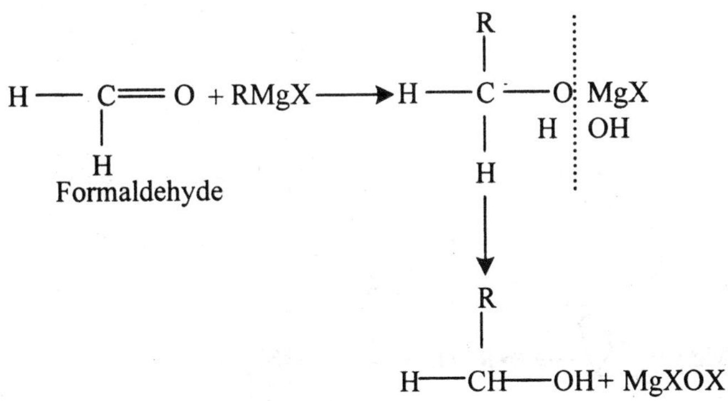

Formaldehyde

(ii)

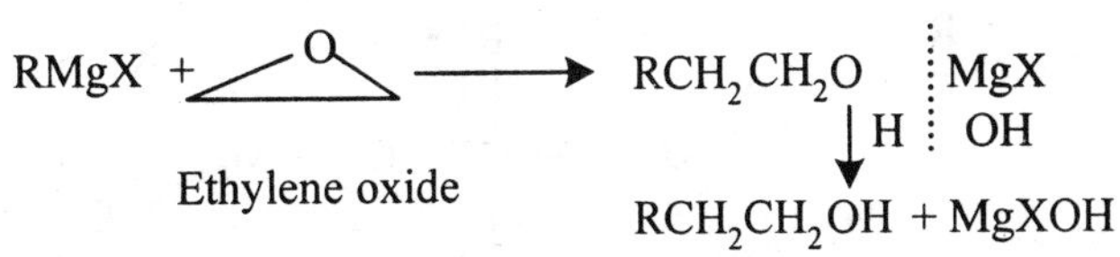

Ethylene oxide

8. *Synthesis of Secondary Alcohols:*

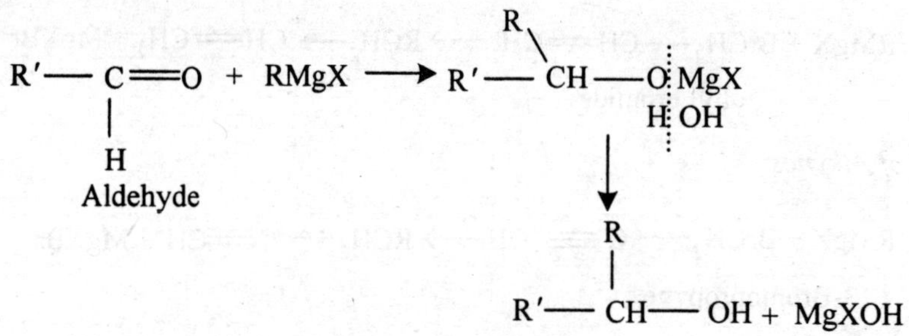

e.g., If $R' = CH_3$, it is acetaldehyde

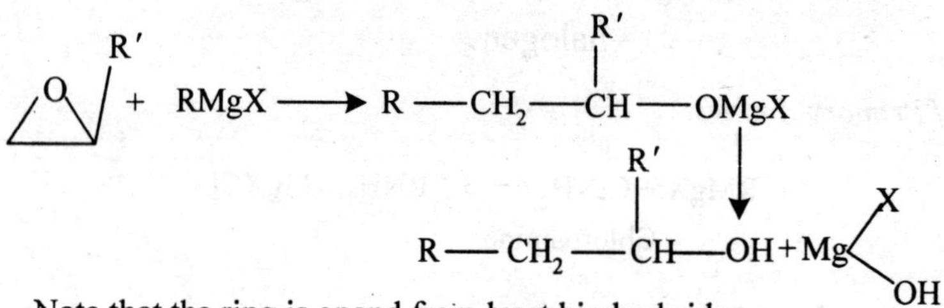

Note that the ring is opend from least hindred side

9. *Synthesis of TertiaryAlcohol:*

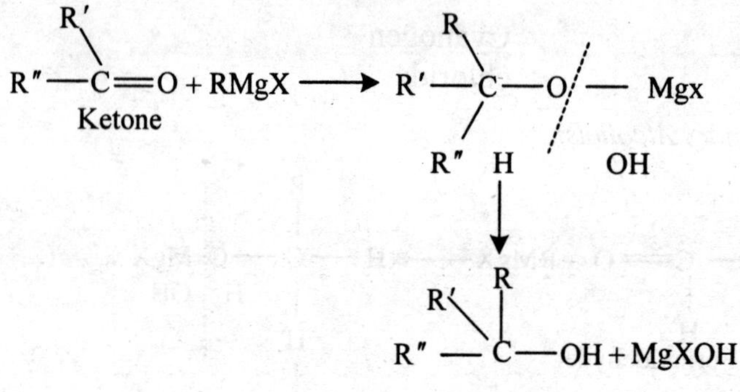

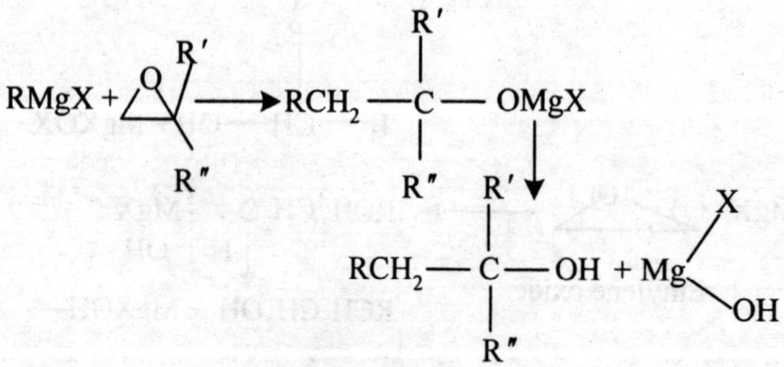

10. *Synthesis of Aldehydes:* There are two methods to prepare an aldehyde

(i)

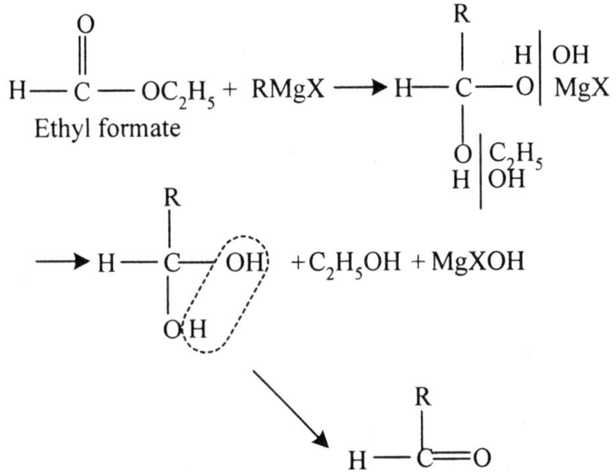

(ii)

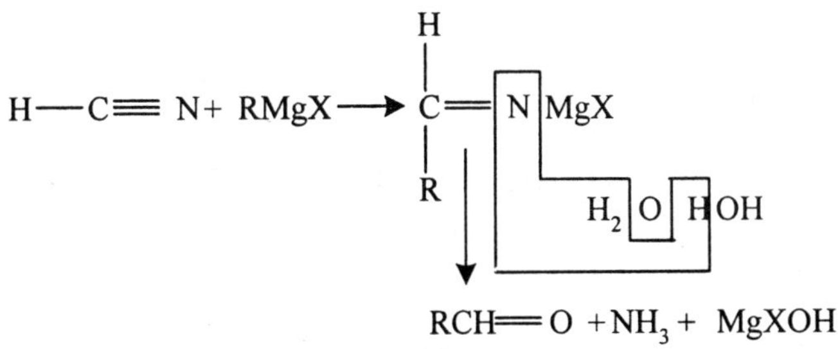

11. *Synthesis of Ketone:* There are three methods to prepare a ketone:

(i)

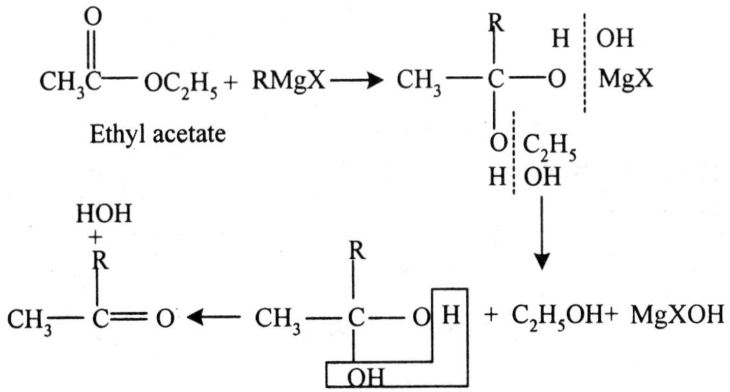

(ii) 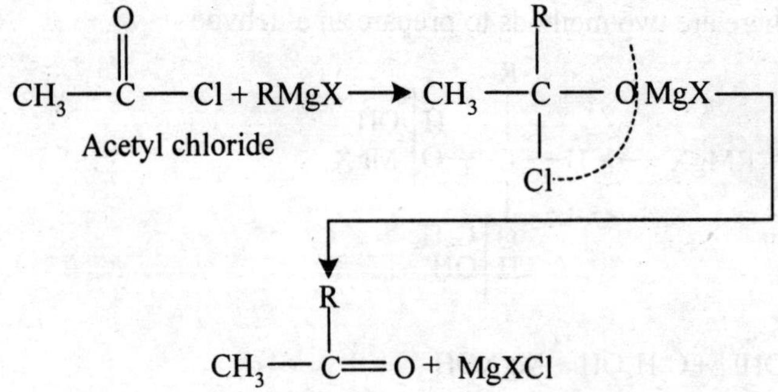

(iii) 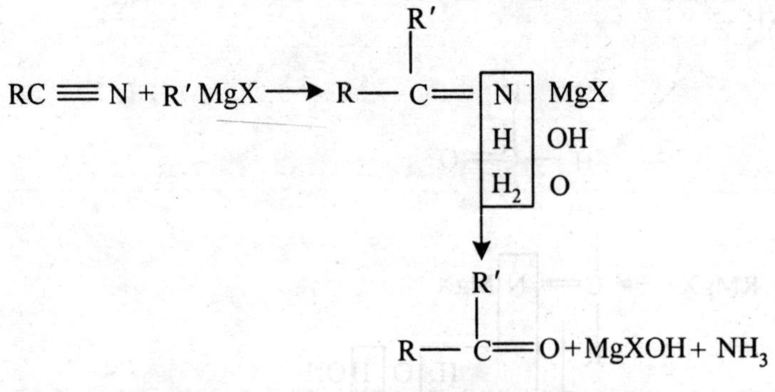

12. *Synthesis of Carboxylic Acids:*

$$RMgX + CO_2 \xrightarrow{0^\circ C} RCOOMgX \xrightarrow{HOH} RCOOH + MgXOH$$

13. *Synthesis of Esters:*

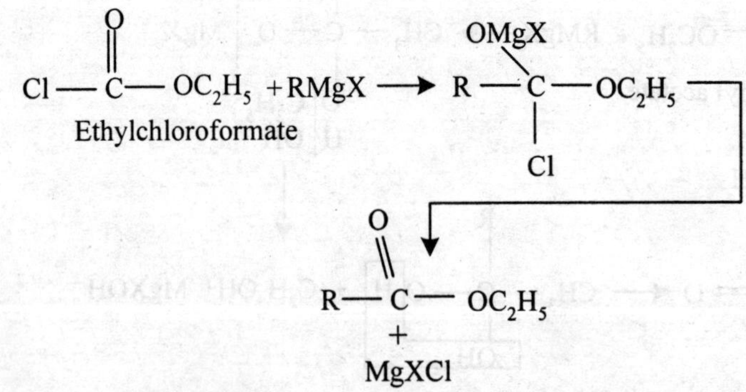

14. *Synthesis of Ethers:*

$$RMgX + Cl\,CH_2CH_2OCH_3 \longrightarrow RCH_2CH_2\,OCH_3 + MgXCl$$

Chloroethylmethyl ether

15. *Synthesis of Mercaptanes:*

$$RMgX + \underset{\text{Sulphur}}{S} \longrightarrow \begin{matrix} RS \\ H \end{matrix} \vdots \begin{matrix} MgX \\ OH \end{matrix} \xrightarrow{H^+} \underset{\text{Mercaptane}}{RSH} + MgXOH$$

16. *Synthesis of Other Organometallic Compounds:* Grignard reagent reacts with a number of halides of less electropositive elements to produce new organometallic compounds. In general,

$$nRMgX + MXn \rightarrow R_nM + nMgX_2$$
(where M is less electropositive than Mg)

e.g.

$$2CH_3MgCl + HgCl_2 \longrightarrow \underset{\text{Dimethylmercury}}{(CH_3)_2Hg} + 2MgCl_2$$

$$2CH_3CH_2MgBr + ZnCl_2 \longrightarrow \underset{\text{Diethyl Zinc (100\%)}}{(CH_3CH_2)_2Zn} + 2MgClBr$$

$$2C_6H_5MgBr + CdCl_2 \longrightarrow \underset{\text{Diphenyl cadmium > 83\%}}{(C_6H_5)_2Cd} + 2MgClBr$$

$$4CH_3CH_2MgCl + SiCl_4 \longrightarrow \underset{\text{Tetraethylsilane (80\%)}}{(CH_3CH_2)_4Si} + 4MgCl_2$$

$$3CH_3CH_2CH_2CH_2MgBr + PCl_3 \longrightarrow \underset{\text{Tributylphosphine (83\%)}}{(CH_3CH_2CH_2CH_2)_3P} + 3MgClBr$$

6.3 STRUCTURE OF GRIGNARD REAGENT

Experiments have shown that there is an equilibrium between an alkyl magnesium halide and dialkyl magnesium.

$$2RMgX \rightleftharpoons R_2Mg + MgX_2$$

A Grignard reagent forms a complex with its ether solvent; the structure of the complex is as follows:

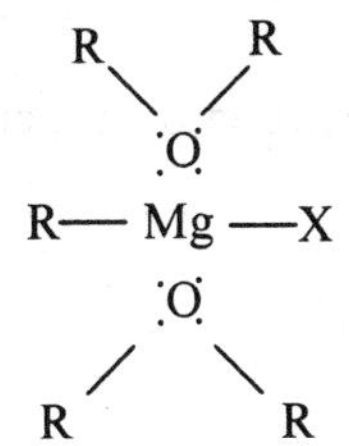

6.4 LIMITATIONS OF GRIGNARD REAGENT

(i) Grignard reagent is very reactive, hence it should be immediately used after synthesis.

(ii) Grignard reagents containing following groups cannot be prepared because it is very powerful base.

—OH, —NH_2, — NHR, — COOH,— SO_3H,— SH,

$=C=CH_2$,— CH_2— $\overset{O}{\overset{\|}{C}}$— R— COOR,

— $CONH_2$,— NO_2, — C≡ N, 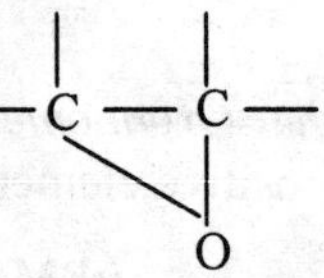

PROBLEMS

1. 8.24 mg of an unknown alcohol (ROH), when treated with methyl magnesium bromide, gives 3.12 ml methane at STP. What is alcohol? Write its isomer.
2. An organic compound, having molecular weight 90–93, when treated with CH_3MgI, thereby confirming the presence of active hydrogen atoms. 0.90 mg of the compound produced 0.67 ml CH_4 at STP. Calculate the number of active hydrogen atoms/mole. If these are present as hydroxyl groups, suggest a possible structure for the compound.
3. An alkyl iodide 'A' forms Grignard reagent with Mg, which when treated with water forms n-pentane, with sodium it forms 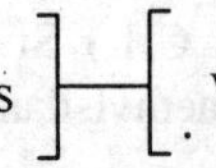 What is 'A'?
4. Identify A to E in the following reactions:

$$\text{EtMeCHCOph} \xrightarrow[\text{(ii) aq. } NH_4Cl]{\text{(i) PhMgBr}} A \xrightarrow{H^+} B \xrightarrow{O_3} C + D$$

$$\text{EtCHO} \xrightarrow{E} C$$

5. Complete the following reactions:

(i) $C_6H_5COOH + CH_3MgI \rightarrow ? + ?$ **(IIT,1993)**

(ii) $CH_3CHO \xrightarrow{CH_3MgBr} ? \longrightarrow CH_3—\underset{}{\overset{CH_3}{\overset{|}{CH}}}— OH$ **(IIT, 1988)**

6. Carry out following conversions using Grignard Reagent:

(i) $CH_3CH_2OH \longrightarrow$ 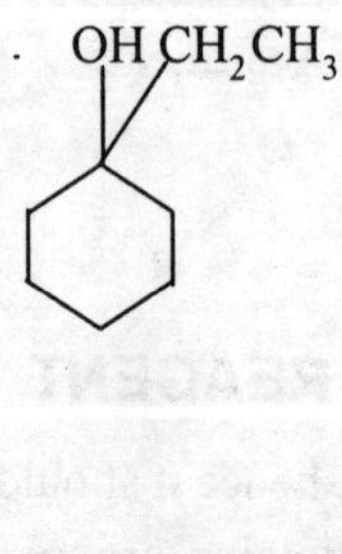

(ii) 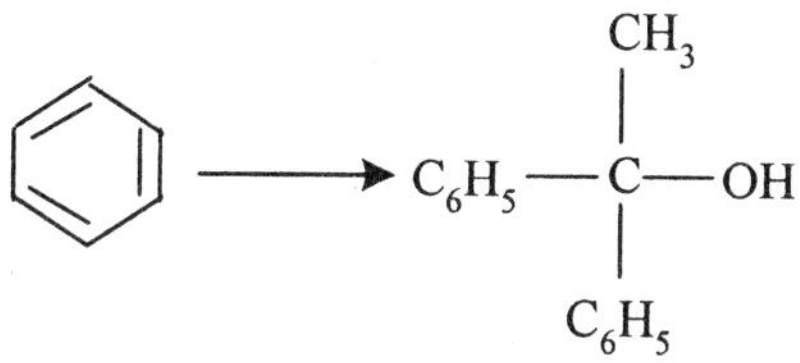

(iii) 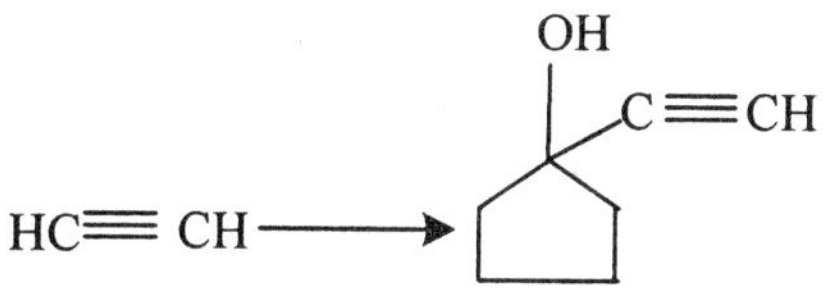

7. 0.037 gm of an alcohol, ROH, was added to CH_3MgI and the gas evolved measured 11.2 ml at STP. What is the molecular weight of ROH? **(R, 1994)**

8. Amongst the four alternative answers given for each question below, one is correct. Choose the correct answer:

(i) Phenyl magnesium bromide reacts with methanol to give: **(AIEEE 2006)**

(a) a mixture of Phenol & Mg (Me) Br

(b) a mixture of anisole & Mg (OH) Br

(c) a mixture of benzene and Mg(OMe) Br

(d) a mixture of toluene & Mg (OH) Br

(ii) Consider following reaction:

$2\ CH_3MgBr +$  $\longrightarrow$ A

'A' is:

(a) CH_4 *(b)* C_2H_6 *(c)* C_2H_5OH *(d)* CH_3OH

(iii) What is the product of following reaction?

$3\ Et\ MgBr + (Eto)_2\ C = O \longrightarrow ?$

(a) $(Eto)_3—C—OH$ *(b)* $Et_3—C—OH$ *(c)* C_2H_6 *(d)* CH_4

(iv) What is the product of following reaction?

$PrMgBr + (MeO)_3C—Me \longrightarrow ?$

(a) 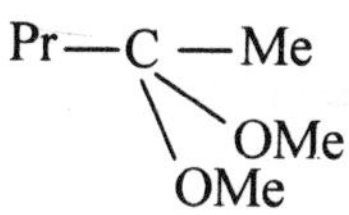 *(b)* 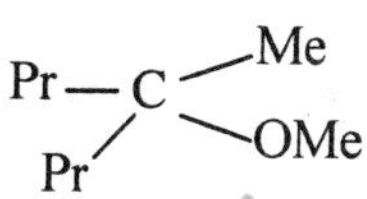 *(c)* 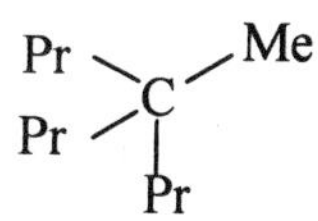

(d) None of the above

ANSWERS

1. $$\text{Moles of alcohol} \times \frac{\text{active H atoms}}{\text{mole}} = \text{moles of alkane}$$

Or $$8.24 \times \frac{10^{-3}}{M \times 1} = \frac{3.12}{22400}$$

Where M is the molecular weight of alcohol: M = 59,
i.e., It is propanol-1 ($CH_3CH_2CH_2OH$)

Possible isomer propanol – 2 $CH_3CH(OH)-CH_3$

2. $$\text{Moles of alcohol} \times \frac{\text{H atoms}}{\text{mole}} = \text{moles of alkane}$$

$$0.909 \times \frac{10^{-3}}{90 \times n} = \frac{0.67}{22400}$$

$n = 2.8$, i.e., 3

So, the number of active H atoms is 3. Possible compound is CH_2OH—CHOH—CH_2OH Glycerol.

3. Since Grignard reagent forms *n*-pentane with water, R is a pentyl group. When 'A' is treated with sodium, it gives 3,4-Diethyl hexane, hence it is 3-iodopentane (Wurtz reaction).

$CH_3CH_2-CH(I)-CH_2-CH_3$

4. 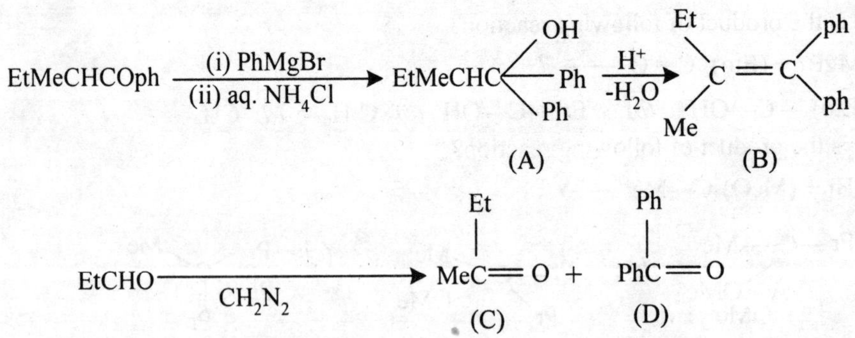

5. *(i)* $CH_4 + C_6H_5COOMgI$.

(ii) $CH_3-CH(OMgBr)-CH_3$, H_2O

6. *(i)* $CH_3CH_2OH \xrightarrow{PBr_3} CH_3CH_2Br \xrightarrow{\text{Mg, ether}} CH_3CH_2MgBr$

(ii) 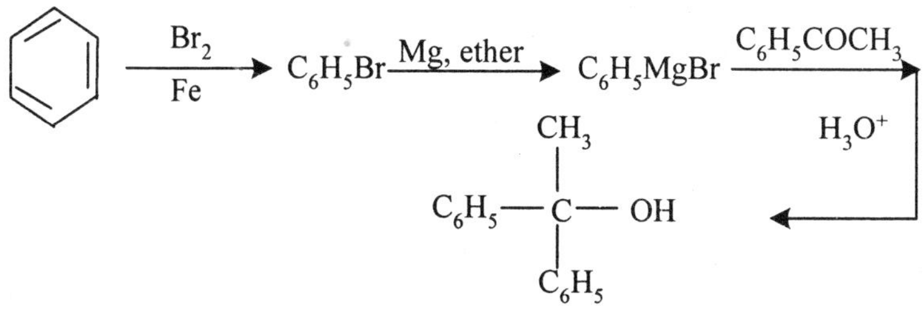

(iii) 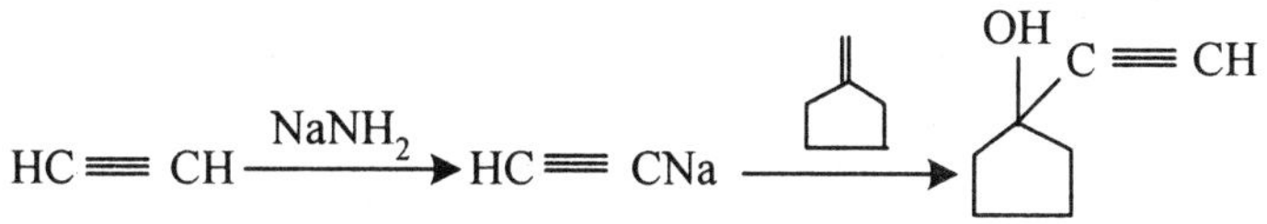

7. Moles of alcohol × active H atoms/mole = moles of alkane $0.037/m \times 1 = 11.2/22400$
$M = 74$.

8. *(i)* (c); *(ii)* (a); *(iii)* (b); *(iv)* (a)

Chapter 7

Isomerism

7.1 INTRODUCTION

Compounds that have the same molecular formula but different properties are called isomers and the phenomenon is known as 'isomerism'. If it is due to different arrangement of atoms in the molecule, it is known as 'structural isomerism'. However, if the arrangement of atoms is different in space it gives rise to stereoisomerism.

7.2 KINDS OF ISOMERISM

Isomerism has been classified as follows:

(a) *Structural Isomerism:* This isomerism arises due to different arrangement of atoms in the molecule.

(b) *Stereoisomerism*: This isomerism arises due to different arrangement of atoms in space. It has been further classified in two categories:

(i) Geometrical Isomerism

(ii) Optical Isomerism.

7.3 STRUCTURAL ISOMERISM

It has been further classified into following categories:

(i) *Chain Isomerism*: Chain Isomers are the compounds that have the same molecular formula but different carbon skeleton and hence different properties. The phenomenon is termed as chain isomerism. It is exhibited by alkanes, e.g., C_5H_{12} has three isomers:

It is clear that as branching increases, the boiling point decreases because the molecule assumes spherical shape, and as a result, van der Waals forces become less. In general

$$\text{Boiling point} \propto \text{no. of carbon atoms}^{1}$$

[1] For example, boiling point of hexane is 68.7 °C.

$CH_3CH_2CH_2CH_2CH_3$

n-pentane
(B.P. 36°C)

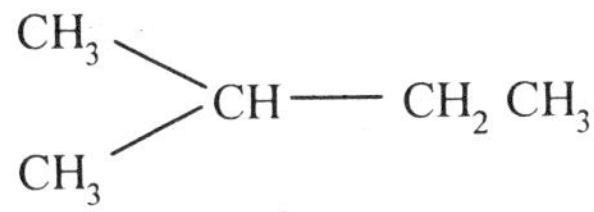

Isopentane
(B.P. 28° C)

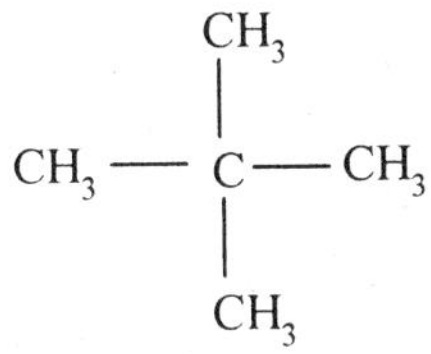

Neopentane
(B.P. 9.5°C)

Or

$$\text{Melting point} \propto \frac{1}{\text{Branching if number of carbon atoms is same}}.$$

(ii) *Functional Isomerism*: Functional isomers are the compounds that have the same molecular formula but different functional groups and hence different properties. The phenomenon is called functional isomerism.

Example:

(a)	Alcohol and Ether	C_2H_5OH Ethyl alcohol	CH_3OCH_3 Dimethyl ether
(b)	Aldehyde and Ketones	C_2H_5CHO Propionaldehyde	CH_3COCH_3 Propanone
(c)	Cyanide and Isocyanide	CH_3CN Methylcyanide	CH_3NC Methylisocyanide
(d)	Cyanate and Isocyanate	CH_3CNO Methylcyanate	CH_3NCO Methylisocyanate

(iii) *Position Isomerism*: Position isomers are the compounds that have the same molecular formula but different position of functional groups or multiple bonds and hence different properties. The phenomenon is called position isomerism.

(a) $CH_3CH_2CH{=}CH_2$ $\quad$ $CH_3CH{=}CHCH_3$

Butene - 1 $\quad$ Butene - 2

Different positions of double bond

(b) $CH_3CH_2{-}C{\equiv}CH$ $\quad$ $CH_3{-}C{\equiv}C{-}CH_3$

Butyne - 1 $\quad$ Butyne -2

Different positions of triple bond

(c) $CH_3CH_2CH_2CH_2OH$ — Butanol - 1 $\qquad$ $CH_3CH_2—CH(OH)—CH_3$ — Butanol - 2

Different positions of — OH groups

(iv) *Metamerism*: Metamers are the compounds that have the same molecular formula but number of carbon atoms on either side of the functional group are different, hence different properties. The isomerism is called metamerism, e.g.,

(a) $C_2H_5OC_2H_5$ Diethylether $\qquad$ $CH_3OC_3H_7$ Methylpropylether

(b) $CH_3COC_3H_7$ Methylpropylketone $\qquad$ $C_2H_5COC_2H_5$ Diethylketone

(v) *Tautomerism*: Interconvertible keto and enol forms of a carbonyl compound are known as 'tautomers', and their interconversion is called tautomerization. This special type of constitutional isomerism is called 'tautomerism'. It is represented by the symbol $\rightleftharpoons$ and this isomerism is shown by a carbonyl compound that has an α-hydrogen atom, i.e.,

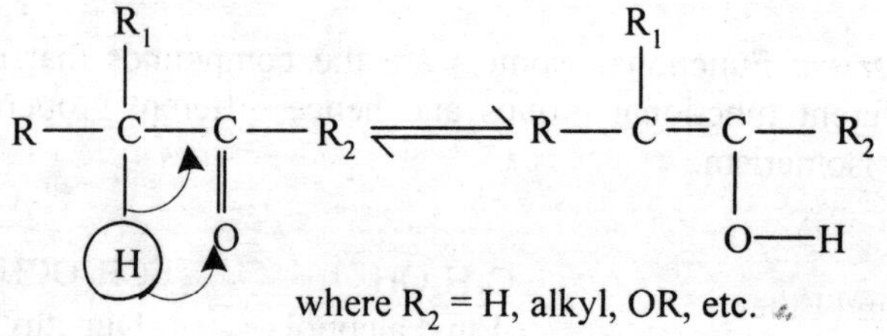

where R_2 = H, alkyl, OR, etc.

Essential condition for an organic compound to exhibit tautomerism is the presence of carbonyl group of aldehyde, ketone or ester with α Hydrogen atom, or alcoholic group in conjugation, e.g.,

(i) $(CH_3)_2C=O$ shows tautomerism but $C_6H_5—CO—C_6H_5$ (benzophenone) does not show tautomerism since it has no α-H atom.

(ii) 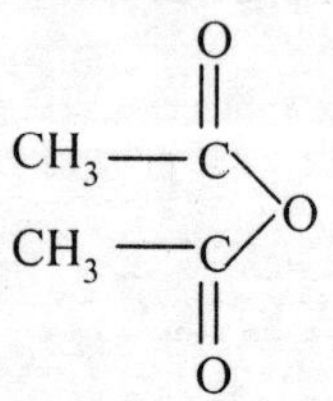 . It does not show tautomerism because it is an acid anhydride.

The keto form has a (C—H, C—C, C=O) bond. Whereas enolic form has a (C=C, C—O, and O—H) bond. The sum is 360 kcal/mole in the former and 345 kcal/mole in the latter. Therefore, keto form is more stable than enolic form by about 15 kcal mol^{-1}. Let us consider following tautomers for the percentage enol content:

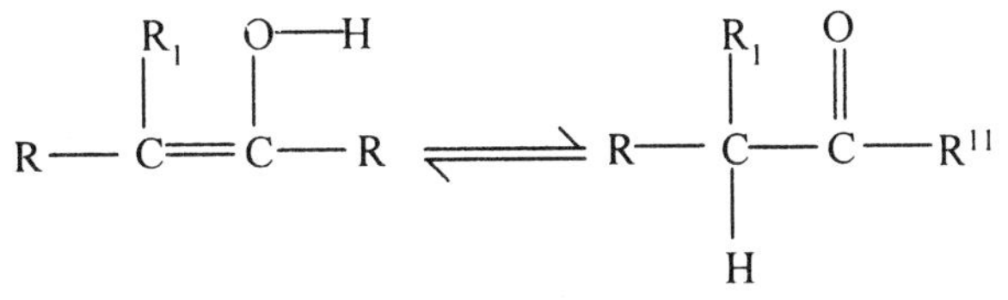

It has been observed that when R contains a multiple bond that it can conjugate with double bond of enolic structure (II), the percentage of enol content is high and it may even be of predominate form. Briefly, the percentage of enol content can be formulated as below:

$$\text{Percentage of enol content} \propto \frac{1}{\text{Temperature}}$$

Conjugation (π bonds)

$$\propto \frac{1}{\text{Hydrogen bonding}}$$

$$\propto \frac{1}{\text{Base concentration}}$$

Percent enol content is given in Table 7.1.

Table 7.1 Percent enol content of various organic compounds

	Compound	*Enolic content %*
(i)	$(CH_3)_2CO$	6×10^7
(ii)	$(CH_3)CHO$	1×10^{-3}
(iii)	$CH_3COCH_2COCH_3$	76.4%
(iv)	$(CH_3)_2CHCHO$	1.3×10^{-2}
(v)	$CH_3COCH_2COO\ C_2H_5$	8
(vi)	$C_6H_5COCH_2\ COCH_3$	89.2
(vii)	$C_2H_5OOC.\ CH_2COOC_2H_5$	7.7×10^3
(viii)	$NCCH_2COOC_2H_5$	0.25

Few more examples are:

(i) $R = R_1 = R_2 = H$, i.e.,

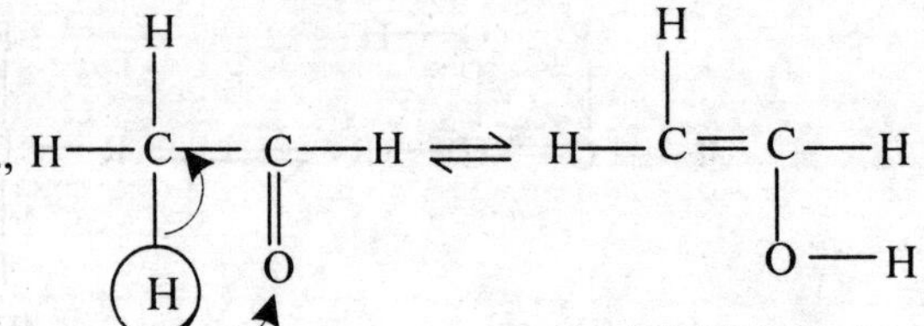

(acetaldehyde 100%) Vinyl alcohol (extremely small)

(ii) $R = R_1 = H$

$R_2 = CH_3$

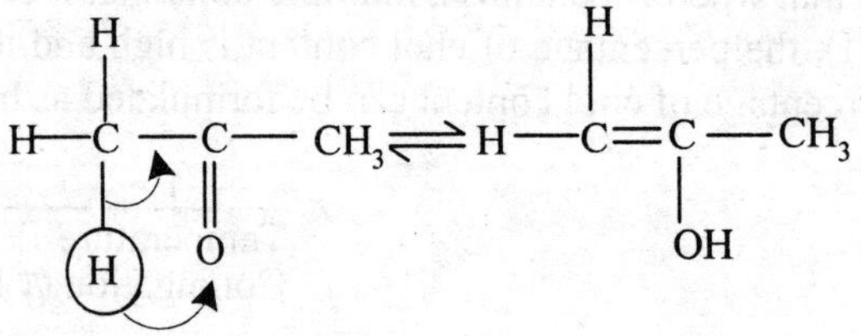

keto form (>99%) enolic form

(acetone)

This isomerism is also shown by molecules that have two carbonyl groups separated by one saturated carbon (called β dicarbonyl compounds) or by cyclohexanone and phenol, For example:

(i)

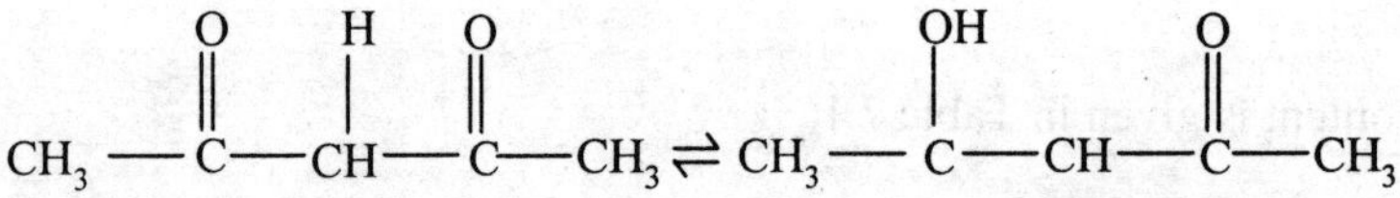

(24 %) keto form (76%) Enol form

(2,4 - Pentandione)

(ii)

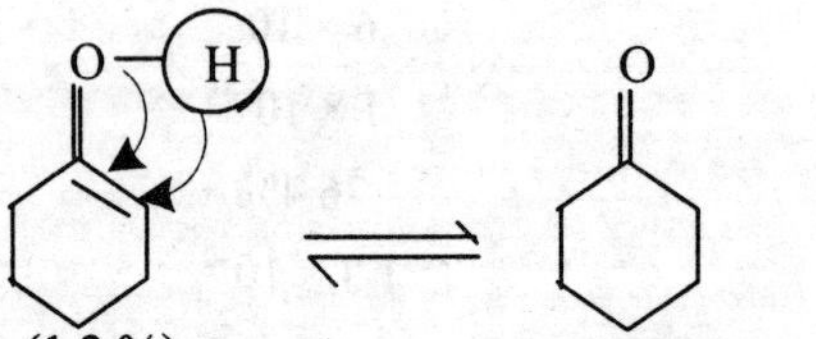

(1.2 %) Cyclohexenol (98.8%) Cyclohexanone

(iii)

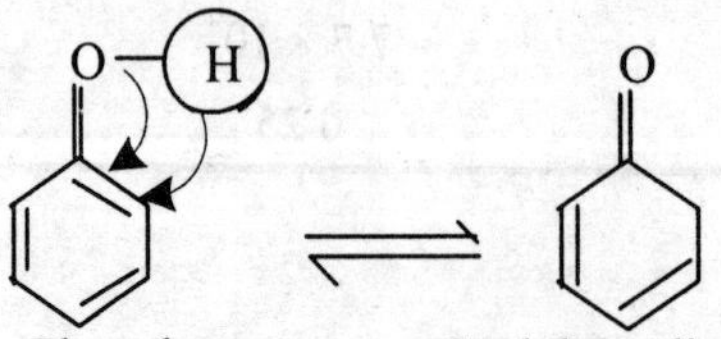

Phenol Cyclohexadienone

(iv) 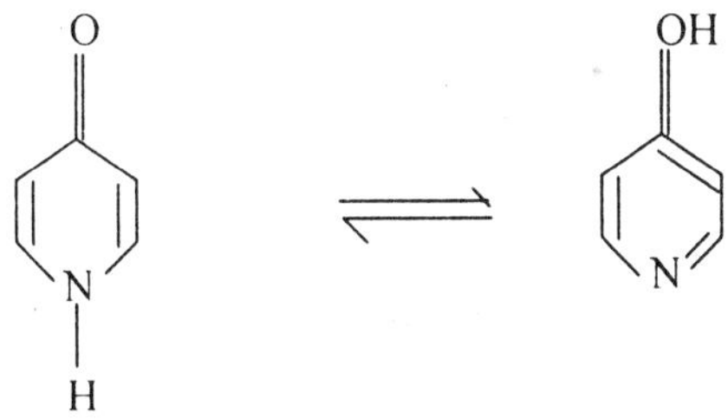

4— pyridone 4 - hydroxypyridine

(v) $R_2CH—N{=}O \rightleftharpoons R_2C{=}N—OH$

Nitroso Oxime

(vi) $R_2CH—CR{=}NR \rightleftharpoons R_2C{=}CR—NHR$

Imine Enamine

7.4 GEOMETRICAL OR *CIS-TRANS*-ISOMERISM

When two carbon atoms are linked by a single bond, there is free rotation about C—C bond. However, when two carbon atoms are linked by a double bond, the rotation is restricted.

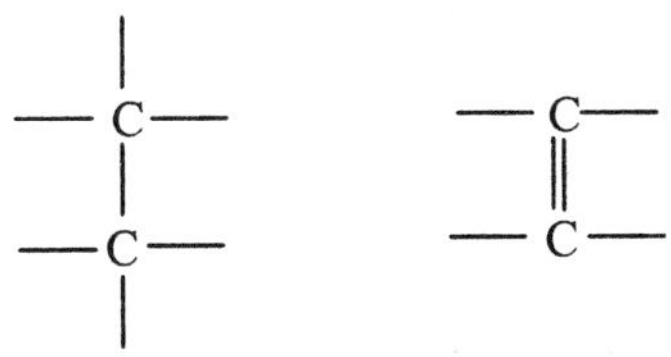

Free Rotation Restricted Rotation

The compounds, in which the rotation is restricted, with different group or atom on same carbon atom generally, exhibit geometrical isomerism. This condition results either from double bond or from rings. Stereoisomers which differ in properties due to restricted rotation are known as geometrical isomers and the phenomenon is called geometrical isomerism. The isomer in which the same groups are present on the same side is called *cis*-isomer and the isomer in which the same groups are on the opposite side is called *trans*-isomer. The essential conditions for geometrical isomerism is as follows:

(i) 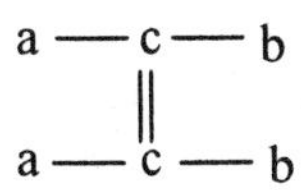

cis

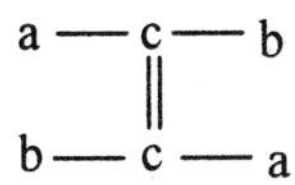

trans

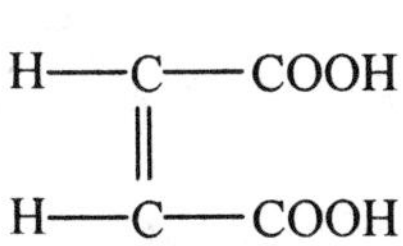

Maleic acid

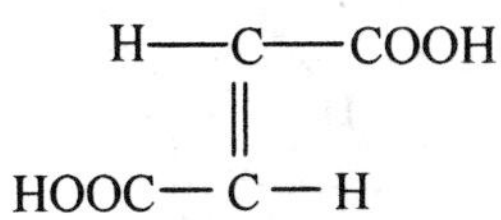

Fumaric acid

(ii)

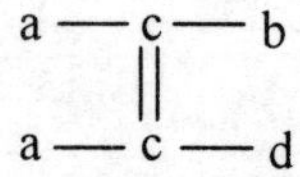

cis

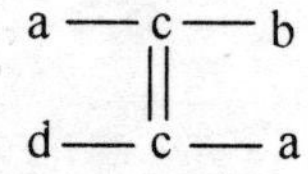

trans

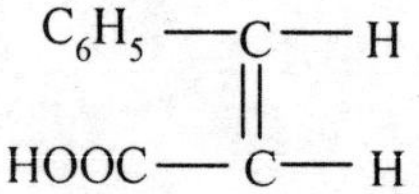

cis-Cinnamic acid

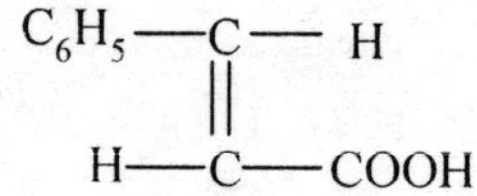

trans -Cinnamic acid

(iii)

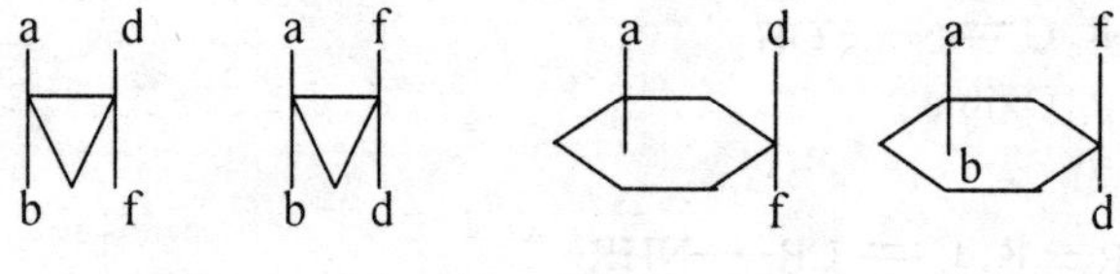

where a = d and b = f but a ≠ b and d ≠ f

The presence of a ring, a double bond, restricts rotation. *cis* and *trans* isomers are possible whenever two carbon atoms on a ring, bears two different groups. It is not essential that the two carbon atoms are adjacent, e.g.,

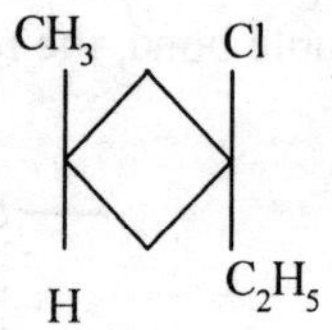

The number of geometrical isomers is calculated by the formula $2n^2$ (where n is the number of double bonds), in case both ends are identical, e.g.,

$CH_3\,CH{=}CH{-}CH_3$;

Butene - 2

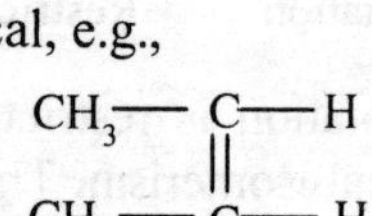

cis

and

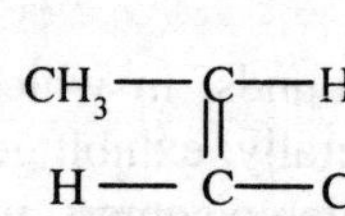

trans

However, if both ends are different, the number of isomers can be calculated by formula $2^{n-1} + 2p - 1$, where $p = \frac{n}{2}$ for even number and $P = \frac{n+1}{2}$ for odd number of double bonds.

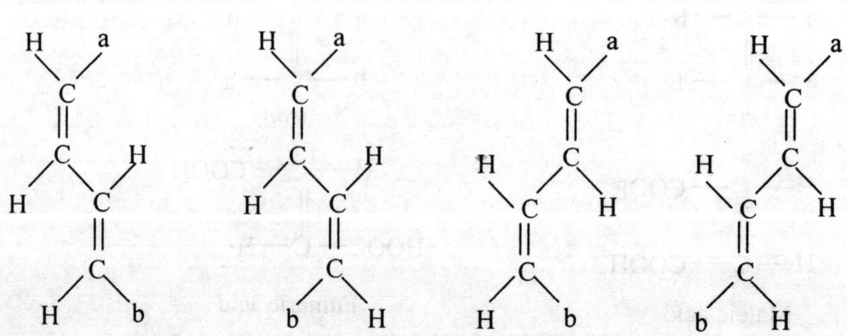

In case of tri or tetra substituted alkene the number of geometrical isomer is equal to number of E-Z forms (Please See Chapter 8).

Cis and *trans* isomers differ in properties (see Table 7.2), some of which are as follows:

(i) The melting point of the *cis*-isomer, as a rule, is always less than that of the *trans*-isomer. It is due to that fact that in *cis*-isomers, the molecules are arranged loosely in crystal lattice.

(ii) The values for other properties like boiling point, solubility, refractive index, disociation constant are higher for *cis*-isomers than for *trans*-isomers.

(iii) In case of maleic acid, anhydride is obtained upon heating which is not obtained in case of fumaric acid.

(iv) The *trans*-isomer has zero dipole moment.

Table 7.2 Differences between properties of *cis* and *trans* isomers

S.no.	*Properties*	*Maleic acid*	*Fumaric acid*
1.	Melting point	130°C	288°C
2.	Solubility in water at 25°C	788 g/litre	7 g/litre
3.	Dissociation constant (Ka) at 25°C	1.5×10^{-2}	1.0×10^{-2}

Stereomutation: Interconversion of *cis* and *trans* form is called stereomutation, e.g.,

$$\text{Maleic acid} \underset{}{\overset{\text{UV radiations, } I_2 \text{ in traces}}{\rightleftharpoons}} \text{Fumaric acid}$$

7.5 OPTICAL ISOMERISM

Any substance that rotates the plane of polarized light is said to be *optically active* and the property by virtue of which it rotates the plane of polarized light is called *optical activity*. A molecule is optically active only when it is non-superimposable upon its mirror image. This property is called *chirality* and the molecule is said to be *chiral*. If it is superimposable on its mirror image, it is achiral.

Stereoisomers which rotate the plane of polarized light in opposite directions, though in equal amounts, are called *optical isomers* and the phenomenon is called *optical isomerism*. The optical isomer can exist in four forms.

(i) *Dextrorotatory (d) or (+) form*: The isomer that rotates the plane of polarized light in (clockwise) direction is the dextrorotatory.

(ii) *Laevorotatory (l) or (—) form*: The isomer that rotates the plane of polarized light to the left (anticlockwise) direction is Laevorotatory.

(iii) *Meso form:* Optical isomer with a plane of symmetry is called meso form. It is optically inactive due to *internal compensation*, i.e., the rotation caused by the upper half of the molecule is neutralized by the lower half of the molecule, e.g., Meso tartaric acid.

(iv) *Racemic form (dl) or (±):* Equimolar mixture of *d* and *l* forms is known as racemic mixture. It is optically inactive due to *external compensation*, i.e., the rotation caused by the *d* form is neutralized by the *l* form, e.g. racemic tartaric acid.

The rotation caused by a molecule depends upon a number of factors like concentration, length of the cell, wavelength of the light, temperature, etc. In order to generalize, a term 'specific Rotation' $[\alpha_{sp}]$ has been used which is given by the following equation:

$$[\alpha_{sp}]_{\lambda}^{Temp} = \frac{\alpha}{lc}$$

where, α = Rotation observed
c = The concentration of solution in g/ml of solution
l = The length of the tube in decimeters.

The percent optical purity of an organic compound can be calculated with the help of following two formulae:

(i) $$\text{Percent optical purity} = \frac{\begin{pmatrix}\text{Moles of one}\\ \text{enantiomer}\end{pmatrix} - \begin{pmatrix}\text{Moles of other}\\ \text{enantiomer}\end{pmatrix}}{\text{Moles of both enantiomers}} \times 100$$

or

(ii) $$\text{Percent optical purity} = \frac{\text{Observed specific rotation}}{\text{Specific rotation of the pure enantiomers}} \times 100.$$

Cause of Optical Isomerism

A compound is said to be optically active if it is *non-superimposable* upon its mirror image due to asymmetry. This condition arises if it contains asymmetric carbon atom or it has no elements of symmetry.

(a) *Asymmetric or Chiral Carbon Atom*

A carbon atom is asymmetric if four different groups or atoms are attached to it, i.e.,

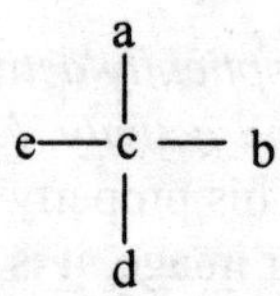
e.g.,
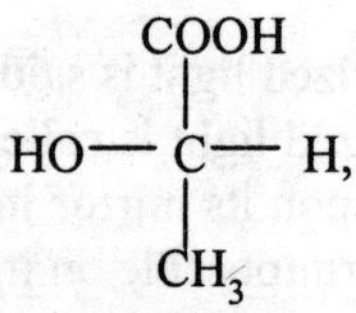
(Lactic acid)
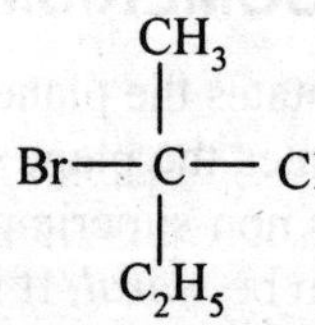
2-Bromo-2-Chlorobutane

Where $a \neq b \neq d \neq e$

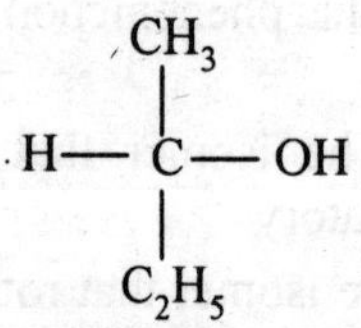
Butanol - 2

All alkanol – 2 & 2-haloalkanes contain asymmetric carbon atom. Number of optical isomers = $2n^2$, where n = Number of asymmetric carbon atoms.

(b) *Elements of Symmetry*

Elements of symmetry are as follows:

(i) *Plane of symmetry*: It may be defined as a plane which divides a molecule in two such parts which are related to each other as an object and its mirror image, e.g.,

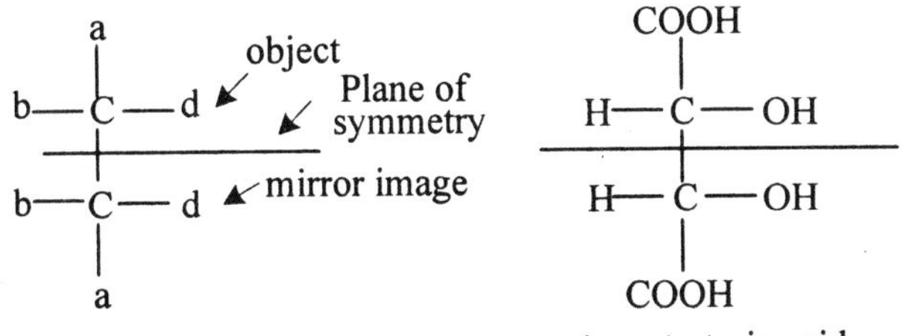

meso-tartaric acid

(ii) *Centre of Symmetry*: It may be defined as a point in the molecule through which if a line is drawn in one direction, then extended to equal distance in opposite direction, it meets another similar group or atom, e.g.,

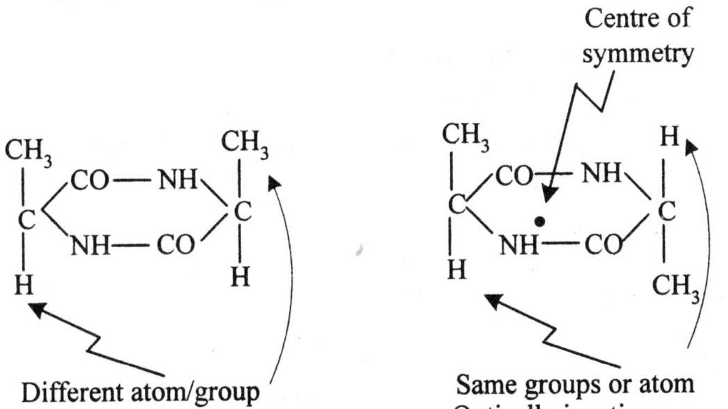

So no centre of symmetry hence optically active

cis-dimetyldiketo-piperzine

trans-dimetyldiketopiperzine

(iii) *Alternating Axis of Symmetry*: A molecule is said to have an alternating axis of symmetry through which when the molecule is rotated by an angle, and its mirror image is seen at an angle, right to the axis, the image is same as the compound was in the original position, e.g.,

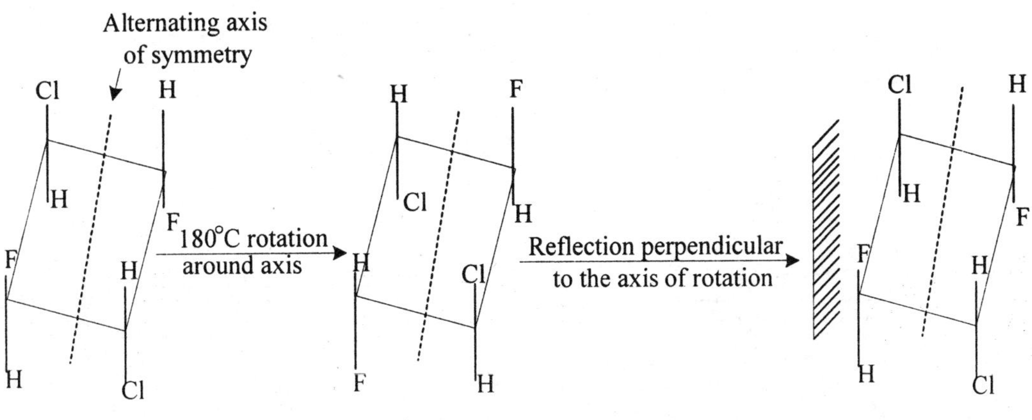

In general, an organic compound is said to have alternating axis of symmetry if it has following structure:

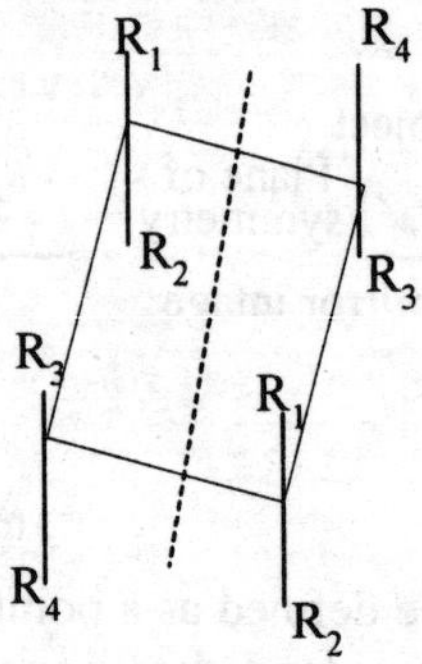

The multiplicity of the alternating axis can be calculated with the help of formula $\frac{360}{\theta}$, where θ is the angle through which molecule has been rotated.

(c) ***Optical Activity even in Absence of Chirality***

Following compounds show optical activity even in the absence of chirality:

(i) 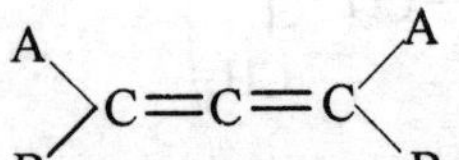

A ≠ B
(Allene)

(ii) 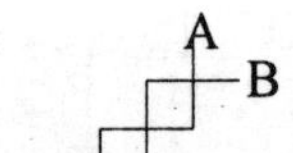

(Spiranes)

(iii) 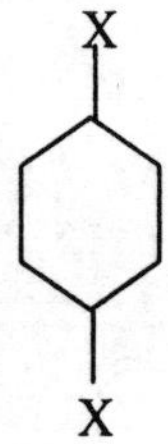

(iv) 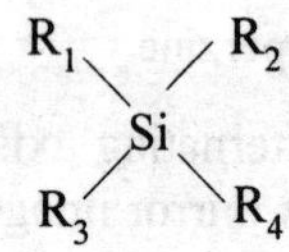

(v) 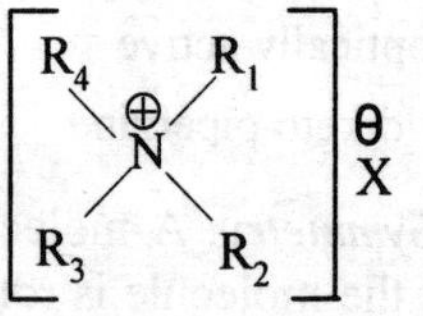

(vi) 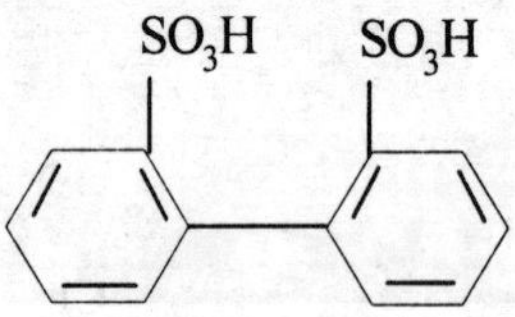

(vii) 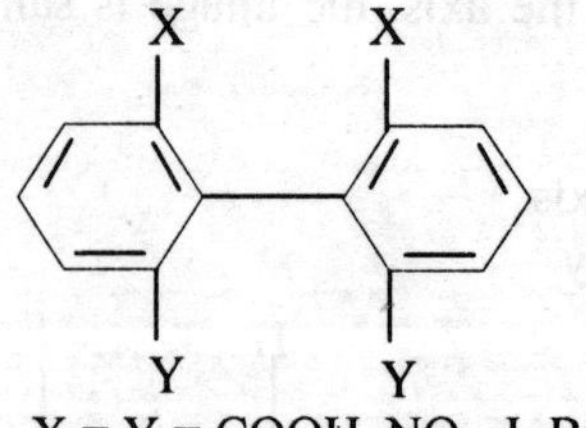

X = Y = COOH, NO_2, I, Br

(viii) 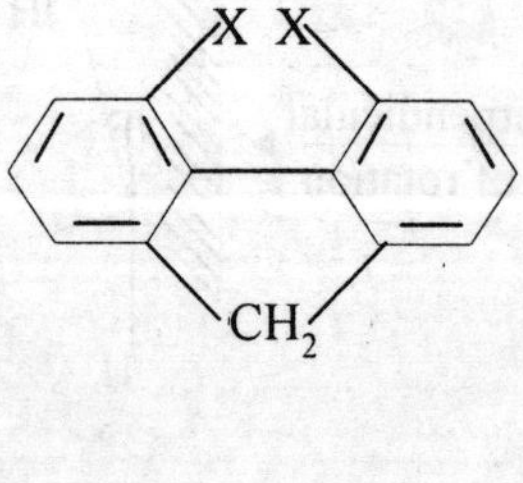

(ix) When even number of cumulated bonds is present, the compound will show optical isomerism, e.g.,

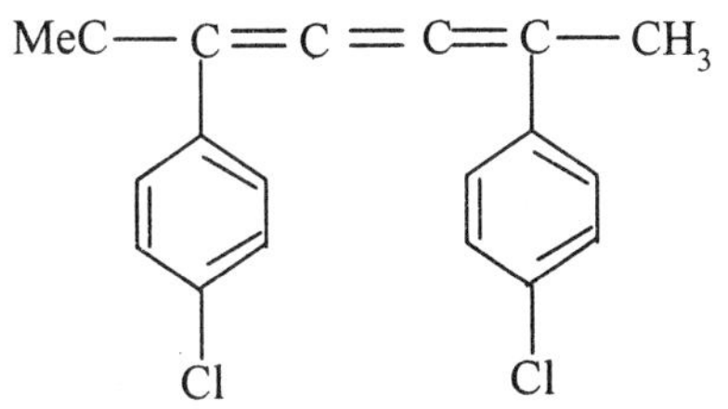

(d) ***Optical Activity of Alicyclic Compounds***

(i) In 1,2-disubstituted cyclopropane only *trans* form is optically active and *cis* is optically inactive (meso form).

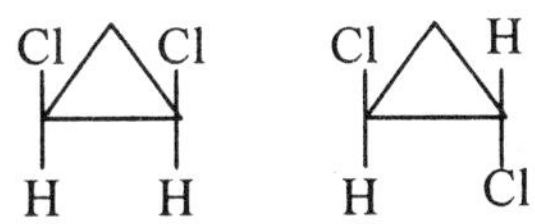

cis *trans*

(ii) *Cis* & *trans* 1, 3-Dimethyl cyclobutanes are optically inactive.

(iii) In case of 1, 2-Dimethyl cyclopentane, the *trans* form exists as enantiomers and *cis* form as meso.

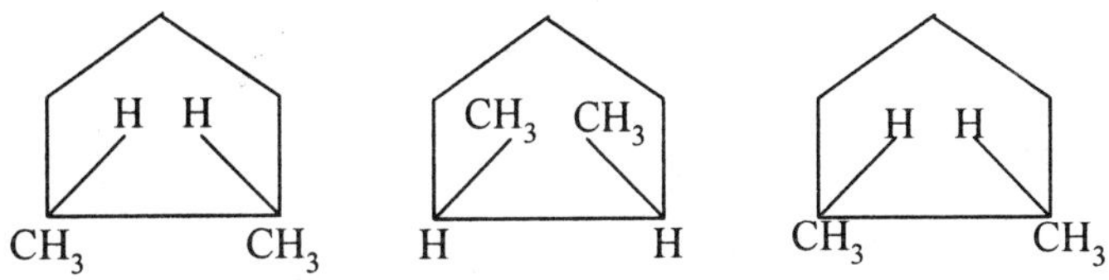

(iv) In 1,2-disubstituted cyclohexane, if $R_1 \neq R_2$ then it will be optically active:

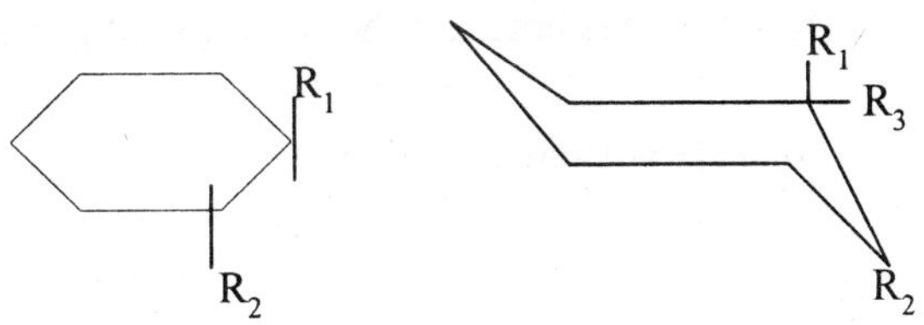

(v) In 1,3-disubstituted cyclohexane

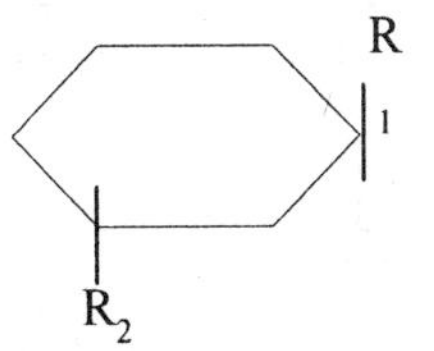

if $R_1 \neq R_2$
The compound will be optically active

(vi) In 1,4-disubstituted cyclohexane, all forms are optically active:

Calculation of optical isomers

Following table can be used for calculating optical isomers:

Compound	*Optically-active form*	*Optically-inactive form*
(i) n different chiral carbon and the molecule cannot be divided into two equal halves	2^n	O
(ii) Even number of chiral carbon atoms but molecule can be divided into equal and similar halves	$2^{(n-1)}$	$2^{\frac{(n-2)}{2}}$
(iii) Odd number of n chiral carbon atoms when molecule can be divided into two equal and similar halves	$2^{(n-1)}$	$2^{\frac{(n-1)}{2}}$

Number of stereoisomers

Number of optical isomers if the chiral carbon atom coincides with terminal responsible for geometrical isomerism.

Examples:

CH_3
|
*CHOH
|
CH=CH—CH_3

$n = 1$
Number of optical isomers = $2^n = 2$
Total stereoisomers = $2 \times 2 = 4$

CHO
|
CHOH
|
CHOH
|
CH_2OH

Number of optical isomers = $2^n = 4$
Number of geometrical isomers = 0
(Since it does not meet the condition of geometric isomerism)
Total stereoisomers = 4 + 0 = 4

COOH
|
*CHOH
|
*CHOH
|
COOH

Number of optical isomers = $2^{n-1} = 2^{(2-1)} = 2$
Number of optically inactive = $2^{(2-2)/2} = 2^{\circ} = 1$
Number of geometrical isomers = 0
Total stereoisomers = 2 + 1 = 3

Enantiomers or Enantiomorphs

Optical isomers, which are related to each other, as an object and its mirror image are known as enantiomers. They have identical physical and chemical properties and same magnitude of rotation in opposite direction. However, they react at a different rate with other optically-active substances, e.g.,

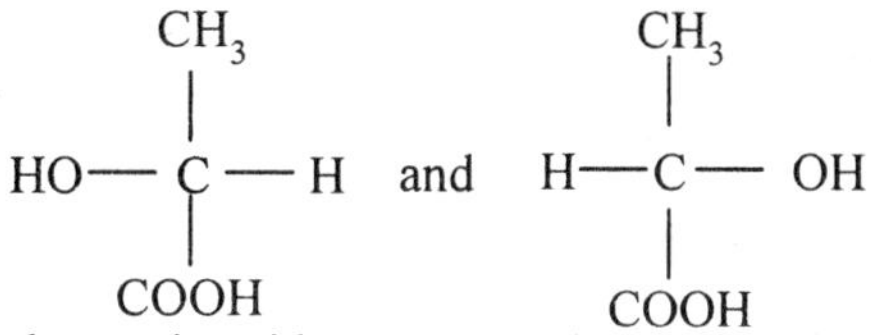

d–Lactic acid and l–Lactic acid

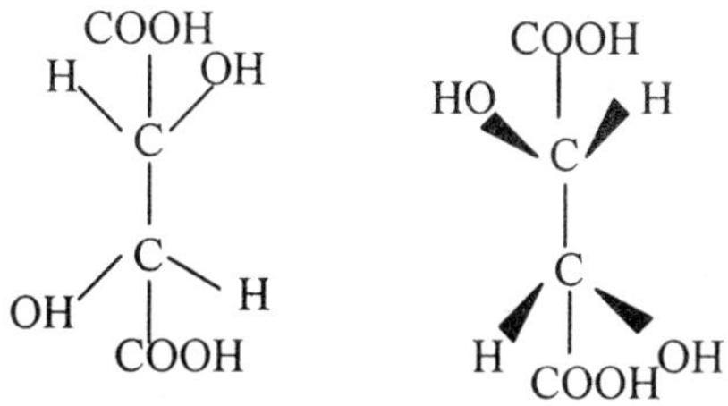

d–Tartaric acid l–Tartaric acid

$$\text{Enantiomeric excess} = \frac{\text{Specific rotation of the mixture} \times 100}{2 \times \text{specific rotation of one enantiomers}}$$

e.g., if specific rotation of a mixture (+) 22° and of *d* enantiomer is (+) 25° than:

$$\text{Enantiomeric excess} \frac{22 \times 100}{2 \times 25} = 44\%$$

It means mixture contain 44% excess (d) form & rest 56% is racemic mixture.

So, the composition of mixture is as follows:

(*d*) form $= 44 + \frac{56}{2} = 72\%$ (*l*) form $= \frac{56}{2} = 28\%$

Diastereoisomers

Optical isomers which are not related to each other as an object and its mirror image are called Diastereoisomers. They have different physical and chemical properties, e.g., d tartaric acid and mesotartaric acid.

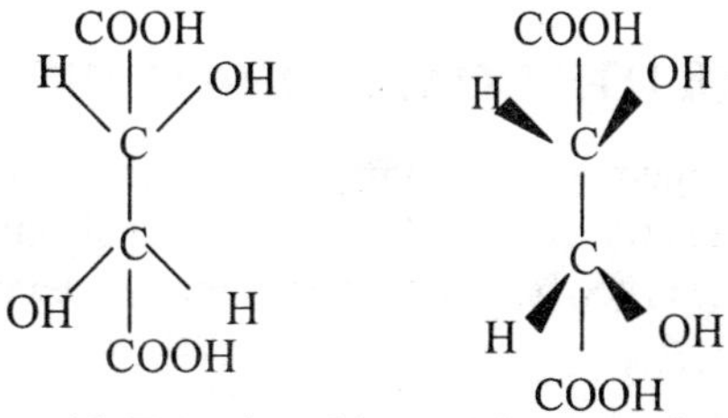

(d) Tartaric acid Mesotartaric acid

7.6 RESOLUTION

Separation of racemic mixture into dextrorotatory (*d*) and laevorotatory (*l*) form is called resolution.

7.7. RACEMIZATION

Conversation of dextro or laevo forms into racemic mixture is called racemization, e.g., SN^1 reactions.

7.8. ASYMMETRIC SYNTHESIS

Synthesis of asymmetric compound (optically active) from symmetric compounds (optically inactive) without resolution is called asymmetric synthesis.

Examples of asymmetric syntheses are as follows:

(i)

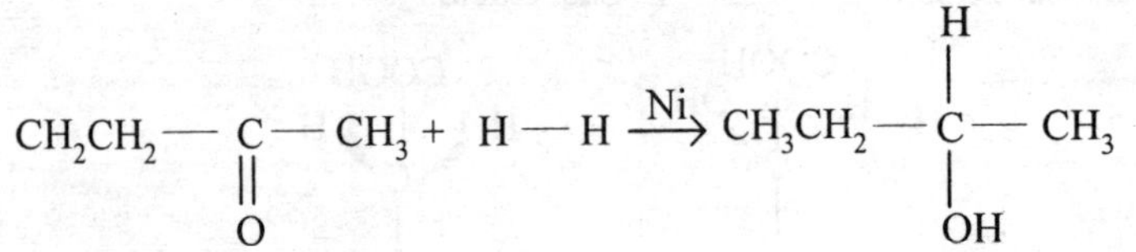

2-Butanone (Achiral molecule) — (+) -2 -Butanol (Chiral molecule)

(ii)

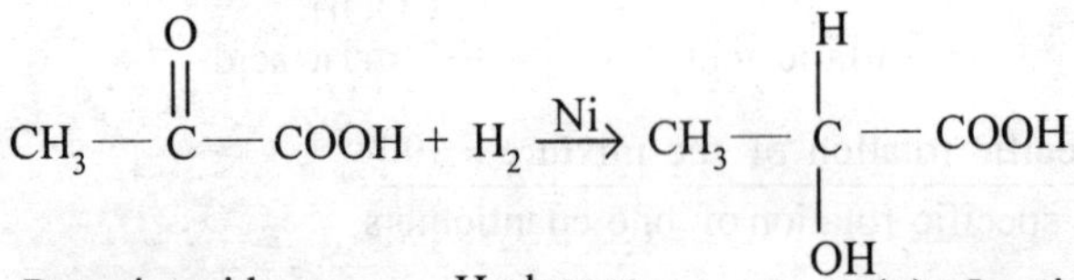

Pyruvic acid (Achiral molecule) — Hydrogen — (+) - Lactic acid (Chiral molecule)

7.9 STEREOSELECTIVE REACTIONS

A stereoselective reaction is a reaction that yields exclusively (or predominantly), only one of a set of stereoisomers, e.g.,

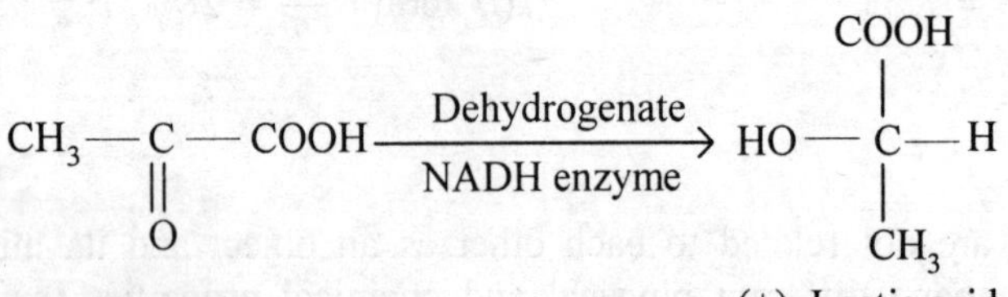

Pyruvic acid — (+) Lactic acid

7.10 RELATIVE AND ABSOLUTE CONFIGURATIONS

Configuration is the particular arrangement of atoms (or groups) which is possible by making and breaking of covalent bond that is characteristic of a given stereoisomer. The configuration at each stereocentre can be designated as (R) or (S). Let us consider the two enantiomers of 2-butanol:

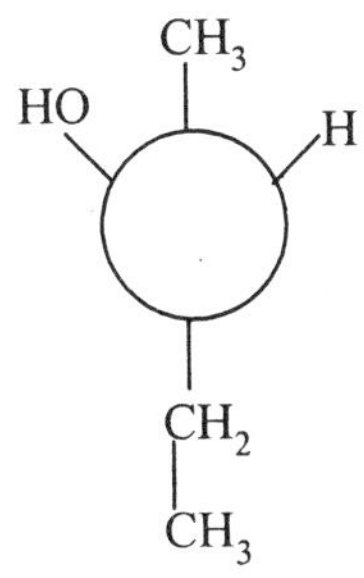

I

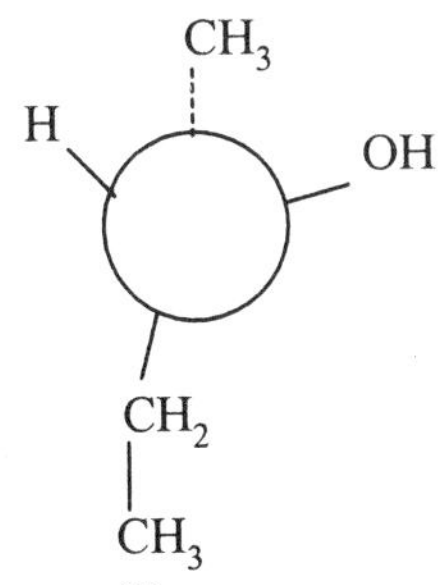

II

Then for the same 2-butanol, one could write structure I or structure II. Three chemists, R.S. Cahn, C.K. Ingold, and V. Prelong devised a system of nomenclature, called (R)-(S) system, to solve the problem. (R)-(S) are from the Latin words *rectus* and *sinster*, meaning right and left respectively. The rules are as follows:

(i) *Assignment of priority to groups/atoms a, b, d or e.* The priority is assigned on the basis of the atomic number of the atom that is directly attached to a stereocentre. The group with the highest atomic number is given highest priority. In the case of isotopes, the isotope of greatest atomic mass has highest priority, e.g., Oxygen is given highest priority because of highest atomic number and Hydrogen is given lowest priority because of lowest atomic number.

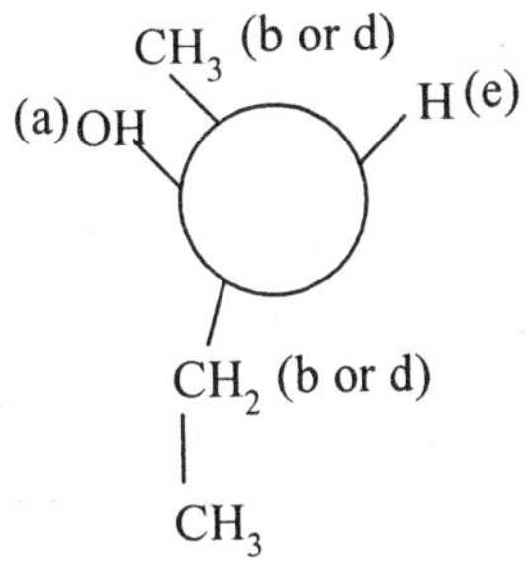

(ii) When first atom is same, then the next sets of atoms in the unassigned groups are examined. The process is continued until we can decide the priority, e.g.,

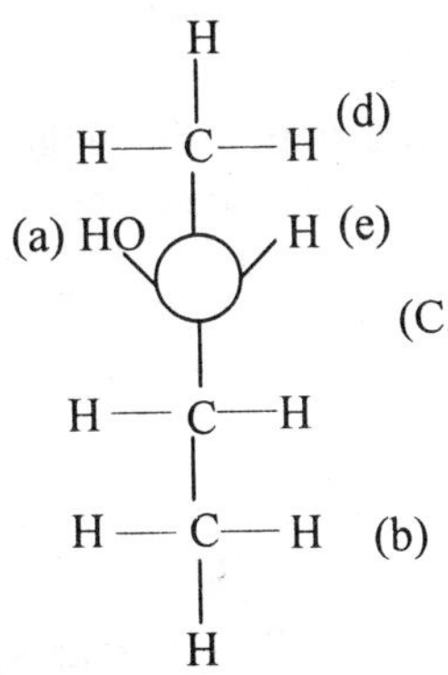

(C, H, H) > (H, H, H)

(iii) Groups containing double or triple bonds are assigned priorities as if both atoms were duplicated or triplicated,

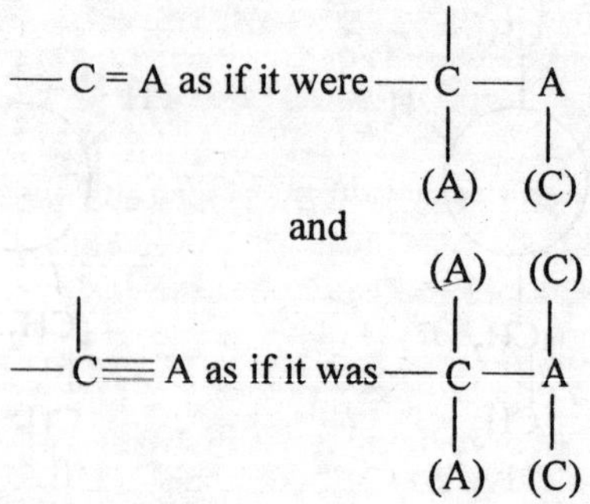

Thus, the vinyl group, —CH =CH_2 has higher priority then the isopropyl group

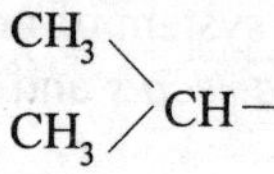

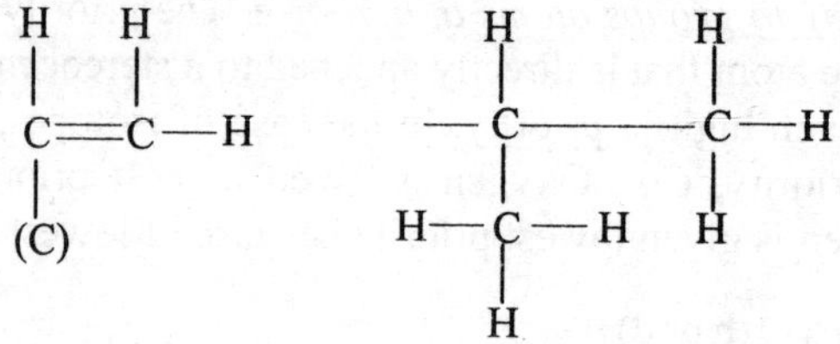

Vinyl group Isopropyl group

(iv) Now rotate the formula (or model) so that the group with lowest priority (e) is directed away from us.

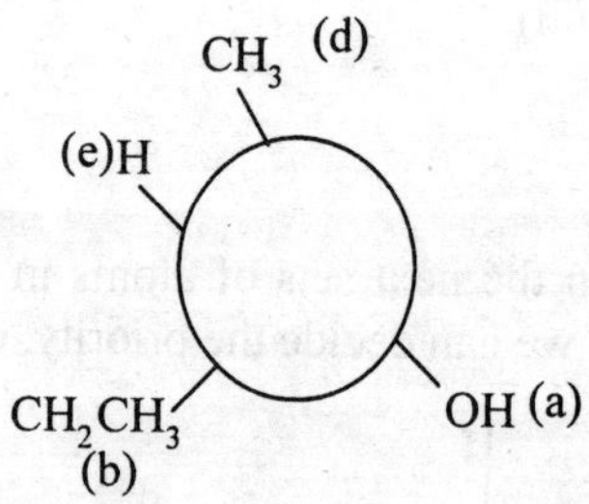

Now we trace a path from a to b to d. If direction of our finger is clockwise, the enantiomer is designed (R). If the direction is anti-clockwise, the enantiomer is designated (S). Thus

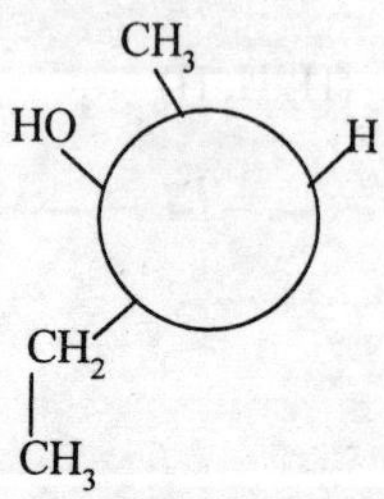

2-butanol enantiomer I

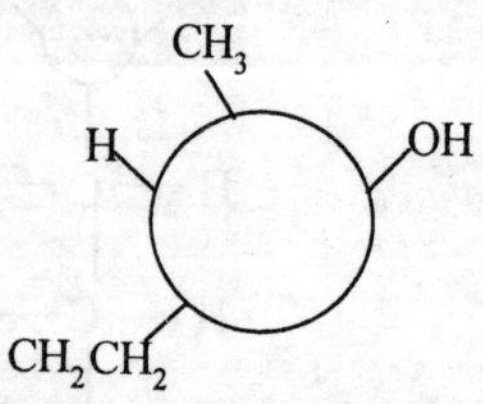

(*h*)-2-butanol

7.11 CONFORMATIONAL ANALYSIS

First let us know some terms:

(i) *Conformation*: The term conformation is used to denote any one of the infinite number of momentary arrangements of the atoms of a molecule in space that results from *rotation about single bond.* Three conformers are possible.

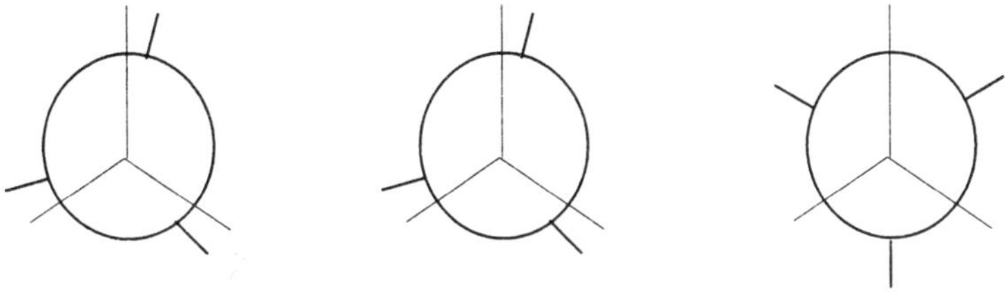

Eclipsed Skew or gauche Staggered

(ii) *Conformational analysis*: Conformational analysis is the analysis of the physical and chemical properties of a compound in terms of its most-preferred conformation in ground state, transition state.

(iii) *Axial and Equatorial bonds*: On each carbon atom one bond is directed up or down and the other more or less in the 'plane' of the ring. The up or down bonds are called axial (a) and the others are called equatorial (e) bonds.

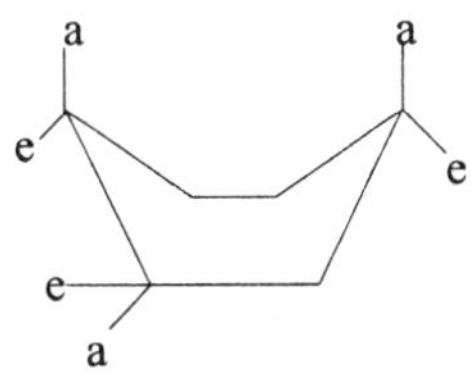

(i) *Conformations of Ethane*

Following six conformers are obtained by first allowing methyl groups to rotate along the C—C axis keeping rest of the molecule undisturbed.

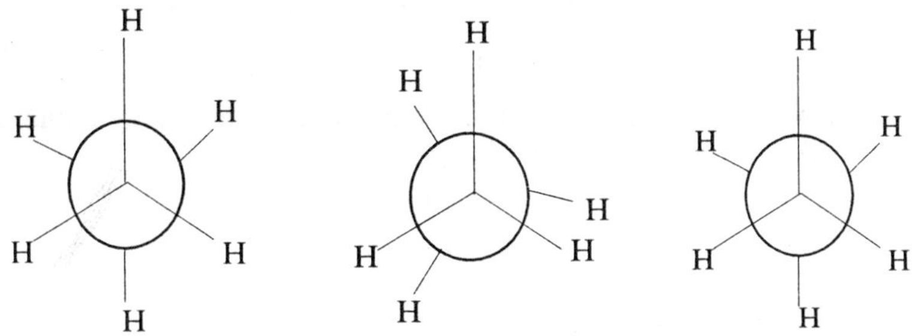

(I) (II) (III)

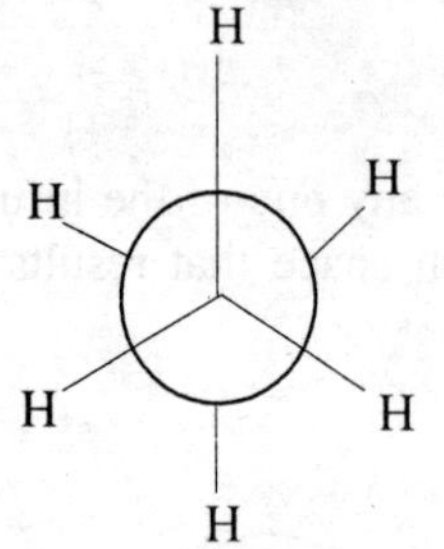

(IV)

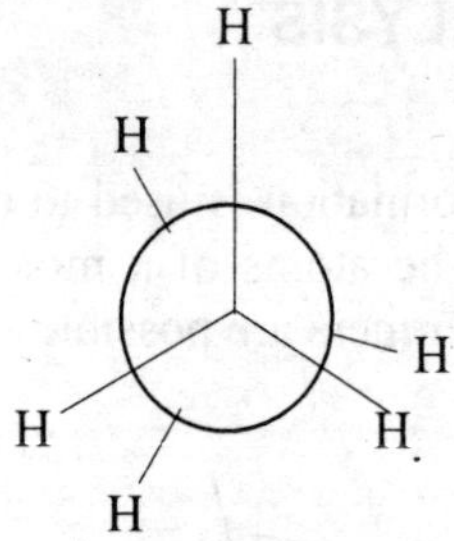

(V)

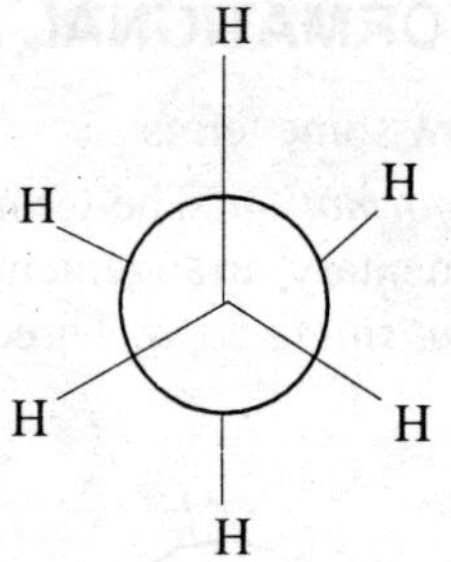

(VI)

Out of six, three are staggered form (I, III, V) in which the 2H atoms on different carbon atoms are as far as possible in staggered or Gauche form there are three. Eclipsed form (II, IV, VI) in which the 2H atoms from different carbon atoms are as closed, together as possible. In skew form the atoms are neither close nor far away.

(ii) ***Conformation of Disubstituted Ethane***

Three conformers exist for disubstituted ethane. These are staggered form, gauche form and eclipsed form. The order of stability is as follows:

{Staggered form or Gauche with minimum energy} > Skew form > {Eclipsed form with maximum energy}

(iii) ***Conformation of Cyclohexane***

For cyclohexane, there are two extreme conformations:

(a) The boat form which is an eclipsed form corresponding to an energy maximum hence less stable.

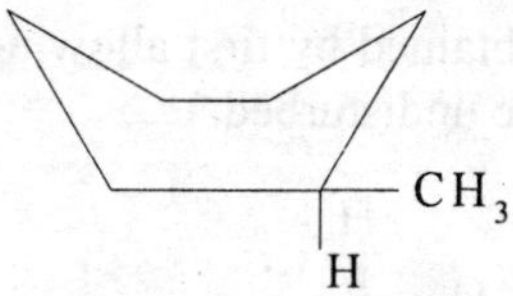

(b) The chair form which is staggered form corresponding to an energy minimum hence more stable.

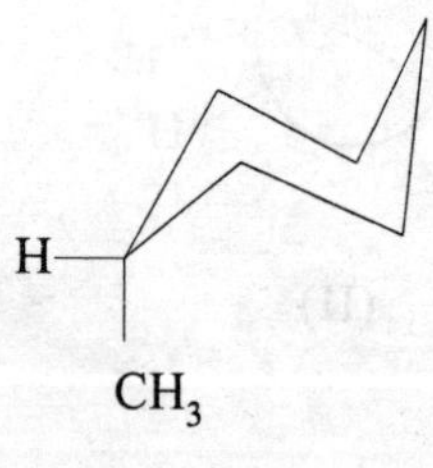

(iv) ***Conformation of Mono Substituted Cyclohexanes***

In mono-substituted cyclohexanes chair conformations are more stable than the boat conformation and further out of the two chair conformations, the equatorial chair form is more stable than axial form, because the substituted group in the equatorial position is subjected to less non-bonded interaction than when it is in axial position.

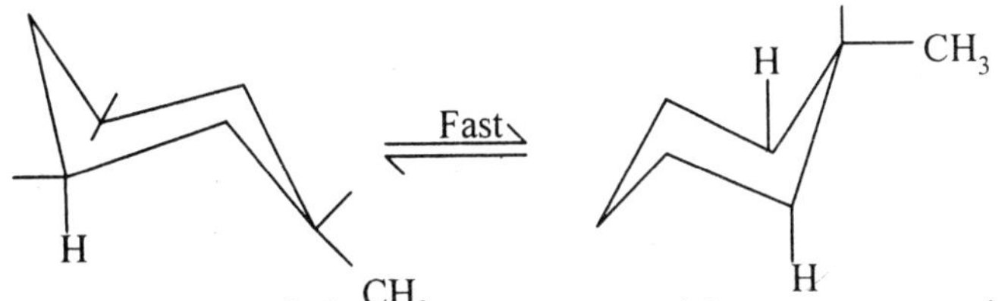

Axial groups crowded Equatorial groups staggered

PROBLEMS

1. Tick the correct answer amongst the following. One or more may be correct:

(i) An optically pure compound A, gave an $[a]^{25}$ D = 30°. A mixture of A and its enatiomer B, gave $[a]^{25}$ D = 15°. The ratio of A to B in the mixture is:

(a) 3 to 1 *(b)* 2 to 1 *(c)* 1 to 3 *(d)* 1 to 2

(ii) If (+) –2-chlorobutane had $[a]^{25}$ D = + 10.6° when pure, the amounts of each isomer in a mixture of (+) and (–) –2-chlorobutane which has a rotation of +7.2° is:

(a) 80° (+) to 20 (–)

(b) 84 (+) to 16 (–)

(c) 80 (–) to 20 (+)

(d) 84 (–) to 16 (+)

(iii) Which of the following compounds is or are optically active?

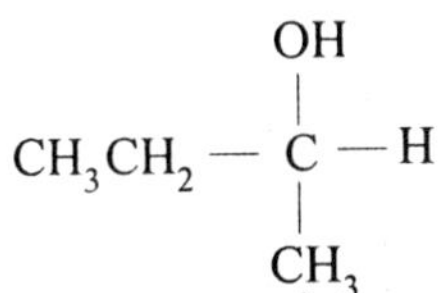

I

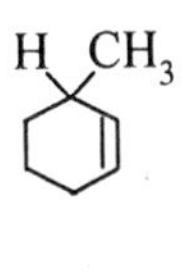

II

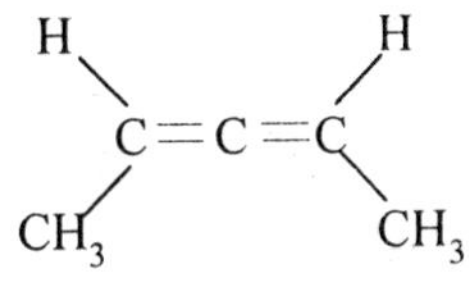

III

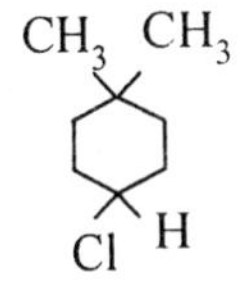

IV

(a) I, III *(b)* I *(c)* I, II, III *(d)* all

(iv) Examine each of the following structure-pairs, and determine whether or not they are identical. If they are different, indicate the relationship they bear with each other.

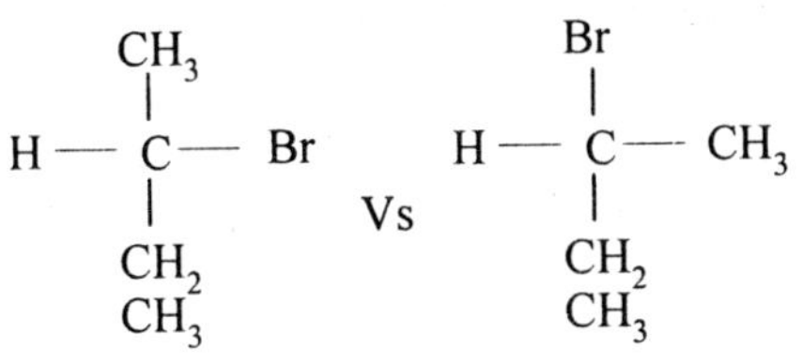

(a)

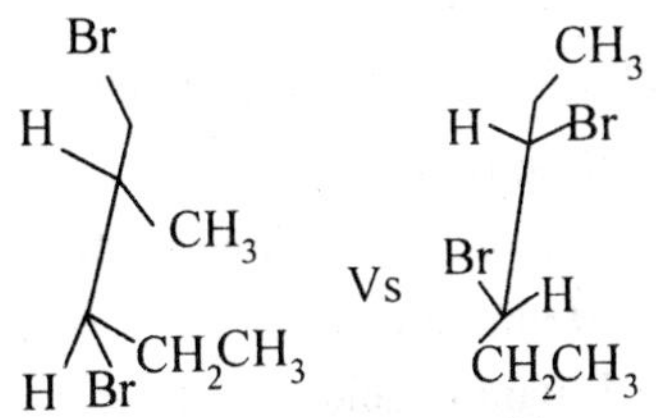

(b)

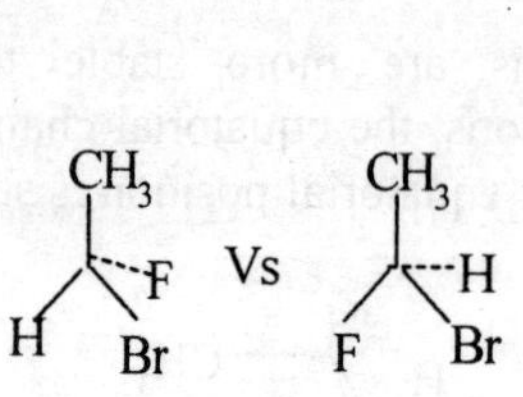

(c)

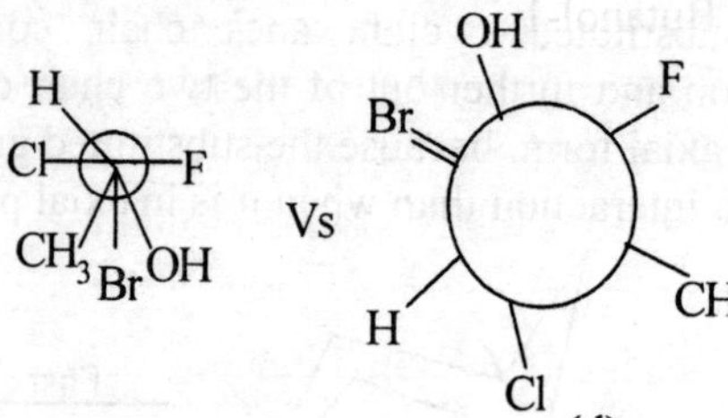

(d)

(v) Which of the following compounds can exist as 'Meso' form?

(i) 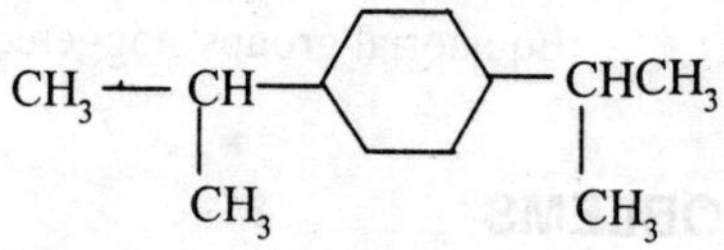

(ii) 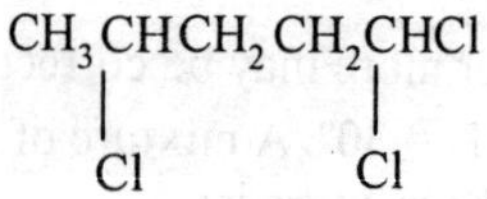

(iii) 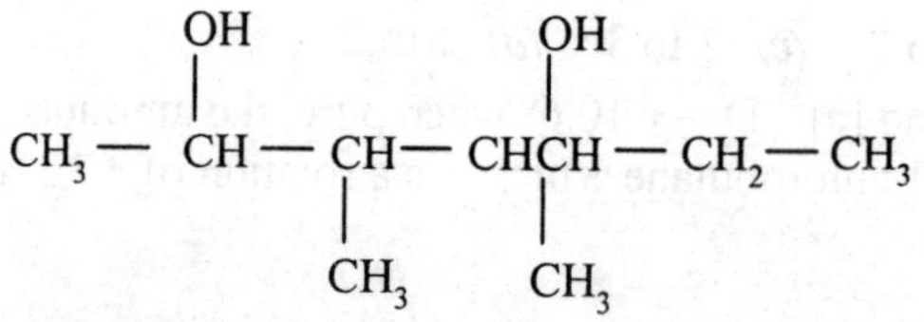

(iv) CHClBrI

(a) (I), (II), (III) *(b)* II, III, IV *(c)* I, III, IV *(d)* I, II, IV

(vi) Compound $MeCHBrCH{=}CH_2$ shows:

(a) Only geometrical isomersim
(b) Only optical isomerism
(c) Both geometrical and optical isomerism
(d) Optical inactivity

(vii) Compound: $CH_3CHOH—CH{=}CH—CH_3$ shows:

(a) Only geometrical isomersim
(b) Only optical isomerism
(c) Both geometrical and optical isomerism
(d) Tautomerism

(viii) Which of the following can be resolved?

(a) Propionic acid
(b) 2-hydroxy propionic acid
(c) Methanoic acid
(d) Ethanoic acid

(ix) The compound that is not isometric with diethyl ether is:

(a) *n*-propylmethyl ether

(b) Butanol-1

(c) 2-methyl propan = 2 = 01

(d) Butanone **(IIT 1986, MNR, 1990)**

(x) The highest boiling point is expected for:

(a) Iso-octane (b) $n-$ octane

(c) 2,2,3,3,-tetramethyl butane (d) n-butane **(IIT, 1986)**

(xi) Keto-enol tautomerism is possible in:

(a) 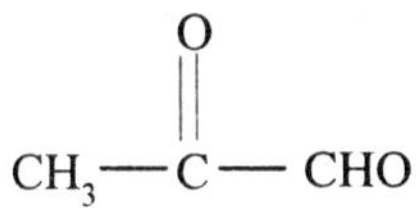

(b) 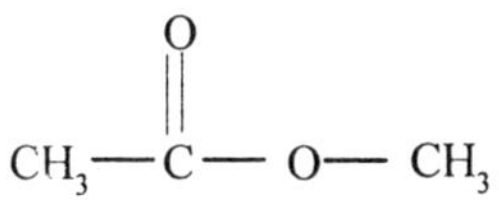

(c) 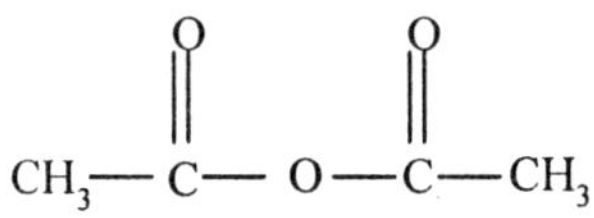

(d) 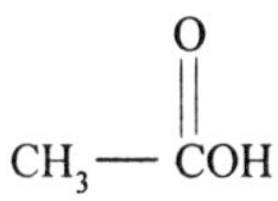 **(IIT, 1998)**

(xii) Total number of possible stereoisomers for the compound:

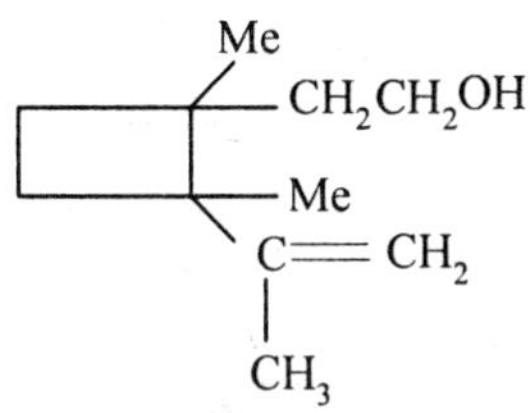

(a) 1 (b) 2 (c) 4 (d) 8

(xiii) Total number of stereoisomers for the compound

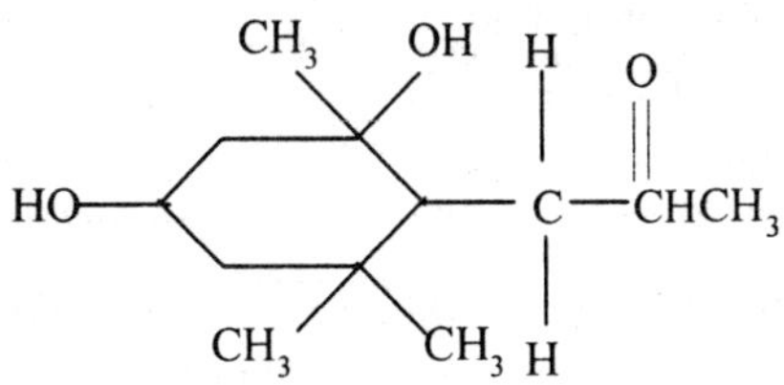

(a) 2 (b) 4 (c) 8 (d) 16

(xiv) A successful resolution of a racemic mixture depends on

(a) the fact that the enantiomers have different physical properties.

(b) conversion of the enantiomers to diastereomers which have different physical properties.

(c) the use of acids or bases to convert the enantiomers to salts.

(d) Fractional distillation of the racemic mixture into enantiomers.

(xv) Isomers which can be converted through rotation around a single bond are:

(a) Conformers (b) Diastereomers

(c) Enantiomers (d) Position Isomers **(IIT, 1992)**

(xvi) Tautomerism is exhibited by:

(a) 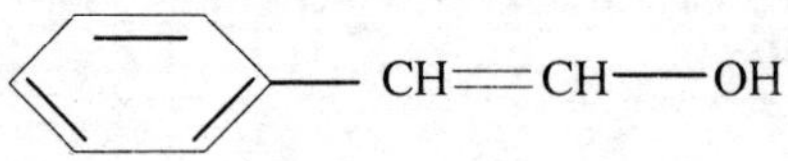

(b) 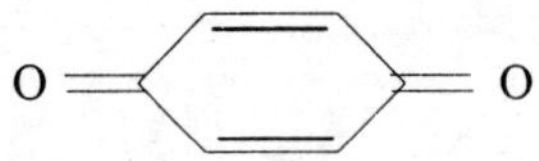

(c) 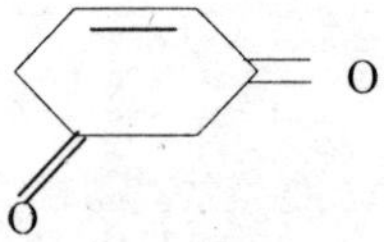

(d) 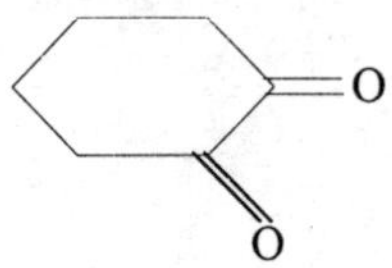

(IIT, 1998)

(xvii) Which of the following compounds will show geometrical isomerism? **(IIT, 1998)**

(a) 2-Butene (b) Propene

(c) 1-Phenylpropene (d) 2-methyl-2-butene

(xviii) The number of isomers for the compound with molecular formula $C_2BrCl FI$ is

(a) 3 (b) 4 (c) 5 (d) 6 **(IIT, 2001)**

(xix) 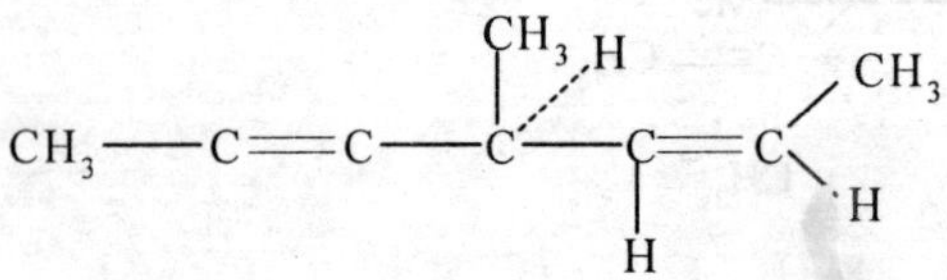

Hydrogenation of the above compound in the presence of poisoned Palladium catalyst gives

(a) an optically-active compound

(b) an optically-inactive compound

(c) a racemix mixture

(d) a diastereomeric mixture **(IIT, 2001)**

(xx) Which of the following compounds exhibits stereoisomerism?

(a) 2-methylbutene –1 (b) 3-methylbutyne – 1

(c) 3-methyl butanoic acid (d) 2-methylbutanoic acid

(xxi) Two chemicals having same chemical formula but different arrangement of atoms are called.

(a) Isotopes (b) Isomers (c) Polymers (d) Epimers

(xxii) The number of geometrical isomers for compound with molecular formula C_2FClBr are: **(IIT, 2004)**

(a) 2 *(b)* 4 *(c)* 6 *(d)* 8

2. Fill in the blanks:

(i) Number of isomers for C_4H_8 ....

(ii) Number of isomers for C_6H_{14}......... **(IIT, 1987)**

(iii) Number of isomers for C_4H_{10} ....

(iv) Number of isomers for $C_{10}H_{22}$.........

(v) Number of isomers for $C_{20}H_{42}$.........

(vi) Number of isomers for $C_{40}H_{82}$.........

(vii) Optically-active alkane with lowest molecular weight is .....

(viii) An optically-active compound 'A' having molecular formula $C_6H_{12}O$ gives optically-inactive compound 'B' on hydrogenation. Compound 'A' and 'B' are. ......and .... respectively.

(ix) Amongst 1-chloropentane, 2-chloropentane, 3-chloropentane, 1-chloro –2-methylpentane, 2-chloro- 2-methylpentane, the chiral compounds are.......

(x) Amongst *n*-pentane, isopentane and neopentane, the boiling point of .... is lowest.

(xi) Possible isomers of *cis*-chloroethene are ...., and ....has zero dipole moment.

3. Write number of isomers for compound with formula C_6H_6.

4. Draw all possible stereoisomers for the following and state which are optically active, inactive and are meso form

(i) MeCHOHCHOHMe

(ii) MeCHOHCHOH

(iii) MeCHOHCOCHOHMe

(iv) MeCHClCOOH

(v) $HO_2C.CHBrCH_2COOEt$

(vi) $CH_2{=}CHMe$

(vii) MeCH=CHEt

(viii) $MeCHBrCH{=}CH_2$

(ix) $CMe_2{=}C{=}CH_2$

(x) CMeCl=C=CHMe

5. Write all possible stereoisomers of the following compounds:

(i) 1,2-dibromopropane

(ii) 3 Chloro-3-bromo-4,4-dimethyl hexane

(iii) 2-Bromo-3-chlorobutane

6. The structure of five isomeric compounds are shown:

(a) $CH_3CH{=}C{=}CHCH_3$

(b) $CH_3HC{=}C{=}CHCH_3$

(c) $CH_2{=}CH{-}HC{=}CHCH_3$

(d) $CH_2{=}CH{-}HC{=}CHCH_3$

(e) $CH_2{=}CH{-}CH_2{-}CH{=}CH_2$

Answer the following:

(i) Do (a) and (b) have same boiling point?

(ii) Do (a) and (b) have same effect on plane polarized light?

(iii) Do (c) and (d) have same effect on plane polarized light?

(iv) Do (c) and (d) have same boiling point?

(v) How many stereoisomers exist for (e)?

ANSWERS

1. (i) (**a**), $\frac{15}{30}$ =50% optical purity or 50% (+) in excess of (+) thus

50% (+)	25% (–)
25% (+)	0% (–)
75% (+)	25% (–)

hence 3 to 1.

(ii) (**b**), $\frac{7.2}{10.6}$ = 68% optical purity, or 68% of (+) in excess of (+). Thus, 68% (+), 32% (+) or

68% (+)	16% (–)
16% (+)	0% (–)
84% (+)	16% (–)

So, 84 parts (+) to 16 parts (–) in the mixture.

(iii) (**c**), If you have doubt about III, think of *sp* hybridization.

(iv) *(a)* are enantiomers *(b)* are diastereomers

(c) are identical *(d)* are diastereomers

(v) (**b**), Structure (I) cannot exist as a meso form since it contains no asymmetric carbon atoms. Rest have two asymmetric carbon atoms.

(vi) (**b**), $CH_3{-}C(H)(Br){-}CH{=}CH_2$ There is one asymmetric carbon atom.

(vii) (**c**),

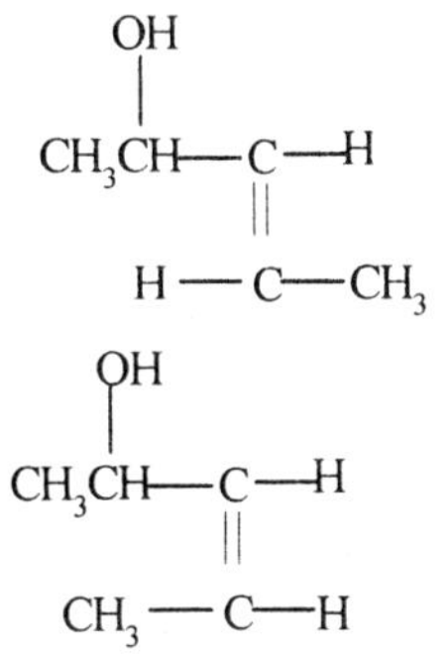

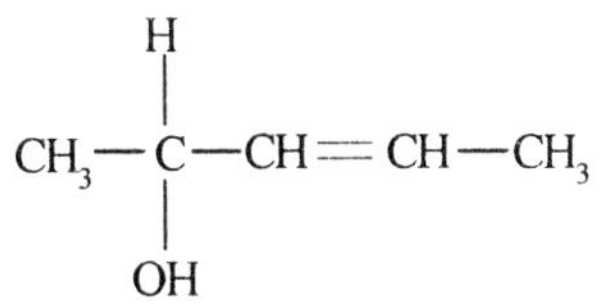

asymmetric C atom,

Hence shows geometrical isomerism.

(viii) (**b**), because it contains one asymmetric carbon atom.

(xi) (**d**), ethers show, metamerism with ether and functional isomerism with alcohols, hence butanone is not an isomer.

(x) (**b**) *n*-octane.

(xi) (**a**), see condition 'Carbonyl compound with a H-atom', I b.c. and d. There is a carbonyl group it is attached to oxygen atom, hence it is not carbonyl compound.

(xii) (**c**), See the uncommon allenic linkage which is a dissymmetric grouping.

(xiii) (**c**), 2 pairs of enantiomers.

(xiv) (**b**), Diastereoisomers should be converted into enantiomers by reacting them with another dissymmetric species (example. xi)

$d, 1 + 2x' \rightarrow \quad dx + 1x'$

enantiomers Diastereoisomers

(xv) (**a**)

(xvi) (**a**), (**b**), (**d**)

(xvii) (**a**), (**c**)

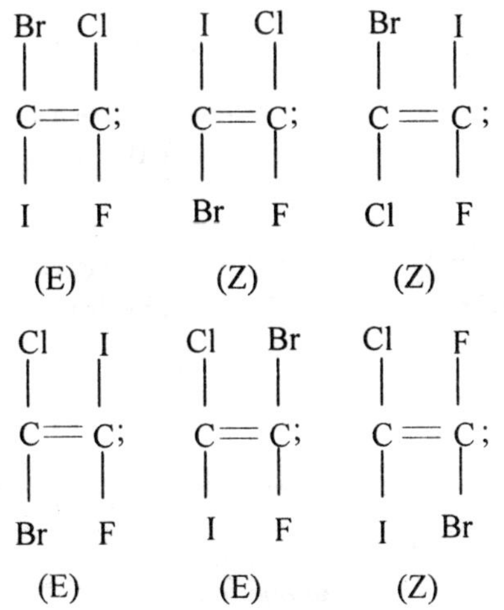

(xviii) (**d**)

(xix) (**b**)

(xx) (**d**), it will show optical isomerism.

(xxi) (**b**)

(xxii) (**c**)

2. (i) (**6**), $CH_3CH_2CH{=}CH_2$

Butene = 1

(I)

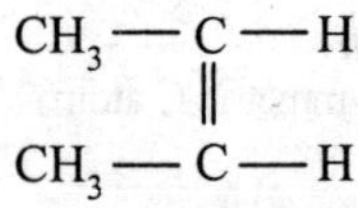

cis Butene-2

(II)

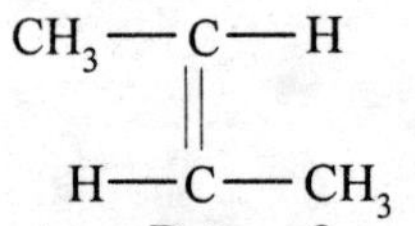

trans Butene-2

III

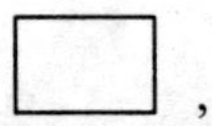

Cyclobutane

IV

and

Methyl cyclopropane

V

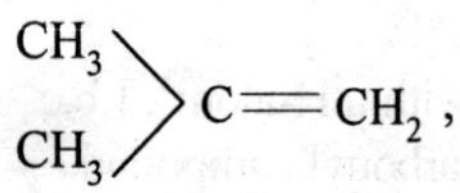

Isobutene

VI

(ii) (**5**) $CH_3CH_2CH_2CH_2CH_2CH_3$

n-hexane

(I)

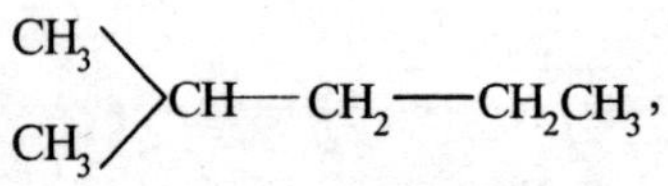

Isohexane

(II)

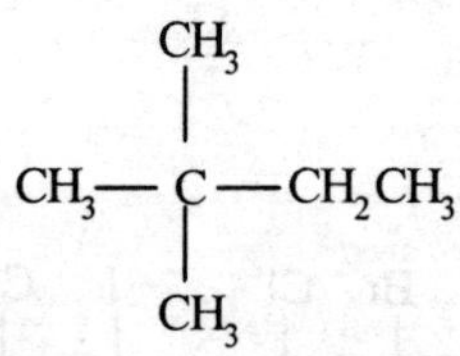

2,2-dimethyl butane

(III)

$(CH_3)_2CH—CH(CH_3)_2$,

2,3-Dimethyl butane

(IV)

$(CH_3)_3C—CH(CH_3)_2$

2,2,3-Trimethyl butane

(V)

(iii) (7), $CH_3CH_2CH_2CH_2OH$

Butanol - 1

(I)

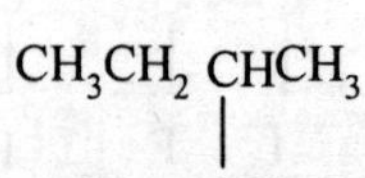

Butanol -2

(II)

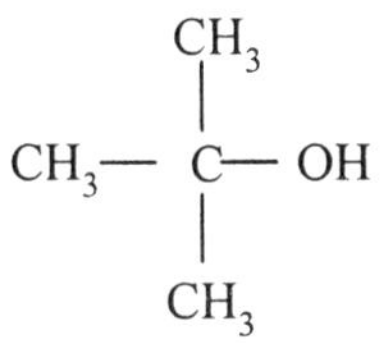

Terbutylalcohol
(III)

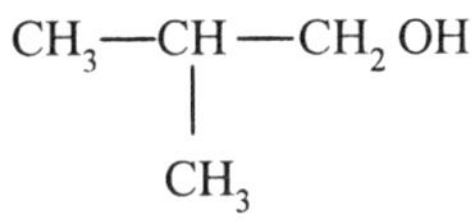

2-methylpropanol
(IV)

$C_2H_5OC_2H_5$

Diethylether
(V)

$CH_3OC_3H_7$
Methylpropyl ether
(VI)

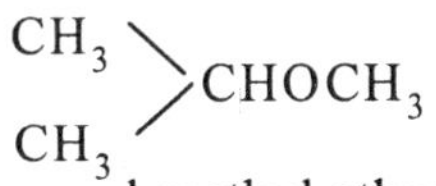

Isopropyl methyl ether
(VII)

(iv) (75)

(v) (10^5) Hypothetical structures. Not possible to draw

(vi) (10^{13})

(vii)

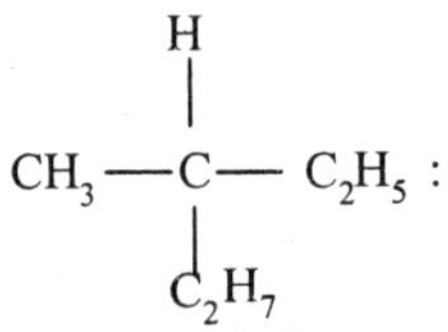

: mol. wt. = 100

3-methylhexane

(viii)

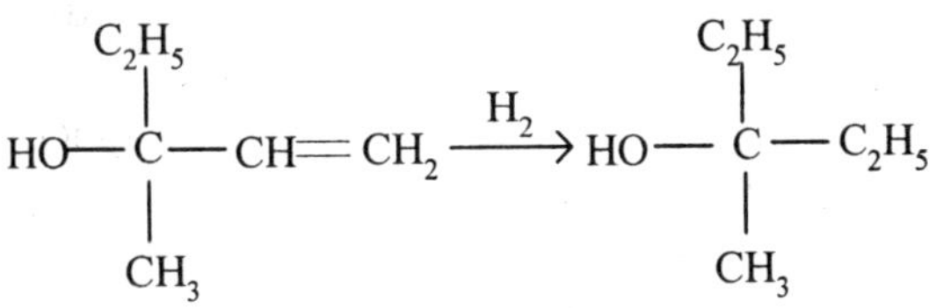

(A) (B)

(ix) 2-chloropentane, 1-chloro-2 methylpentane contain asymmetric carbon atom, hence chiral, e.g.,

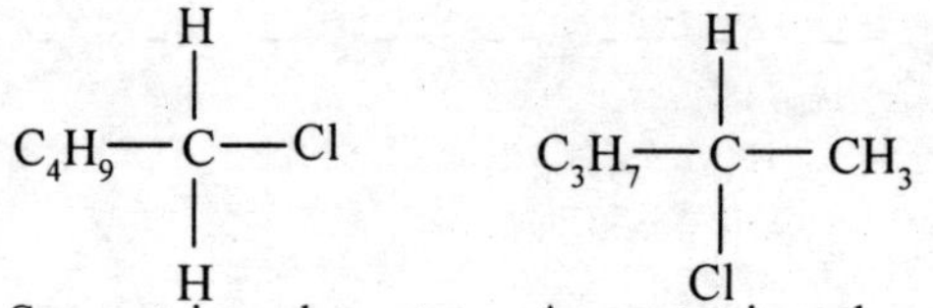

Symmetric carbon atom 1-chlorobutane   Asymmetric carbon atom 2-chlorobutane

(x) Neopentane

(xi) (2) *trans* form:

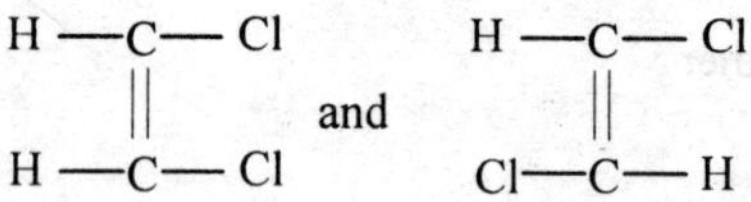

cis   *trans*

3. (7) The various structures are as follows:

(i) 

(ii) $HC\equiv C-CH_2CH_2-CH=CH$

(iii) $CH_2=CH-CH=CH-C\equiv CH$

(iv) $CH_3CH_2-CH=CH-C\equiv C-H$

(v) $HC\equiv C-CH_2-C\equiv C-CH_3$

(vi) 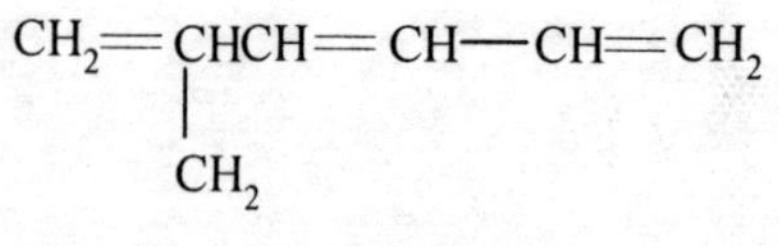

(vii) 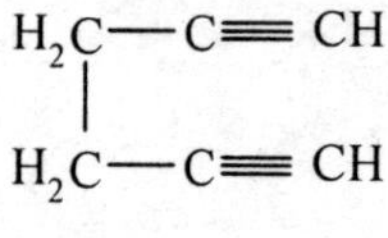

4. (i) 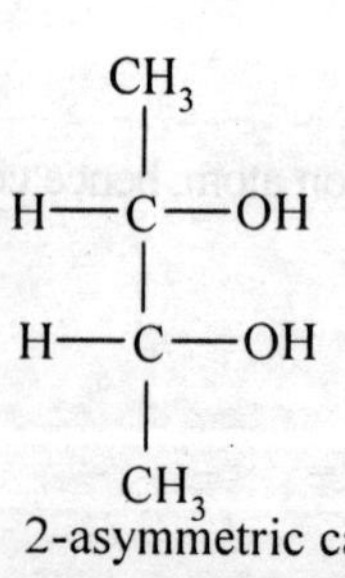

2-asymmetric carbon atoms
hence 4-stereoisomers d, 1, dl, meso

(ii) 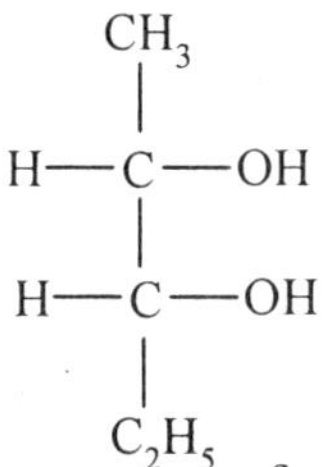

same as for (i)

(iii) 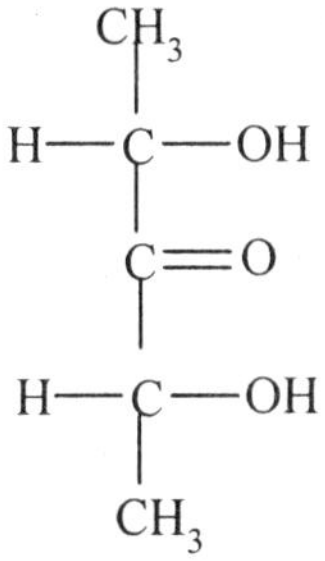

same as for (i)

(iv) 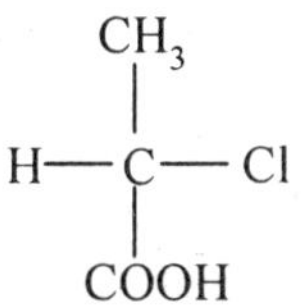

(v) 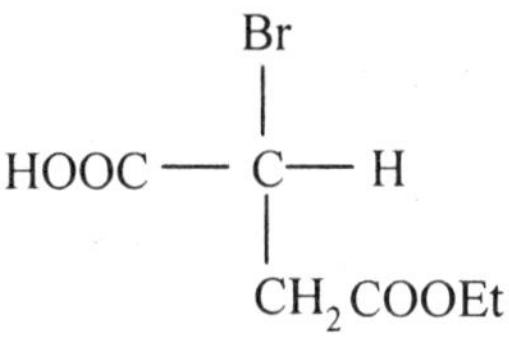

same as (iv)

(vi) No stereoisomers

1-asymmetric carbon atom, hence 2-stereoisomers

(vii) 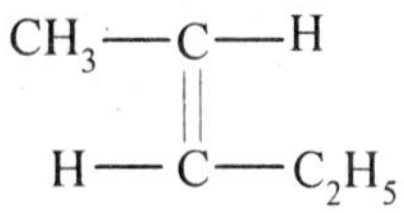

Two geometrical isomers

(viii) 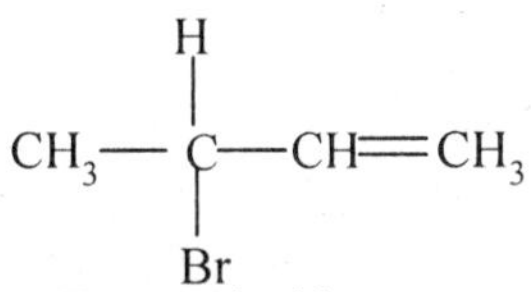

Two optical isomers

(ix) Stereoisomers

(x) No stereoisomer.

5. (i)

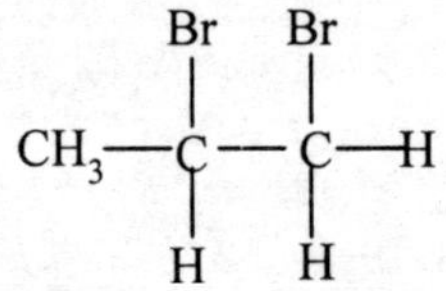

One asymmetric C atom hence 2-optical isomers

(ii)

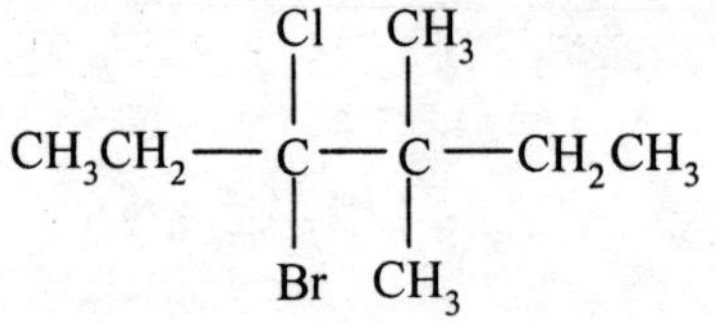

(iii)

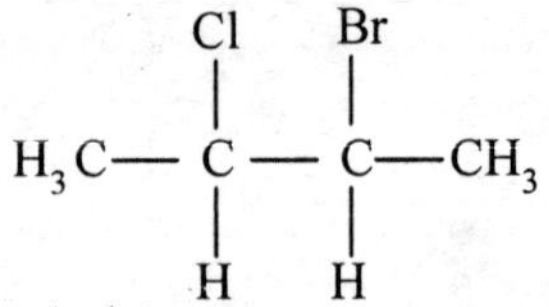

(iv) Same as (i).

6. (i) No, they have different boiling points since (a) and (b) are different compounds.

(ii) (a) and (b) have same effect on plane polarized light because both are optically inactive.

(iii) (c) and (d) have same effect since both are optically inactive.

(iv) (c) and (d) have different boiling points because (c) is *cis* and (d) is *trans* isomer.

(v) No stereoisomer.

Chapter 8

Nomenclature and Classification of Organic Compounds

Two general systems, (i) Trivial (ii) IUPAC, of naming organic compounds have been used, but we shall discuss IUPAC nomenclature here. The basic rules of IUPAC nomenclature are as follows:

8.1 IUPAC NOMENCLATURE

Following rules should be used as a basic guideline for the IUPAC nomenclature of organic compounds.

(1) Selection of Parent Chain

(a) Locate the longest continuous chain of carbon atoms as parent chain, e.g.,

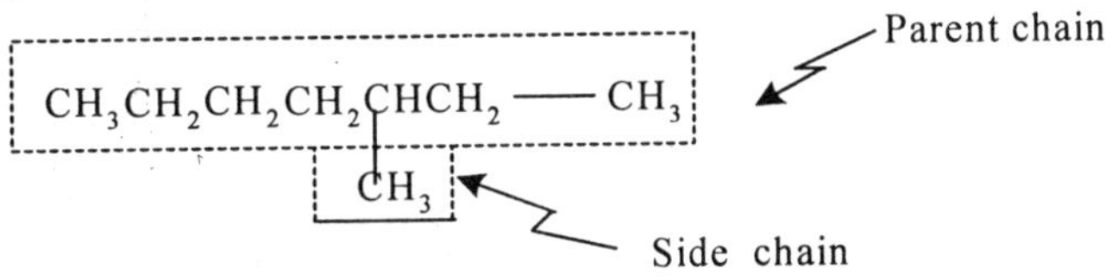

(b) The parent chain must include functional group/multiple bond, although it may not be longest, e.g.,

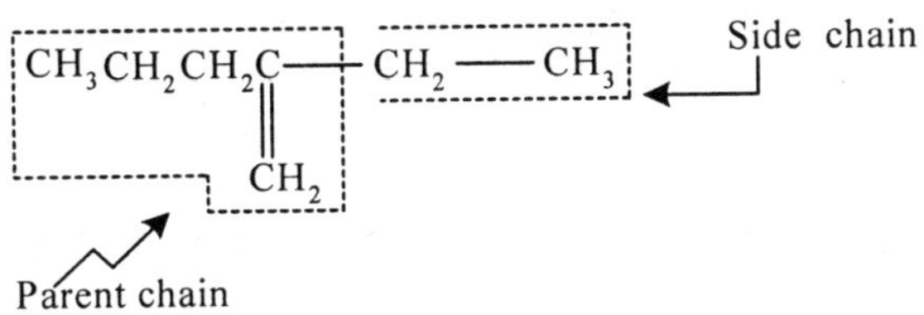

(although not longest but includes double bond)

(c) When two chains of equal length compete for selection as the parent chain, choose the chain with the greater number of substituents, as parent chain. e.g.,

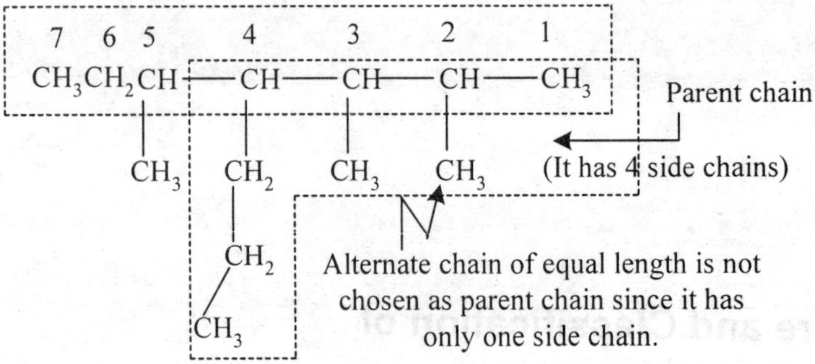

2,3,5-trimethyl-4-propylheptane (not 4-sec. butyl-2,3-dimethyl heptane)

(2) Numbering of Carbon Atoms of Parent Chain

The carbon atoms of the parent chain are numbered on the basis of following rules:

(a) The locant (carbon atom carrying functional group/multiple bond substitutent) must get lowest number, e.g.,

(i)

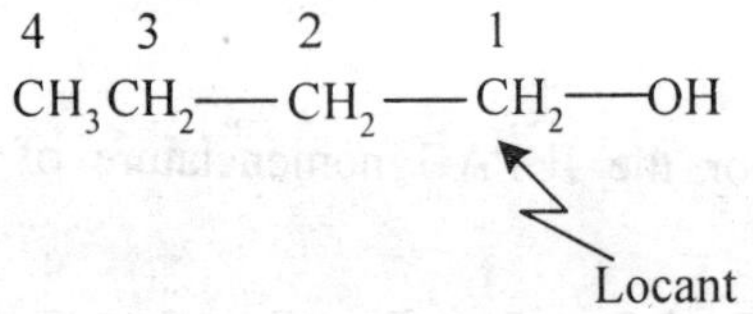

(ii)

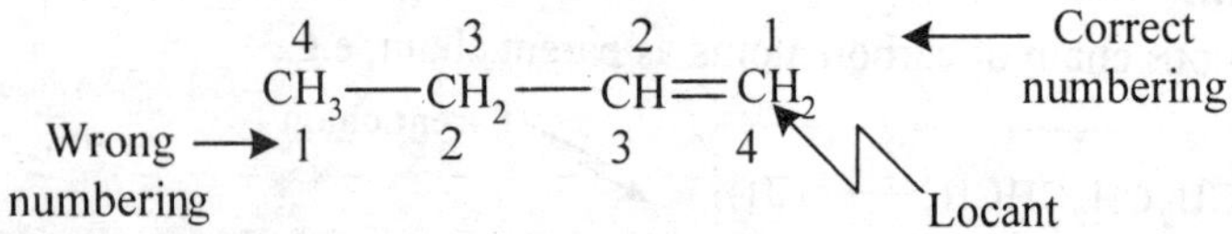

(b) If two or more substituents are present, one of them is regarded as senior group and rest junior groups. The locant containing senior group must get lowest number. The order of seniority is as follows:
Acids > Acid derivatives excluding nitriles > Aldehydes > Nitriles > Ketones > Alcohols > Amines > Ethers > => ≡> substituents.

(c) Lowest sum rule: The sum of numbers used to indicate the position of substituents should be minimum, e.g.,

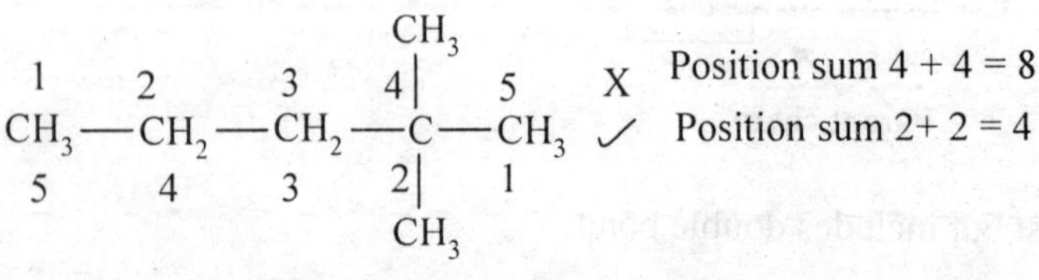

(d) The carbon in chain terminating functional groups like —COOH, —$CONH_2$, must always be given number 1.

(3) Other Rules

(i) If there are more than one substitutents in side chains, they should be prefixed in an alphabetical order, e.g.,

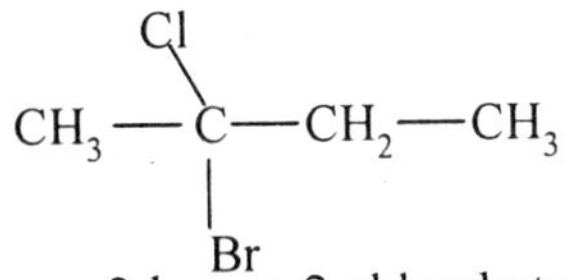

2-bromo-2-chlorobutane and not
2-chloro-2-bromobutane

(ii) When 2 or more identical substituents/side chains are present, numerical prefixes di, tri-, etc., are used before the names of such substituents/side chain. However, these numerical prefixes are not considered while arranging the substituents/side chains alphabetically.

(iii) The IUPAC name of functional groups are shown in Table 8.1.

Table 8.1

S.No.	*Functional group*	*Common/Trivial name*	*IUPAC*	*Examples*
1.	—	Saturated hydrocarbon	Alk*ane*	C_2H_6 (Ethane)
2.	—C=C—	Unsaturated hydrocarbon containing double bond	Al*kene*	C_2H_4 (Ethene)
3.	—C≡C—	Unsaturated hydrocarbon containing triple bond	Alk*yne*	C_2H_2 (Ethyne)
4.	RX	Alkyl halide	*Halo-alkane*	C_2H_5 Br (Brmoethane)
5.	R—OH	Alcohol	Alk*anols*	C_2H_5OH (Ethanol)
6.	R—CHO	Aldehyde	Alkan*al*	CH_3CHO (Ethanal)
7.	R—C=O R (below C)	Ketone	Alkan*one*	CH_3COCH_3 (Propanone)
8.	R—COOH	Carboxylic acid	Alkan*oic* Acid	CH_3COOH (Ethanoic Acid)
9.	ROR′ (R′ > R in case of mixed ether)	Ether	Alk*oxy*-(R) Alk*ane* (R′)	$CH_3OC_2H_5$ (Methoxy ethane)

10.	RCOOR′	Ester	Alkyl-alkanoate	$HCOOCH_3$ Methyl methanoate $C_2H_5COOCH_3$ Methyl propionate $C_2H_5COOC_3H_7$ Propyl propionate
11.	R — CO O R — CO	Acid anhydride	Alkanoic acid anhydride	CH_3— CO O CH_3— CO Ethanoic acid anhydride
12.	RCOCl	Acid chloride	Alkanoyl chloride	CH_3COCl Ethanoyl chloride
13.	$RCONH_2$	Alkyl amide	Alkanamide	CH_3CONH_2 Ethanamide
14.	R—CN	Alkyl cyanide	Alkane Nitrile	CH_3CN Ethane-nitrile
15.	R—NC	Alkyl isocyanide	Alkyl carbylamine	CH_3NC Methylcarbylamine
16.	R—NO_2	Nitro alkane	Nitro alkane	CH_3NO_2 Nitromethane
17.	R—SH	Alkyl mercaptan	Alkanethiol	C_2H_5SH Ethanethiol
18.	R—NH_2	Primary amines	Amino-alkane	$C_2H_5NH_2$ Aminoethane
19.	R—NHR	Secondary amine	N-alkyl-amino alkane	$CH_3NHC_2H_5$ N-methyl amino ethane
20.	R–NR | R	Tertiary amine	N,N-dialkylamino alkane	C_2H_5—N—C_2H_5 | C_2H_5 N, N-diethylamino ethane

8.2 E-Z NOMENCLATURE

The term *cis* and *trans* are unambiguous only when applied to disubstituted alkenes. However, in case of trisubstituted or tetrasubstituted alkene, the term *cis* and *trans* are either ambiguous or do not apply at all, e.g.,

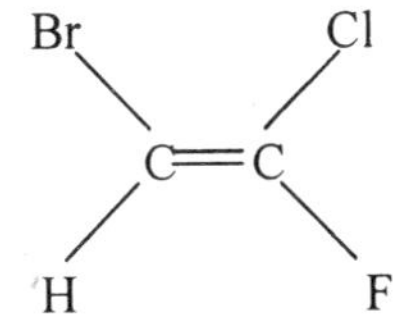

A new system called the E-Z system, has been introduced. This system is based on Cahn-Ingold-Prelog convention. The steps are as follows:

(i) Examine the two groups attached to one carbon atom of the double bond and decide which has higher priority on the basis of higher atomic number. Now repeat the exercise at the other carbon atom.

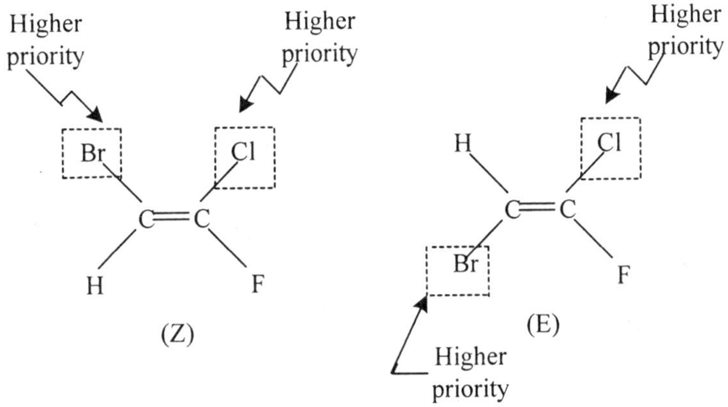

(ii) If two groups of higher priority are on the same side of the double bond, the alkene is designated Z (German word Zusammen, meaning together). However, if two groups of higher priority are on the opposite side of the double bond, the alkene is designated E (German word Entgegen meaning opposite).

(iii) Most compounds that would normally be designated *cis* designated Z. Likewise *trans* forms are designated E.

Other example:

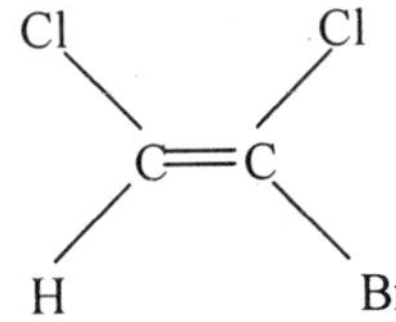

Cl > H
Br >Cl

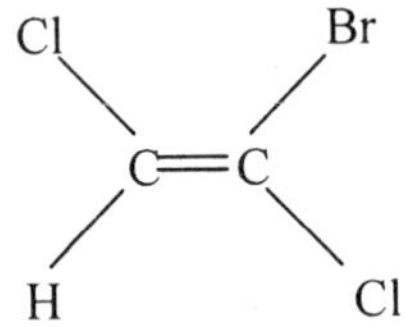

(E) (Z)

8.3 IUPAC NOMENCLATURE OF HETEROCYCLES

The 'Heteroatom' is indicated by **'oxa', 'aza' and 'thia' for oxygen, nitrogen and sulphur** respectively and size of ring from the Table 8.2.

Table 8.2

No. of atoms in ring including heteroatom	Saturated ring		Unsaturated ring	
	X = N	X = O or S	X = N	X = O or S
3	-iridine	-irane	-irine	-irene
4	-etidine	-etane	-ete	-ete
5	-olidine	-olane	-ole	-ole
6	-a	-ane	-ine	-in

Some examples are:

(i)

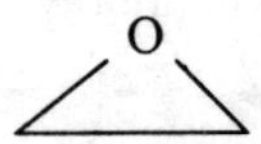

(Saturated ring containing oxygen— three membered)

Oxa + irane = Oxirane

(ii)

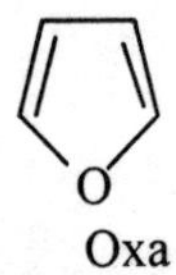

(Unsaturated righ containing oxygen— five membered)

Oxa + ole = Oxole

(iii)

Thia + ole = Thiole

(iv)

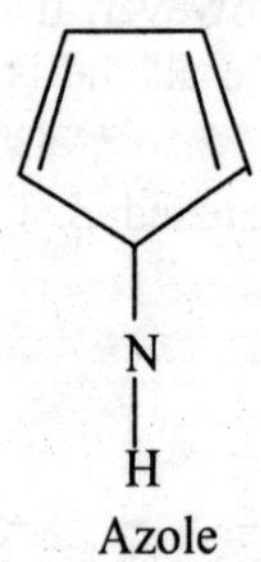

Azole

(v)

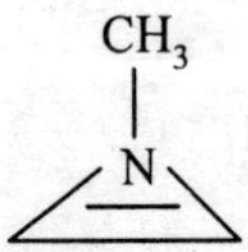

N-methyl azirine

(vi)

Dioxane

8.4 IUPAC NOMENCLATURE OF ALICYCLIC COMPOUNDS

Add the word cyclo, and rest like aliphatic compounds, e.g.,

Methylcyclopropane

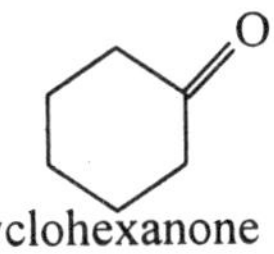

Cyclohexanone

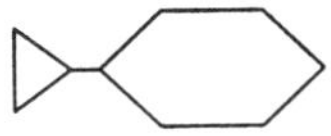

Cyclopropylcyclohexane

Chlorocyclopentane

8.5 IUPAC NAME OF BICYCLIC ALKANE

(i) We name alkane containing two fused or bridged rings as bicycloalkanes and name corresponding to the total number of carbon atoms in the rings as the base number.

(ii) Now introduce in the name an expression in brackets that denotes the number of carbon atoms in each bridge in descending order.

Bicyclo [Insert number of carbon atoms of three carbon bridges in descending order] alkane

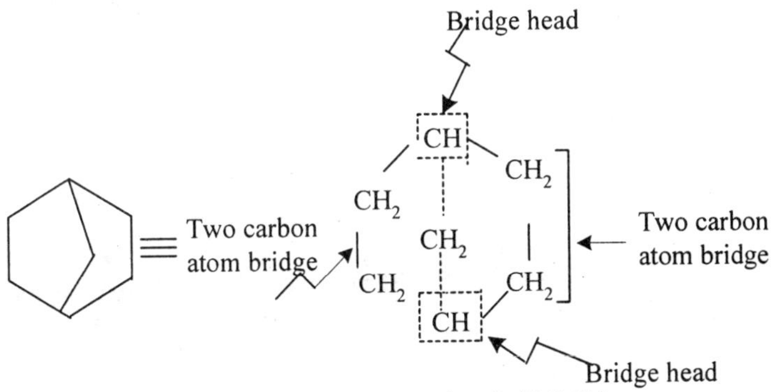

Bicyclo [2.2.1] heptane

8.6 NOMENCLATURE OF AROMATIC COMPOUNDS

(i) If two groups/substituents are present on adjacent alternate or parallel position; these are called ortho (*o*), meta (*m*) and para (*p*) positions respectively.

(ii) Number the carbon atoms of benzene ring and assign the position of groups or substituent.

Some examples are as follows:

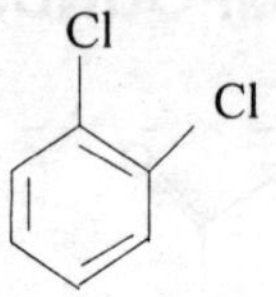

o-Dichlorobenzene or 1,2-Dichlorobenzene

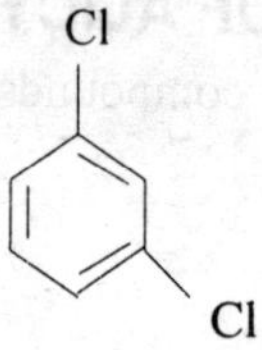

m-Dichlorobenzene or 1,3-Dichlorobenzene

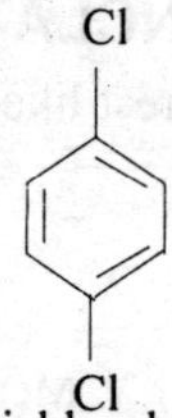

p-Dichlorobenzene or 1,4-Dichlorobenzene

8.7 SPIRO COMPOUNDS

Spiro [2.2] pentane

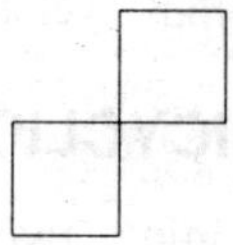

Spiro [3.3] heptane

8.8 CLASSIFICATION OF ORGANIC COMPOUNDS

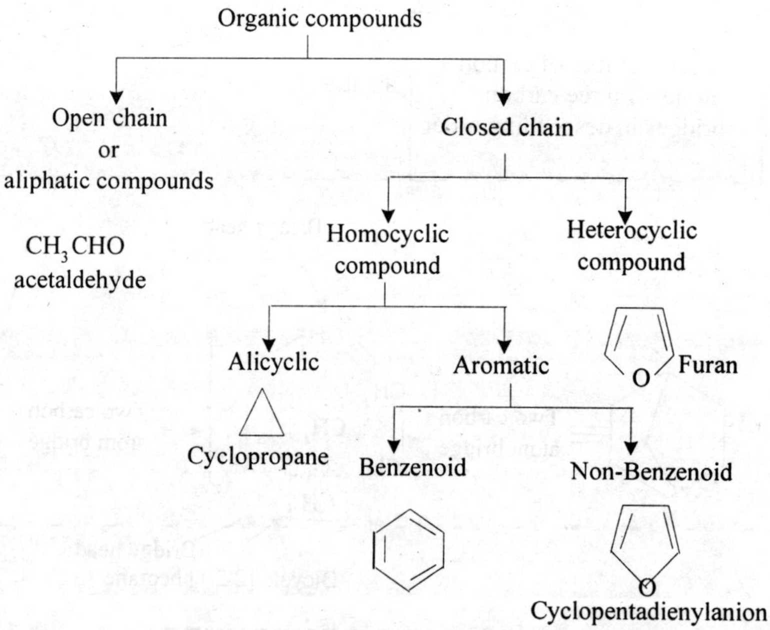

PROBLEMS

1. Write IUPAC names of the following compounds:

(i) 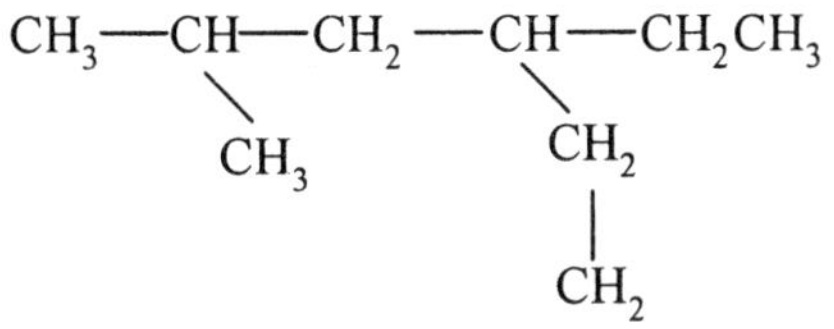

(ii) 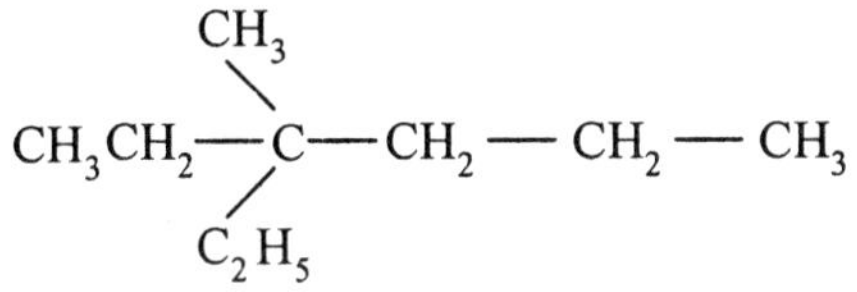

(iii) 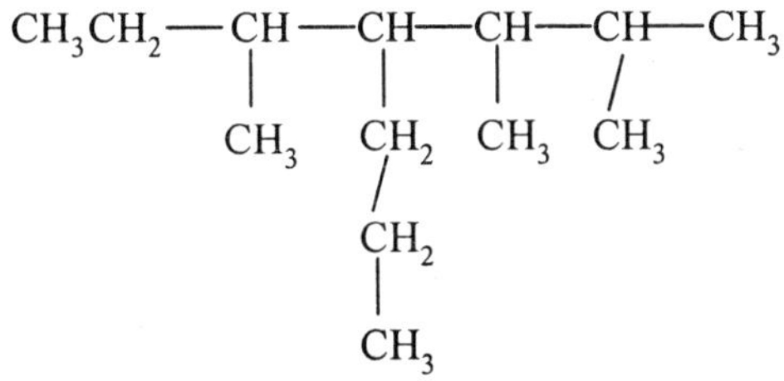

(iv) 

(v) 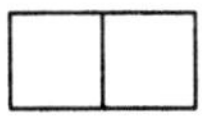

(vi) 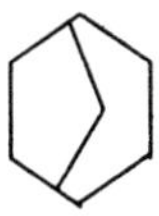

(vii) 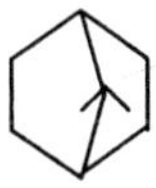

(viii) 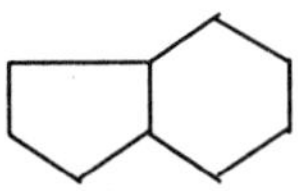

(ix) 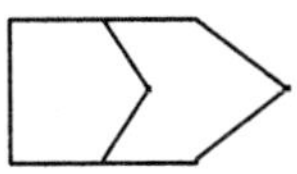

(x) 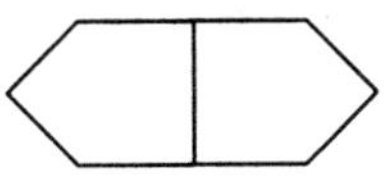

(xi) 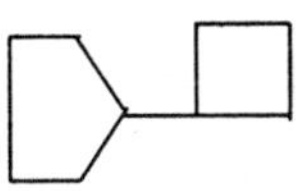

(xii) 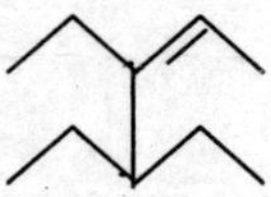

(xiii) 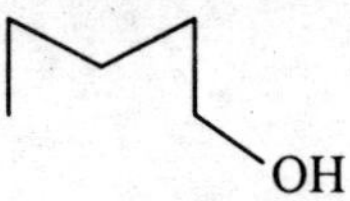

(xiv) 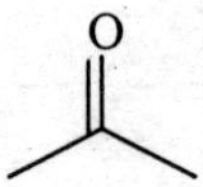

(xv) 

(xvi) 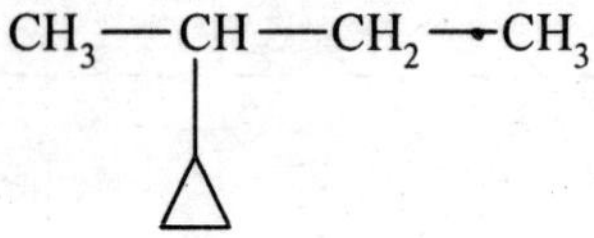

(xvii) 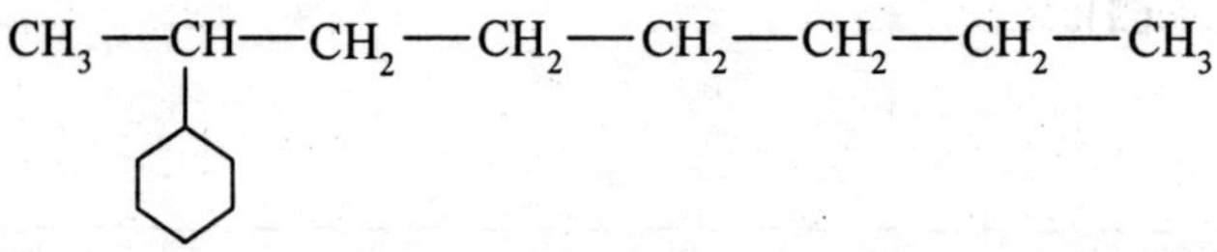

(xviii) 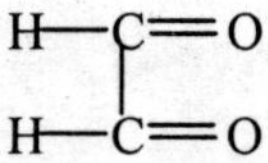

(xix) 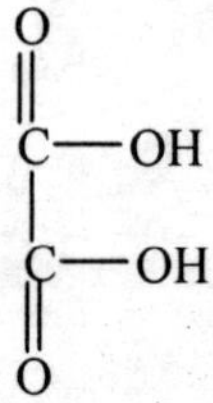

(xx) 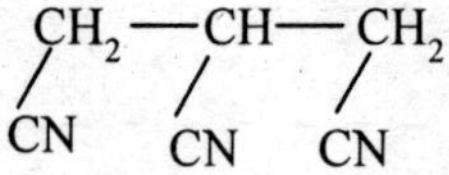

(xxi) 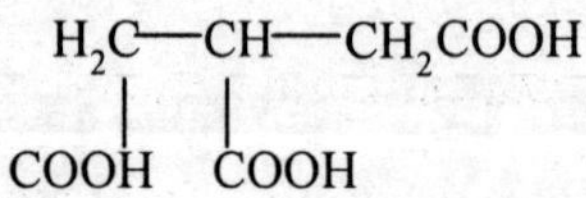

(xxii) 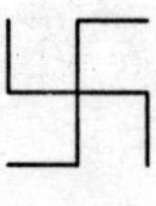

(xxiii) $CH_3CH_2CH{=}CHCOOH$

(xxiv) 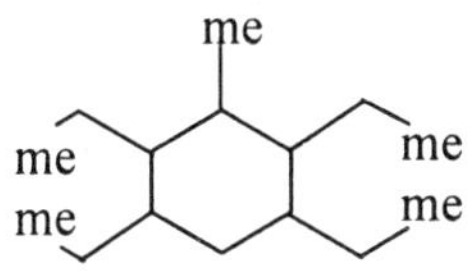

(xxv) $H_2C{-}NH{-}\underset{}{\overset{CH_3}{CH}}{-}CH_2CH_3$

(xxvi) Succinic acid **(IIT, 1994)**

(xxvii) $ClCH_2\,\underset{CH_3}{\underset{|}{CH}}{-}CHO$ **(GATE, 1992)**

2. Write structure of the following

(i) Bicyclo [1.1.0] butane

(ii) 2-chlorobicyclo [3.2.0] heptane

(iii) Propanone

(iv) Ethyl-ethanoate

(v) 4-chloro-2-pentene **(IIT, 1988)**

3. Amongst the following, one or more answers may be correct. Write the correct answer:

(i) The IUPAC name of the compound having the formula.

$$CH_3{-}\underset{CH_3}{\underset{|}{\overset{\overset{CH_3}{|}}{C}}}{-}CH{=}CH_2 \text{ is}$$

(a) 3,3,3-Trimethyl-1-1-propene

(b) 1,1,1-Trimethyl-1-2-propene

(c) 3,3-Dimethyl-1-1-butene

(d) 2,2-Dimethyl-3-butene **(IIT, 1984)**

(ii) The IUPAC name of the compound $CH_2{=}CH\text{-}CH(CH_3)_2$ is

(a) 1,1-Dimethyl-1-2-propene

(b) 3-Methyl-1-butene

(c) 2-Vinylpropane

(d) 1-Isopropyl ethylene **(IIT, 1987; MNR, 1995)**

(iii) The IUPAC name for the following compound is

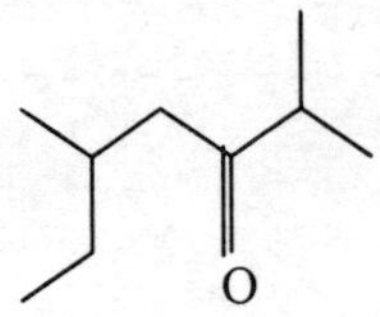

(a) 5-Ethyl-2-methylhexan-3-one
(b) 2,5-Dimethylheptan-3-one
(c) 3,6-Dimethylheptan-5-one
(d) 2-Ethyl-5-methylhexane-4-one. **(GATE, 1994)**

(iv) The IUPAC name of

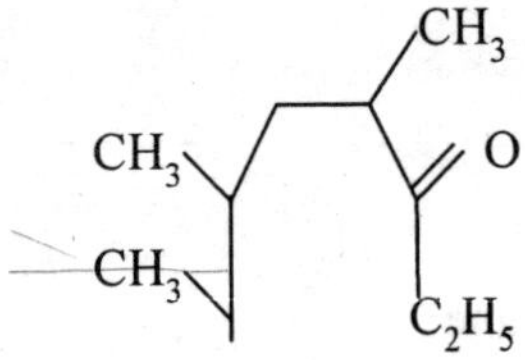

(a) Z-4,6-Dimethyloctan-4-ene-one
(b) E-4,6-Dimethyloctan-4-ene-one
(c) Z-3,5-Dimethyloctan-4-ene-one
(d) E-3,5-Dimethyloctan-4-ene-6-one. **(GATE, 1993)**

(v) C-60 isotope is known as
(a) Fullerene (b) Graphite (c) Diamond (d) None above

ANSWERS

1. (i) 4-Ethyl-2-methylhexane
(ii) 3-Ethyl-3-methylhexane
(iii) 2,3,5—Trimethyl-4-propylheptane
(iv) Bicyclo [1.1.0] butane
(v) Bcyclo [2.2.0] hexane
(vi) Bicyclo [2.2.1] heptane
(vii) 7.7 ′-Dimethylbicyclo [2.2.0] heptane
(viii) 8-Methylbicyclo [4.3.0] nonane
(ix) 8-Methylbicyclo [3.2.1] octane
(x) Bicyclo [3.3.0] octane
(xi) Cyclobutyl cyclopentane

(xii) 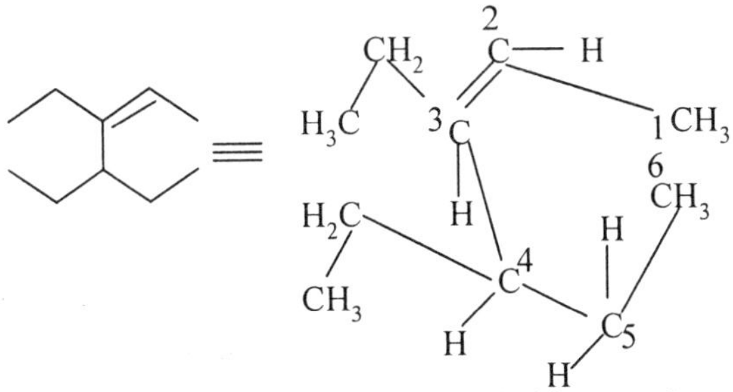

3,4-Diethyl hexene-2

(xiii) 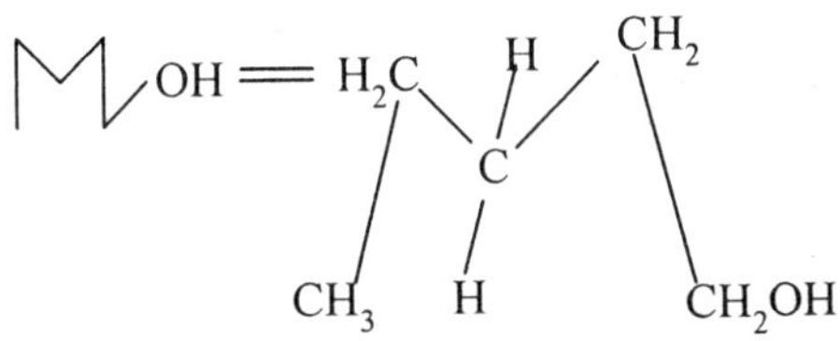

Pentanol-1

(xiv) 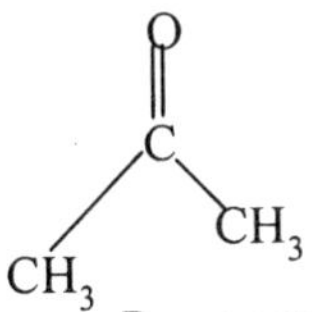

Propanone

(xv) 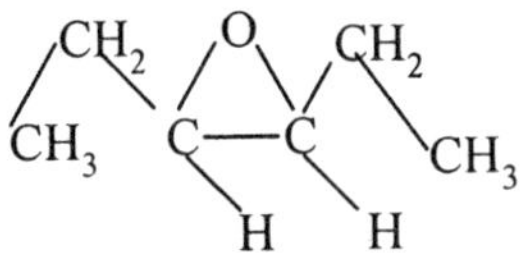

2,3-diethyl Oxirane

(xvi) 2-Cyclopropyl butane ( is cyclopropane so 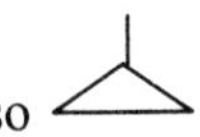 is cyclopropyl)

(xvii) 2-Cyclohexyloctane

(xviii) Ethandinal

(xix) Ethandioic acid

(xx) 1,2,3-Tricyanopropane or 2-Cyanopropan-1, -5-dinitrile

(xxi) 3-Carboxylicpent-1,4-dioic acid

(xxii) 3,3′-Diethylpentane

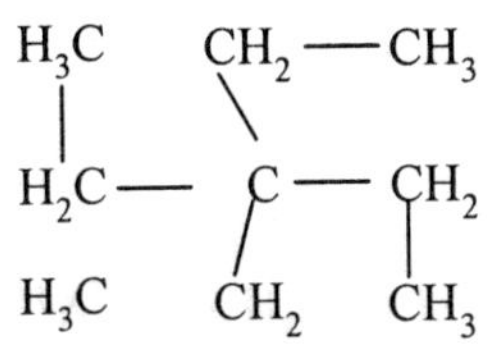

(xxiii) Pent-2-enoic acid-1

(xxiv) 5,6-Diethyl 1-3-methyldec-4-ene

(xxv) 3-(N, N-Dimethylamine) 3-methylpentane

(xxvi) 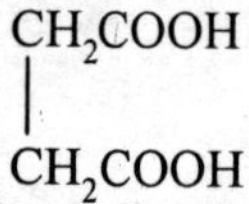

Butandioic acid

(xxvii) 2-Methyl-3-chloropropanal.

2. (i) 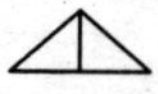

(ii) 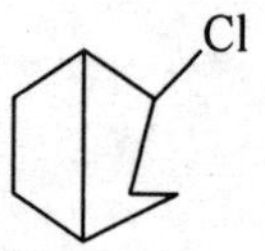

(iii) 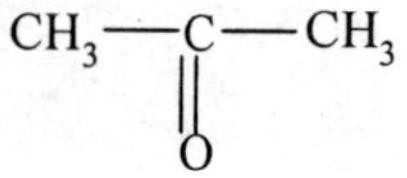

(iv) $CH_3COOC_2H_5$

(v) 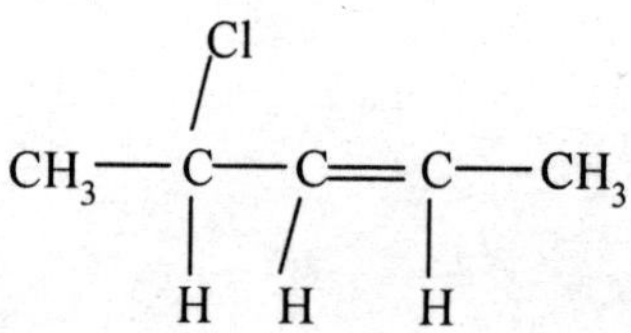

3. (i) **(c)**

(ii) **(b)**

(iii) **(b)**

(iv) **(b)**

(v) **(a)**

Chapter 9

Aliphatic Hydrocarbons

Aliphatic hydrocarbons are organic compounds containing carbon and hydrogen elements. These can be classified into following groups based on the type of bond that exists between the individual carbon atoms:

1. Alkanes
2. Alkenes
3. Alkynes

9.1 ALKANES

Those hydrocarbons in which all of the carbon-carbon bonds are single bonds, are called alkanes. Alkanes have general formula C_nH_{2n+2}. Alkanes in which all or some of the carbon atoms are arranged in a ring are known as *Cycloalkanes*. Cycloalkanes with single ring have the general formula C_nH_{2n}. For example:

$$CH_3—CH_2—CH_3;$$

Propane

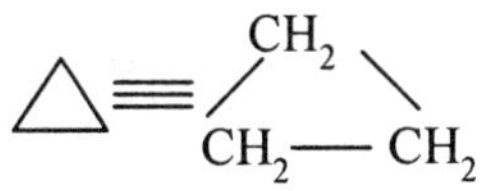

Cyclopropane

The chief source of alkanes is petroleum. Alkanes show structural isomerism. They contain four types of carbon atoms which are as follows:

(i) A primary carbon atom is the one that is joined to only one carbon atom.
(ii) A secondary carbon atom is the one which is attached to two carbon atoms.
(iii) A tertiary carbon atom is the one which is attached to three carbon atoms.
(iv) Neo carbon atom is the one which is attached to four carbon atoms.

Hydrogen atoms joined to primary, secondary, and tertiary carbon atoms are known as primary, secondary, and tertiary hydrogen atoms respectively: All the carbon and hydrogen atoms are shown in following structure:

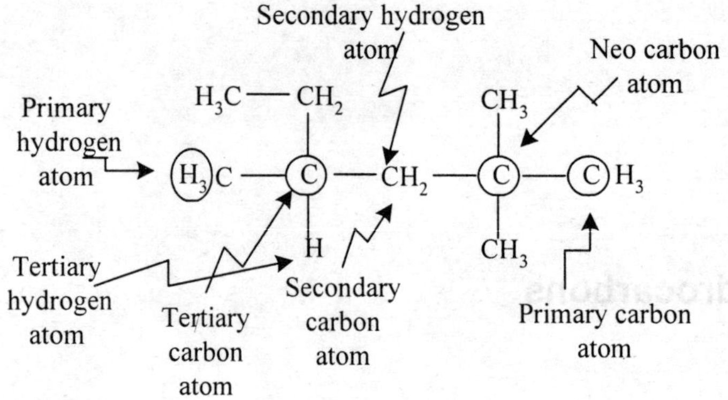

When one hydrogen atom of an alkane is removed, an alkyl group is obtained. The structures of some common alkyl groups are as follows:

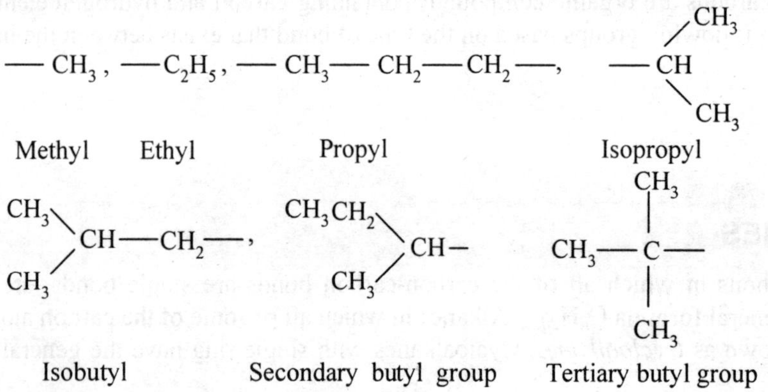

(a) Preparation

(i) This method involves reduction of alkene into alkanes using Raney-Ni as catalyst. The reaction is as follows:

$$C_nH_{2n} \xrightarrow[H_2,\ 200\text{-}300^\circ C]{\text{Raney-Ni}} C_nH_{2n+2}$$

Alloy containing equal amounts of Ni and Al is digested with NaOH. The Al is dissolved away, and the residual, very finely-divided Ni is washed and stored under water or ethanol.

(ii) From Grignard Reagent (see Chapter 6).

(iii) Alkanes can also be prepared by reduction of alkyl halide with zinc and aqueous acid or triphenyl antimonyhydride (Ph_3SnH). The general reaction is:

$$2R{-}X + Zn + 2H^+ \longrightarrow 2R{-}H + ZnX_2$$

e.g.,

$$2CH_3CH_2\underset{\underset{Br}{|}}{CH}—CH_3 + Zn + 2H^+ \longrightarrow 2CH_3CH_2\underset{\underset{H}{|}}{CH}—CH_3 + ZnBr_2$$

$LiAlH_4$ and $NaBH_4$ can also be used as reducing agent for reduction of primary, secondary, and tertiary alkyl halides, respectively.

(iv) Recently, alkyl halides have been converted into alkanes by reaction with lithium dialkyl cuperate and the raction is called **Corey-Posner, white side house synthesis**. The lithium dialkyl cuperate is called **Gillman's catalyst**.

$$\boldsymbol{R}'_2\,CuLi + R\text{-}X \longrightarrow R-\boldsymbol{R}' + RCu + LiX$$

e.g.,

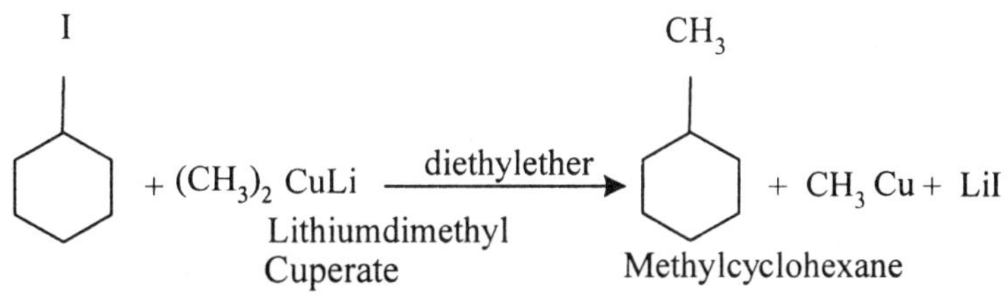

(v) *Wurtz Reaction*: It involves reaction of an alkyl halide with sodium to form the alkane. The general reaction is as follows:

$$Rx + 2Na + \boldsymbol{R}'x \longrightarrow R-R + R-\boldsymbol{R}' + \boldsymbol{R}'-\boldsymbol{R}' + 2NaX$$

$$\text{e.g., } CH_3Br + 2Na + C_2H_5Br \longrightarrow C_2H_6 + C_4H_{10} + C_3H_8 + 2NaBr$$

It gives best yield for "even carbon alkanes" of higher molecular weight, reaction gives poor yield for a secondary alkyl halide and fails with tertiary alkyl halide.

(vi) *Kolbe's electrolytic method*: Electrolysis of the salts of the mono-carboxylic acids results in the formation of alkanes.

$$RCOOK + RCOOK + 2H_2O \longrightarrow R-R + 2CO_2 + H_2 + 2KOH$$

e.g.,

$$CH_3COOK + CH_3COOK + 2H_2O \longrightarrow CH_3CH_3 + 2CO_2 + H_2 + 2KOH$$

(vii) When alcohols, ketones or fatty acids are reduced with red phosphorus and hot hydroidic acid in a sealed tube, alkanes are obtained.

(a) $$ROH + 2HI \xrightarrow[150\text{-}200^\circ]{\text{Red phosphorus}} RH + H_2O + I_2$$

e.g.,

$$C_2H_5OH + 2HI \longrightarrow C_2H_5 + H_2O + H_2$$

(b) $RCOR + 4HI \longrightarrow RCH_2R + H_2O + I_2$

e.g.,

$CH_3COH_3 + 4HI \longrightarrow CH_3CH_2CH_3 + H_2O + 2I_2$

(c) $RCOOH + 6HI \longrightarrow RCH_3 + 2H_2O + 3I_2$

e.g.,

$CH_3COOH + 6HI \longrightarrow CH_3CH_3 + 2H_2O + 3I_2$

(viii) Alkanes can also be synthesized from its element. For example,

$$C + 2H_2 \xrightarrow[1200^{\circ}C \text{ methane}]{\text{Electric arc}} CH_4$$

(ix) Alkanes can be prepared by decarboxylation of carboxylic acid. The method involves strong heating of sodium salt of carboxylic acid with soda lime (NaOH + CaO). The mechanism involves carbanion.

$$RCOOH + NaOH \longrightarrow RCOONa + H_2O$$
$$RCOONa + NaOH \longrightarrow RH + Na_2CO_3$$

e.g.,

$$CH_3COOH + NaOH \longrightarrow CH_3COONa + H_2O$$
$$\underset{\text{sodium acetate}}{CH_3COONa} + NaOH \longrightarrow CH_4 + Na_2CO_3$$

Properties

Physical: The alkanes containing up to four carbon atoms are colourless gas; and alkanes containing carbon atoms from five to eighteen are colourless liquids. The alkanes containing eighteen or more carbon atoms are colourless solid.

The density and boiling point of alkanes are directly proportional to the number of carbon atoms while solubility in solvents of low polarity like benzene, chloroform is inversely proportional to the number of carbon atoms (Table 9.1). As already mentioned in Chapter 7, the boiling point of branched alkane is always less than that of isomeric normal alkanes.

Table 9.1 Physical constants of unbranched alkanes

Number of carbon atoms	*Name*	*bp (°C) (1atm)*	*Density* d^{20} *(gml*$^{-1}$*)*
1	Methane	–161.5	–
2	Ethane	–88.6	–
3	Propane	–42.1	–
4	Butane	–0.5	–
5	Pentane	36.1	0.626
6	Hexane	66.2	0.659
7	Heptane	98.4	0.684

Similar pattern is also observed for cycloalkanes (Table 9.2).

Table 9.2 Physical constants of cycloalkanes

Number of carbon atoms	*Name*	*bp (° C) (1atm)*	*Density (d^{20}) (gml^{-1})*
3	Cyclopropane	–33	–
4	Cyclobutane	13	–
5	Cyclopentane	49	0.751
6	Cyclohexane	81	0.779

Chemical: Alkanes undergo substitution. Ease of substitution is tert H > see H > Pri H.

(i) Halogenation: Let us consider following case and answer questions:

$$CH_3—CH_2—CH_2—CH_3 + Cl_2 \rightarrow ?$$

(a) Kind of monochloro derivatives
(b) 1st monochloro derivative obtained
(c) Relative population of all monochloro derivatives

The Answer is as follows:

(1) Kinds of monochloro derivatives = Kinds of hydrogen atoms, e.g., Kinds of monochloro derivative = Two kinds of hydrogen atoms, that is, primary and secondary.

(2) First monochloro derivative is in the order of ease of substitution, e.g.,

$$\underset{\displaystyle\overset{|}{Cl}}{} CH_3—\underset{\underset{\displaystyle Cl}{|}}{CH}—CH_2—CH_3$$ is the 1st monochloro derivative.

(3) Relative population is directly proportional to that of population of primary, secondary, and tertiary hydrogen atoms, e.g.,

Relative population[1]

Reactivity order $F_2 > Cl_2 > Br_2 > I_2$

Fluorination is explosive, iodination is reversible hence oxidizing agents like HNO_3, HIO_3, HgO are used.

(ii) Nitration: Number of mononitro derivative = Kinds of hydrogen atoms + All possible fractions, e.g.,

---

[1] for chlorination: Pri H × 1 : Sec H × 3.8 : Tert H × 5
for brominating: Pri H × 1 : Sec H × 82 : Tert H × 1600

$$CH_3CH_2CH_2CH_3 \xrightarrow[400^{\circ}C]{HNO_3}$$

$$CH_3CH—CH_2CH_2NO_2 + CH_3CH_2CH_2CH_2NO_2 +$$

1st product

$$CH_3NO_2 + CH_3CH_2CH_2NO_2 + C_2N_5NO_2$$

(iii) Isomerization: *n*-alkane $\xrightarrow[300^{\circ}C]{Alcl_3}$ Isoalkane

e.g., n – butane $\xrightarrow[300^{\circ}C]{AlCl_3}$ Isobutane.

(iv) Cracking : Alkane $\xrightarrow[\text{without catalyst + alkanes}]{\text{400-600 with or}}$ H_2 + smaller alkanes.

9.2 ALKENES

Alkenes or olefins are hydrocarbons that contains a carbon-carbon double bond and have general formula C_nH_{2n}. Alkenyl groups are obtained by the removal of one hydrogen atoms from alkenes. The carbon atom with free valence is always number 1. For example:

$$\overset{3}{C}H_2 = \overset{2}{C}H — \overset{1}{C}H_2—$$

Allyl or 2-propenyl group

$$\overset{2}{C}H_2 = \overset{1}{C}H—$$

Vinyl group

(a) Preparation

(i) From Grignard reagent (see Chapter 4).

(ii) Dehydrohalogenation of an alkyl halide

$$RCH_2CH_2Br + \text{alc. KOH} \longrightarrow RCH = CH_2 + KBr + H_2O$$

e.g., $CH_3CH_2Br + \text{alc. KOH} \longrightarrow CH_2 = CH_2 + KBr$

(iii) Dehydration of alcohol

$$R— CH_2CH_2OH \xrightarrow{H^+} RCH = CH_2$$

e.g., $CH_3CH_2OH \xrightarrow{H_2SO_4} CH_2 = CH_2 + H_2O$

(iv) By selective reduction of an alkyne

$$RC \equiv CH + H_2 \xrightarrow{\text{Lindlar's catalyst}} RCH = CH_2$$

$$\text{e.g., } CH_3 - C \equiv CH + H_2 \xrightarrow{\text{Lindlar's catalyst}} CH_3CH = CH_2$$

Propyne → Propene (97% cis & 3% trans)

Lindlar's catalyst is prepared from Palladium deposited on $CaCO_3$ followed by the treatment with lead salts which moderate the activity of the catalyst. It is symbolized as $Pd/CaCO_3$-PbO or $Pd/CaCO_3$-$Pb(OOCCH_3)_2$.

$$CH_3 - C \equiv C - CH_3 + H_2 \xrightarrow{\text{Wilkinson's catalyst*}} CH_3 - CH = CH_3$$

Butyne → 100% *cis* - Butene-2

(v) By dehalogenation of vicinal dihalides by heating with Zinc dust in ethyl alcohol

$$R - \underset{\substack{|\\Br}}{CH} - \underset{\substack{|\\Br}}{CH_2} + Zn \xrightarrow[\Delta]{\text{alcohol}} RCH = CH_2 + ZnBr_2$$

$$\text{e.g., } CH_3\underset{\substack{|\\Br}}{CH} - \underset{\substack{|\\Br}}{CH_2} + Zn \longrightarrow CH_3CH = CH_2 + ZnBr_2$$

(vi) Wittig Reaction[2]

$$Ph_3P + CH_3Br\,PhLi \longrightarrow Ph_3P = CH_2 + PhH + LiBr$$

$$Ph_3P = CH_2 + \underset{R}{\overset{R'}{>}}C = O \longrightarrow Ph_3PO + \underset{R}{\overset{R'}{>}}C = CH_2$$

(Where, $R' = R = H$ or $R = R' = CH_3$).

(vii) By cracking of alkanes: When alkanes are heated at 500–700°C in the absence of air, alkenes are obtained, e.g.,

$$CH_3 - CH_3 \xrightarrow{600\,°C} CH_2 = CH_2 + CH_4 + H_2$$

(viii) Cope reaction

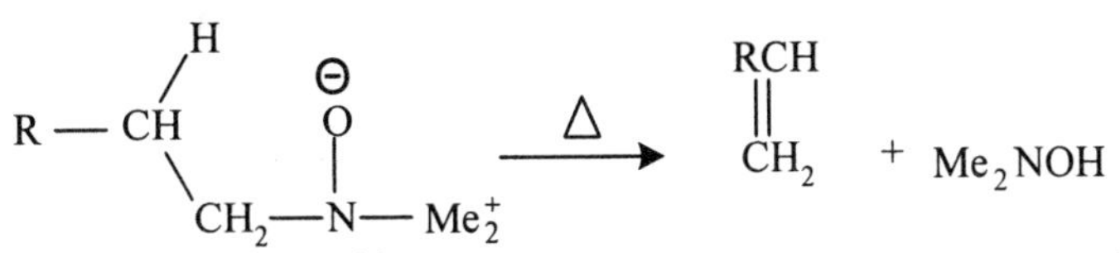

Amine oxide

---

* Wilkinson's catalyst is tris-triphenyl chloro Rhodium II $[Ph_3P)_3RhCl]$

[2] Ph means phenyl group.

Properties

Physical: Alkenes containing carbon atoms up to four are colourless gases; while those containing carbon atoms from five up to seventeen are colourless liquids. Alkenes containing 18 or more carbon atoms are colourless solids. Like alkanes, the boiling point and density of alkenes are directly proportional to the number of carbon atoms while solubility in solvents of low polarity like benzene, chloroform, is inversely proportional to the number of carbon atoms (Table 9.3).

Table 9.3 Physical constants of alkenes

Name	*Formula*	*bp(°C)*	*Density* $d^{20}(ge^{-1})$
Ethene	$CH_2{=}CH_2$	–104	0.384*
Propene	$CH_3CH{=}CH_2$	–47	0.514
1-Butene	$CH_3CH_2CH{=}CH_2$	–6.3	0.596
(Z)-2-butene	$CH_3CH{=}CHCH_3$ (*cis*)	3.7	0.621
(E)-2-butene	$CH_3CH{=}CHCH_3$ (*trans*)	0.9	0.604
1-Pentene	$CH_3CH_2CH_2CH{=}CH_2$	30	0.641

*Density at –10°C.

Chemical:

1. Alkenes undergo addition reactions[3], In general, it can be written as:

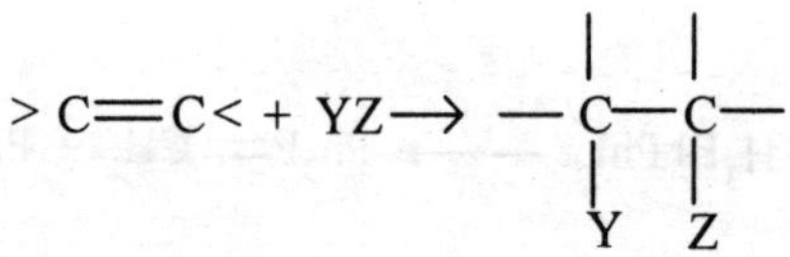

e.g., Y=H Z=H Product alkane
Y=H Z=X Haloakane (halogen)
Y=X Z=X Dihalide

2. *Oxymercuration–demercuration:* It is a method to covert alkene into alcohol. Addition occurs according to Markovnikov's Rule

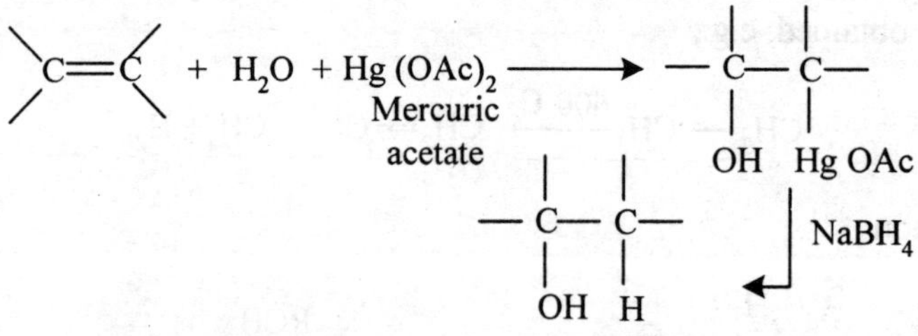

[3] For details see Chapter 5.

e.g.,

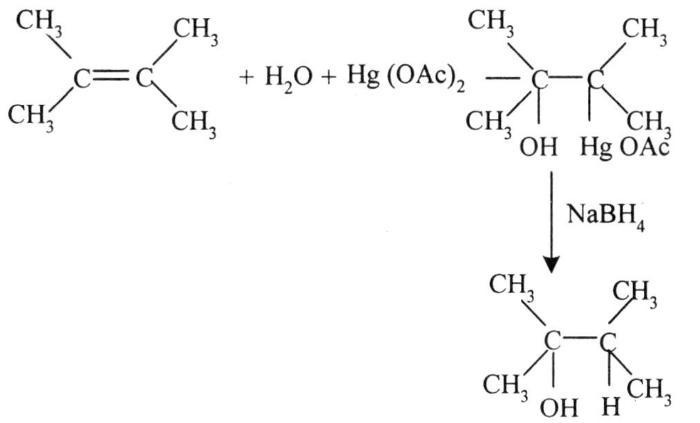

2,3-Dimethylbutanol-2

3. *Hydroboration*

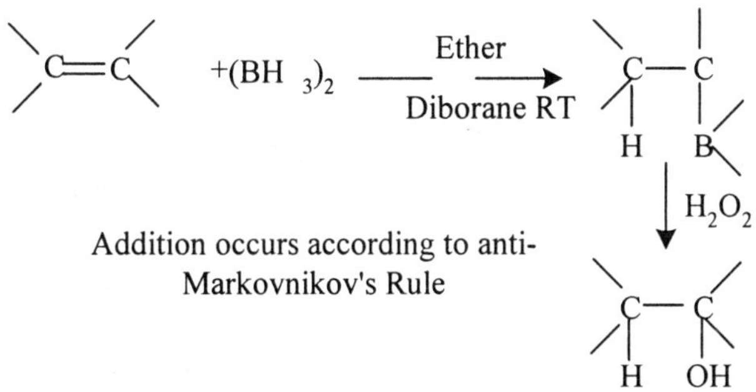

Addition occurs according to anti-Markovnikov's Rule

4. *Polymerization*: See Chapter 12.
5. *Isomerization*

$$\text{Alkene-1} \xrightarrow[500 - 700^\circ]{AlCl_3} \text{Alkene-2}$$

e.g., $CH_3CH_2CH_2\,CH{=}CH_2 \xrightarrow{AlCl_3} CH_3CH_2CH{=}CHCH_3$

Pentene-1 Pentene-2

6. *Ozonolysis*: It is an important method for the determination of the position of double bond in an alkene. The reaction, called *ozonolysis*, consists of passing ozonized air into alkene, when ozonide is obtained. It is hydrolysed using $ZnCl_2/H_2O$ or Zinc powder/H_2O. The general picture is as follows:

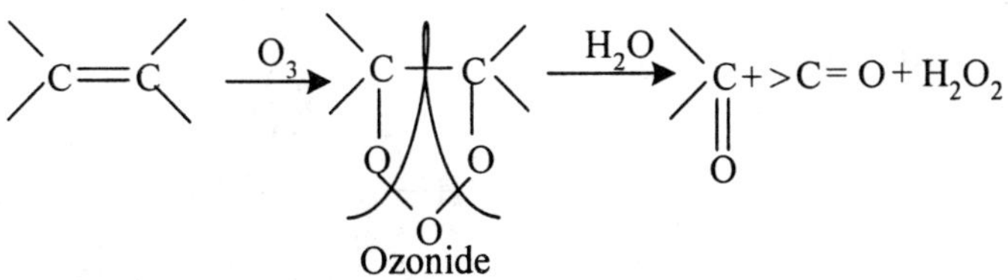

Ozonide

Remember that number of carbon atoms in the products must be equal to that of alkene.

Now there can be two kinds of problems:

(i) *To find out ozonolysis product of a given alkene.*

Following examples will clear the way to find out oznolysis product.

(i) Ozonolysis of butene-1

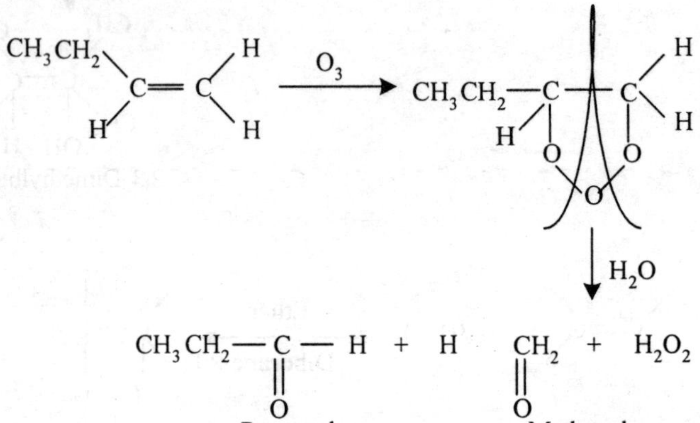

Propanal Methanal

(ii) Ozonolysis of butene-2

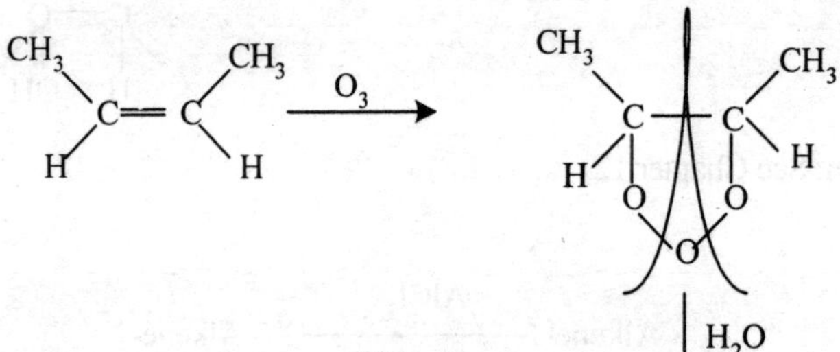

$CH_3 — CHO + OHC — CH_3 + H_2O_2$

Ethanal Ethanal

(iii) Ozonolysis of cyclobutene

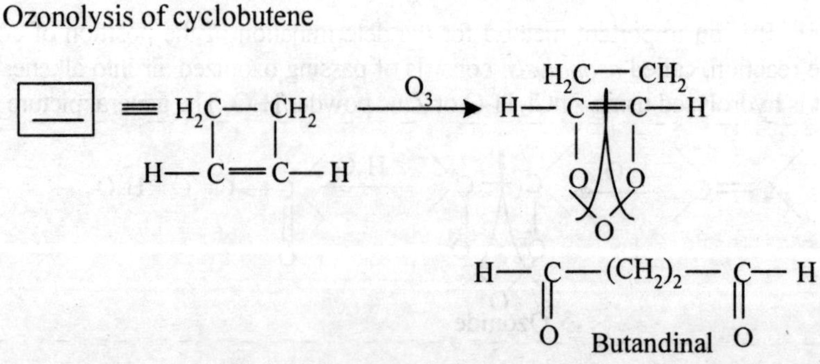

Butandinal

(iv) Ozonolysis of benzene

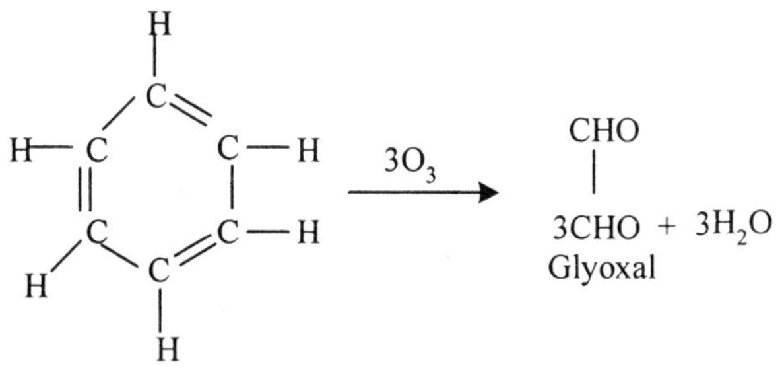

(ii) To find out the structure of alkene on the basis of ozonolysis product.

Such problems should be solved by counting the carbon atoms of ozonolysis product and then writing the general scheme of ozonolysis as shown above. Few examples are as follows:

Write structure of alkene which upon ozonolysis yields acetone only.

(IIT, 1995)

Since acetone is the only ozonolysis product, therefore, two molecules of acetone must have been formed. Therefore, total number of carbon atoms in alkene must be six. Now with general scheme of ozonolysis and move for ozonolysis.

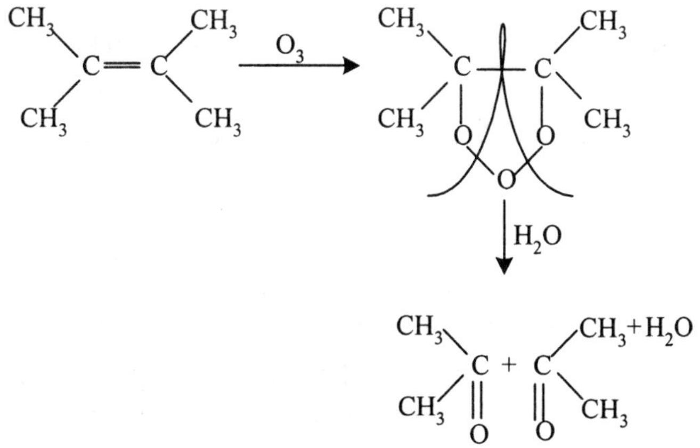

Hence, alkene is 2,3, -Dimethyl butene-2.

7. *Epoxidation*

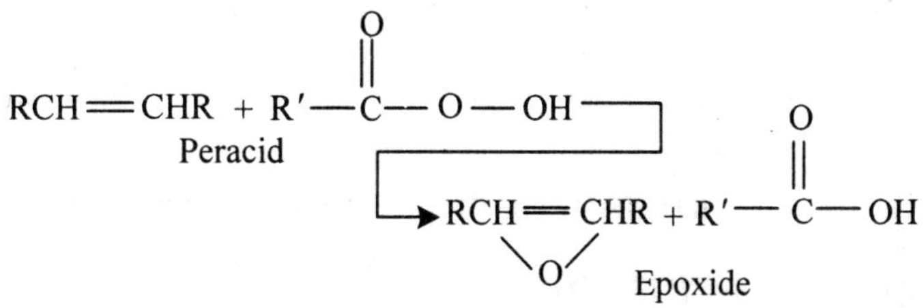

If R′ = H, it is performic acid

$R' = CH_3$, it is peracetic acid

8. *Oxidation*: The general picture for oxidation is as follows:

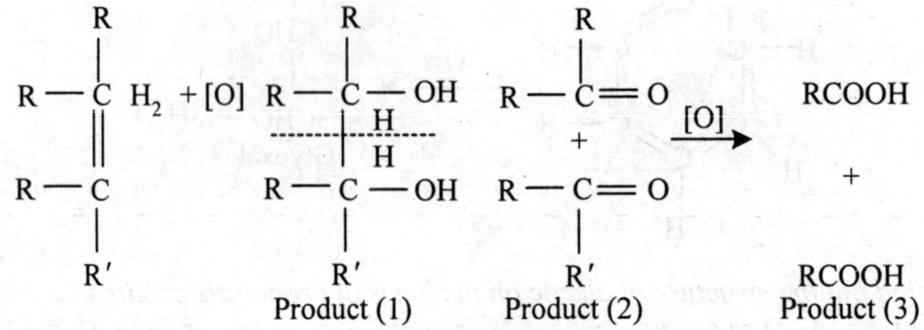

End product depends upon the oxidizing agent. See Table 9.4.

Table 9.4

	Reagent	*Product*
1.	Alk. $KMnO_4$(Baeyer's Reagent)	1. Diol
2.	OSO_4	2. Aldehyde or Ketone
3.	HIO_4 or Boil with $KMnO_4$	3. Acid
4.	Lemineux reagent (Sodium per iodate and traces of $KMnO_4$	4. Acid

Examples:

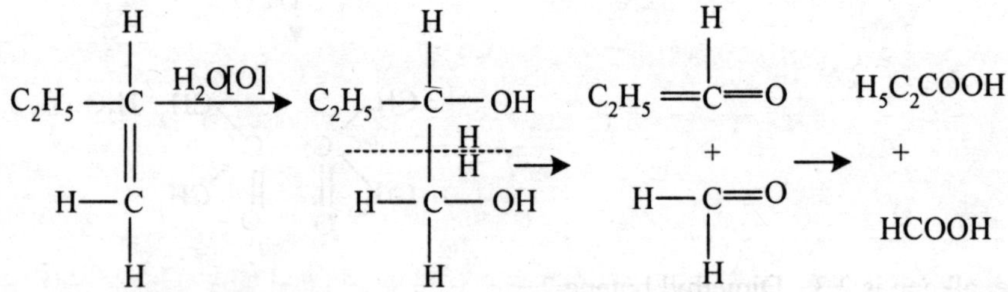

Butene-1

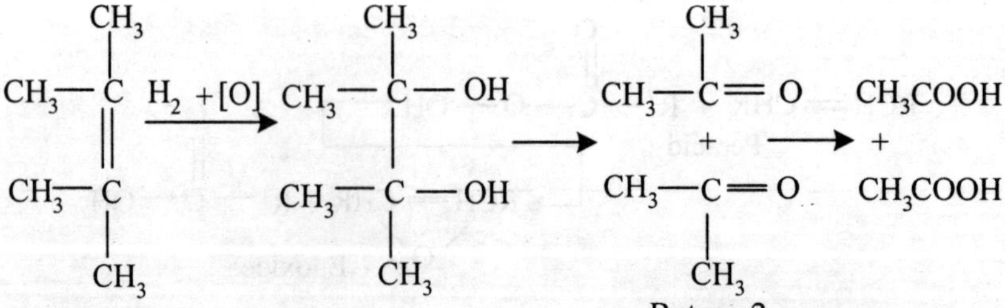

Butene-2

When alkenes are treated with Cl_2 or Br_2 at high temperature, substitution occur

$$CH_2 = CH_2 + Cl_2 \xrightarrow{400^\circ C} CH_2 = CH - Cl.$$

Test for Unsaturation: Alkene decolorizes alkl. $KMnO_4$ or Baeyer's reagent. It is a test for unsaturation.
Witting Reaction: The properties of cycloalkenes are similar to alkenes as shown in the following chart:

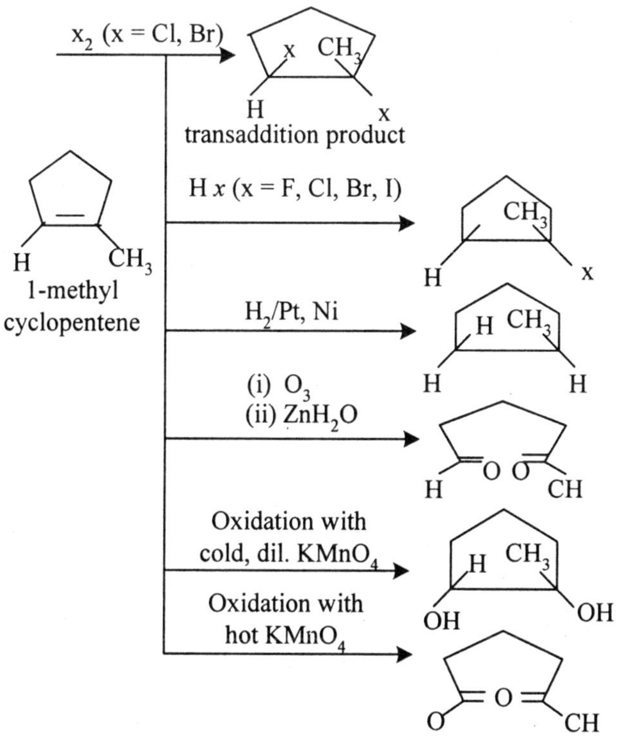

9.3 ALKYNES

Alkynes are hydrocarbons that contain a carbon-carbon triple bond and have general formula C_nH_{2n-2}.

(a) Preparation

(1) Alkenes can be converted into alkynes via formation of vicinaldihalide

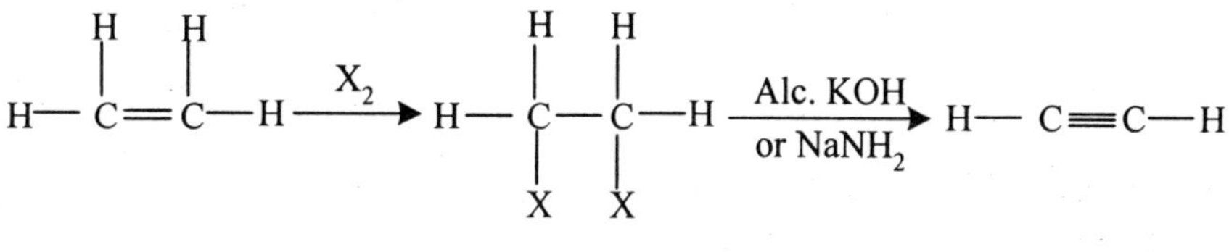

(2) Acetylene or alkyne-1 can be converted into higher alkynes

$$H—C\equiv C—H \xrightarrow{LiNH_2} H—C\equiv C—Li \xrightarrow{Rx} H—C\equiv C—R + LiX$$

(3) By reaction of calcium carbide with water

$$CaC_2 + H_2 \rightarrow C_2H_2 + Ca(OH)_2$$

(b) Properties

Physical: Physical properties are similar to alkenes as shown in Table 9.5.

Table 9.5 Physical constants of alkynes

Name	*Formula*	*bp(°C)*	*Density (d^{20}(gml^{-1})*
Ethyne	$HC\equiv CH$	–84	—
Propyne	$CH_3—C\equiv CH$	–23.2	—
1-Butyne	$CH_3CH_2—C\equiv CH$	8.1	—
2-Butyne	$CH_3—C\equiv C—CH_3$	27	0.691
1-Pentyne	$CH_3CH_2CH_2\equiv C—CH_3$	39.3	0.6952
2-Pentyne	$CH_3—CH_2CH_2C\equiv CH$	55.5	0.71

Chemical reactions:

(i) Alkynes undergo addition reaction, as follows

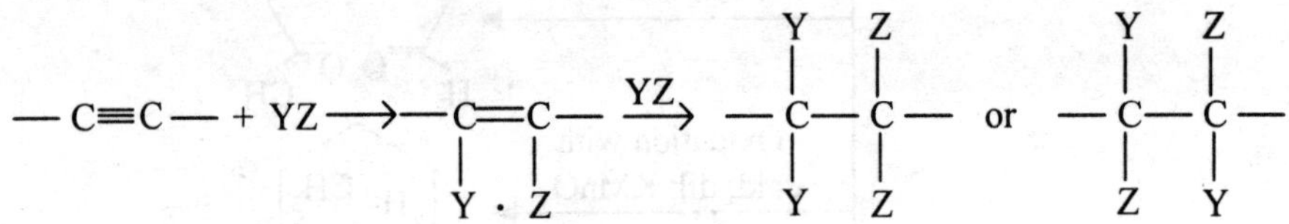

If Y = Z = Cl, product is tetrachloro alkane.
If Y = H, Z = Halogen like Cl, product is gem-dihalide

(ii) Addition of water

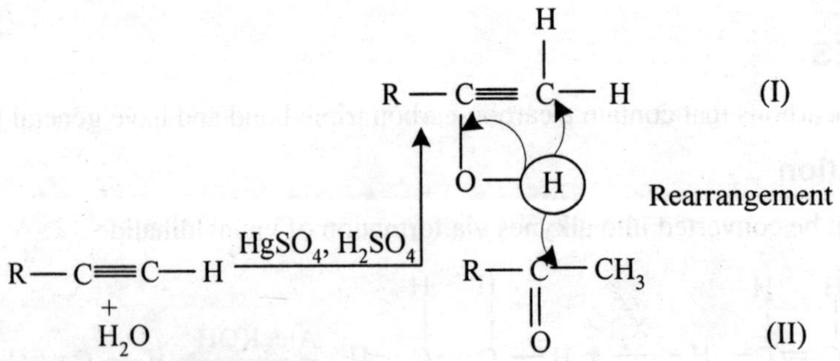

For example:

$$\underset{\text{Propyne}}{CH_3-C\equiv C-H_3} + H_2O \xrightarrow{HgSO_4,\ H_2SO_4} \underset{\text{Propanone}}{CH_3-\overset{\overset{\large O}{\|}}{C}-CH_3}$$

(IIT, 1983)

(iii) (a) Formation of acetylides

$$R-C\equiv C-H + nBuLi \longrightarrow \underset{\text{Lithium acetylide}}{RC\equiv C-Li} + nBuH$$

(b) $$R-C\equiv C-H + Cu(NH_3)_2 + \overset{\ominus}{O}H \xrightarrow{H_2O} \underset{\substack{\text{Copper acetylide}\\\text{(Brown ppt.)}}}{RC\equiv C-Cu}$$

(c) $$RC\equiv CH + Ag(NH_3)_2 + OH^- \xrightarrow{H_2O} \underset{\substack{\text{Silver acetylide}\\\text{(White ppt.)}}}{RC\equiv C-Ag}$$

(d) $$R-C\equiv C-H \xrightarrow{NaNH_2 \text{ in } NH_3} \underset{\text{Sodium acetylide}}{RC\equiv C-Ag}$$

Note that alkyne-1 forms acetylide. Rest do not.[4]

(iv) Some important addition reactions of acetylene are as follows:[5]

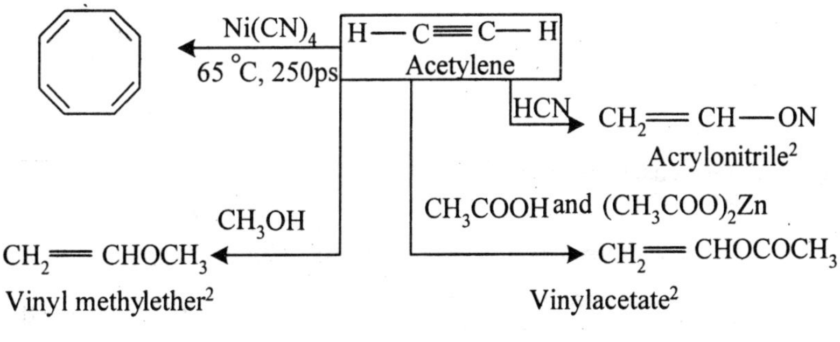

(v) Isomerization: $$\underset{\text{Butyne-1}}{CH_3CH_2\equiv CH} \xrightarrow[NaNH_2]{\text{Alc. KOH}} \underset{\text{Butyne-2}}{CH_3-C\equiv C-CH_3}$$

(vi) Ozonolysis: Alkyes react with O_3 to produce ozonide which upon hydrolysis, yields glyoxal or its derivatives depending upon structure of alkyne, e.g.,

---

[4] See Chapter 1: Hybridization.

[5] These are used as monomers (see Chapter 12)

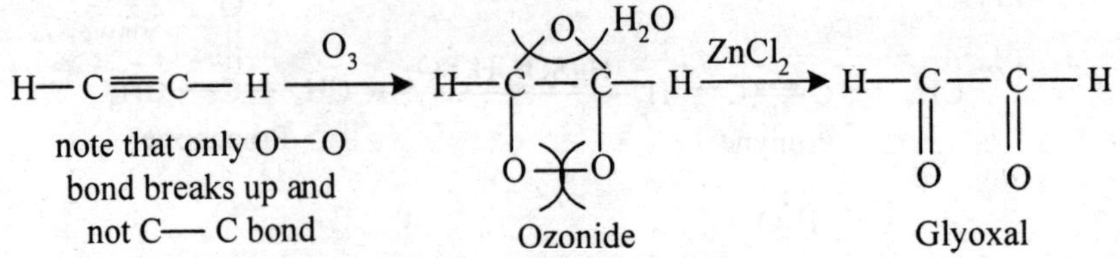

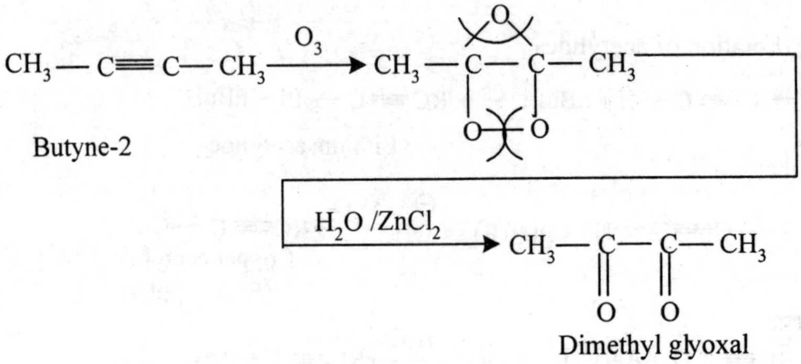

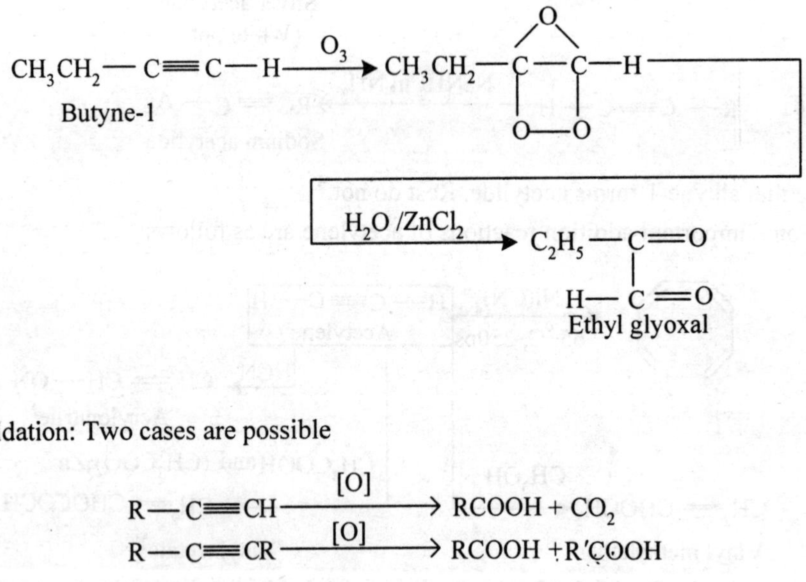

(vii) Oxidation: Two cases are possible

$$R-C\equiv CH \xrightarrow{[O]} RCOOH + CO_2$$
$$R-C\equiv CR \xrightarrow{[O]} RCOOH + R'COOH$$

Note that the ≡CH group of a terminal alkyne is oxidization to CO_2. and RC≡ is oxidized to RCOOH

(viii) Alkynes containing methylene groups adjacent to the $-C\equiv C^-$ bond undergo, 3-carbon tautomeric shifts to the allene and to conjugated diene.

$$CH_3CH_2CH_2-C\equiv CH \xrightarrow{Base} CH_3CH_2CH=C=CH_2$$
$$\longrightarrow CH_3CH=CH-CH=CH_2$$

PROBLEMS

1. Amongst the four alternate answers given for the subquestion below, one answer is correct. Write correct answer:

(i) The reaction conditions leading to the best yields of C_2H_5Cl are:

(a) $C_2H_6 \text{ (excess)} + Cl_2 \xrightarrow{\text{UV light}}$

(b) $C_2H_6 + Cl_2 \xrightarrow[\text{Room temperature}]{\text{dark}}$

(c) $C_2H_6 + Cl_2 \text{ (excess)} \xrightarrow{\text{UV light}}$

(d) $C_2H_6 + Cl_2 \xrightarrow{\text{UV light}}$ **(IIT, 1986)**

(ii) Baeyer's reagent is

(a) alkaline permanganate solution

(b) Acidified permanganate solution

(c) Neutral permanganate solution

(d) Aqueous bromine solution **(IIT, 1984)**

(iii) *n*-propyl bromide on treatment with ethanolic KOH produces:

(a) propane (b) propene

(c) propyne (d) propanol **(IIT, 1987)**

(iv) Addition of one equivalent of bromine to 1,3-pentadiene produces:

(a) 4,5-dibromopentene

(b) 1,4-dibromo-2-pentene

(c) 5,4-dibromopentene

(d) 3,4-dibromo-2-pentene **(GATE, 1992)**

(v) Consider the following:

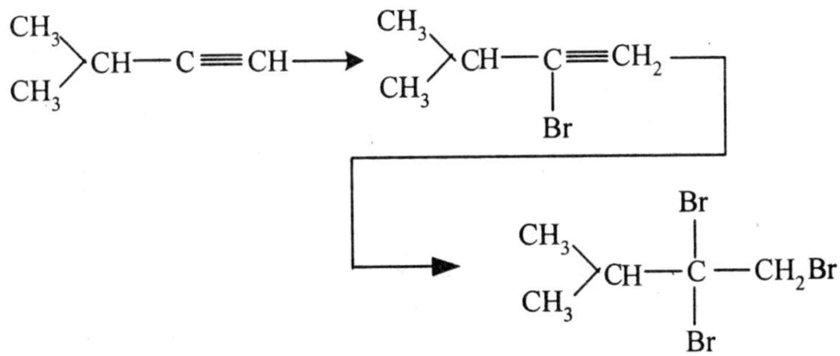

What reagent would you support for step 1 and 2, respectively?

(i) HBr then Br_2

(ii) HBr + peroxide then HBr

(iii) Br_2 then HBr

(iv) HBr, then HBr + peroxide

(vi) Which of the following hydrocarbons has the lowest dipole moment? **(IIT 2002)**

(a) 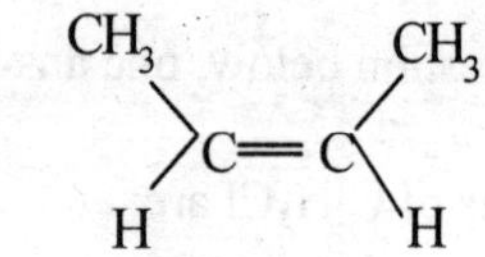 (b) $CH_3C{\equiv}C{-}CH_3$

(c) $CH_3CH_2C{\equiv}CH$ (d) $CH_2{=}CH{-}C{\equiv}CH$

(vii) HBr reacts with $CH_2{=}C{=}OCH_3$ under anhydrous conditions at room temperature to give

(a) $CH_3-CHBr-OCH_3$ (b) CH_3CHO & CH_3Br
(c) $BrCH_2CHO$ & CH_3OH (d) $BrCH_2-CH_2-OCH_3$ **(AIEEE 2006)**

2. Complete following reaction:

(i) $MeCH_2CH_2CMe{=}CHMe \xrightarrow{SeO_2}$

(ii) $\begin{matrix}CH_3\\CH_3\end{matrix}\!\!>C{=}CH_2 \xrightarrow{HIO_4}$

(iii) $\begin{matrix}CH_3\\CH_3\end{matrix}\!\!>C{=}CH_2 \xrightarrow[NaBH_4]{(CH_3COO)_2Hg}$

3. State whether True or False:
"Addition of HBr to 1-hexene in the presence of benzoyl peroxide generate 1-bromohexane".
(IIT, 1994)

4. Suggest appropriate structures for the missing compounds:

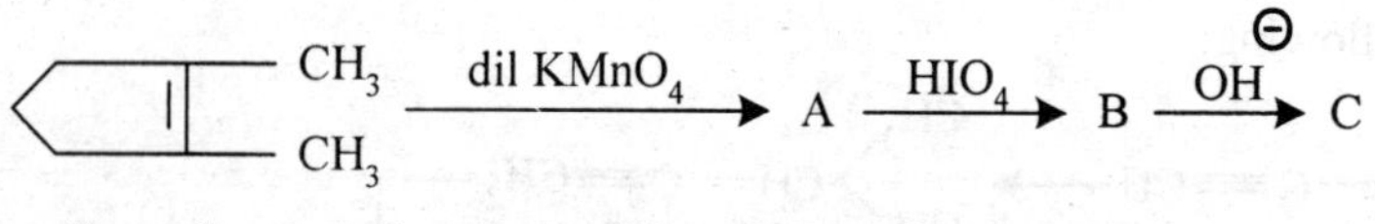 **(GATE, 1996)**

5. An alkane has a molecular weight of 72, the monochlorination produces only one product. What is the compound?

6. A hydrocarbon C_8H_{18} gives only one monochloro derivative. What is the hydrocarbon?
(GATE, 1992)

7. Write the structure of an alkane/cycloalkane which

(i) has lowest molecular weight and optically active
(ii) contains secondary butyl group and isobutyl group
(iii) has one neopentyl group
(iv) has molecular weight 86 and has 4 methyl group
(v) has molecular weight 100 and has pri, tert and neo carbon atom
(vi) has molecular formula C_5H_{12} and has only one tertiary H-atom
(vii) has molecular formula C_5H_{10} and has only secondary H-atoms
(viii) has molecular formula C_6H_{14} and has only pri and tert H-atoms
(ix) has molecular formula C_8H_{14} that has 12 secondary and 2 tertiary hydrogen atoms
(x) has molecular formula C_6H_{12} that has only primary and secondary hydrogen atoms

8. n-Butane is produced by the monobromination of ethane followed by Wurtz reaction. Calculate the volume of ethane at STP required to produce 55 g n-butane, if the bromination takes place with 90% yield and the Wurtz reaction with 85% yield. **(IIT, 1989)**

9. n-pentane has higher boiling point than neopentane why? **(MNR, 1995)**

10. Convert CaC_2 into C_2Cl_6.

11. It the relative reactivities of primary, secondary and tertiary hydrogen atoms are in the ratio of 4:6:10, calculate two amounts of primary, secondary and tertiary monochloro derivative, obtained by chlorination of 72 g of 2-methyl butane.

12. Give the structure of alkene that yields on ozonolysis:

(i) $CH_3CH_2CH_2CHO$ and HCHO only

(ii) 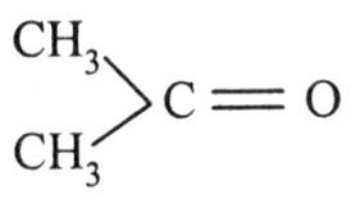

(iii) 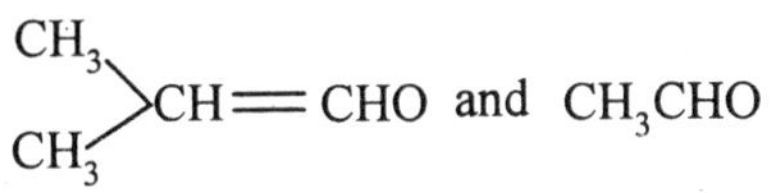

(iv) CHO—CH_2—CH_2—CH_2—CHO

(v) CHO—CH_2—CHO, HCHO, CH_3CHO

(vi) CHO—CH_2—CHO

13. Predict the ozonolysis products of

(i) Cyclopentene (ii) Toluene (iii) 1-Methylcyclopentene (iv) o-Xylene

(v) Benzene

14. Identify a compound $C_{10}H_{16}$ that absorbs 3 moles of H_2 to form $C_{10}H_{22}$. Upon ozonolysis, it yields:

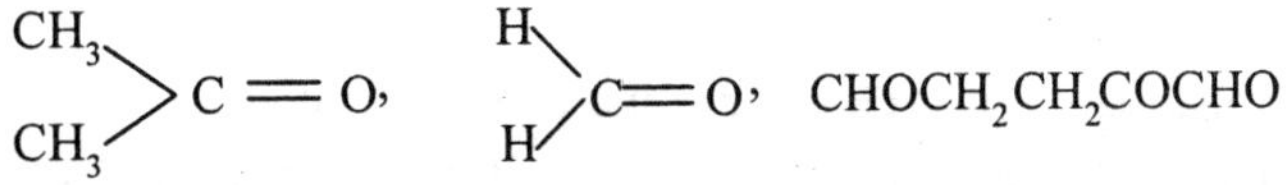

15. A compound A (C = 87.1/H = 12.8) on catalytic reduction gives B(C = 84.1/, H = 15.9). Ozonolysis of A gives acetic acid, acetone and pyruvic acid ($CH_3COCOOH$). What are A and B?

16. An alkene has molecular weight 80–85. 10.02 milligram of its sample took up 8.40 ml of hydrogen gas measured at 0°C and 760 mm pressure. Upon ozonolysis, it yielded followings:

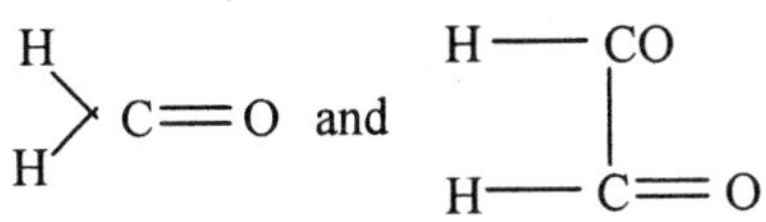

What is alkene?

17. Compound A, B and C all have the same molecular formula C_6H_8. Compound A yields three different products on ozonolysis. Compound B yields two different products, and compound C yields a single product. Identify A, B, and C.

18. Give structures of the major products expected from dehydrohalogenation of:

(i) 2-Bromohexane

(ii) 2-Bromo-2-methylpentane

(iii) 3-Bromo-2-methylpentane

(iv) 3-Bromo-2, 3-dimethylpentane

19. Arrange the compounds of each set in order of reactivity towards dehydrohalogenation by strong base.

(i) 2-Bromo-2-methylbutane and 1-Bromopentane

(ii) 2-Bromopentane and 1-Bromopentane

20. Which alkyl halide would yield each of the following alkenes upon dehydrohalogenation by strong base?

(i) Isobutylene *(ii)* Pentene-1

21. Arrange following alkenes in order of increasing stability

(i) 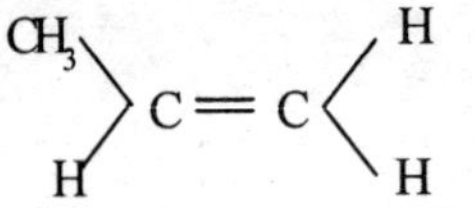

(ii) 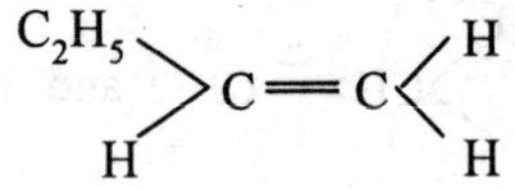

(iii) 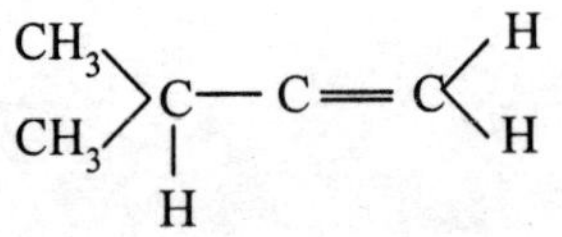

22. Write oxidation products of most stable alkene using alk.$KMnO_4$ (hot).

23. Treatment of C $(CH_2Br)_4$ with Zn yields a hydrocarbon of molecular formula C_5H_8 in which all the H-atoms are equivalent. Suggest a structure for C_5H_8.

24. The following statement is true only under some specific conditions. Write the condition in not more than two sentences. "2-methyl propene can be converted into isobutyl bromide by hydrogen bromide". **(IIT, 1984)**

25. How will you distinguish between cyclohexane and cyclohexene? **(IIT, 1988)**

26. A (C_4H_8) adds on HBr in the presence and absence of peroxide to give same product C_4H_9Br. Identify A. **(IIT, 1993)**

27. An organic compound 'A' C_8H_{12} on oxidation with $KMnO_4$ yields only HOOC $(CH_2)_6COOH$. What is 'A'?

28. An organic compound 'A' on oxidation yields and CH_3CH_2COOH. What is 'A'?

29. How will you distinguish between

(i) Butyne-1 and Butyne-2

(ii) CH_4 and C_2H_2 **(MNR, 1995)**

(iii) C_2H_4 and C_2H_2 **(MNR, 1989; 1995)**

30. Carry out following conversion:

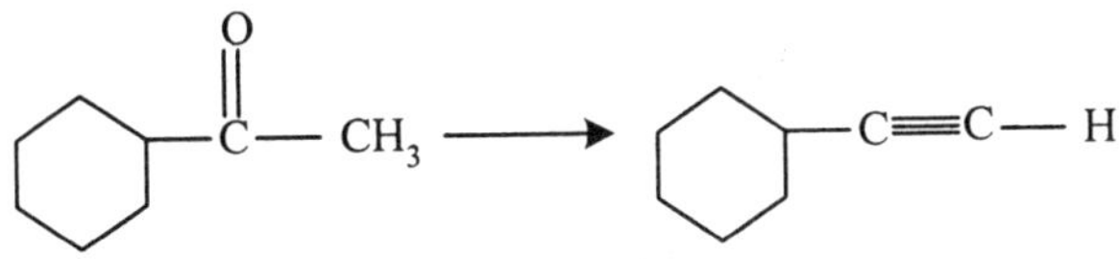

Methyl cyclohexyl ketone   Cyclohexyl acetylene

31. Write down the structure of the stereoisomers formed when cis-2-butene is reacted with bromine. **(IIT, 1995)**

32. Complete the following equations:

(i) $RC{\equiv}C{-}R \xrightarrow[\text{Linder Catalyst}]{H_2} ?$

(ii) $MeCH_2C{\equiv}CH \xrightarrow{Na/NH_3} ? \xrightarrow{E\,tBr}$ **(R, 1993)**

(iii) 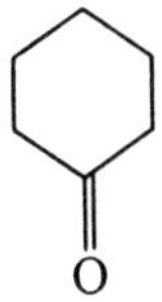 + $Ph_3P{=}CH_2 \longrightarrow$?

(IIT, 1997)

33. How will you distinguish between:

(i) CH_4 and C_2H_2 **(MNR, 1995)**

(ii) C_2H_4 and C_2H_2 **(MNR, 1989; 1995)**

34. What is Wilkinson's catalyst? What is its significance?

35. What is P-200 Catalyst? What is its significance?

36. Complete the following: **(IIT, 1999)**

(i) 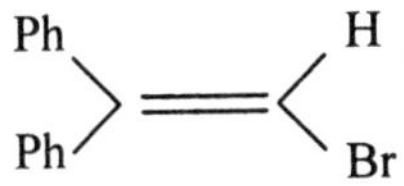 + $KNH_2 \longrightarrow$ A

(ii) 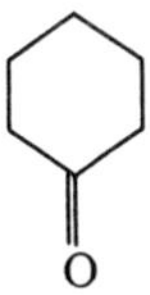 + $Ph_3P{=}CH_2 \longrightarrow$ E

(iii) 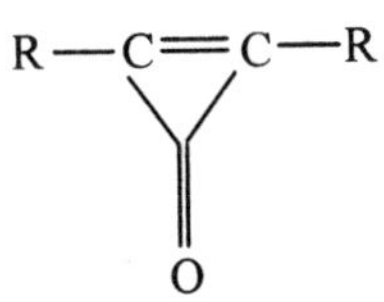 + $HClO_4 \longrightarrow$ G

ANSWERS

1. *(i)* **(a)**

(ii) **(a)**

(iii) **(b)**

(iv) **(a)**, See stability of alkene formed.

(v) **(i)**, In the presence of peroxide (see peroxide effect on page 66)

(vi) **(b)**

(vii) **(d)**

2. *(i)*

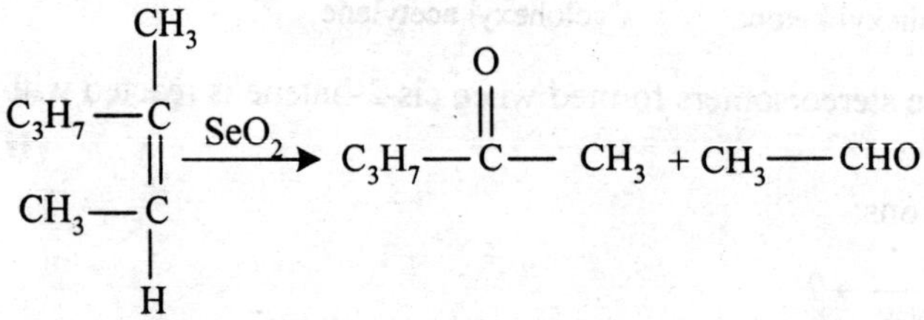

(ii)

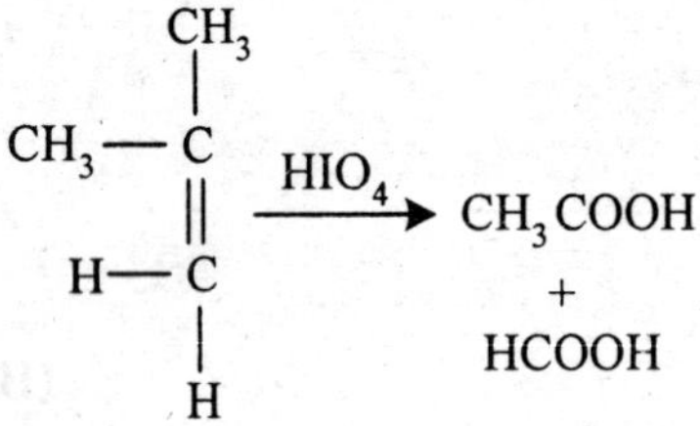

(iii)

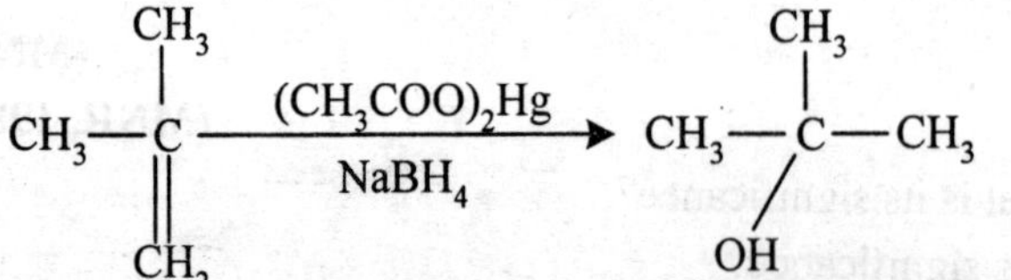

2-Methylpropanol-2

3. True.(See peroxide effect).

4.

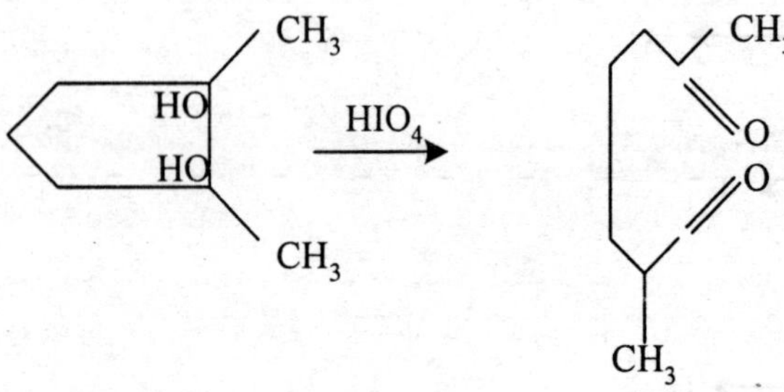

(A) (B)

5. Three alkanes are possible:

$CH_3CH_2CH_2CH_2CH_3$

n-pentane

(It contains primary and secondary H-atoms, hence two monochloro derivatives)

Isopentane

(It contains primary secondary and tertiary H-atoms, hence two monochloro derivatives)

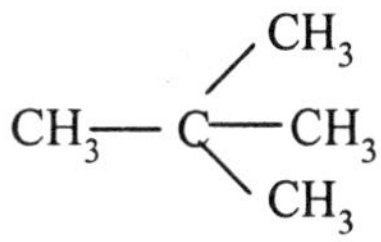

Neopentane

(It contains only primary H-atoms, hence one monochloro derivative)

Hence, correct answer is neopentane.

6. It is 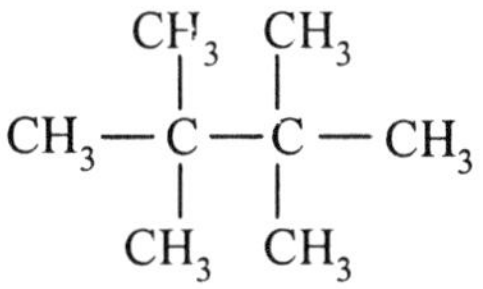

2,2,3,3-Tetramethylbutane

7. *(i)* 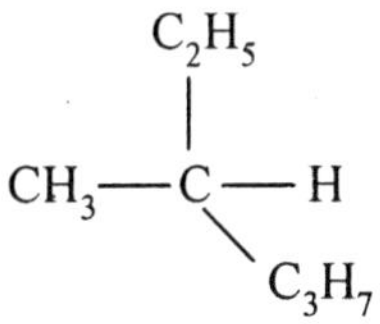

Asymmetric carbon atom, hence optically active molecular weight 100.

(ii) 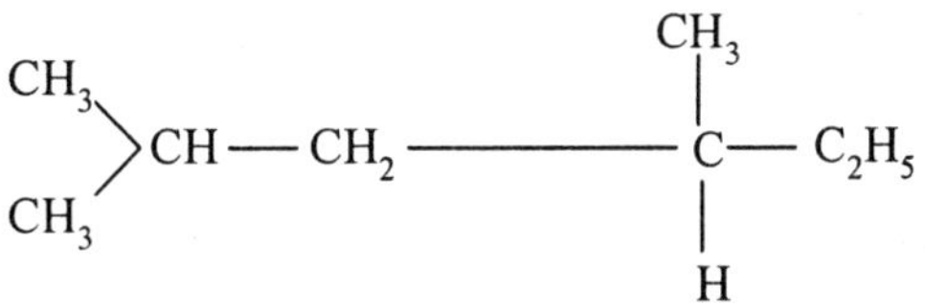

Isobutyl gruop Secondary butyl group

2,4-Dimethylhexane

(iii) 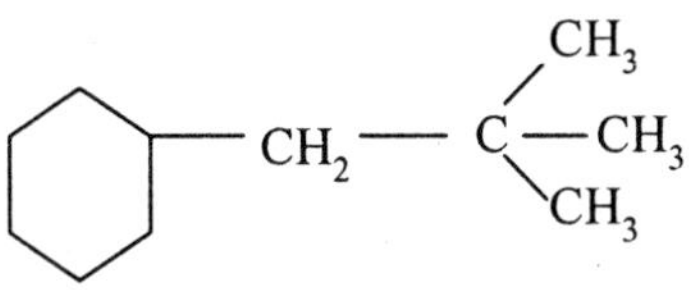

(iv) 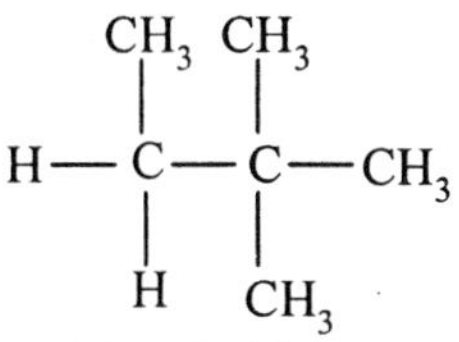

2,2-Dimethyl butane

(v)

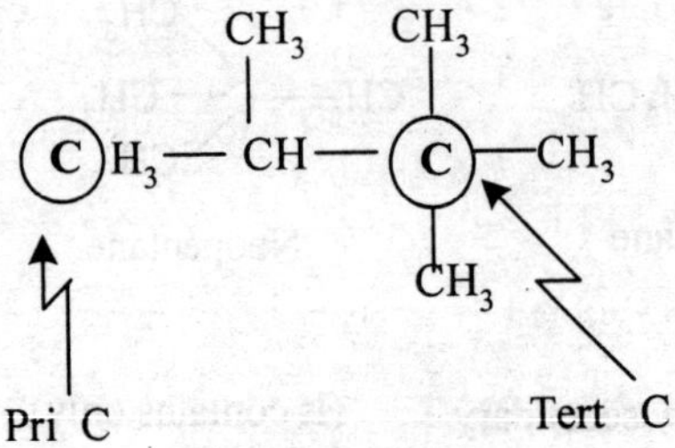

2,2,3-Trimethylbutane

(vi)

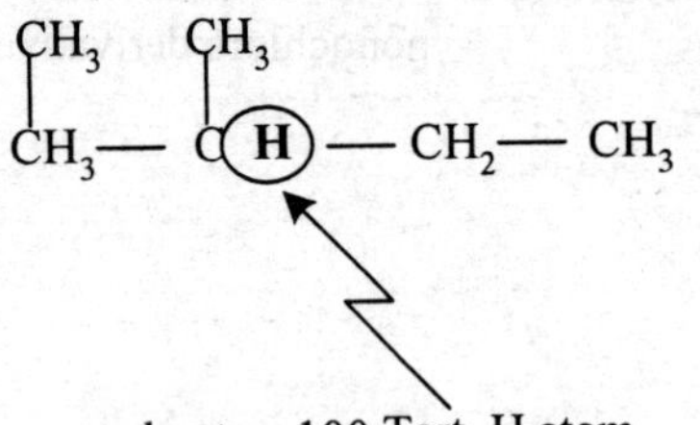

mol. wt. = 100 Tert. H atom
2-Methylbutane atom

(vii)

Cyclopentane

(viii)

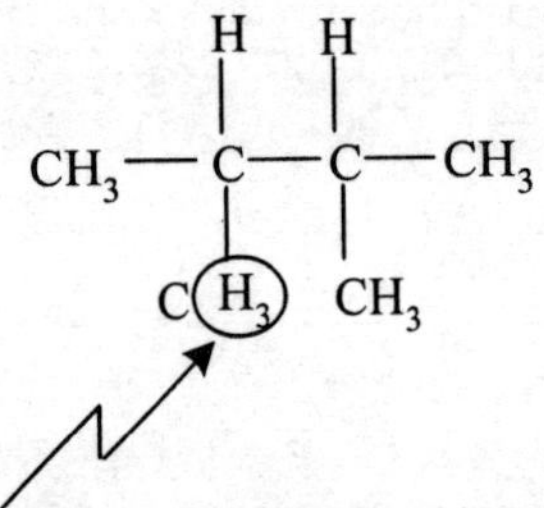

Pri. H atom 2,3-Dimethyl butane

(ix)

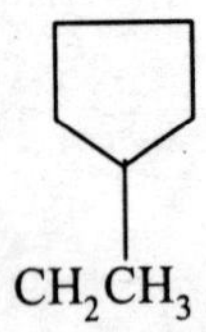

Ethylcyclopentane

(x) Cyclopropyl cyclopentane -

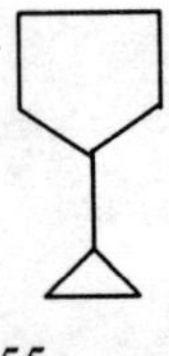

8. Moles of *n*-butane to be produced $\frac{55}{58} = 0.948$.

Wurtz reaction is $2C_2H_5Br + 2Na \longrightarrow C_4H_{10} + 2NaBr$
2 Moles 1 Moles

However, yield is 85 percent. So, 0.948 moles of C_4H_{10} would be obtained from $2\times\frac{0.948}{0.85} = 2.23$ moles of C_2H_5Br

Monobromination reaction is $C_2H_6 + Br_2 \rightarrow C_2H_5Br + HBr$
1 Mole 1 Mole

But the yield is 90 percent.

So, 2.23 moles of C_2H_5Br would need $\frac{1}{0.9} \times 2.23 = 2.478$ moles of $C_2H_6 = 2.478 \times 22.4 = 55.5$ litres.

9. As branching increases, the molecular surface becomes spherical, so the van der Waal's forces become less, hence *n*-pentane has higher boiling point.

10. 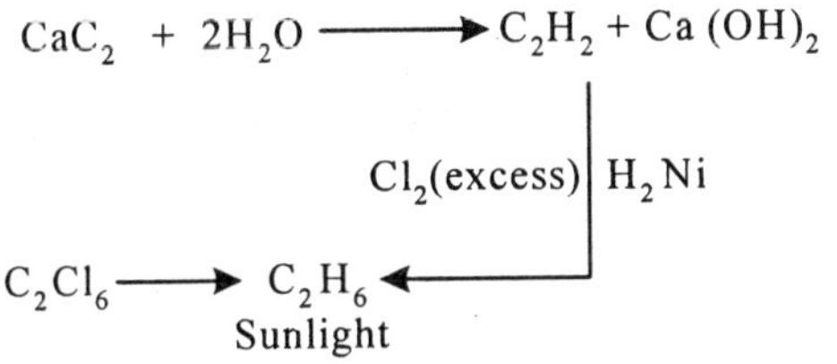

11. The alkane has structure

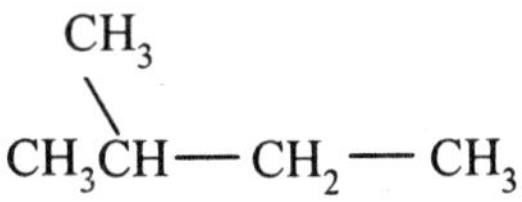

Number of primary H-atoms = 9
Number of secondary H-atoms = 2
Number of tertiary H-atoms = 1
The relative ratio of chlorides = 9 × 4:2 × 6:1 × 10 = 36:12:10
Weight of monochloro derivative from 72 gm of alkane = 106.5 g
Amount of primary chloride = 36 × 106.5/58 = 66.12 g
Amount of secondary chloride = 12 × 106.5/588 = 21.99 g
Amount of tertiary chloride = 10 × 106.5/58 = 18.37 g.

12. *(i)* $CH_3CH_2CH_2CH{=}CH_2$
Pentene – 1

(ii) 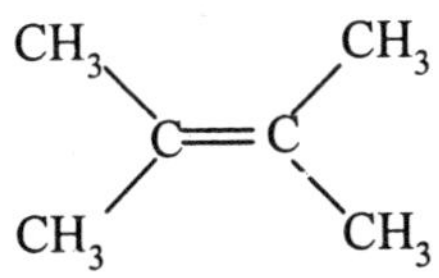

2,3-Dimethyl butene -2

(iii) 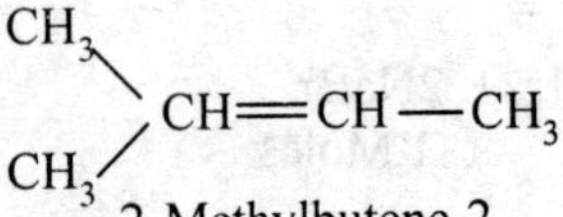

2-Methylbutene-2

(iv) 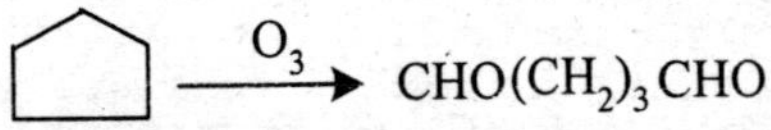

Cyclopentane

(v) $CH_2 = CH - CH_2 - CH = CH - CH_3$

1,4-hexadiene

(vi) 

Cyclopropene

13. *(i)* 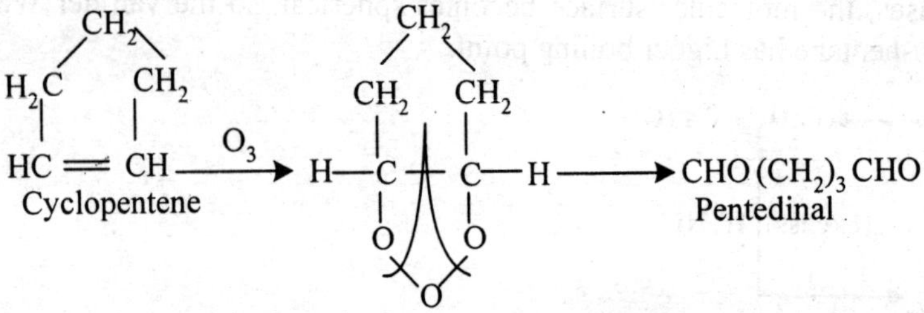

Cyclopentene → Pentedinal

(ii) 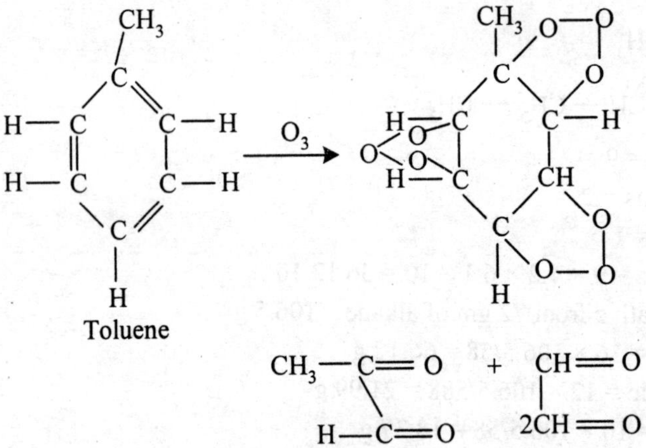

Toluene

Methyglyoxal Glyoxal

(iii)

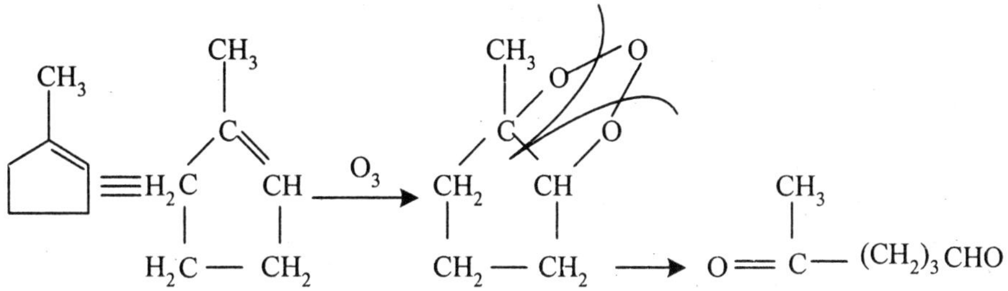

2-Ketohexanal

(iv) In ozonolysis of *o*-xylene, we have to consider both the Kekulé structures. Hence, three ozonolysis products are obtained.

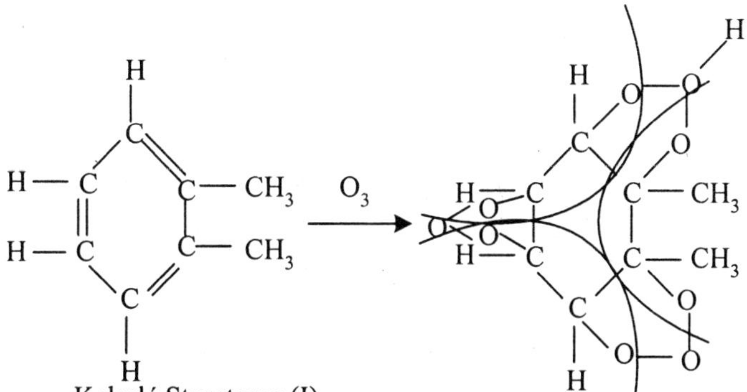

Kekulé Structures (I)

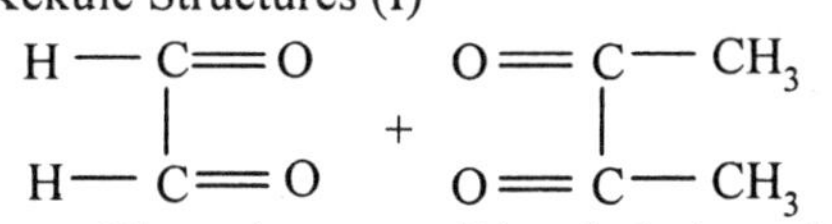

Glyoxal   Dimethyl glyoxal

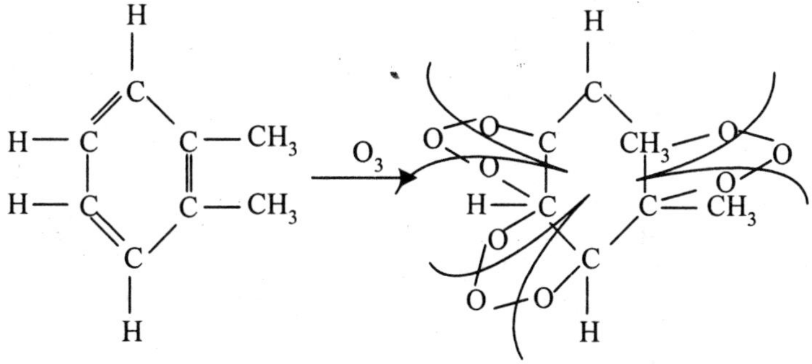

Kekulé structure (II) (see that position of double bond is changed)

Glyoxal   Methyl glyoxal

14. Since it absorbs 3 moles of hydrogen hence it must contain three double bonds. Since alkene contains ten carbon atoms, two moles of HCHO will be formed. The formation of two HCHO is possible when alkene must have one double bond at each terminal position. Therefore, the alkene is as follows:

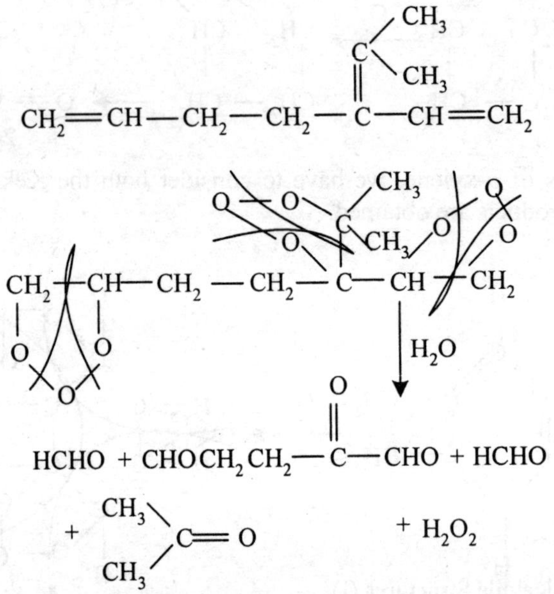

15. Since ozonolysis results this products, it indicates presence of two unsaturated sites. Hence, it is alkdiene.

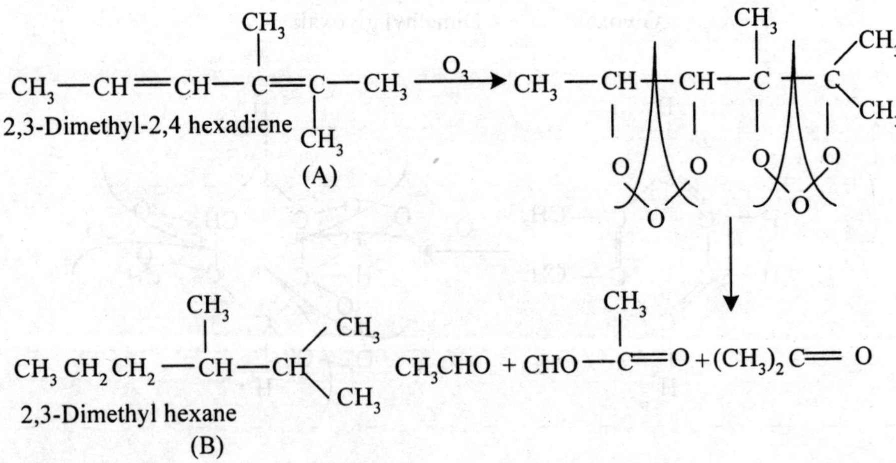

16. 1,3,5-hexatriene.

Volume of hydrogen taken up STP = 8.4 ml

Weight of hydrogen = $2 \times 8.4/22400$ gm

Moles of hydrogen = $2 \times 8.4/2 \times 22400$

10.02×10^{-3} g of the compound took up $2 \times 8.4/2 \times 22400$ moles of hydrogen.

Moles of hydrogen taken up = 80 × 2 × 8.4 × 10PT3PT/2 × 22400 × 10.02 by 80 g of the compound = 3 moles.

Hence, alkene contains 3 double bonds.

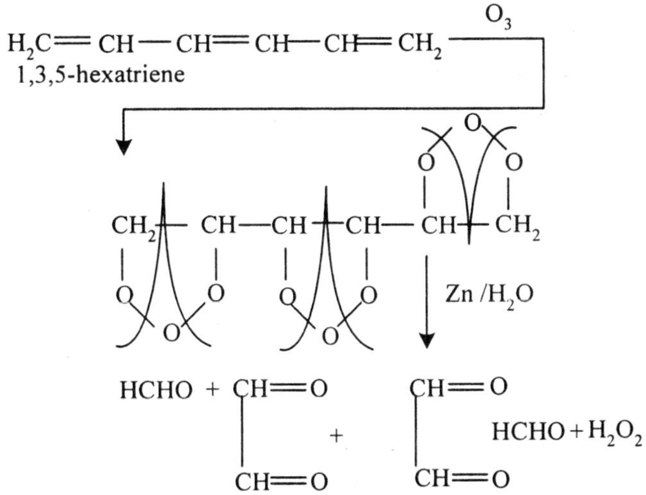

17. $CH_2{=}CH{-}CH{=}CH{-}CH{=}CH_2$ or 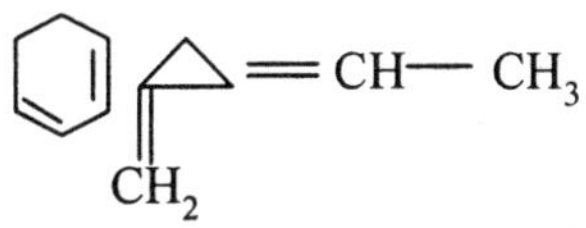

(B)(C)

18. See Saytzeff Rule to solve these problems: (Page 91)

(i) 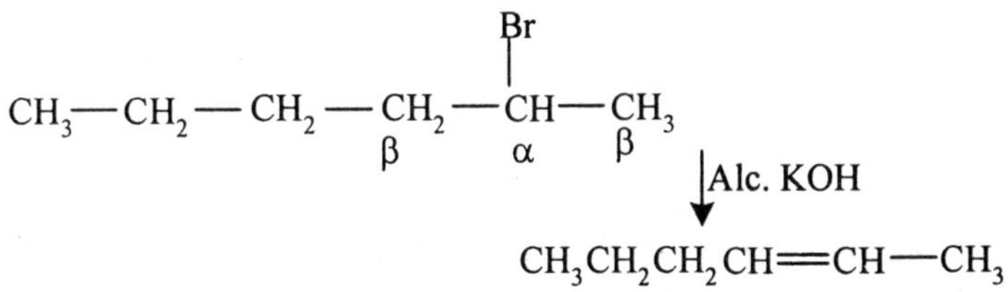

(*cis* and *trans* mixture) pentene-2

(ii) 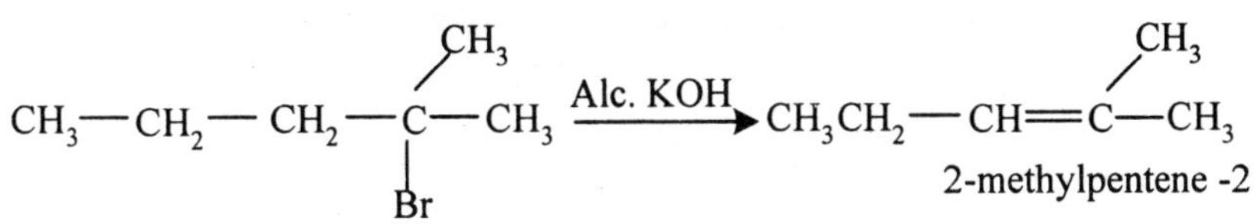

2-methylpentene -2

(iii) 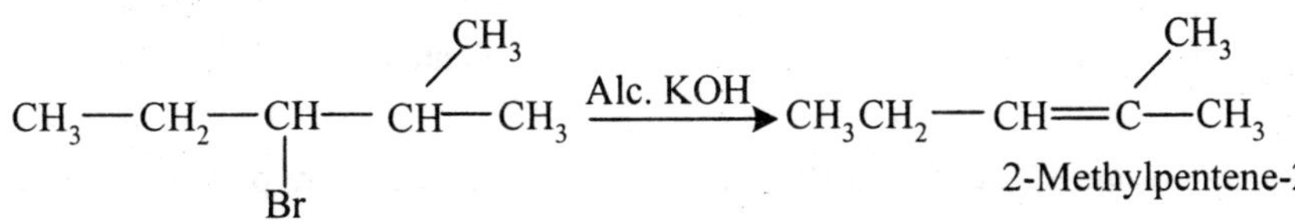

2-Methylpentene-2

(iv) 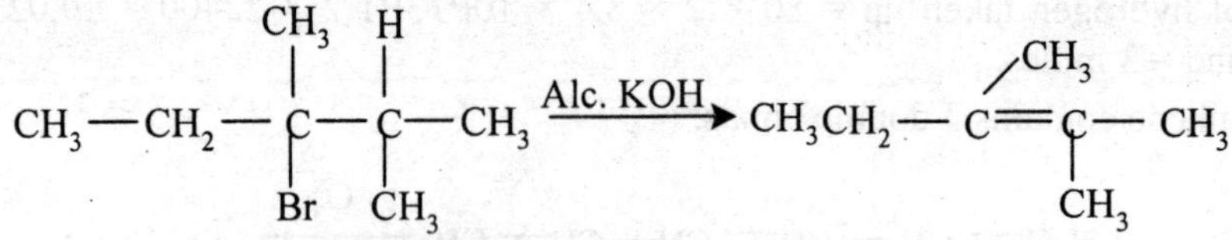

2-Bromo-2-Methylbutane 2,3-Dimethylpentene-2

19. *(i)* Examine the stability of 2-methylbutene = 2, the compound formed from each alkyl halide

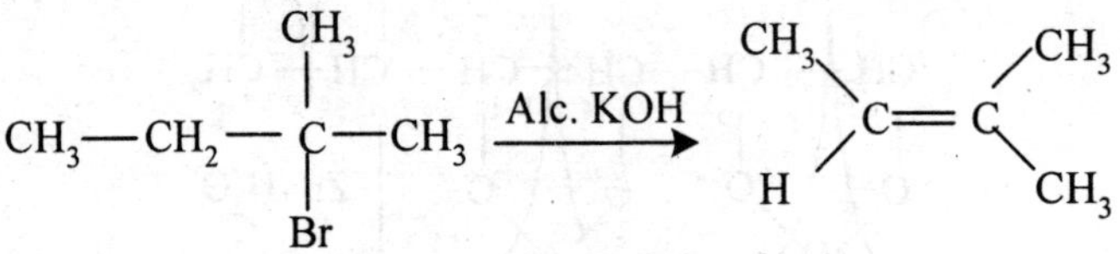

9 Hyperconjugating structures, hence more stable

$$CH_3CH_2CH_2CH_2—CH_2Br \xrightarrow{\text{Alc. KOH}} CH_3CH_2CH_2CH=CH_2$$

2 Hyperconjugating structures, hence less stable

(ii) 2-Bromopentane Reason is same as above

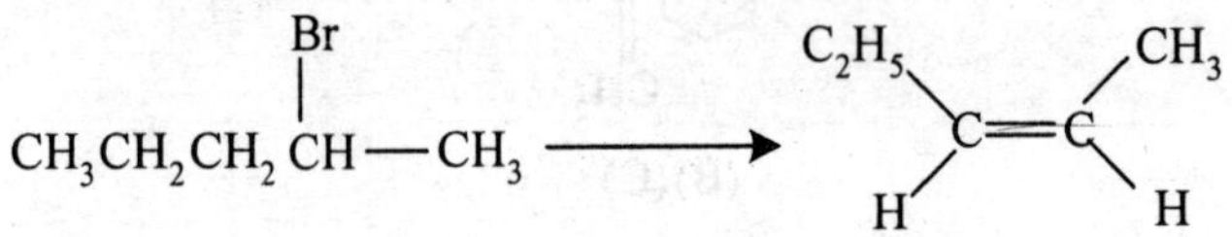

2-Bromopentane 5 Hyperconjugating structures

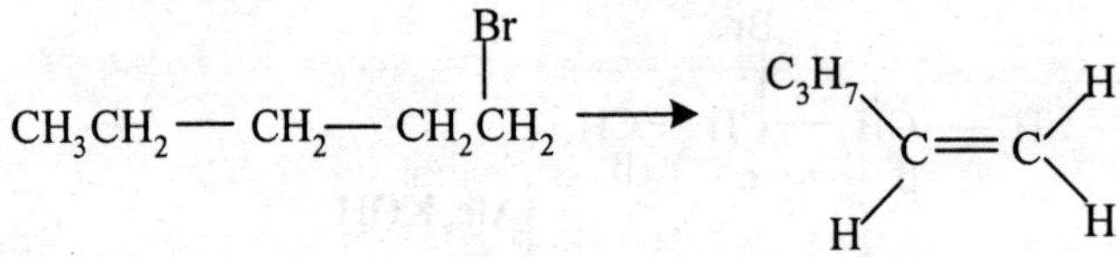

1-Bromopentane Isobutylene

20. *(i)* 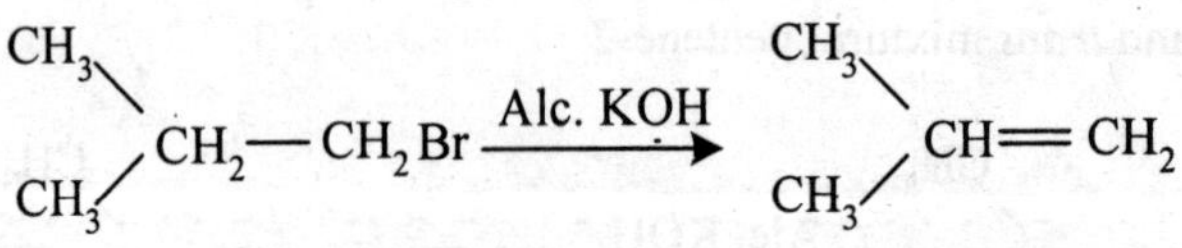

2-Methyl-1-Bromopropane Isobutylene

(ii) $$CH_3CH_2CH_2CH_2CH_2Br \xrightarrow{\text{Alc. KOH}} CH_3CH_2CH_2CH=CH_3$$

Pentene – 1

21. Increasing order of stability is III < II < I. See the number of hyperconjugation structures:

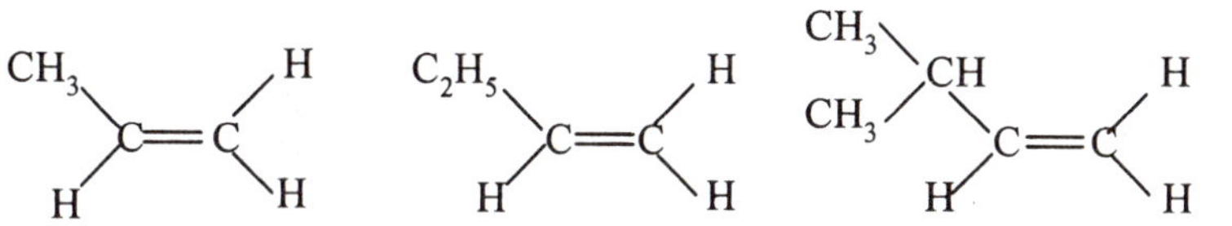

3 Hyperconjugating structures due to methyl group

2 Hyperconjugating structures due to ethyl group

1 Hyperconjugating structures due to isopropyl group

22. 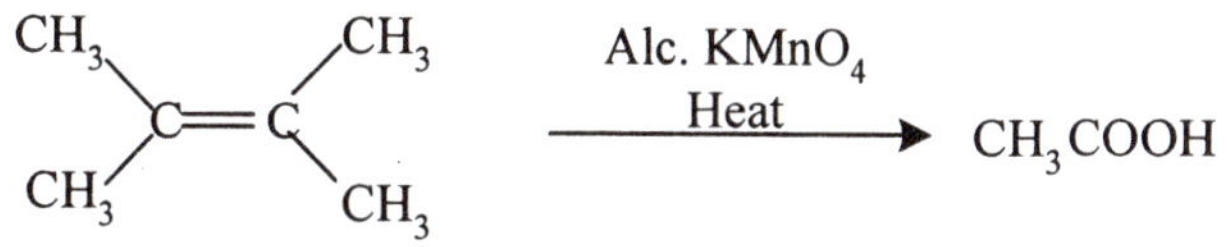

12 Hyperconjugating structure (Most stable alkene)

23. 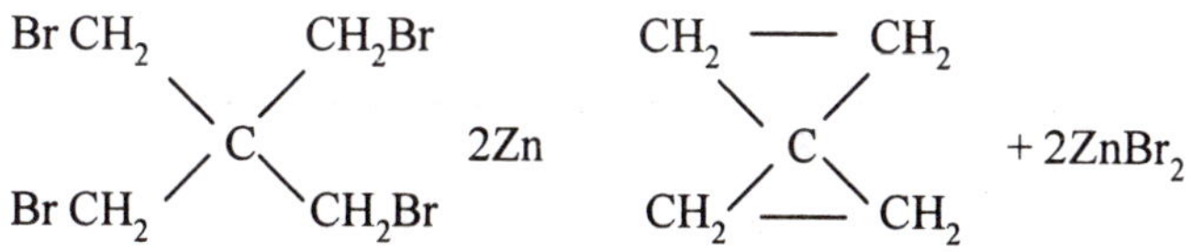

24. (i) in the presence of a peroxide (See peroxide effect).

25. Cyclohexane does not reduce Baeyer's reagent and bromine water, whereas cyclohexene reduces both the reagents.

26. Since it gives same product in the presence and absence of peroxide, it should be a symmetrical alkene. So, A is $CH_3—CH{=}CH—CH_3$ (for details see the Chapter on Mechanism).

27. 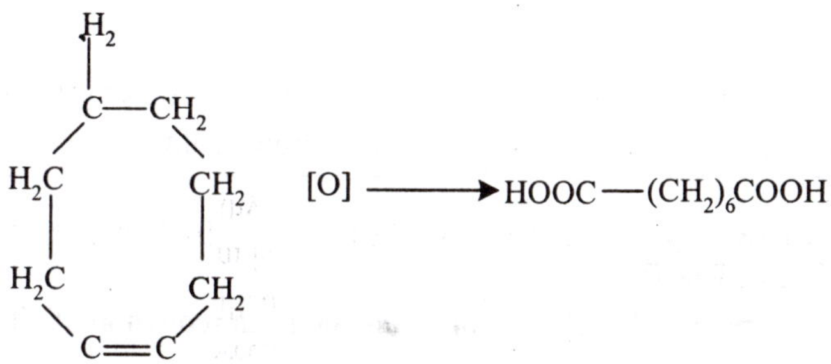

Cycloctyne

28. 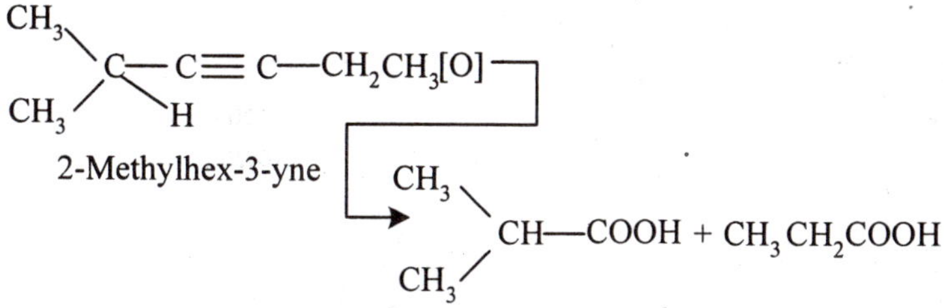

29. Butyne-1 contains acidic H-atom, and hence forms white ppt of silver acetylide with Amm. $AgNO_3$ and brown ppt of copper acetylide with Amm. CuCl.

30. 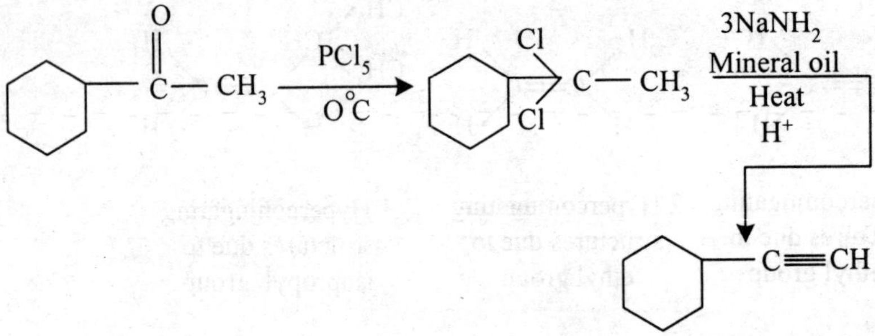

31. 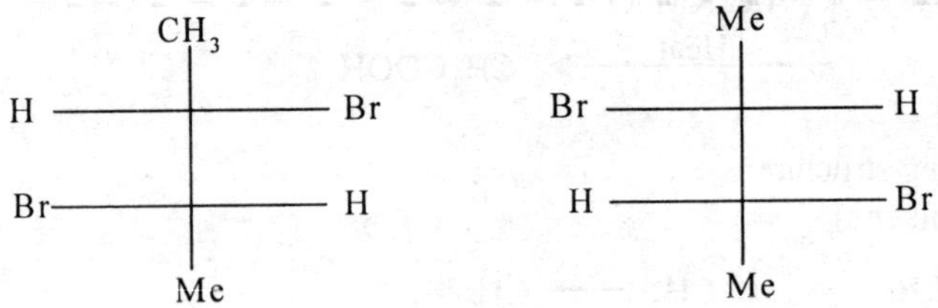

32. *(i)* RCH═CHR (Lindlar's catalyst is useful for partial reduction)

(ii) $MeCH_2C{\equiv}CH \xrightarrow{Na/NH_3} MeCH_2C{\equiv}CNa \xrightarrow{EtBr} MeCH_2—C{\equiv}C—Et$

(iii) 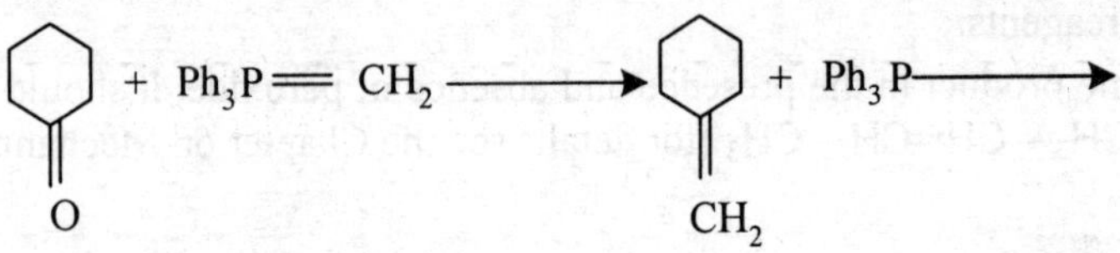

33. *(i)* CH_4 does not decolour Baeyer's reagent and bromine water, whereas C_2H_2 does so.

(ii) C_2H_4 does not give precipitate with ammonical cuprous chloride or ammonical $AgNO_3$ whereas C_2H_2 does so.

34. Wilkinson's catalyst is *Tris* (triphenyl phosphine) Chloro Rhodium II. It is used to covert alkyne to *cis*-alkene alony [$Ph_3P)_3RhCl$].

35. It is nickel boride. It is used to obtain 100% Z form of an alkene from reduction of alkyne.

36. *(i)* A is Ph—C≡C—Ph

(ii) E is 

(iii) G is RCOOH

Chapter 10

Derivatives of Aliphatic Hydrocarbons

Derivatives of aliphatic hydrocarbons include:

(1) Alkyl halides; (2) alcohols; (c) Ethers; (4) Aldehydes and Ketones; (5) Acids and their Derivatives; (6) Nitro compounds; (7) Nitrosoalkanes; (8) Cyanides; (9) Isocyanides; and (10) Alkyl amines.

Let us study them one by one.

10.1 ALKYL HALIDES

Aliphatic compounds in which halogen atom(s) is attached to an alkyl group are called alkyl halides. Alkyl halides are of four types, which are as follows:

(1) Monohalogen alkanes;

(2) Dihalogen alkanes;

(3) Trihalogen alkanes; and

(4) Tetrahalogen alkanes.

10.1.1 Monohalogen Alkanes

These have general formula C_nH_{2n+1} X, e.g., Ethyl bromide (C_2H_5Br).

(a) ***Preparation***

(i) By reaction of alkanes (excess) with halogens:

$$\underset{\text{Excess}}{RH} + X_2 \xrightarrow{\text{Sunlight}} RX + HX$$

$$\text{For example, } C_2H_6 + Br_2 \xrightarrow{\text{Sunlight}} C_2H_5Br + HBr$$

(ii) By reaction of alcohols with PCl_5, PCl_3

$$ROH + PCl_5 \rightarrow RCl + HCl + POCl_3$$

For example, $C_2H_5OH + PCl_5 \longrightarrow C_2H_5Cl + HCl + POCl_3$

Ethyl alcohol   Ethyl chloride

$$3ROH + PCl_3 \longrightarrow 3RCl + H_3PO_3$$

Phosphorus trichloride

$$ROH + SOCl_2 \longrightarrow RCl + SO_2 + HCl$$

Thionyl chloride

$$ROH + HX \longrightarrow RX + H_2O$$

Halogen acid

For example, $C_2H_5OH + HI \longrightarrow C_2H_5I + H_2O$

(b) *Properties*

Physical: In general alkyl chlorides, bromides and iodides are all liquids and tend to have boiling point near those of alkanes of similar molecular weight. For example, methyl-iodide is the only monohalomethane that is a liquid. Ethyl chloride is a gas (bp, 13°C) whereas ethyl bromide (bp, 38°C) and ethyl iodide (bp, 72°C) both are liquids. The propyl chlorides, bomides and iodides, are all liquids.

Chemical :Monohalogen alkanes undergo nucleophilic substitution reaction. The chemical properties can be summarized as below:

(i) It undergoes hydrolysis[1] with aqueous KOH to yield alcohol:

$$RCH_2CH_2X + aq.\ KOH \rightarrow RCH_2CH_2OH$$

(ii) It undergoes dehydrohalogenation[2] with alcoholic KOH to yield alkene:

$$RCH_2CH_2X + alc.\ KOH \rightarrow RCH=CH_2 + KX + H_2O$$

(iii) It eacts with ammonia to form amine:

$$RCH_2CH_2X + H-NH_2 \rightarrow RCH_2CH_2NH_2$$

(iv) It reacts with magnesium in ether to form Grignard reagent[3]

$$RCH_2CH_2X + Mg \rightarrow RCH_2CH_2MgX$$

(v) It reacts with potassium cyanide to form alkyl cyanide

$$RCH_2CH_2X + KCN \rightarrow RCH_2CH_2CN$$

(vi) It reacts with silver cyanide to form alkyl isocyanide

$$RCH_2CH_2X + AgCN \rightarrow RCH_2CH_2NC$$

(vii) It reacts with potassium hydrogen sulphide to form mercaptane

---

[1] See Chapter 5 for details.

[2] See Chapter 5 for details.

[3] See Chapter 6 for details.

$$RCH_2CH_2X + KSH \longrightarrow RCH_2CH_2SH + KX$$

Alkylmercaptane or Alkanthiol

For example,

$$C_2H_5Br + KSH \longrightarrow C_2H_5SH + KBr$$

Ethyl bromide — Ethyl mercaptane or Ethanthiol

(viii) It reacts with sodium to form alkane (Wurtz reaction)

$$2RCH_2CH_2X + 2Na \longrightarrow RCH_2CH_2CH_2CH_2R +$$

10.1.2 Dihalogen Alkanes

When two hydrogen atoms of an alkane are replaced by two halogen atoms, the compound are classified into two categòries, viz., vicinal (*vic*-) and geminal (*gem*-) dihalides on the basis of the attachment of two halogen atoms to two adjacent carbon atoms and, to one carbon atom respectively, e.g.,

$$\begin{array}{l} CH_2-Cl \\ | \\ CH_2-Cl \end{array} \qquad CH_3-CHCl_2$$

Ethylene dichloride (*vic*) Ethylidene dichloride (*gem*)

(a) *Preparation*

vic-dihalides can be prepared either by addition of halogens to alkenes or by the reaction of PCl_5 with glycols.

$$CH_3-CH=CH_2 + Br_2 \longrightarrow CH_3-CH(Br)-CH_2(Br)$$

1,2-Dibromopropane

$$\begin{array}{l} CH_2OH \\ | \\ CH_2OH \end{array} + PCl_5 \longrightarrow \begin{array}{l} CH_2Cl \\ | \\ CH_2Cl \end{array} + 2POCl_3 + 2HCl$$

1,2-Dichloroethane (Ethylenedichloride)

The *gem*-dihalides are prepared either by addition of hydrogen halides to alkynes or by reaction of PCl_5 with aldehydes and ketones:

$$CH_3-C\equiv CH + HBr \longrightarrow CH_3-\underset{}{\overset{Br}{\overset{|}{C}}}=CH_2 \xrightarrow{HBr} CH_3-\overset{Br}{\overset{|}{\underset{Br}{\underset{|}{C}}}}-CH_3$$

Propyne $\qquad$ 2,2-Dichloropropane

$$\begin{matrix}CH_3 \\ CH_3\end{matrix}\!\!>C=O + PCl_5 \longrightarrow (CH_3)_2CCl_2 + POCl_3$$

Propanone $\qquad$ 2,2-Dichloropropane

(b) ***Properties***

Physical: Dihaloalkanes are sweet smelling liquids.

Chemical: The chemical properties of dihaloalkanes are similar to those of monohaloalkanes, e.g.,

(1)

$$\begin{matrix}CH_2Cl \\ | \\ CH_2Cl\end{matrix} + \text{alc. } 2KOH \rightarrow \begin{matrix}CH_2OH \\ | \\ CH_2OH\end{matrix} + 2KCl$$

1,2 – Dichloroethane

$$CH_3CHCl_2 + \text{aq. } 2KOH \longrightarrow CH_3CH\begin{matrix}OH \\ OH\end{matrix} \xrightarrow{-H_2O} CH_3CHO + 2KCl$$

Ethylidene chloride

(2)

$$\begin{matrix}CH_2Cl \\ | \\ CH_2Cl\end{matrix} + \text{alc. } 2KOH \longrightarrow \begin{matrix}C-H \\ ||| \\ C-H\end{matrix} + 2KCl + 2H_2O$$

$$CH_3CHCl_2 + \text{alc. } 2KOH \longrightarrow CH\equiv CH + 2KCl + 2H_2O$$

10.1.3 Trihalogen Derivatives

These include important compounds like chloroform, iodoform, Westron, etc.

Chloroform (Trichloromethane, $CHCl_3$)

(a) ***Preparation***

(1) By heating bleaching powder, ethyl alcohol, and water together. The reactions are as follows:

(i) $CaOCl_2 \xrightarrow{\Delta} CaO + Cl_2$

(ii) $CaO + HOH \rightarrow Ca(OH)_2$

(iii) $CH_3CH_2OH + Cl_2 \longrightarrow CH_3CHO + 2HCl$

(Alcohol is oxidized by chlorine in above reaction)

(iv) $CH_3CHO + 3Cl_2 \rightarrow \underset{\text{Chloral}}{CCl_3CHO} + 3HCl$

(v)

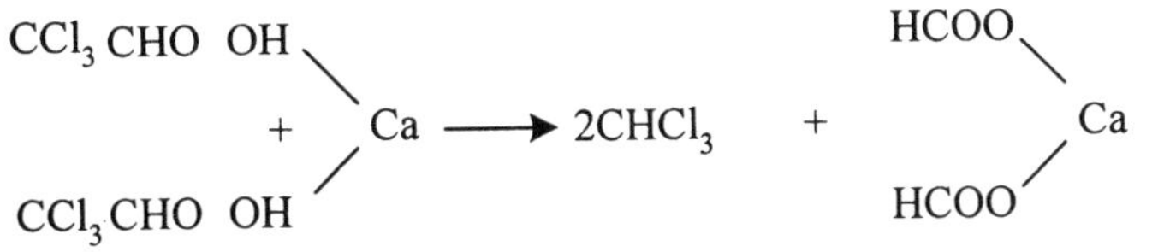

Chloroform Calcium formate

(2) By heating acetone, bleaching powder, and water together. Steps (i) and (ii) are same as above and reset are as follows:

(i)

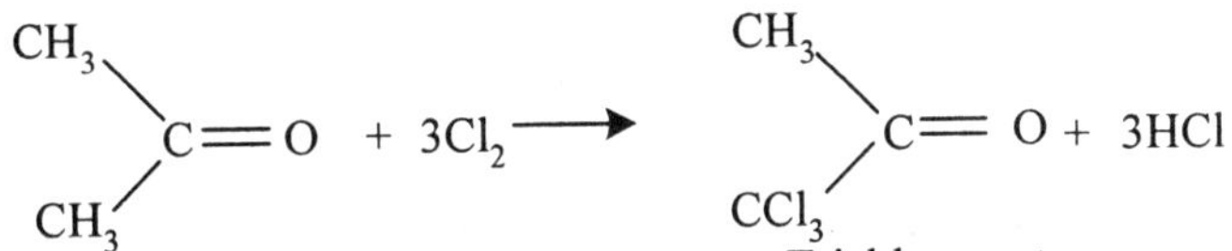

Trichloroacetone

(ii)

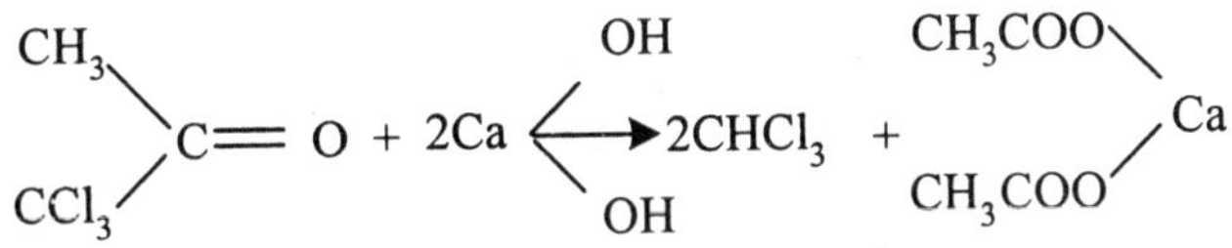

Chloroform Calcium acetate

(3) Laboratory method: By heating chloral hydrate with NaOH:

$$\underset{\text{Chloral hydrate}}{CCl_3CHO.H_2O} \xrightarrow{\text{NaOH}} CHCl_3 + HCOONa + H_2O$$

(4) Commercial method

$$CCl_4 + 2[H] \xrightarrow{Fe/H_2O} CHCl_3 + HCl$$

(b) *Properties*

Chemical: *(i)* *Oxidation*: Chloroform is oxidized in the presence of air and sunlight to form a poisonous gas, Phosgene.

$$CHCl_3 + \underset{\text{air}}{[O]} \xrightarrow{\text{Sunlight}} \underset{\text{Phosgene}}{COCl_2} + HCl$$

Therefore, following precautions are taken while storing the chloroform for medical purpose.

(a) The b ottle is filled up to the neck to avoid air.

(b) The chlor oform is kept in dark bottle in dark places to avoid sunlight.

(c) One percent ethyl alcohol is added, which converts harmful phosgene gas into harmless ethyl carbonate.

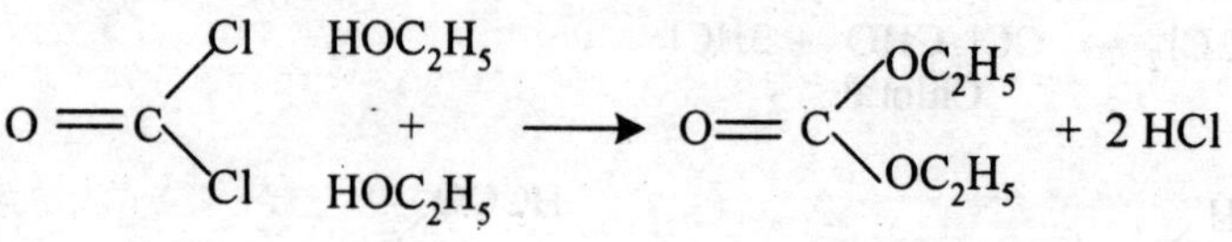

Phosgene Ethyl carbonate

(ii) *Hydrolysis*

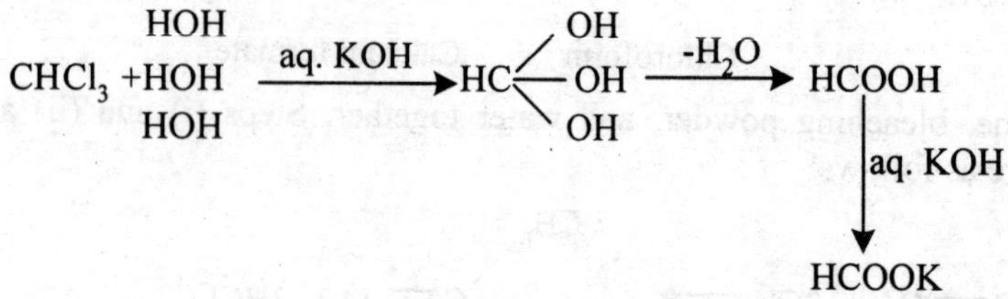

Potassium formate

Note that *4 molecules of aq. KOH are consumed and potassium formate is formed.*

(iii) *Reaction with nitric acid*

$$\underset{}{CHCl_3} + \underset{\text{Nitric acid}}{HONO_2} \rightarrow \underset{\text{Chloropicrin}}{CCl_3NO_2} + H_2O$$

(iv) *Reaction with acetone*

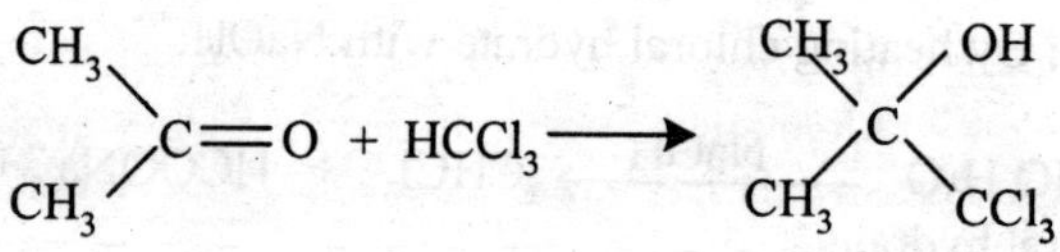

Chloretone (Hypnotics)

(v) *Reaction with silver*:

$$2CHCl_3 + 6Ag \rightarrow \underset{\text{Acetylene}}{CH \equiv CH} + 6AgCl$$

(vi) *Carbylamine reaction*: When primary amine is treated with $CHCl_3$ and KOH, it forms phenyl isocyanide, having bad smell:

$$RNH_2 + CHCl_3 + 3KOH \rightarrow \underset{\text{Alkyl isocyanide}}{RNC} + 3KCl + 3H_2O$$

(vii) *Reimer-Tiemann reaction*: When phenol reacts with $CHCl_3$ and KOH, it forms salicylic aldehyde as main product:

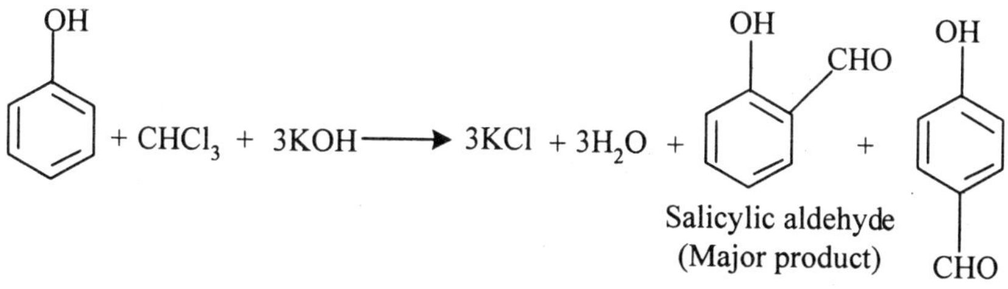

10.1.4 Tetrahalogen Derivatives

Some examples of these compounds are as follows:

CF_2Cl_2	$CF_2— CF_{2n}$	$CHCl$
Freon	Teflon	‖
		CCl_2
		Westrosol

Freons have general formula $CF_x\ Cl_y$. Where $X + Y = 4$

10.2 ALCOHOLS

Alcohols are compounds whose molecules have a hydroxyl group attached to a saturated carbon atom.[4] Alcohols have been classified into three categories, viz., Monohydric alcohols, Dihydric alcohols and Trihydric alcohols, where one, two, and three hydroxyl groups(s) are present respectively.

CH_3OH	CH_2OH	CH_2OH
Methanol	|	|
	CH_2OH	$CHOH$
	Glycol	|
		CH_2OH
		Glycerol
(Monohydric Alcohol)	(Dihydric Alcohol)	(Trihydric Alcohol)

(a) Preparation

(i) From Grignard reagent. See Chapter 6.
(ii) From alkenes. See Chapter 9.
(iii) Hydrolysis of alkyl halide. See Section 10.1.
(iv) By reduction of carbonyl compounds. See Section 10.4.
(v) Industrial preparation of methanol.

---

[4] f—OH is attached to an unsaturated carbon atom,i.e., (C=C), they are called enols.

$$CO + 2H_2 \xrightarrow[200-300\,\text{atm.}]{300-400^\circ C} CH_3OH$$
$$CHO + Cr_3O_3$$

(vi) Industrial preparation of ethanol

Molasses containing:

(i) Yeast diluted to 10%;

(ii) $C_6H_{12}O_6$, cane sugar.

$$C_6H_{12}O_6 \xrightarrow{\text{Zymase enzyme}} 2C_2H_5OH + 2CO_2$$

(vii) $$\underset{\text{alkyl boride}}{R_3B} + CO \longrightarrow R_2-B-\overset{\overset{O}{\|}}{C}-R \xrightarrow{LiBH_4} R_2BCH\genfrac{}{}{0pt}{}{\diagup OH}{\diagdown R} \xrightarrow[H_2O]{OH^-} RCH_2OH$$

(b) Properties

Physical properties: Alcohols are liquids having high boiling point (as compared to ether) due to intermolecular H-bonding.

Chemical properties: Chemical properties of alcohols can be discussed under following three categories:

(i) Reaction in which —O—H bond ruptures;

(ii) Reaction in with $-\overset{|}{\underset{|}{C}}-OH$ bond ruptures; and

(iii) Miscellaneous reactions.

(i) Reactions involving fission of —O—H bond Reactivity of alcohols CH_3OH > Pri > Sec > Tert. Following reactions are the examples of this category:

(1) Reaction with active metals

$$\underset{\text{1 mole}}{RO{-}H + M} \rightarrow \underset{\text{Metal alkoxide}}{ROM} + \underset{\text{0.5 mole}}{1/2\,H_2}$$

where, M = Na, K Mg, Al

$$\text{e.g., } CH_3OH + Na \longrightarrow \underset{\text{Sodium methoxide}}{CH_3ONa} + 1/2H_2$$

This reaction is of significance because it shows weak acidic nature of alcohols. The relative acidity and basicity is as follows:

Relative acidity

$$H_2O > ROH > RC \equiv CH > NH_3 > H_2 > RH$$

Relative basicity

$$OH^- < RO^- < RC \equiv C^- < NH_2^- < H^- < R^-$$

(2) *Ester formation*: Alcohol reacts with acid to form ester. Reaction is called esterification

$$R-\overset{\overset{\displaystyle O}{\|}}{C}-OH + HOR' \xrightarrow{H^+} \underset{\text{ester}}{R-\overset{\overset{\displaystyle O}{\|}}{C}-OR'}$$

(ii) *Reaction involving breaking of C—O bond*: The reactions are as follows:

Reactivity order of alcohol: allyl > benzyl > tert > sec > pri. Reaction generally involves carbonium ion as reaction intermediate, therefore, carbonium ion may rearrange.[5]

(1) *Reaction with hydrogen halides*

$$R-OH + HX \xrightarrow{ZnCl_2} RX + H_2O$$

Reaction order: HI > HBr > HCl > HF (generally not reactive). The reason is lowest bond association energy for HI and highest for HF (See Peroxide effect in Chapter 9). HX can be generated *in situ* by the reaction of sodium halide with H_2SO_4. The reaction is acid-catalyzed. The function of the acid is to help the formation of a carbonium ion in case of secondary and tertiary alcohols. However, with primary alcohols the function of the acid is to produce as substrate in which the leaving group is a weakly basic water molecule rather than a strongly basic hydroxide ion ($-OH^-$). It is important to point out that hydrogen chloride does not react with primary or secondary alcohol unless $ZnCl_2$ or similar Lewis acid is used. Zinc chloride forms a complex with the alcohol. This provides a better leaving group for the reaction than water.

It is further important to note that reactions of the following type do not occur to any appreciable extent

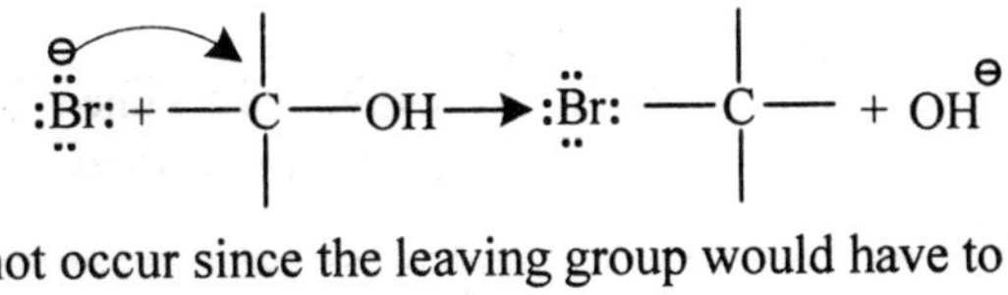

does not occur

They do not occur since the leaving group would have to be a strongly basic hydroxide ion.

(2) *Reaction with* PCl_5, PBr_3 *or* $SOCl_2$: The reaction of an alcohol with PBr_3, PCl_5 or $SOCl_2$ does not involve the formation of a carbonium ion, and normally takes place without rearrangement of the carbon skeleton. Hence, these reagents are preferred for the conversion of an alcohol to the corresponding alkyl halide.

$$R-OH + PCl_5 \rightarrow RCl + POCl_3 + HCl$$
$$3ROH + PBr_3 \rightarrow 3RBr + H_3PO_3$$
$$ROH + SOCl_2 \rightarrow RCl + HCl + SO_2$$

[5] See Chapter 4 for carbonium ions.

(3) *Dehydration of alcohol*: Alcohols on dehydration[6] yield alkenes, e.g.,

$$RCH_2CH_2OH \xrightarrow{H^+} RCH{=}CH_2$$

(4) *Reaction with hot-reduced copper*: The primary alcohol forms aldehyde and secondary alcohol forms ketone. However tertiary alcohol forms alkene.

$$RCH_2OH \xrightarrow{Cu,\ 300\ ^\circ C} RCHO$$

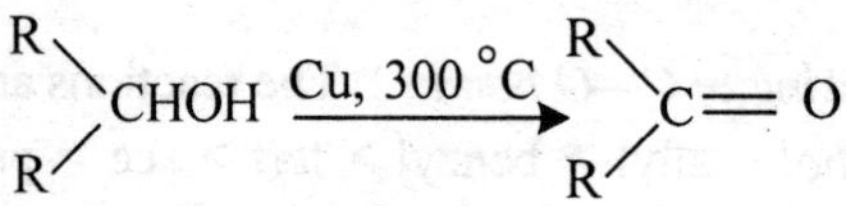

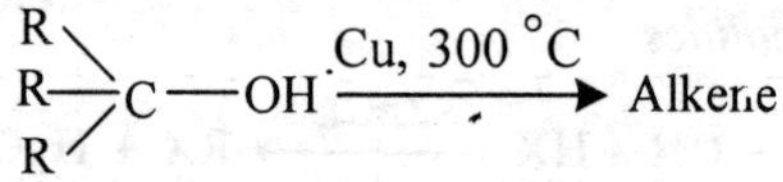

e.g.,

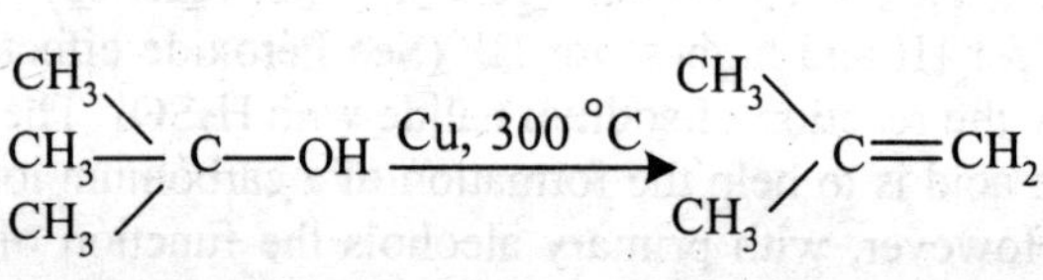

Tertiary butyl alcohol 2-Methyl propene

(iii) *Miscellaneous properties*

(1) *Oxidation*: **Number of alpha (α) H atom ∝ oxidation rate ∝ product yield.** Primary alcohol, containing 2 αH atoms undergo oxidation faster and the yield of aldehyde is greater than that of the ketone obtained from oxidation of secondary alcohol (1 αH atom). A tertiary alcohol does not contain any α atom and hence, is not oxidized easily, e.g.,

$$RCH_2OH + [O] \longrightarrow RCHO \xrightarrow{[O]} RCOOH$$

Primary alcohol — acid with the same no. of carbon atoms

e.g.,

$$CH_3CH_2OH + [O] \longrightarrow CH_3CHO \xrightarrow{[O]} CH_3COOH$$

Ethyl alcohol — Acetaldehyde — Acetic acid

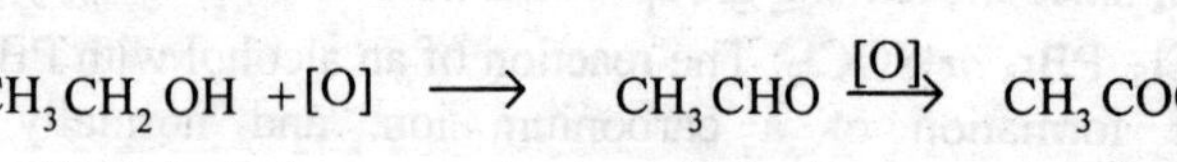

Secondary alcohol — acid with lesser no. of carbon atoms.

[6] For details, see Chapter 5.

e.g.,

$$(CH_3)_2CHOH + [O] \rightarrow (CH_3)_2CO \xrightarrow{[O]} CH_3COOH$$

Isopropyl alcohol → Acetone → Acetic acid

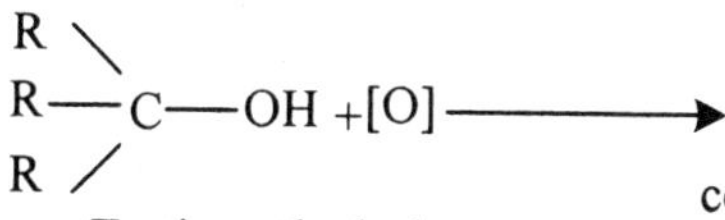

does not occur under ordinary conditions, but under drastic conditons gives ketones and acids

Tertiary alcohol

i.e.,

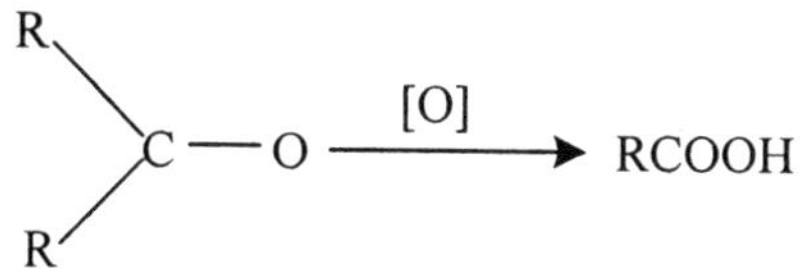

$$(CH_3)_3COH + [O] \longrightarrow CH_3COCH_3 + HCHO$$

Tertiary butyl alcohol → Acetone + Formaldehyde

$$\downarrow [O] \qquad \downarrow [O]$$

$$CH_3COOH + HCOOH$$

Acetic Acid + Formic Acid

(2) *Iodoform Reaction*: Compounds containing following structure, when treated with iodine and alkali (NaOH or Na_2CO_3) form Iodoform (Yellow solid, melting point, 119°C).

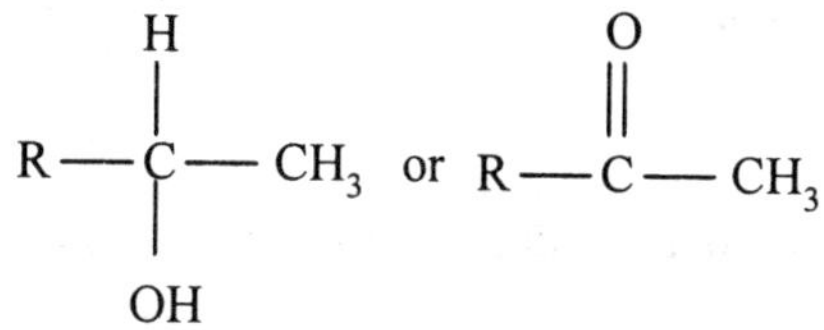

where R = H, alkyl, aryl group

All Alkanol-2 and methylketones gives Iodoform test.

Compound giving positive Iodoform test	*Compounds giving negative Iodoform test*
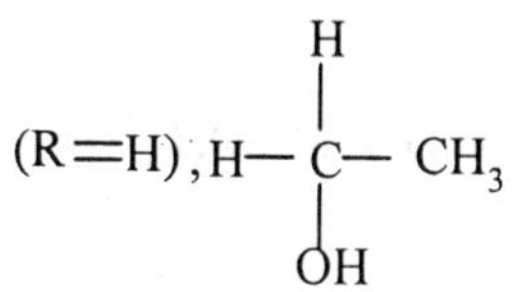 Ethanol	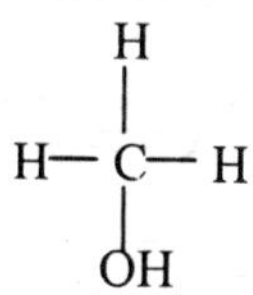 Methanol Note the absence of CH_3 group

$(R{=}CH_3)$, $CH_3-\underset{OH}{\overset{CH_3}{C}}-H$ Propanol-2	$CH_3CH_2-\underset{OH}{\overset{H}{C}}-H$ Propanol-1
$(R{=}H)$, $H-\overset{O}{\overset{\Vert}{C}}-CH_3$ Acetaldehyde	$H-\overset{O}{\overset{\Vert}{C}}-H$ Formaldehyde
$(R{=}CH_3)$, $CH_3-\overset{O}{\overset{\Vert}{C}}-CH_3$ Propanone	$C_2H_5-\overset{O}{\overset{\Vert}{C}}-C_2H_5$ Butanone-3
$(R{=}C_6H_5)$, $C_6H_5-\overset{O}{\overset{\Vert}{C}}-CH_3$ Acetophenone	$C_6H_5-\overset{O}{\overset{\Vert}{C}}-C_6H_5$ Benzophenone

The reaction involves oxidation and halogenation. The various reaction steps are as follows:

$$R-\underset{CH_3}{\overset{H}{C}}-OH + NaOI \longrightarrow R-\underset{CH_3}{\overset{O}{\overset{\Vert}{C}}} + NaI + H_2O$$

$$\begin{matrix} R \\ CH_3 \end{matrix}\!\!>C{=}O + NaOI \longrightarrow \begin{matrix} R \\ CI_3 \end{matrix}\!\!>C{=}O + 3NaOH$$

$$\begin{matrix} R \\ CI_3 \end{matrix}\!\!>C{=}O + NaOH \longrightarrow RCOONa + \underset{\substack{\text{Iodoform} \\ \text{m.p. } 119^\circ C \\ \text{(yellow ppt.)}}}{CHI_3}$$

(3) Reaction with Lucas reagent (Anhy. $ZnCl_2$ + conc. HCl):

Alcohols (of not more than six carbons) are soluble in the Lucas reagent; the corresponding alkyl chlorides are insoluble. Formation of a *chloride* from an alcohol is indicated by the cloudiness that appears when the chloride separates. It is chosen as a criterion for identification of alcohols.

- Allyl or tertiary alcohol + Lucas Reagent → White turbidity *at once*
- Secondary alcohol + Lucas Reagent → White turbidity *within 5 minutes*
- Primary alcohol + Lucas reagent → *No Turbidity*

Distinction between primary, secondary and tertiary alcohols Following methods have been used:

(1) Oxidation method
(2) Lucas reagent
(3) Iodoform method
(4) Victor = Meyer Test

} See above

(4) *Victor-Meyer Test*:

(i)

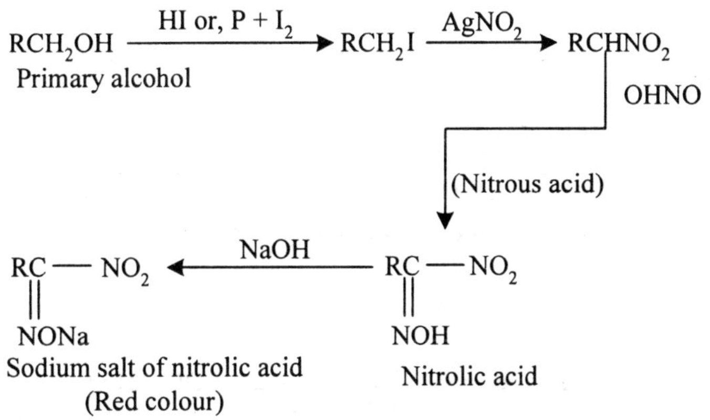

(ii)

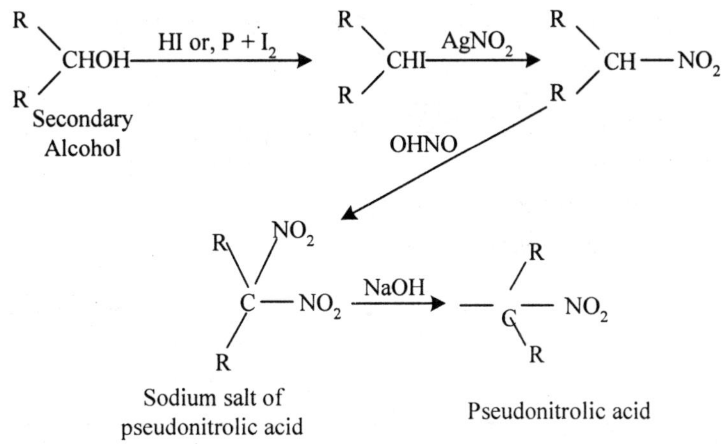

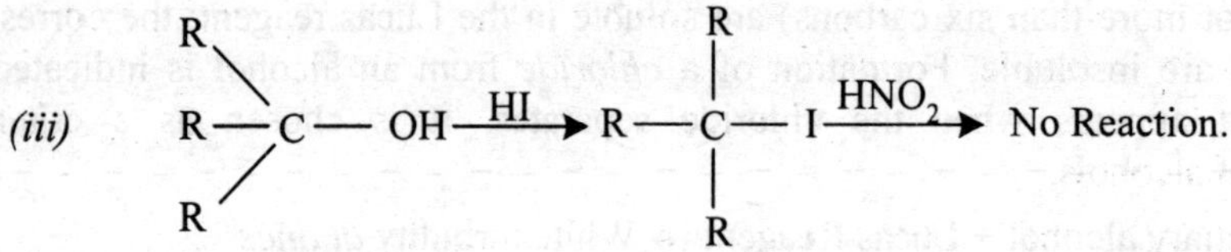

10.3 ETHERS

Ethers are organic compounds in which two alkyl groups are attached together through oxygen atom. They have general formula ROR′ (where R and R′ are alkyl groups). If R=R′, it is a simple ether; and if R ≠ R′, it is a mixed ether. Ethers show metamerism. The structure of an ether is

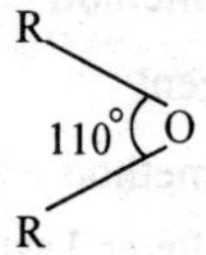

(a) Preparation

(i) Williamson's synthesis: It is a good method for the preparation of ethers, both simple as well as mixed. The reaction is as follows:

$$\underset{\text{Sodium alkoxide}}{R'ONa} + RX \rightarrow ROR' + NaX$$
$$(X = Br, I, R'OSO_2)$$

For primary alkyl halides it is SN2 mechanism; for secondary and tertiary alkyl halides, it is SN1 mechanism.

(ii) By intermolecular dehydration of alcohols:

$$R-OH + HO-R \xrightarrow[H_2SO_4,\ 180\ ^\circ C]{H^+} ROR + H_2O$$

e.g., $$\underset{\text{Ethyl alcohol}}{CH_3CH_2OH} \longrightarrow CH_2 = CH_2 + H_2O$$

e.g., $$\underset{\text{Ethyl alcohol}}{CH_3CH_2OH} \xrightarrow{H_2SO_4, 140^\circ C} \underset{\text{Diethyl ether}}{CH_3CH_2OCH_2CH_3} + H_2O$$

Secondary alcohols generally do not form ether.

(iii) *Solvomercuration–Demercuration Reaction:*

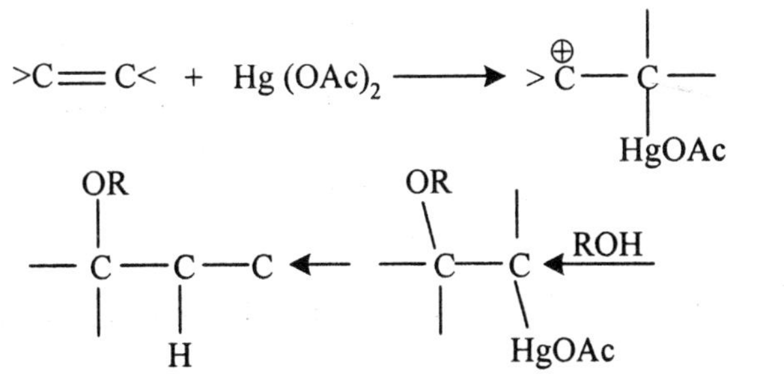

Ether Alkoxy alkyl mercuric acetate

(iv) By passing alcohol vapours over alumina:

$$C_2H_5OH + HOC_2H_5 \xrightarrow[250\ °C]{Al_2O_3} C_2H_5OC_2H_5 + H_2O$$

(v) By heating alkyl halides with dry silver oxide:

$$2CH_3CH_2Br + Ag_2O \rightarrow C_2H_5OC_2H_5 + 2AgBr$$

(vi) By reaction of alcohol with diazomethane:

$$CH_3CH_2OH + CH_2N_2 \xrightarrow{HB_4} CH_3CH_2OCH_3 + N_2$$

(b) Properties

Physical:

(1) Dimethyl ether and methylethyl ether are gases. Rest are sweet-smelling liquids with low boiling point in comparison to isomeric alcohol.

(2) They are slightly soluble in water because of formation of hydrogen bonding (less extent) with water.

Chemical: Ethers are chemically inert due to absence of functional group. Therefore, these are frequently used as inert solvent in the organic synthesis. Some important reactions are as follows:

(1) Reaction with HI

(a) $ROR' + HI \rightarrow ROH + R'I$

(where R > R′)

e.g., $C_2H_5OCH_3 + HI \rightarrow C_2H_5OH + CH_3I$

(b) $ROR' + \underset{\text{hot}}{2HI} \rightarrow RI + R'I + H_2O$

The latter forms the basis of *Zeisel method* for the estimation of —OCH_3 (methoxy) groups. The compound containing methoxy group is heated with HI, when the methoxy group is converted into CH_3I. It is digested with $AgNO_3$ when yellow ppt. of AgI is obtained, from which the number of methoxy group can be calculated. The reaction can be written as:

$$\underset{\text{1 mole}}{—OCH_3} \xrightarrow{HI} \underset{\text{1 mole}}{CH_3I} \xrightarrow{AgNO_3} \underset{\text{1 mole}}{AgI}$$

(2) Like alkanes, ethers also undergo substitution reaction. For example:

$$CH_3CH_2OCH_2CH_3 \xrightarrow[\text{dark}]{Cl_2} CH_3CH_2OCH(Cl)CH_3 \xrightarrow[\text{dark}]{Cl_2} CH_3CH(Cl)OCH(Cl)CH_3$$

α-chlorodiethyl ether   αα′-Dichloro diethyl ether

(3) Unlike alkanes, ethers react with cold concentrated HCl or H_2SO_4 to form oxonium salt.

$$R—OR' + H_2SO_4 \longrightarrow [R—\overset{\oplus}{\underset{..}{O}}(H)—R]HSO_4^{\ominus}$$

e.g., $$CH_3CH_2\ddot{O}CH_2CH_3 + H_2SO_4 \longrightarrow [CH_3CH_2\ddot{O}(H)CH_2CH_3]HSO_4^{\oplus}$$

Oxonium salt of diethyl ether

(4) Ethers, like alcohols, react with PCl_5 to form alkyl halides.

$$R—OR' + PCl_5 \xrightarrow{\Delta} RCl + R'Cl + POCl_3$$

e.g., $$C_2H_5OC_2H_5 + PCl_5 \xrightarrow{\Delta} 2C_2H_5Cl + POCl_3$$

Ethyl chloride

(5) When ethers are heated with dil. H_2SO_4, alcohols are obtained.

$$ROR + H_2O \xrightarrow{H^{\oplus}} 2ROH$$

e.g., $$C_2H_5OC_2H_5 + H_2O \xrightarrow{H} 2C_2H_5OH$$

(6) Ethers react with acetyl chloride in presence of anhydrous zinc chloride to form alkyl halide and ester.

$$ROR + CH_3COCl \xrightarrow[\Delta]{ZnCl_2} RCl + CH_3COOR$$

e.g., $$CH_3OCH_3 + CH_3COCl \xrightarrow[\Delta]{ZnCl_2} C_2H_5Cl + CH_3COOC_2H_5$$

Ethyl chloride   Ethyl acetate

Epoxides: Epoxides are cyclic ethers with three membered rings.

(c) Preparation

(i)

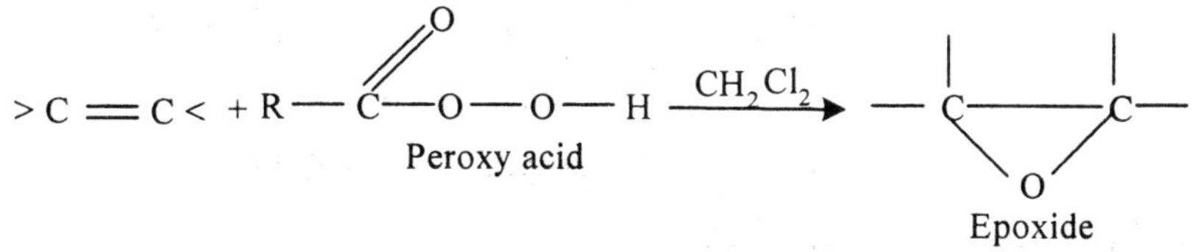

R = C_6H_5, Perbenzoic acid
= CH_3, Peracetic acid

(ii)

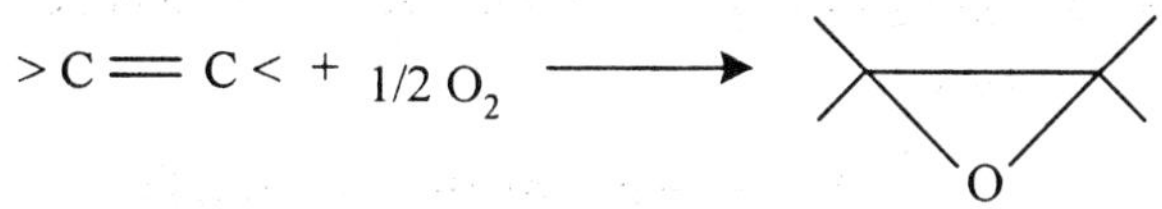

(d) Properties

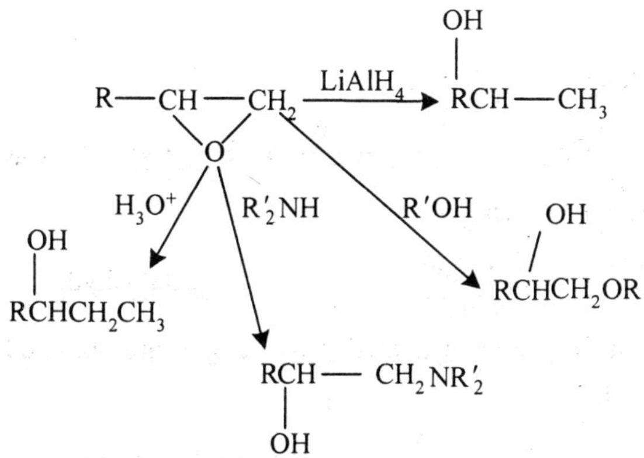

(Where R′ = H, acyl group)

(e) Crown ether

The Crown ether is a heterocyclic polyether, usually minimum with four oxygen atoms. The first of two numbers in the name of Crown ether is the total number of atoms in the ring, the second number of oxygen atoms e.g.,

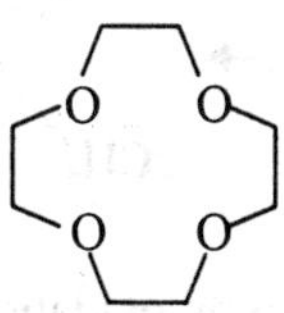

12-Crown 4

10.4 ALDEHYDES AND KETONES

Aldehydes are organic compounds in which at least one H-atom is attached to a carbonyl group (> C = O). In ketones, the carbonyl group is situated between two carbon atoms (alkyl or aryl groups).

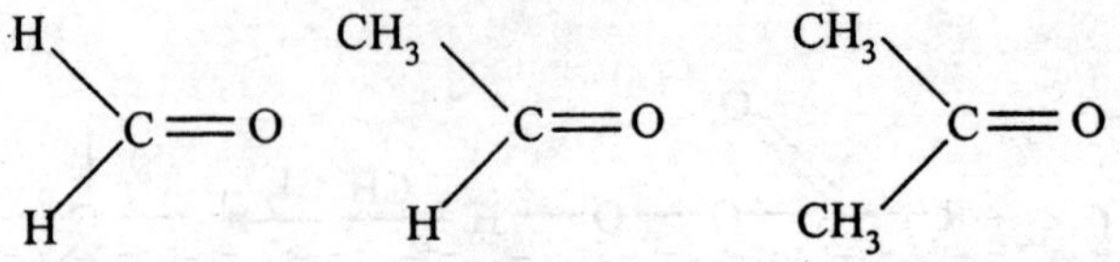

Formaldehyde Acetaldehyde Acetone

(a) Preparation of aldehydes and ketones

Aldehydes and ketones can be prepared by following methods:

(i) Aldehydes can be prepared by heating calcium salt of fatty acid with calcium formate.

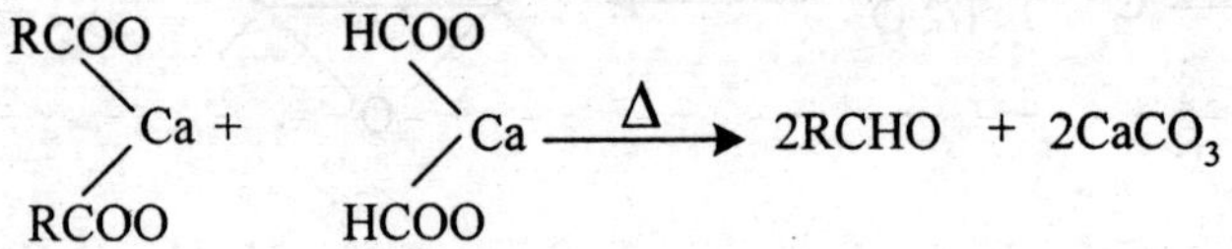

e.g.,

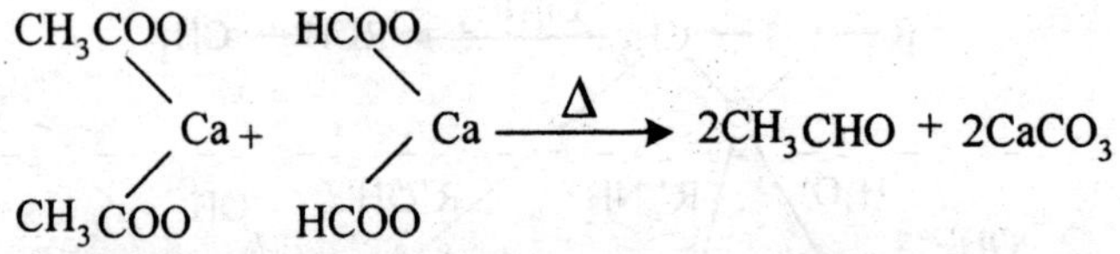

Calcium acetate Calcium formate acetaldehyde

(ii) When calcium salt of fatty acids is heated, only ketones are obtained.

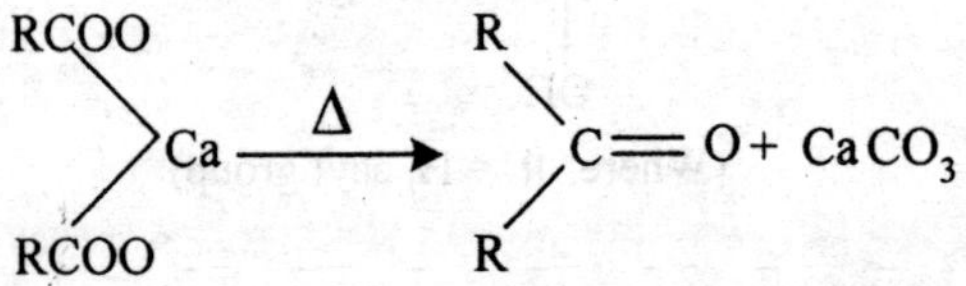

e.g.,

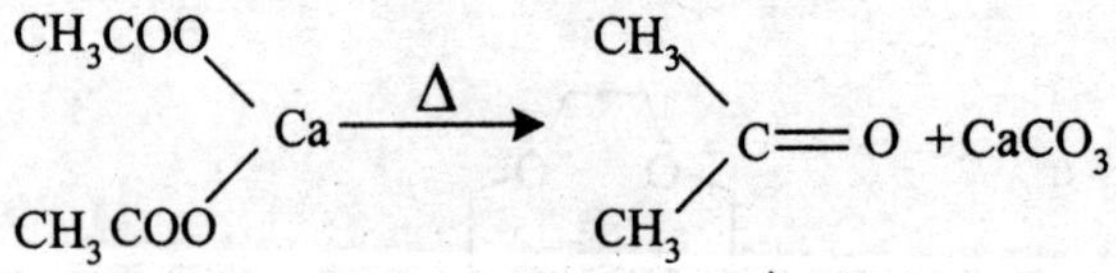

Acetone

(iii) Aldehydes and ketones can be prepared by oxidation of primary alcohols and secondary alcohols, respectively. Pyridiniumchlorochromate in dichloromethane K2Cr2O$_7$ or H_2SO_4 are used as oxidizing agents.

(a) $RCH_2OH + [O] \rightarrow RCHO + H_2O$
Primary alcohol

e.g.,

$$\underset{\text{Ethylalcohol}}{CH_3CH_2OH} + [O] \rightarrow \underset{\text{Acetaldehyde}}{CH_3CHO} + H_2O$$

(b)

$$\underset{\text{Secondary alcohol}}{R_2CHOH} + [O] \longrightarrow \underset{\text{Ketone}}{R_2C{=}O} + H_2O$$

e.g.,

$$\underset{\text{Isopropyl alcohol}}{(CH_3)_2CHOH} + [O] \longrightarrow \underset{\text{Acetone}}{(CH_3)_2C{=}O} + H_2O$$

(iv) Aldehydes and ketones can be prepared from Grignard reagent (see Chapter 6).

(v) Only aldehydes can be prepared by reaction of acid chlorides with water using lithium-tert-butoxyaliuminium-hydride. LiAlH $[OT\text{-}Bu]_3$ in ether at 78°C.

$$R{-}\overset{O}{\overset{\|}{C}}{-}Cl + H_2O \xrightarrow[\text{in ether at } 78^{\circ}C]{LiAlH+[OT\text{-}Bu]_3} RCHO$$

Ketones can be prepared by reaction of acid chloride with lithium dialkyl cuperate in ether at −78°C

$$R'_2\,CuLi + RCOCl \rightarrow R{-}COR'$$

(vi) Only aldehydes can be obtained by reaction of ester with diisobutylaluminiumhydride $(Me_2CHCH_2)_2$ AlH

$$RCOOR' + H_2O \xrightarrow[\text{at } -78\,^{\circ}C]{(Me_2CHCH_2)_2\text{ AlH in hexane}} RCHO$$

(vii) Both aldehydes and ketones can be obtained from alkynes by reaction with $HgSO_4$ and H_2SO_4:

$$H{-}C{\equiv}CH + H_2O \xrightarrow{1\%\ HgSO_4,\ H_2SO_4} CH_3CHO$$

$$CH_3{-}C{\equiv}CH + H_2O \xrightarrow{1\%\ HgSO_4,\ H_2SO_4} CH_3\,COCH_3$$

However, when alkynes are treated with disiamyl borane

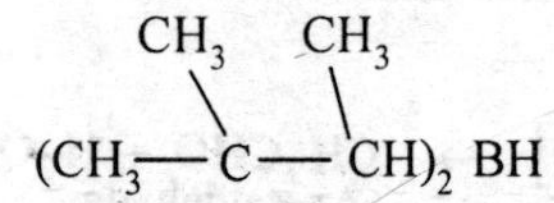

(Siamyl is common name for the 2,3,-dimethylpropyl group)

$$RC{\equiv}CH \xrightarrow[\text{ether } H_2O_2 \text{ and } OH^-]{\text{Disiamyl borane in}} RCHO$$

(viii) Both aldehyles and ketones can be prepared by ozonolysis of alkenes (see Chapter 9).

(ix) The industrial process for preparation of only aldehydes is known as oxoreaction or hydroformylation. The reaction is as follows:

$$RCH{=}CH_2 + CO + H_2 \xrightarrow[\text{2000 psi (pressure), 1000°C}]{\text{Cobalt carbonyl [HCo (CO)}_4]} RCH_2CH_2CHO + \text{Isomers}$$

(x) Carboxylic acids can be converted only into ketones by reaction with alkyl lithium. The reaction is as follows:

$$R{-}\overset{O}{\overset{\|}{C}}{-}OH + 2R'Li \longrightarrow R{-}\overset{O}{\overset{\|}{C}}{-}R' + R'H + 2LiOH$$

(b) Properties

Physical properties: They are all liquids with high boiling point than non-polar compounds of same molecular weight. However, they have lower boiling points than alcohols and carboxylic acids of same molecular weight, due to absence of hydrogen bonding. Lower members are soluble in water due to hydrogen bonding.

Chemical properties: Chemical properties can be studied under following points:

- *(i)* Nucleophilic addition to the carbon-oxygen double bond.
- *(ii)* Removal of oxygen atom in the form of water.
- *(iii)* Miscellaneous reactions.

(i) *Nucleophilic addition reactions* (common to both aldehydes and ketones)

1.

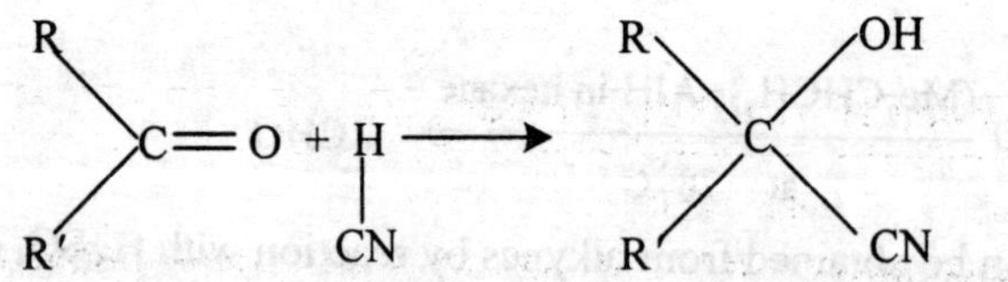

Cyanohydrin addition product

2. 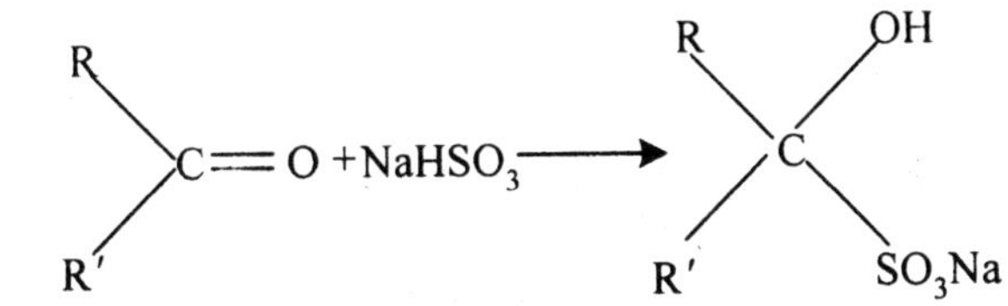

Sodium bisulphite addition product (white ppt.)

This reaction is used to identify carbonyl compounds.

3. 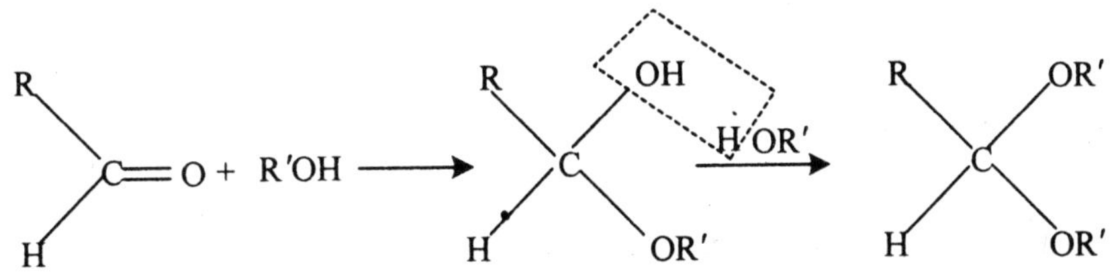

Aldehyde Hemiacetal Acetal

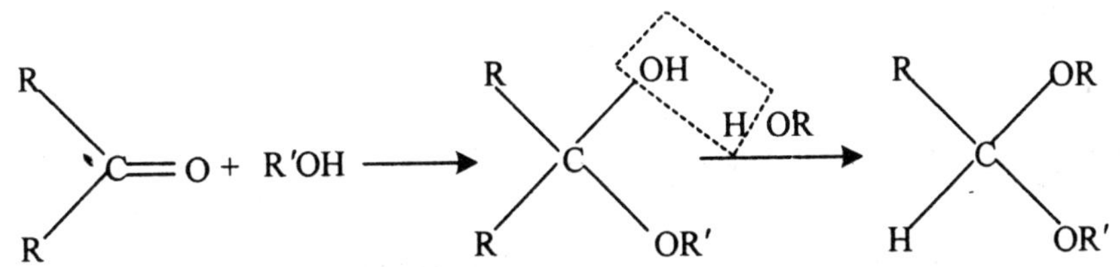

Ketone Hemiacetal Ketal

If we place the acetal in water and add a small amount of acid, all the steps are reversed. Thus, the acetal undergoes hydrolysis:

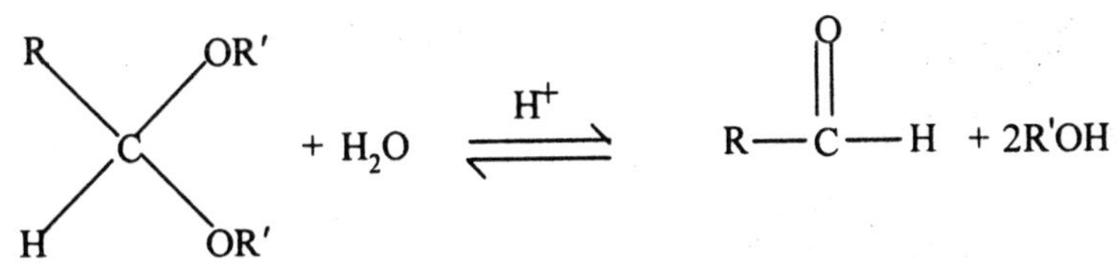

(ii) *Removal of oxygen atom of carbonyl group as water:* General reaction is as follows:

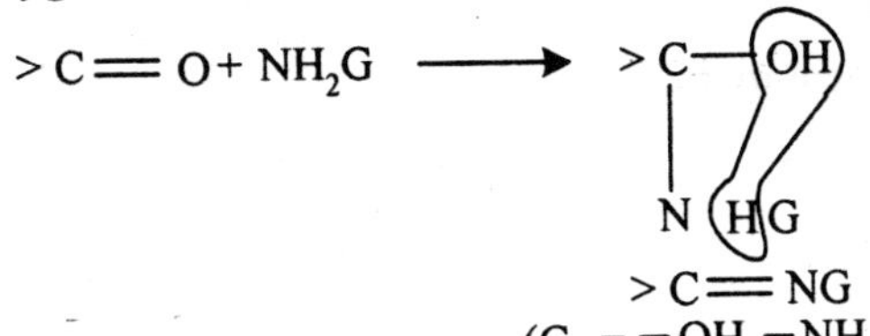

(G $= -OH, -NH_2, -NHC_6H_5$)

Note: first step is addition reaction and second step is elimination reaction.

e.g.,

1. 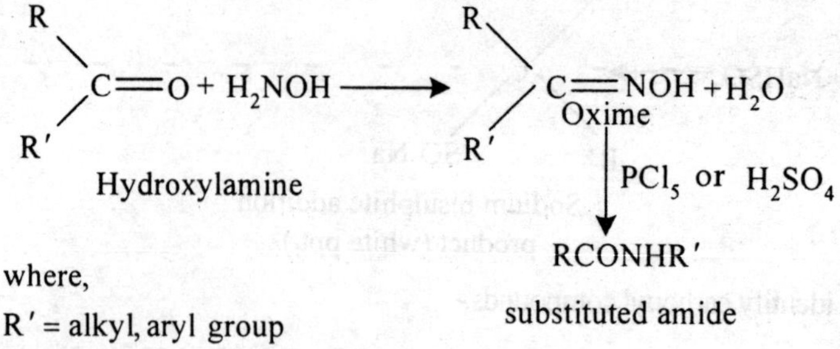

where,
R′ = alkyl, aryl group
R = alkyl, aryl group, H

This is called Beckmann rearrangement

2. 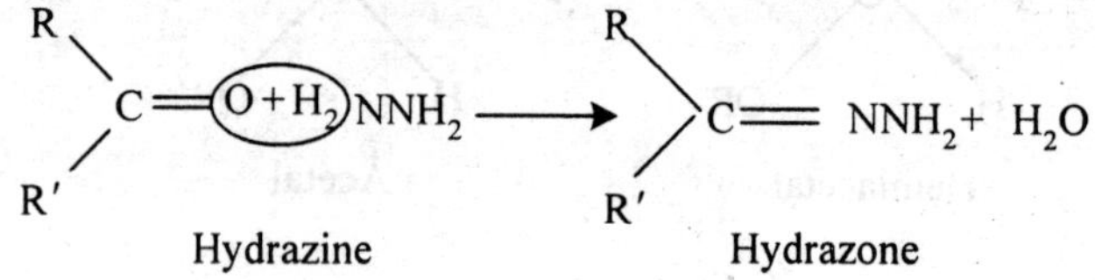

3. 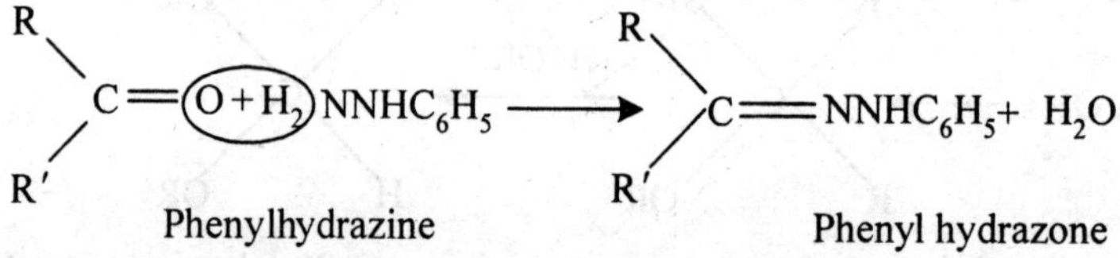

4. 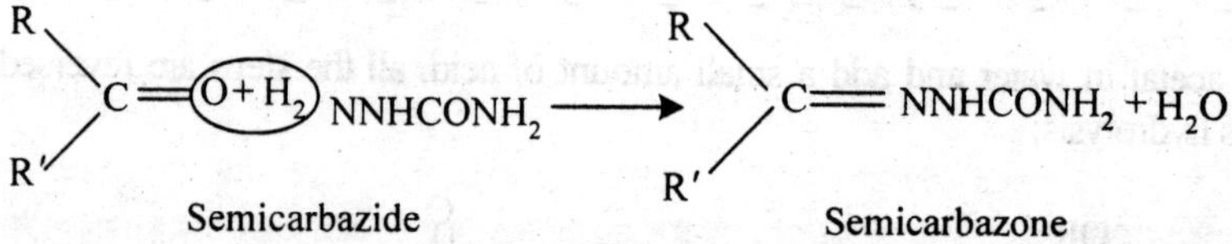

(iii) *Miscellaneous reactions*

1. 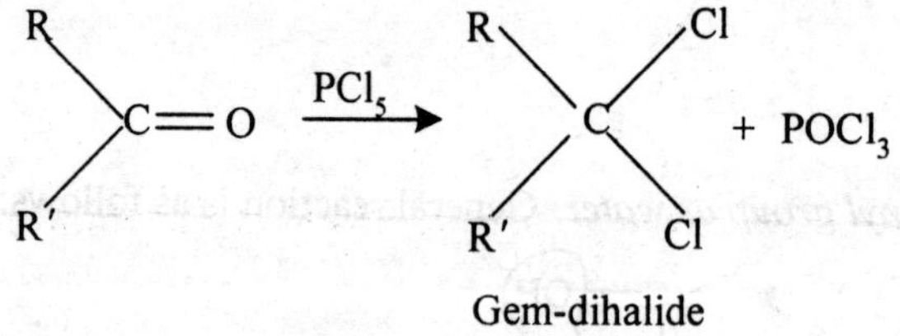

2. Addition of NH_3: Reactions of NH_3 with aldehydes and ketones are as follows:

(1) $$\underset{\text{Formaldehyde}}{6CH_2O} + 4NH_3 \longrightarrow \underset{\text{Hexamethylene tetramine or Urotropine (orange colour solid)}}{(CH_2)_6N_4} + 6H_2O$$

(2)

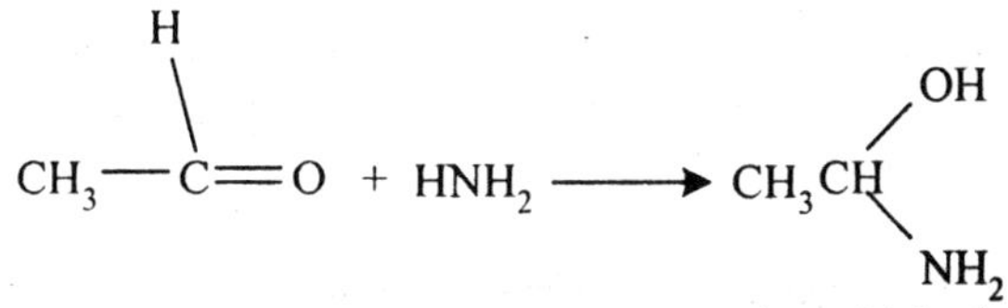

Acetaldehyde ammonia

(3)

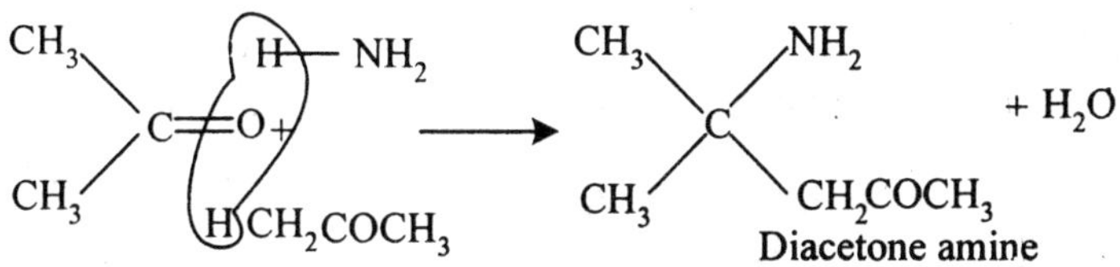

Diacetone amine

(4)

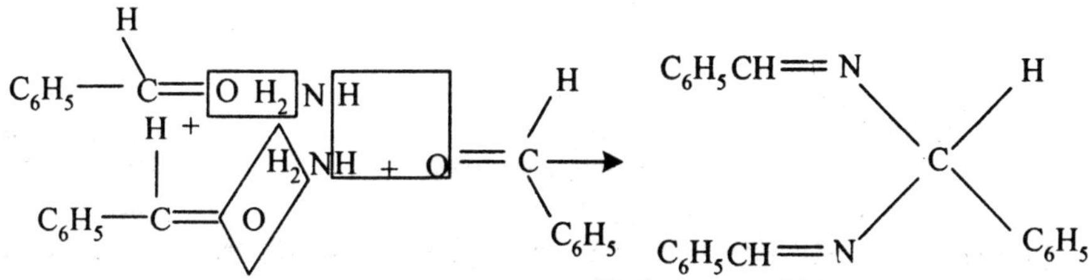

Hydrobenzamide

(5) *Reaction with alkali*:

(i) *Aldehydes with α-H atom(s)*: Aldehydes like ethanal, propanal undergo aldol condensation (for details, see Chapter 5).

(ii) *Aldehydes without α-H atom: Cannizzaro reaction:* Aldehydes which *have no α-hydrogen atom* when treated with alkali undergo *cannizzaro reaction* in which two molecules of aldehyde react, one is oxidized to form sodium or potassium salt of fatty acid and other is reduced to primary alcohol.

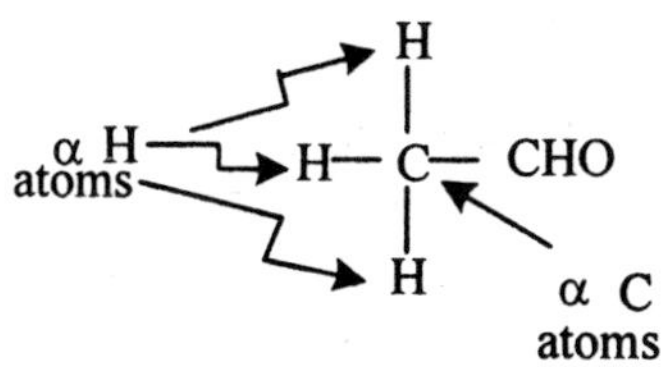

Ethanal
Aldehyde with
α Hydrogen atom

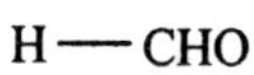

Methanal
Aldehyde without
α Carbon atom so
α Hydrogen atom
is absent

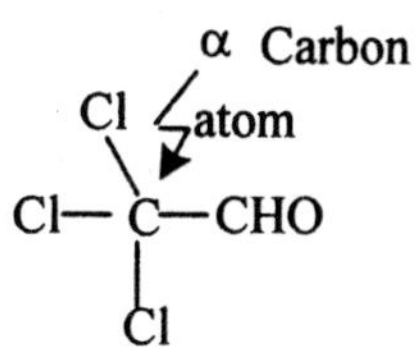

Trichloro ethanal
Aldehyde with
α Carbon atom but
without α Hydrogen atom

In general, the reaction may be represented as below:

$$\underset{}{RCHO} \xrightarrow{NaOH} \underset{\text{Sodium salt of fatty acid}}{RCOONa} + \underset{\text{Primary alcohol}}{RCH_2OH}$$

The Cannizzaro reaction of various aldehydes with alkali and reaction products are given in Table 10.1. Mechanism involves hydride ion ($\overset{\ominus}{H}$).

Table 10.1 Reaction products of the Cannizzaro reaction of various aldehydes

Value of R	*Aldehyde*	*Product*		
R=H,	Formaldehyde	HCOONa Sodium formate	+	HCH_2OH Methanol
R=$(CH_3)_3C$	Trimethyl acetaldehyde	$(CH_3)_3$ C.COONa Trimethyl Sodium acetate	+	$(CH_3)_3$ $C.CH_2OH$ Trimethyl ethyl alcohol
R=C_6H_5	Benzaldehyde	C_6H_5COONa Sodium benzoate	+	$C_6H_5CH_2OH$ Benzyl alcohol

(iii) If HCHO and C_6H_5CHO reacts together in presence of an alkali, it is known as *crossed Cannizzaro reaction*, e.g.,

$$\underset{\text{Formaldehyde}}{HCHO} + \underset{\text{Benzaldehyde}}{C_6H_5CHO} \xrightarrow{NaOH} \underset{\text{Sodium formate}}{HCOONa} + \underset{\text{Benzyl alcohol}}{C_6H_5CH_2OH}$$

(HCHO is more reactive, hence oxidized, and benzaldehyde is reduced)

(iv) If an aldehyde contains two aldehyde groups, then one —CHO group is oxidized while other is reduced. It is known as internal-Cannizzaro reaction. e.g.,

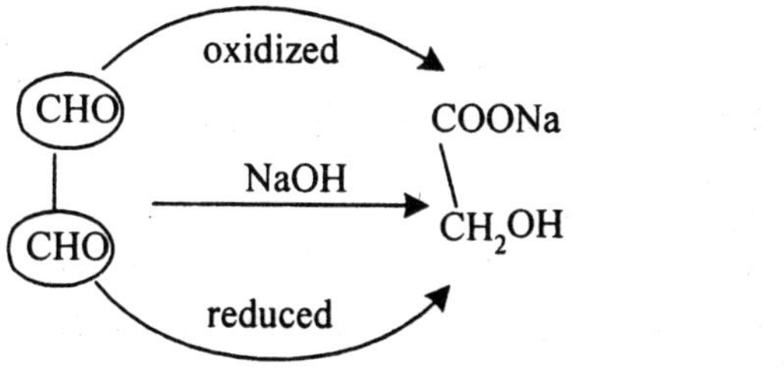

Glyoxal α -Hydroxy sodium acetate

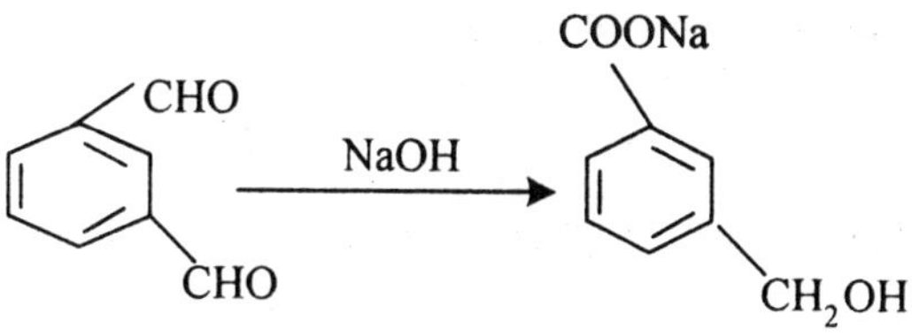

(v) If acetaldehyde reacts with formaldehyde in presence of alkali, an addition product results.

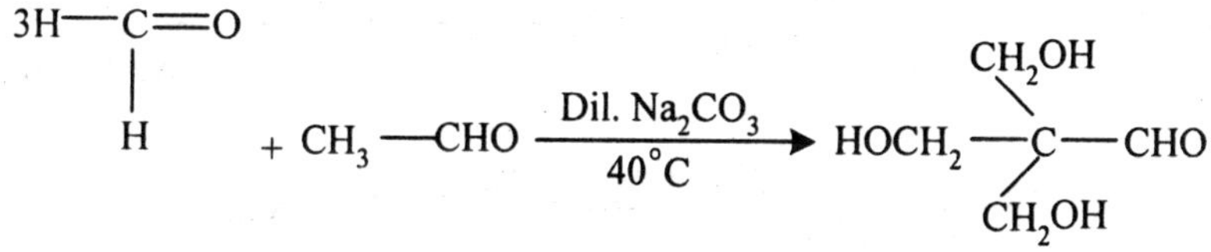

(82%)

(6) *Oxidation*: Aldehydes on oxidation yield carboxylic acid with same number of carbon atoms. Ketones on oxidation yield carboxylic acid with less number of carbon atoms.

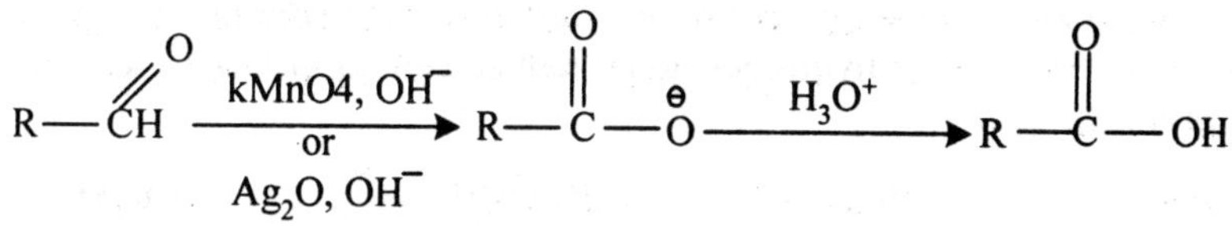

e.g.,

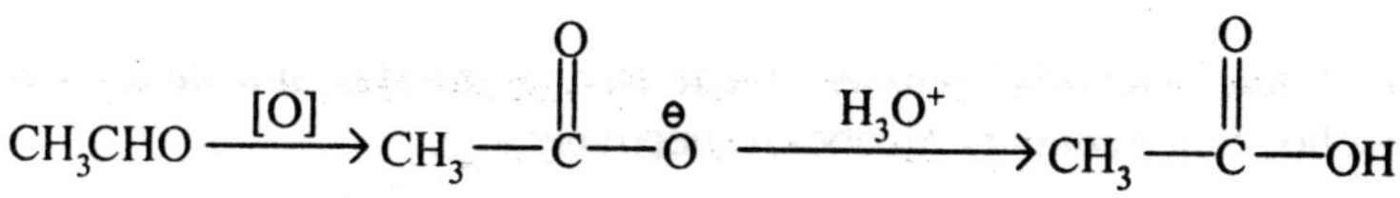

Aldehydes are oxidized much more easily as compared to ketones. Oxidation of both aldehydes and ketones can be affected by a number of reagents, e.g.,

(i) Both aldehydes and ketones can be oxidized by peroxy acids and the reaction is called *Baeyer-Villiger Oxidation*. The reaction product is carboxylic ester, e.g,

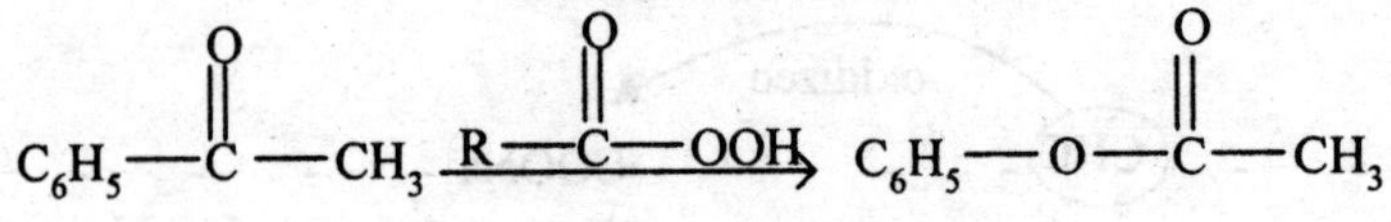

(ii) *Tollen's Test (Silver Mirror Test)*: Tollen's reagent is prepared by mixing aqueous silver nitrate with aqueous ammonia. The reagent contains the diammonio silver (I) ion, $Ag(NH_3)_2^+$. The ion oxidized aldehydes to carboxylate ions, as the silver is reduced from the +1 oxidation state to metallic silver. The metallic silver deposits on the walls of the test tube as a mirror. All ketones except α-hydroxy ketones give negative test with Tollen's reagent:

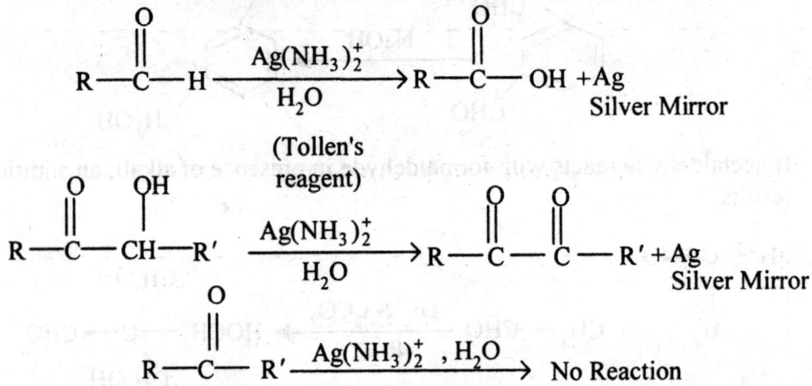

(iii) *Fehling solution*: Fehling solution is obtained by mixing Fehling solution 'a' (copper sulphate) and Fehling solution 'b' (sodium potassium tartrate and conc. NaOH). When aldehydes are heated with Fehling solution (blue colour), a brown precipitate of cuprous oxide is obtained. *All ketones and benzaldehydes do not reduce Fehling solution.* However α-hydroxy ketones reduce Fehling solution (see Chapter 13, Carbohydrates).

$$\underset{}{RCHO} + \underset{\substack{\text{Fehling solution}\\ \text{(Blue Colour)}}}{2CuO} \rightarrow RCOOH + \underset{\substack{\text{Cuprous oxide}\\ \text{(Brown ppt.)}}}{Cu_2O}$$

(7) *Reduction of aldehydes and ketones:* Generally, aldehydes and ketones on reduction yield primary alcohol and secondary alcohol respectively, e.g.,

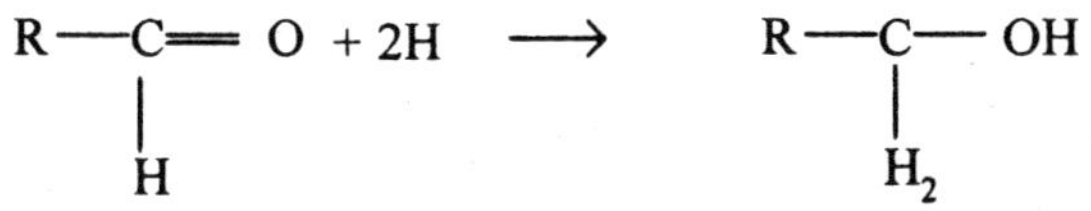

Primary alcohol

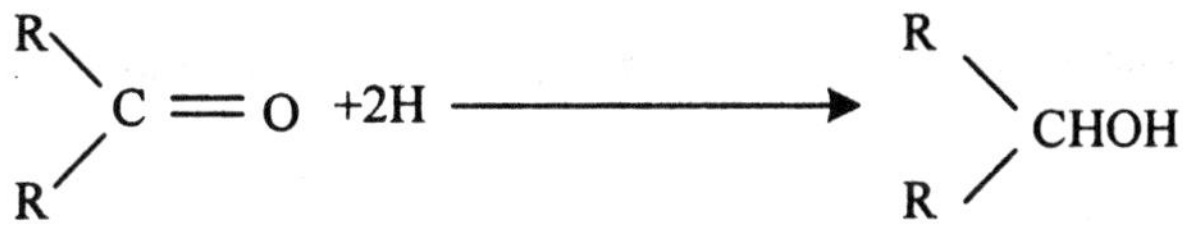

Secondary alcohol

However, the nature of product also depends upon the reducing agent. Few examples are as follows:

(i) *Clemmensen's reduction*: This involves conversion of carbonyl group (both aldehydes and ketones) to methylene group (>CH_2) using zinc amalgam and HCl.

$$>C{=}O + 4H \xrightarrow{Zn/Hg+HCl} >CH_2 + H_2O$$

(ii) *Wolff-Kishner reduction*: When carbonyl group is treated with hydrazine in the presence of a base (alc. KOH), it is also converted into methylene group.

$$\underset{\text{Hydrazine}}{>C{=}O + H_2NNH_2} \xrightarrow[\Delta]{base} \underset{\text{Hydrazone (not isolated)}}{[>C{=}N{-}NH_2]} + H_2O$$

$$\downarrow$$

$$>CH_2 + N_2$$

(iii) *Reformatsky reaction*: When an aldehyde or ketones is heated with α-bromoester in the presence of zinc metal, using benzene as solvent, α-hydroxy ester is obtained.

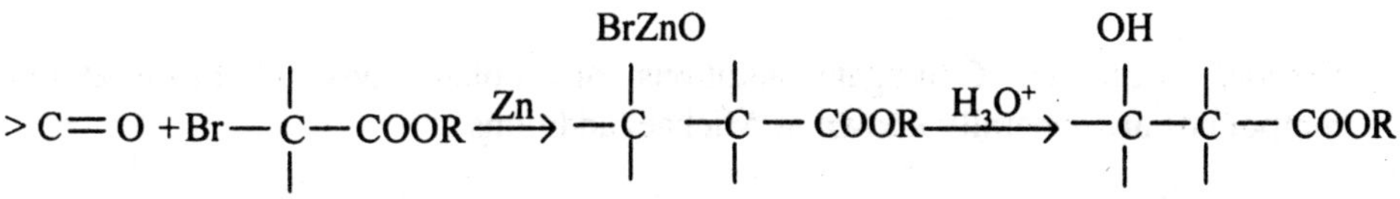

Aldehyde or ketone Benzene α -bromoester β-Hydroxy ester

(8) *Condensation reactions of acetone:*

(i) When acetone reacts with HCl, it forms mesityloxide and phorone

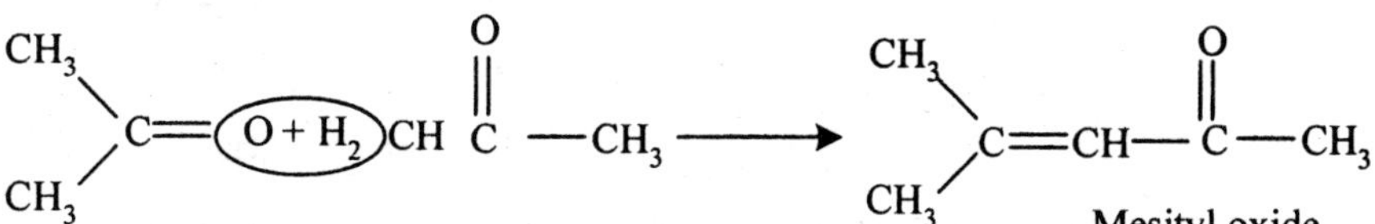

Mesityl oxide

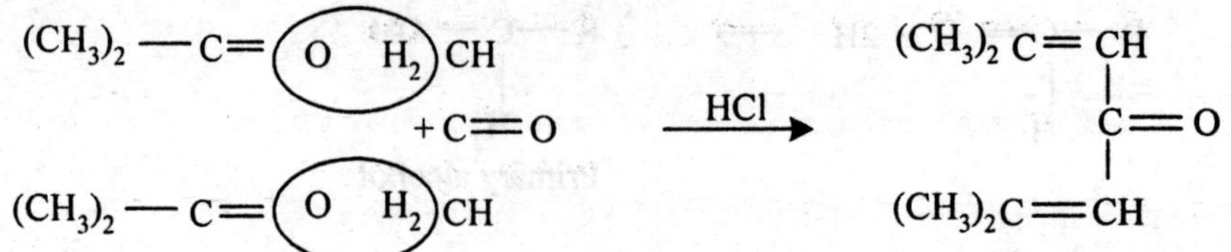

Phorone

(ii) When acetone reacts with conc. H_2SO_4, it forms mesitylene.

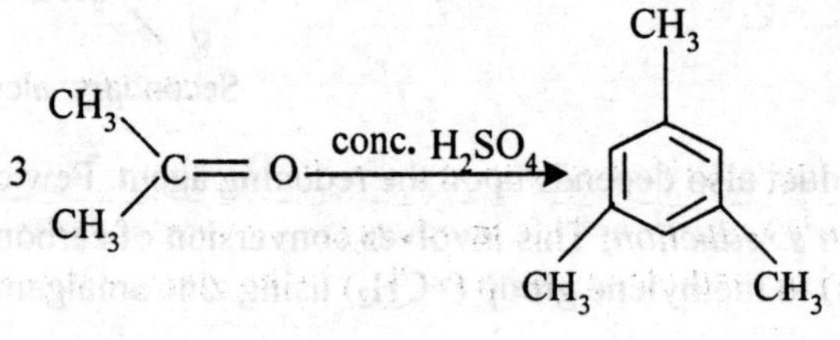

Mesitylene

(iii) When acetone is treated with HCHO in the presence of an alkali, it forms addition product.

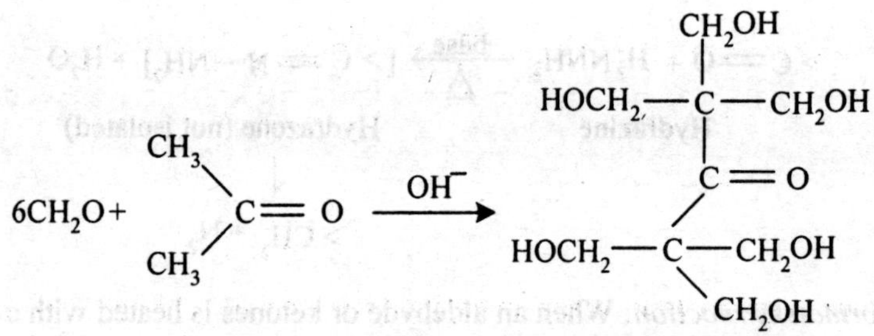

Addition product

(9) *Michael Additions:* Conjugate additions of enolate ions to α,β-unsaturated carbonyl compounds are generally known as Michael additions.

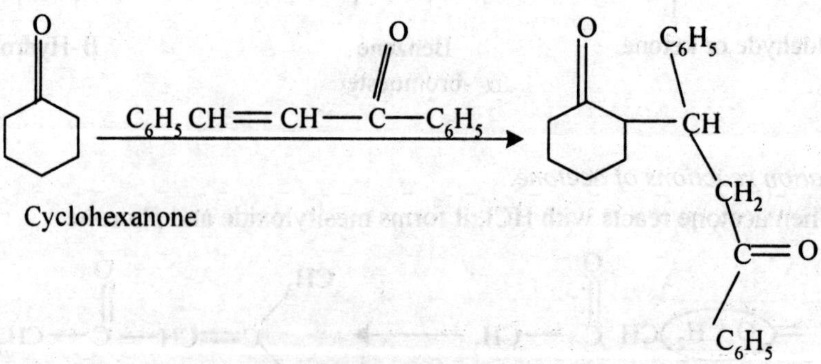

Cyclohexanone

(10) *Tischenkos reaction:* When aldehyde reacts with Aluminium iso-propoxide, it forms ester,

$$RCHO \xrightarrow{\text{Aluminium isoproporide}} RCOOCH_2R$$

e.g.,

$$CH_3CHO \xrightarrow{Al(OC_2H_5)_3} CH_3COOCH_2CH_3$$

(11) *Aldehydes store pink colour with Schiffs reagent but Ketones do not.*

10.5 CARBOXYLIC ACIDS AND DERIVATIVES

General formula for Carboxylic acid is RCOOH or more conveniently C_nH_{2n+1} COOH or RCOOH (where R may be H, alkyl or aryl group, saturated or unsaturated).

(a) Preparation

(i) By oxidation of primary alcohols or aldehydes using potassium dichromate and concentrated H_2SO_4

$$RCH_2OH \xrightarrow{K_2Cr_2O_7/H_2SO_4} RCOOH$$

$$RCHO \xrightarrow{K_2Cr_2O_7/H_2SO_4} RCOOH$$

e.g.,

$$\underset{\text{Ethyl alcohol}}{C_2H_5OH} \xrightarrow{[O]} \underset{\text{Acetaldehyde}}{CH_3CHO} \xrightarrow{[O]} \underset{\text{Acetic acid}}{CH_3COOH}$$

(ii) From Grignard reagent

$$RMgX + CO_2 \xrightarrow{0\,^\circ C} RCOOMgX \xrightarrow{H_3O^+} RCOOH + MgXOH$$

(iii) By hydrolysis of alkane nitriles

$$R—C\boxed{N + H}OH \xrightarrow{H^+} RCOOH + NH_3$$

$$H_2O$$

e.g.,

$$C_2H_5CN + 2H_2O \xrightarrow{H^+} C_2H_5COOH + NH_3$$

(iv) $$CO_2 + H_2 \xrightarrow{Pd} HCOOH$$

(v) $$CO + NaOH \xrightarrow{210\,^\circ C/6\text{-}10\,atm.} HCOONa$$

$$\downarrow NaHSO_4$$

$$\underset{\text{(formic acid)}}{HCOOH} + Na_2SO_4$$

(vi) Carboxylic acids can also be prepared by oxidation of alkene (see Chapter 9).

(vii) Carboxylic acids can be prepared on large scale by reaction of an alkene with carbonmonoxide and steam under pressure in the presence of phosphoric acid at 400°C. The reaction is known as Koch reaction.

$$\underset{\text{Ethylene}}{CH_2 = CH_2} + CO + H_2O \xrightarrow{H_3PO_4/400^\circ C} \underset{\text{Propionic acid}}{CH_3CH_2COOH}$$

(viii) Carboxylic acids can also be prepared by boiling an ester with concentrated aqueous sodium hydroxide followed by treatment with dilute HCl

$$\underset{\text{Ethylacetate}}{CH_3-\overset{O}{\overset{\|}{C}}-OC_2H_5} + NaOH \longrightarrow \underset{\text{Sodium acetate}}{CH_3-\overset{O}{\overset{\|}{C}}-ONa} + \underset{\text{Ethyl alcohol}}{C_2H_5OH}$$

(b) Properties

Physical:

(1) The first three acids are liquids with pungent odour, miscible with water due to formation of hydrogen bonding (like alcohols). These are polar compounds, hence less soluble in solvents of low polarity like ether, alcohol, and toluene. The solubility in water decreases as the molecular weight of carboxylic acid increases. The boiling point of the carboxylic acid is higher than that of the alcohol of comparable molecular weight (e.g., propionic acid b.p., 141°C and *n*-butyl alcohol, b.p., 118°C) due to the fact that the strength of hydrogen bonding is greater in acids than in alcohol, i.e.,

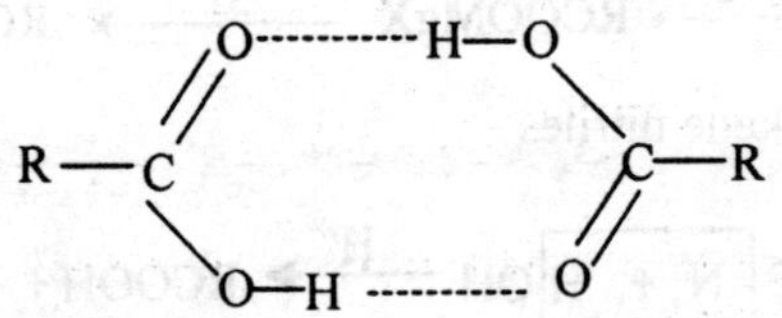

Acids with four to nine carbon atoms are liquids and those having more than ten carbon atoms are solids.

(2) The melting points have an 'even' acid is higher than that of an 'odd' acid immediately above and below it in the series.

Chemical Properties:

1. *Acidity*: Carboxylic acids are much weaker than the strong mineral acids like sulphuric, hydrochloric and nitric acids but are more acidic than alcohols, acetylene and water. It is because the acids form salts with electropositive elements as well as with an alkali.

$$\underset{\text{Stronger acid}}{RCOOH + Na} \rightarrow \underset{\text{Weaker acid}}{RCOONa} + H_2O$$

$$RCOOH + NaOH \rightarrow RCOONa + H_2O$$

e.g.,

$$CH_3COOH + NaOH \rightarrow CH_3COONa + H_2$$

2. The acids evolve bubbles of CO_2 with $NaHCO_3$ solution

$$RCOOH + NaHCO_3 \rightarrow RCOONa + H_2O + CO_2$$

3. *Hell Volhard Zelinsky Reaction*: Carboxylic acids having α-*H atom*, react with P and Cl_2/I_2, forming α-halogen carboxylic acid. The reaction is called *Hell Volhard Zelinksky reaction* (H.V.Z. reaction), e.g.,

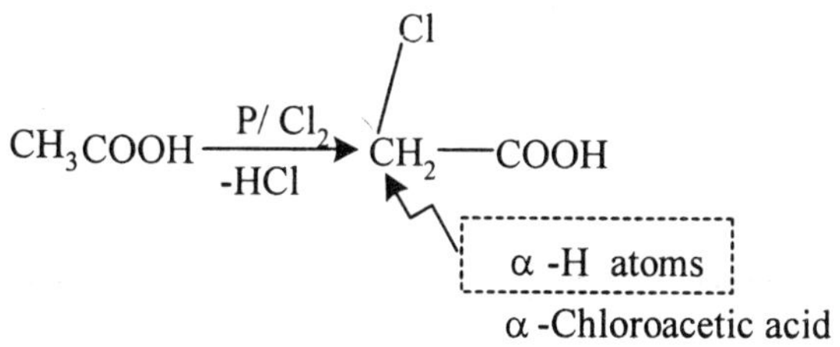

The reaction is not given by following acids due to lack of α-H atom.

HCOOH, CCl_3COOH, C_6H_5COOH

4. Due to presence of $H-\overset{O}{\overset{\|}{C}}-$ group, formic acid $(H-\overset{O}{\overset{\|}{C}}-OH)$ is the only acid that reduces Fehling solution as well as Tollen's reagent.

$$HCOOH + 2Ag(NH_3)_2OH \rightarrow 2Ag + 4NH_3 + 2H_2O + CO_2$$

$$\underset{\text{Fehling solution}}{2CuO} + HCOOH \xrightarrow{\Delta} \underset{\text{Brown ppt.}}{Cu_2O}$$

5. Formic acid reacts with mercuric chloride

$$2HgCl_2 + HCOOH \rightarrow Hg_2Cl_2 + 2HCl + CO_2$$

6. Carboxylic acid react with alcohols in the presence of acids like H_2SO_4 to form ester. The reaction is reversible and is called esterification.

$$R-\overset{O}{\overset{\|}{C}}-OH + H-\overset{H}{\overset{|}{O}}R^1 \xrightarrow{H^{\oplus}} R-\overset{O}{\overset{\|}{C}}-OR^1 + H_2O$$

e.g.,

$$\underset{\text{Propionic acid}}{CH_3CH_2COOH} + \underset{\text{Ethanol}}{HOC_2H_5} \xrightarrow{-H_2O} \underset{\text{Ester}}{CH_3CH_2COOC_2H_5}$$

7. Carboxylic acids react with ammonia to form salts of fatty acids, which when heated with a dehydrating agent like P_2O_5, forms amides and then alkyl cyanides.

$$R—COOH + NH_3 \longrightarrow \underset{\text{Ammonium salt of acid}}{R—COONH_4} \xrightarrow{P_2O_5} \underset{\text{Acidamide}}{RCONH_2} + H_2O$$

$$RCONH_2 \xrightarrow{\Delta,\ P_2O_5} RCN + H_2O$$

e.g.,

$$CH_3COOH + NH_3 \longrightarrow \underset{\text{Ammonium acetate}}{CH_3COONH_4} \xrightarrow{P_2O_5} \underset{\text{Acetamide}}{CH_3CONH_2} + H_2O$$

$$CH_3CONH_2 \xrightarrow{P_2O_5} \underset{\text{Methyl cyanide}}{CH_3CN} + H_2O$$

8. Carboxylic acids react with PCl_5, $SOCl_2$ or PCl_3 to form acid chloride derivatives.

$$R—\overset{O}{\overset{\|}{C}}—OH + PCl_5 \longrightarrow R—\overset{O}{\overset{\|}{C}}—Cl + POCl_3 + HCl$$

$$R—\overset{O}{\overset{\|}{C}}—OH + SOCl_2 \longrightarrow R—\overset{O}{\overset{\|}{C}}—Cl + SO_2 + HCl$$

$$3R—\overset{O}{\overset{\|}{C}}—OH + PCl_3 \longrightarrow 3R—\overset{O}{\overset{\|}{C}}—Cl + H_3PO_3$$

e.g.,

$$CH_3COOH + PCl_5 \rightarrow \underset{\text{Acetyl chloride}}{CH_3COCl} + PoCl_3 + HCl$$

$$CH_3COOH + SOCl_2 \rightarrow \underset{\text{Acetyl chloride}}{CH_3COCl} + SO_2 + HCl$$

$$3CH_3COOH + PCl_3 \rightarrow 3CH_3COCl + H_3PO_3$$

9. When acids are heated with dehydrating agent like phosphorus pentaoxide, it forms acid anhydride.

$$\begin{matrix} RCOOH \\ + \\ RCOOH \end{matrix} \xrightarrow{P_2O_5} \underset{\text{Acid anhydride}}{\begin{matrix} RCO \\ RCO \end{matrix}\!\!>O} + H_2O$$

e.g.,

$$\begin{matrix} CH_3COOH \\ + \\ CH_3COOH \end{matrix} \xrightarrow{P_2O_5} \begin{matrix} CH_3CO \\ \\ CH_3CO \end{matrix}\!\!>O + H_2O$$

Derivatives of carboxylic acids

The derivatives of carboxylic acids include:

(1) Acid halides;

(2) Acid anhydrides;

(3) Amides; and

(4) Esters.

The order of reactivity of the derivatives is as follows:

Acid chloride > Acid anhydride > Ester > Amide

10.5.1 Acid Halides

These are derivatives of acids in which –OH group of acid is replaced by halogen. These are also called acyl halides.

(a) Preparation

Acid halides are prepared by reaction of carboxylic acids with PCl_5, $SOCl_2$, PCl_3, as discussed above.

(b) Properties

Physical: These are colourless liquids having pungent smell.

Chemical: acid chlorides are very reactive compounds and undergo Nucleophilic substitution reactions, e.g.,

(1) Acid chlorides react with water to form acid

$$\underset{\text{Acetyl chloride}}{CH_3COCl} + HOH \rightarrow \underset{\text{Acetic acid}}{CH_3COOH} + HCl$$

(2) Acid chlorides reacts with alcohols to form esters.

$$\underset{\text{Acetyl chloride}}{CH_3COCl} + HOC_2H_5 \rightarrow \underset{\text{Ethyl acetate}}{CH_3COOC_2H_5} + HCl$$

(3) Acid chlorides react with ammonia to form amides.

$$\underset{\text{Acetyl chloride}}{CH_3COCl} + HNH_2 \rightarrow \underset{\text{Acetamide}}{CH_3CONH_2} + HCl$$

(4) Acid chloride, when reduced with Pd suspended in $BaSO_4$, aldehydes are formed. The reaction is called ***Rosenmund reaction***.

$$CH_3-COCl + H_2 \xrightarrow{Pd/BaSO_4} CH_3CHO + HCl$$

$BaSO_4$ acts as poison to inhibit reduction of aldehyde to alcohol.

(5) Acid chlorides react with benzene in the presence of anhydrous $AlCl_3$ to from aromatic ketones. The reaction is known as Friedel-Craft acylation.

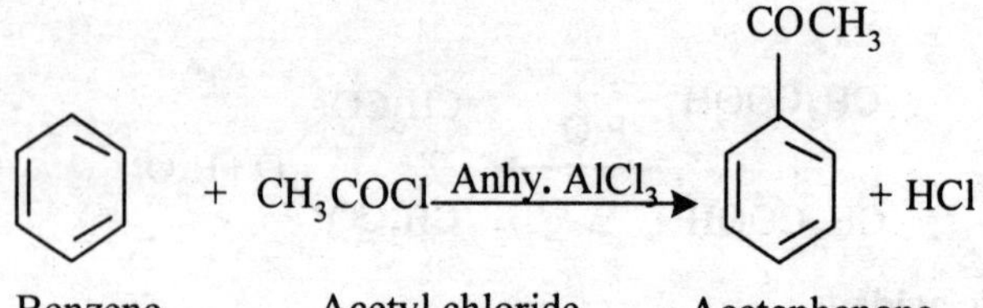

Benzene Acetyl chloride Acetophenone

(6) Acid chlorides react with sodium salts of carboxylic acids to form acid anhydrides

$$\underset{\text{Acetyl chloride}}{CH_3COCl} + \underset{\text{Sodium acetate}}{CH_3COONa} \rightarrow \underset{\text{Acetic anhydride}}{(CH_3CO)_2O} + NaCl$$

10.5.2 Acid Anhydrides

These are derived by loss of one water molecule from two molecules of carboxylic acids. They have general formula $(RCO)_2O$.

(a) *Preparation*

Acid anhydrides can be prepared from acids or Acid chlorides, as mentioned above.

$$\underset{\text{Acetic acid}}{\begin{matrix} CH_3COOH \\ + \\ CH_3COOH \end{matrix}} \xrightarrow{P_2O_5} \underset{\text{Acetic anhydride}}{\begin{matrix} CH_3CO \\ \quad\quad > O \\ CH_3CO \end{matrix}}$$

$$\underset{\text{Acetyl chloride}}{CH_3COCl} + \underset{\text{Sodium acetate}}{CH_3COONa} \rightarrow (CH_3CO)_2O + NaCl$$

(b) *Properties*

Physical: The lower members are colourless liquid with a pungent odour.

Chemical:

(1) Acid anhydrides undergo hydrolysis to form acids

$$(CH_3CO)_2O + H_2O \rightarrow \underset{\text{Acetic acid}}{2CH_3COOH}$$

(2) Acid anhydrides react with alcohols to from ester and acids

$$\underset{\text{Acetic anhydride}}{\begin{matrix} CH_3CO \\ \quad\quad > O \\ CH_3CO \end{matrix}} + \underset{\text{Ethyl alcohol}}{H{-}OC_2H_5} \longrightarrow \underset{\text{Acetic acid}}{CH_3COOH} + \underset{\text{Ethylacetate}}{CH_3COOC_2H_5}$$

(3) Acid anhydrides react with ammonia to form amides and acids

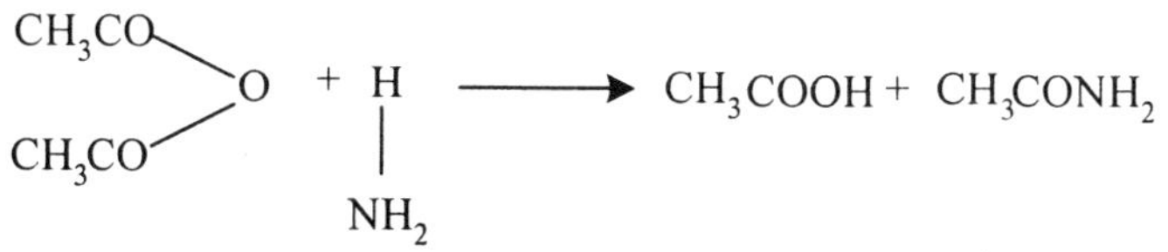

Acetic anhydride Ammonia Acetic acid Acetamide

10.5.3 Amides

These are the derivatives of carboxylic acids in which the —OH function of carboxylic group is replaced by an amino group ($-NH_2$). They have general formula $RCONH_2$.

(a) *Preparation*

(1) By reaction of acid chlorides with ammonia.

$$\underset{\text{Acetyl chloride}}{CH_3COCl} + \underset{\text{Ammonia}}{2HNH_2} \rightarrow \underset{\text{Acetamide}}{CH_3CONH_2} + NH_4Cl$$

(2) By heating ammonium salt of fatty acids

$$\underset{\text{Ammonium acetate}}{CH_3COONH_4} \xrightarrow[\Delta]{P_2O_5} \underset{\text{Acetamide}}{CH_3CONH_2} + H_2O$$

(b) *Properties*

Physical: Amides are almost colourless, white crystalline salts, soluble in water.

Chemical:

(1) Amides undergo hydrolysis under alkaline/acid conditions.

$$\underset{\text{Acetamide}}{CH_3CONH_2} + NaOH \rightarrow \underset{\text{Sodium acetate}}{CH_3COONa} + NH_4Cl$$

$$\underset{\text{Acetamide}}{CH_3CONH_2} + HCl + H_2O \rightarrow \underset{\text{Acetic Acid}}{CH_3COOH} + NH_4Cl$$

(2) Amides react with nitrous acid to form carboxylic acids.

$$\underset{\text{Acetamide}}{CH_3\text{—}CONH_2} + \underset{\text{Nitrous acid}}{OHNO} \rightarrow \underset{\text{Acetic acid}}{CH_3COOH} + N_2 + H_2O$$

(3) Amides can be reduced to primary amines using lithium aluminum hydride ($LiAlH_4$)

$$\underset{\text{Acetamide}}{RH_3CONH_2} \xrightarrow{LiAlH_4} \underset{\text{Ethyl amine}}{CH_3CH_2NH_2}$$

(4) Amides when treated with bromine and sodium hydroxide produce primary amine with one carbon atom less. The reaction is known as *Hofmann Bromide reaction.*

$$CH_3CONH_2 + Br_2 + 4NaOH \rightarrow CH_3NH_2 + 2NaBr + Na_2CO_3 + 2H_2O$$

The reaction occurs in following steps:

(i) $CH_3CONH_2 + Br_2 \rightarrow CH_2CONHBr + HBr$

(ii) $CH_3CONHBr + NaOH \rightarrow CH_3NCO + NaBr + H_2O$

(iii) $\underset{\text{Methyl cynate}}{CH_3CNO} \xrightarrow{\text{Isomerization}} \underset{\text{Methyl isocynate}}{CH_3NCO}$

(iv) $CH_3NCO + 2NaOH \rightarrow CH_3NH_2 + Na_2CO_3$

10.5.4 Esters

Esters are organic compounds with general formula R—COOR′.

(a) ***Preparation***

(i) By esterification reaction

$$R-\overset{\overset{\large O}{\|}}{C}-OH + HO-R' \underset{}{\overset{H^{\oplus}}{\rightleftharpoons}} RCOOR' + H_2O$$

(ii) Esters can be prepared by trans-esterification:

$$\underset{\text{High boiling ester}}{R-\overset{\overset{\large O}{\|}}{C}-OR'} + \underset{\text{Low boiling alcohol}}{R''-OH} \underset{}{\overset{H^+\ \Delta}{\rightleftharpoons}} \underset{\text{Higher boiling ester}}{R-\overset{\overset{\large O}{\|}}{C}-OR'} + \underset{\text{High boiling ester}}{R'-OH}$$

e.g.,

$$\underset{\text{Methyl acrylate}}{CH_2{=}CH-COOCH_3} + \underset{\text{Butyl alcohol}}{CH_3CH_2CH_2CH_2OH} \xrightarrow{H^{\oplus}} \underset{\text{Butyl acrylate}}{CH_2{=}CH-COOCH_2CH_2CH_3} + CH_3OH$$

(b) ***Properties***

Physical: Esters are liquid with pleasant small.

Chemical:

(i) Esters undergo hydrolysis to form acid and alcohol

$$R-\overset{\overset{\large O}{\|}}{C}-OR' + H_3O^{\oplus} \rightleftharpoons R-\overset{\overset{\large O}{\|}}{C}-OH + R'OH$$

(ii) With NaOH, esters undergo saponification.

$$R-\overset{\overset{O}{\|}}{C}-OR' + NaOH \longrightarrow R-\overset{\overset{O}{\|}}{C}-ONa + R'OH$$

10.6 ALIPHATIC COMPOUNDS CONTAINING NITROGEN

Organic compounds in which a nitro group (—NO_2) is attached to an alkyl group are called nitro compounds, which have general formula RNO_2.

(a) ***Preparation***

(i) By nitration of alkanes[7]

$$RH + HNO_3 \rightarrow \underset{\text{Nitro alkane}}{RNO_2} + H_2O$$

(ii) By heating an alkyl halide with aqueous ethanolic solution of silver nitrite:

$$RX + AgONO \rightarrow RNO_2 + RONO + AgX$$

Reactive order: I > Br > Cl

Reactivity of alkyl group: tert > sec > pri halide

(iii) By boiling halogeno-acetic acid with aqueous solution of sodium nitrite

$$ClCH_2COONa + NaNO_2 \longrightarrow \underset{NO_2}{\overset{CH_2-COOH}{|}} + NaCl + CH_3NO_2 + CO_2$$

(iv) $ClCH_2COOH + NaNO_2 \rightarrow NO_2CH_2COOH + NaCl$

(b) ***Properties***

(1) *Reduction:* It takes place under different conditions:

(a) Acidic medium $RNO_2 + 6H \xrightarrow[\text{or metal in acid}]{LiAlH_4} RNH_2 + 2H_2O$

(b) Neutral medium $RCH_2NO_2 \xrightarrow{Zn/NH_4Cl} \underset{\text{N − alkylhydroxyl amine}}{RNHOH} + H_2O$

---

[7] Nitration can be carried out in two ways:

(i) *Liquid-phase nitration:* the hydrocarbon is heated with conc. HNO_3 under pressure at 140°C.

(ii) *Vapour-phase nitration:* the hydrocarbon is heated with conc. HNO_3 under pressure at 150–450°C.

For details, see Chapter 9, Alkanes.

(c) $$RCH_2NO_2 \xrightarrow{SnCl_2/HCl} RCH_2NHOH + \underset{\text{Oxime}}{RCH{=}NOH}$$

(2) Hydrolysis:

$$\underset{\substack{\text{Primary}\\ \text{nitrocompound}}}{RCH_2NO_2} + H_2O \xrightarrow{H_2SO_4 \text{ or } HCl} RCOOH + NH_2OH$$

$$\underset{\substack{\text{Secondary}\\ \text{nitrocompound}}}{2R_2CHNO_2} \xrightarrow{HCl} \underset{\text{ketone}}{2R_2CO} + N_2O + H_2O$$

Tertiary nitrocompounds are unaffected by HCl acid.

(3) Reaction with nitrous acid:

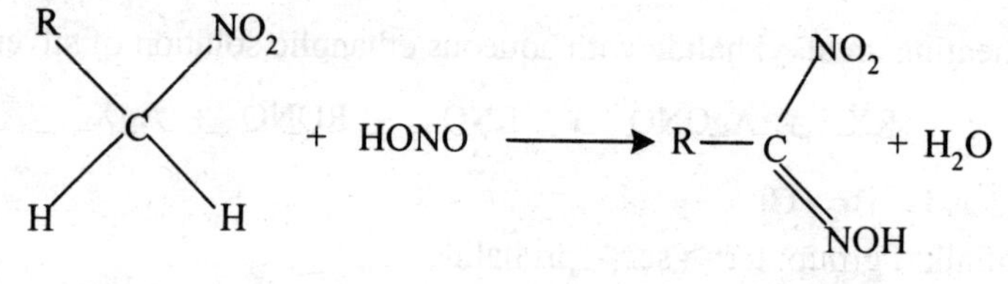

Primary nitrocompounds Nitrolic acids

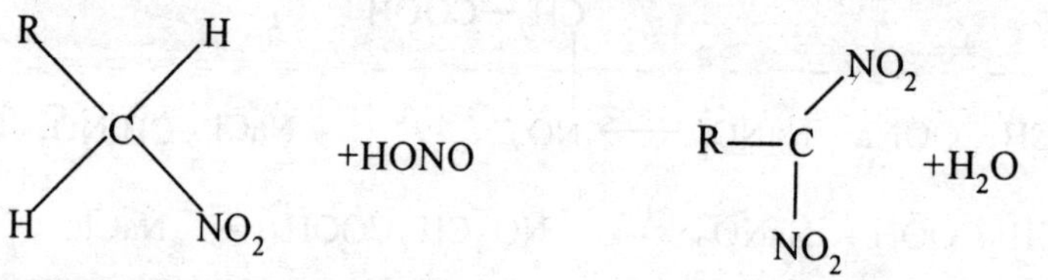

Secondary nitrocompounds Pseudonitrates

Tertiary nitrocompounds do not react with nitrous acid as they lack α-hydrogen atom

(4) Primary and secondary nitrocompounds, due to presence of α-hydrogen atoms, undergo condensation with aldehydes, e.g.,

$$CH_3NO_2 \xrightarrow[C_6H_5CHO]{\overset{\ominus}{O}H} C_6H_5{-}CH(OH){-}CH_2NO_2 \xrightarrow{-H_2O} \underset{\omega\text{-nitrostyrene}}{C_6H_5CH{=}CHNO_2}$$

10.7 NITROSOALKANES

Organic compounds is which a nitroso group —N=O is directly attached to a carbon atom are called nitrosoalkane.

(a) ***Preparation***

(i)

$$\underset{\text{alkene}}{RCH{=}CH_2} + \underset{\text{Nitrosyl chloride}}{NOCl} \rightarrow ClCH_2CH_2NO_2$$

(ii) By oxidation of primary amine containing tertiary alkyl group.

$$R_3CNH_2 + 2[O] \xrightarrow{H_2SO_5} R_3CNO + H_2O$$

(iii) N-substituted hydroxylamines, on oxidation, yield nitrosocompounds

$$RNHOH \xrightarrow{[O]} RNO$$

(b) ***Properties***

(1) Usually blue or green liquids which dimerises to colourless solids.

(2) Nitrosocompounds exist as such only when nitroso group is attached to a tertiary carbon atom. However, if it is attached to primary or secondary C atom, it rearrange to the oxime.

$$\underset{\text{Blue unstable}}{R_2CHNO} \rightarrow \underset{\text{Colourless stable}}{R_2C{=}NOH}$$

(3) On oxidation, it yields nitro compounds

$$R_3CNO \rightarrow R_3CNO_2$$

(4) On reduction, it yields primary amines

$$R_3CNO \xrightarrow{Sn/HCl} R_3CNH_2$$

10.8 ALKYL CYANIDES

Alkyl cyanides are organic derivatives of hydrocyanic acid and have general formula RCN.

(a) ***Preparation***

(1) By passing a mixture of carboxylic acid and ammonia over alumina at 500°C

$$RCOOH + NH_3 \longrightarrow RCOONH_4 \xrightarrow[-H_2O]{P_2O_5} RCONH_2 \xrightarrow{\Delta\ P_2O_5\ -H_2O} RCN$$

(2) By ammoxidation of alkanes:

$$2RCH_3 + 2NH_3 + 3O_2 \xrightarrow[500-600\,°C]{\text{Catalyst}} 2RCN + 6H_2O$$

(3) By dehydration of aldoxime with P_2O_5 or acetic anhydride

$$RCH{=}NOH \xrightarrow{(CH_3CO)_2O} RCN + H_2O$$

(4) By heating alkyl halide with KCN

$$RX + KCN \rightarrow RCN + KX$$

The order of yield is primary > secondary > tertiary (very poor yield)

(5) From Grignard reagent:

$$\underset{}{RMgCl} + \underset{\text{Cyanogen chloride}}{ClCN} \rightarrow RCN + MgCl_2$$

This method gives best yield of tertiary alkyl cyanides.

(b) Properties

(1) Upon hydrolysis by acid or alkali, these yield acids.

$$R{-}C{\equiv}N + HOH \rightarrow \underset{\text{Intermediate}}{[R{-}\overset{OH}{\underset{NH}{C}}]} \xrightarrow{HOH} RCOOH + NH_3$$

i.e.,

$$\begin{matrix} R{-}C{\equiv}N \\ + \\ H_2O \end{matrix} \quad HOH \rightarrow RCOOH + NH_3$$

(2) When a solution of alkyl cyanide in an alcohol is heated with conc. H_2SO_4 or HCl, the ester is obtained.

$$R'CN + R''OH + H_2O \rightarrow R'COOR'' + NH_3$$

(3) When alkyl cyanide is reduced by sodium and alcohol or $LiAlH_4$, it forms primary amine.

$$\begin{matrix} R{-}C{\equiv}N \\ + \\ 2N \quad 2H \end{matrix} \rightarrow RCH_2NH_2$$

or

$$RC{\equiv}N + 2H \rightarrow \underset{\substack{\text{Imine} \\ \text{(Reaction} \\ \text{intermediate)}}}{RCH{=}NH} \xrightarrow{2H} RCH_2NH_2$$

(4) Alkyl cyanide reacts with Grignard reagent to form ketones

$$R'-C\equiv N + R-MgX \rightarrow R''-\underset{\underset{R'}{|}}{C}=NMgX \xrightarrow{H^+} R'COR''$$

(5) Alkyl cyanide undergo condensation reactions in the presence of sodium in ether.

$$CH_3CH_2-C\equiv N + H-\overset{\overset{CH_3}{/}}{C}HCN \xrightarrow{Na/ether} CH_3CH_2-\overset{\overset{NH}{\|}}{C}-\overset{\overset{CH_3}{|}}{C}H-CN$$

The reaction is called *Thorpe nitrile condensation*

(6) Alkyl cyanides also condense with esters, e.g.,

$$CH_3CN + NaNH_2 \xrightarrow{CH_3CH_2COOC_2H_5} CH_3CH_2COCH_2CN + C_2H_5OH$$

10.9 ALKYL ISOCYANIDE

These are organic derivatives of hydrocyanic acid with general formula RNC.

(a) Preparation

(i) By reaction of alkyl iodide with silver cyanide

$$RI + AgCN \rightarrow RNC + AgI$$

(ii) By carbylamine reaction

$$RNH_2 + CHCl_3 + 3KOH \rightarrow RNC + 3KCl + 3H_2O$$

The mechanism is as follows:

$$RNH_2 + \underset{\text{Dichloro carbene}}{:CCl_2} \rightarrow RNH_2-CCl_2 \xrightarrow{-H^+} RNH-CCl_2$$

(Reaction intermediate RN=C, C—Cl⁻, RN=CCl)

(iii) By dehydration of N-substituted phosphoryl chloride in pyridine.

$$RNHCHO \xrightarrow{PoCl_3} RNC$$

(b) Properties

Physical: Poisonous, most unpleasant smelling liquids with lower boiling point than that of cyanides (of same molecular weight), less soluble in water.

Chemical:

(1) Hydrolysis

$$RNC + HOH \rightarrow RNH_2 + HCOOH$$

$$+$$

$$H_2O$$

(2) Reduction: Alkyl isocyanides are reduced to secondary amine by dissolving metal or by catalytic reduction.

$$RN \equiv C \rightarrow RNHCH_3$$

$$H + 3H$$

(3) $RNC + S \rightarrow RNCS$
Alkyl isothiocyanate

(4) $RNC + X_2 \rightarrow RNCX_2$

(5) $RNC + HgO \rightarrow RNCO + Hg$

(6) When alkyl isocyanide is heated for long time, it isomerises to alkyl cyanide.

$$RNC \rightarrow RCN$$

10.10 ALKYL AMINES

Alkyl amines are organic compounds in which an amino group ($-NH_2$) is attached to an alkyl group. Alkyl amines are of three types.

(i) Primary amines (RNH_2): e.g., $CH_3CH_2NH_2$
Ethyl amine
(ethanamine)

(ii) Secondary amines (R_2NH): e.g., $CH_3NHCH_2CH_3$
Ethyl methyl amine
(N – methyl ethanamine)

(iii) Teritary amines (R_3N): e.g., $(CH_3CH_2)_3N$
Triethylamines
(N, N′ – diethylethanimine)

(a) Preparation

(1) By reduction of nitro compounds

$$R\text{—}NO_2 + 6H \text{ metal, } \xrightarrow[\text{Catalyst}]{H^+ \text{ or } H_2} \underset{\text{primary amine}}{RNH_2} + 2H_2O$$

e.g.,

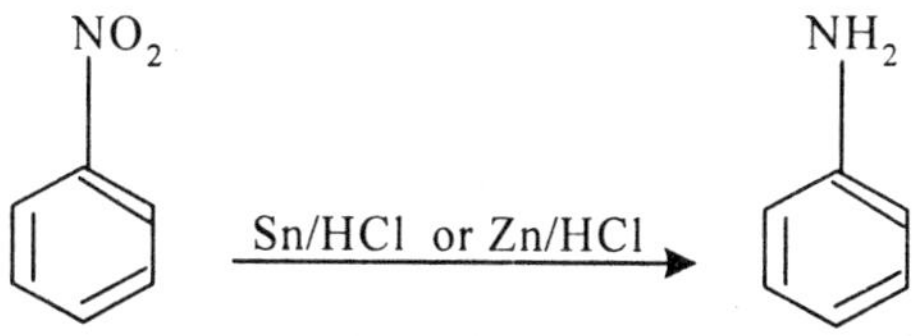

Nitrobenzene Aniline

For industrial preparation Fe/CH_3COOH is used because this combination is economic.

(2) By reaction of alkyl halide with ammonia or amines or with azide ion:

(i)

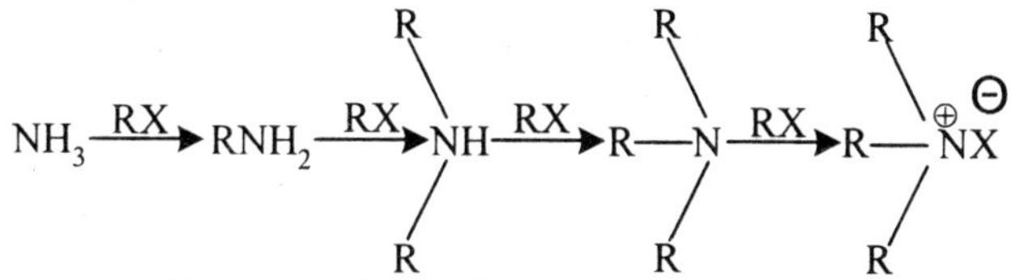

Primary amine Secondary amine Tertiary amine Quarternary salt

If ammonia is used in excess, multiple alkylations can be minimized.

(ii) $R-X + N_3^- \xrightarrow{SN^2} R-N=N=N^- \xrightarrow{Na/alcohol} RNH_2$

Azide ion Alkyl azide

(3) By reduction of cyanides, isocyanides and oximes

(i) $R-C\equiv N \longrightarrow RCH_2NH_2$

+

2H 2H Primary Amine

(ii)

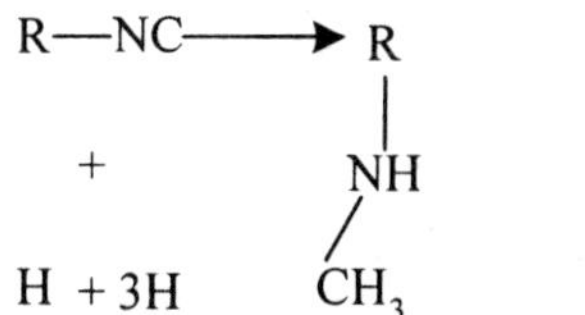

Secondary amine

(iii) $RCH{=}NOH + 2H \longrightarrow RCH_2NH_2 + H_2O$

+

2H

Oxime Primary Amine

(4) By Hoffmann bromide reaction: When amides are treated with OBr^- (in the form of NaOBr obtained from bromine and NaOH), the amine is obtained.

$$RCONH_2 \xrightarrow{Br_2 + KOH} RNH_2 + K_2CO_3$$

The reaction takes place in the following steps

(i) $$R{-}\overset{O}{\overset{\|}{C}}{-}NH_2 + Br_2 \longrightarrow R{-}\overset{O}{\overset{\|}{C}}{-}NHBr$$

(ii) $$R{-}\overset{O}{\overset{\|}{C}}{-}NHBr + KOH \longrightarrow \underset{\text{Alkyl cyanate}}{R{-}CNO} + KBr + H_2O$$

(iii) $$RCNO \xrightarrow{\text{Isomerises}} \underset{\text{Alkyl isocyanate}}{RNCO}$$

(iv) $$\underset{\substack{+\\ KOH}}{RNCO + KOH} \longrightarrow RNH_2 + K_2CO_3$$

(i) Aldehydes and ketones can be converted to amines through catalytic or chemical reduction in the presence of ammonia or an amine. For example:

(i) $$R_2C{=}O + NH_3 \xrightarrow{H} \underset{\text{Primary amine}}{R_2CH{=}NH_2}$$

(ii) $$R_2C{=}O + RNH_2 \xrightarrow{H} \underset{\text{Secondary amine}}{R_2CHHNR}$$

(iii) $$R_2C{=}O + R_2NH \xrightarrow{H} \underset{\text{Tertiary amine}}{R_2CH{-}NR_2}$$

(b) Properties

Physical: Amines are polar liquid compounds. Primary and secondary amines can form intermolecular hydrogen bonds, hence have higher boiling points than non-polar compounds of the same molecular weight, but lower boiling points than alcohols or carboxylic acids. All amines are water soluble due to the formation of hydrogen bonds with water. Amines are also soluble in less polar solvents like ether, alcohol, benzene. Aromatic amines like aniline are colourless when pure, however, are often encountered coloured by oxidation products.

Chemical:

1. Amines are basic[8] in nature, hence react with acids to form salt,

$$\text{e.g., } RNH_2 + HCl \rightarrow \underset{\text{Amine hydrochloride}}{RNH_2.HCl \text{ or } RN^+H_3Cl^-}$$

2. Reaction with acid chloride: Primary amines form N-substituted amide, secondary amines form N, N-disubstituted amide and tertiary amines do not react.

$$RNH_2 \xrightarrow{R'COCl} \underset{\text{N – substituted amide}}{RNHCOR'}$$

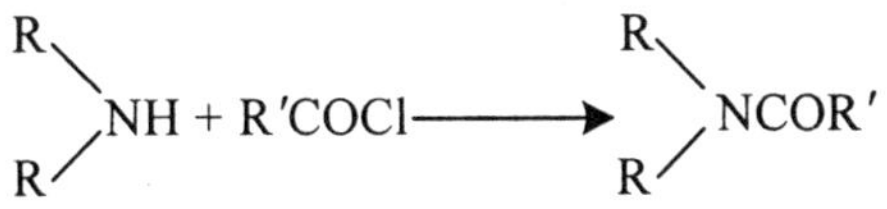

N, N-disubstituted amide

$$R_3N + R'COCl \rightarrow \text{No reaction}$$

3. Reaction with nitrous acid

$$\underset{\text{Primary amine}}{RNH_2 + OHNO} \rightarrow ROH + N_2 + H_2O$$

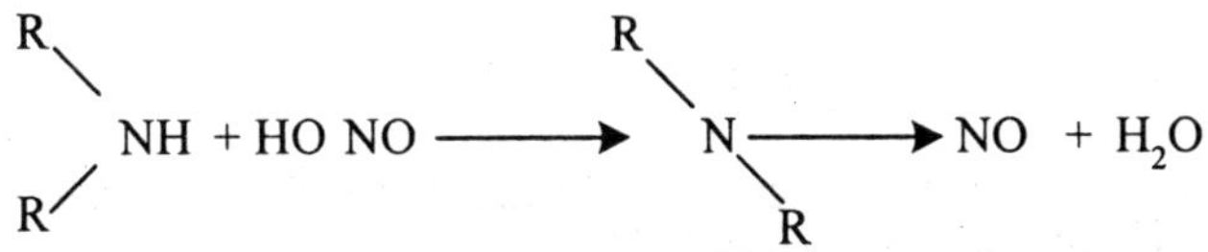

N, nitrosoamine
(yellow oily liquid)

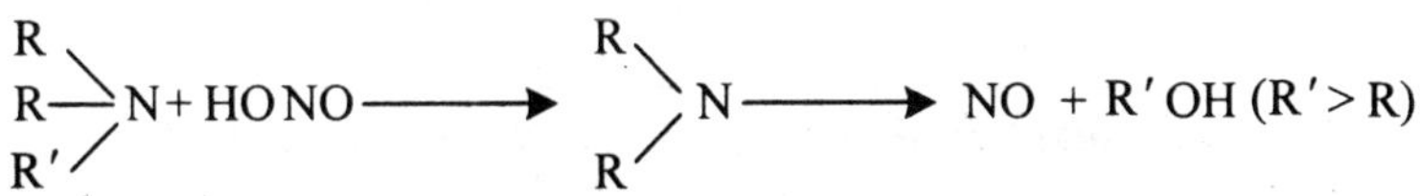

Nitrosoamine

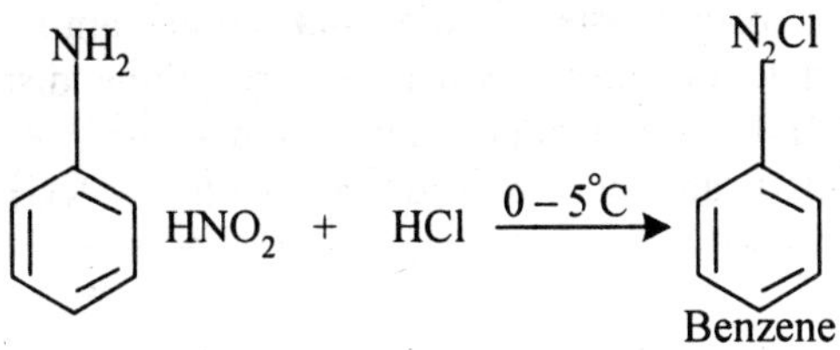

Benzene
diazonium chloride

This reaction is called *Diazotization.*

---

[8] See chapter 3 for details.

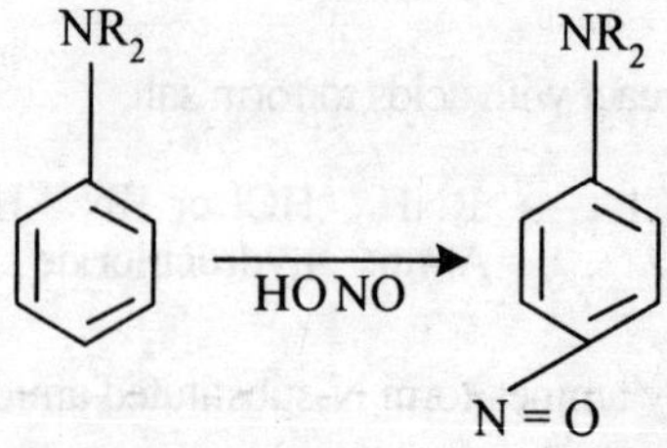

p-nitroso compound

4. Reaction with Benzene sulphonyl chloride (Heinsberg reagent)

$$RNH_2 + C_6H_5SO_2Cl \longrightarrow \underset{\text{N-substituted sulphonamide}}{RNHSO_2C_6H_5} \xrightarrow{\text{aq. KOH}} \underset{\text{Water soluble}}{RNKSO_2C_6H_5}$$

$$R_2NH + C_6H_5SO_2Cl \longrightarrow \underset{\text{N,N-disubstituted sulphonamide}}{R_2NHSO_2C_6H_5} \xrightarrow{\text{aq. KOH}} \underset{\text{Water insoluble}}{R_2NKSO_2C_6H_5}$$

$$R_3N + C_6H_5SO_2Cl \rightarrow \text{No reaction.}$$

5. Carbylamine Reaction: When primary amine (aliphatic or aromatic) is treated with $CHCl_3$ and KOH; phenyl isocyanide of unpleasant odour is formed.

$$RNH_2 + CHCl_3 + 3KOH \rightarrow \underset{\text{Isocyanide}}{RNC} + 3KCl + 3H_2O$$

Separation of primary, secondary and tertiary amines: There are two methods.

(i) *Hofmann method*: The mixture of primary, secondary and tertiary amines is treated with diethyloxalate when primary amine forms monoalkyl oxamide (solid), secondary amine forms dialkyl oxamic ester (liquid) and tertiary amine does not react. So, the content is subjected to distillation when tertiary amine distills off. The remaining content is filtered and solid oxamide and liquid oxamic esters are separated. The primary and secondary amines are recovered with aq. KOH. The reactions are as follows:

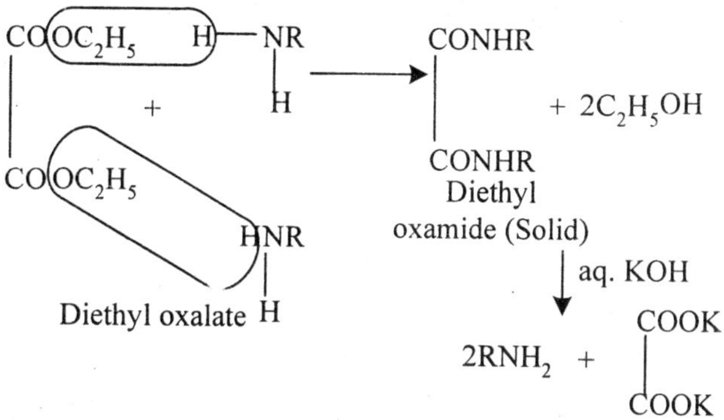

(ii) *Heinsberg method*: Heinsberg reagent is benzene sulphonyl chloride ($C_6H_5SO_2Cl$). The mixture of primary, secondary and tertiary amines is treated with Heinsberg reagent (HR) when primary amine forms monoalkyl sulphonamide, which is soluble in aq. KOH whereas secondary amine forms dialkyl sulphanamide which is also soluble in aq. KOH. The tertiary amine does not react.

$$\underset{\text{Benzene sulphonyl chloride}}{C_6H_5SO_2Cl} + \underset{\text{Primary amine}}{HNHR} \xrightarrow{HCl} \underset{\text{Alkyl mono = sulphonamide}}{C_6H_5SO_2HNR} \xrightarrow[-H_2O]{KOH} \underset{\text{Potassium salt (water soluble)}}{C_6H_5SO_2NKR}$$

$$\underset{\text{Secondary amines}}{C_6H_5SO_2Cl + HNR_2} \rightarrow \underset{\text{Dialkyl sulphonamide (insoluble in water)}}{C_6H_5SO_2NR_2} \xrightarrow{KOH} \text{No reaction}$$

$$C_6H_5SO_2Cl + \underset{\text{Tertiary amine}}{NR_3} \rightarrow \text{No reaction}$$

The whole content is extracted with ether when tertiary amine and dialkyl-sulphonamide passes into ether layer. The aqueous and ether layers are separated. The aqueous layer is acidified with dil. HCl, then heated with conc. HCl. This is then subjected to distillation with NaOH, when primary amine is regenerated:

$$C_6H_5SO_2NKR + HCl \xrightarrow[\text{Dil.}]{} C_6H_5SO_2NHR \xrightarrow[\text{conc. HCl}]{H_2O} C_6H_5SO_2OH + RNH_2HCl$$

$$RNH_2HCl \xrightarrow{\text{aq. NaOH}} RNH_2 + NaCl + H_2O$$

The ethereal layer is fractionally distilled when tertiary amine passes over. The remaining layer is hydrolyzed by conc. HCl and then distilled with NaOH to set free secondary amine.

$$C_6H_5SO_2NR_2 + H_2O \xrightarrow{\text{conc. HCl}} C_6H_5SO_2OH + R_2NH.HCl$$

$$R_2NH.HCl \xrightarrow{\text{aq. NaOH}} \underset{\text{sec. amine}}{R_2NH} + NaCl + H_2O$$

10.11 SOAPS

In a strictly chemical sense, any compound formed by the reaction of a water-insoluble fatty acid with a metallic radical or with an organic base may be called a "soap". Practically, however, the soap industry is concerned mainly with those water-soluble soaps which result from the interaction between fatty acids and alkali metals. In addition to the fatty acids proper, abietic acid and the naphthenic acids are also capable of forming salts with alkali metals, and these compounds are water soluble and display "soap-like" properties. Salts of fatty acids with heavy metals or with alkaline earth metals are water insoluble and are, as a rule termed "metallic soaps". They possess no detergent or soap-like properties.

Detergents

Many definitions of synthetic detergents have been proposed, all of which are very wide. The Comite International de Derives Tensio Actifs has, after several years of deliberation, agreed on the following definitions:

Detergent Product, the formulation of which is specially devised to promote the development of detergency. A detergent is a formulation comprising essential constituents (surface active agents) and subsidiary constituents (builders, boosters, fillers and auxiliaries).

Rancidity Due to presence of unsauration the fatty acids of triglyceride molecule, oxidation takes place at double bond. Further under the conditions of storage the oxidized products get decomposed into aldehyde and ketones. This develops the bad odour in the vegetable oil. This is rancidity.

Saponification Value The saponification value gives the amount of KOH (mg), required to saponify 1 gm of oil or fat. This value still does not give directly the actual amount of caustic soda required. This is simply calculated by multiplying the saponification value by 40/56.1. If the composition of the fatty material is compared with saponification value, it is found that the higher the percentage of fatty acids with lower molecular weight, the higher is the saponification value. This accounts for the fact that coconut oil needs more alkali for saponification than does tallow.

Iodine Value The hardness of the soap depends on the iodine value of the fats and on the molecular weight of the fatty acid: the lower the iodine value and the higher the molecular weight, the harder the soap will be. In order to produce a hard soap, a fat charge with low iodine value and low saponification value should be selected. Coconut and palm kernel oils have high saponification values, but owing to their extremely low iodine value they still produce hard, even brittle, soaps. Solubility and lathering property are generally higher in soaps from oils with higher iodine values and high saponification values. Again, coconut oil has an extremely low iodine value, but a very high saponification value, so it furnishes a special kind of hard soap which dissolves easily and lathers well in cold water.

PROBLEMS

1. How will you distinguish between ethylidene chloride and ethylene chloride.
2. Amongst following answers, one or more may be correct. Write the correct ones: **(IIT, 1984)**

   (i) *n*-propyl bromide reacts with ethanolic KOH to form
   (a) Propane *(b)* Propene
   (c) Propyne *(d)* Propanol **(IIT, 1987)**

   (ii) The reaction between a primary amine, chloroform and a few drops of alcoholic KOH is known as:
   (a) Cannizzaro's reaction
   (b) Frankland's reaction
   (c) Wurtz's reaction
   (d) Carbylamine reaction **(MNR, 1987)**

   (iii) The products of the reaction of alcoholic silver nitrate with ethylbromide:
   (a) Ethene *(b)* Nitroethane *(c)* Ethyl alcohol
   (d) Ethyl nitrite **(Delhi PMT, 1985)**

   (iv) When chloroform is left open in air and sunlight it forms:
   (a) Phosgene gas
   (b) Chlorine
   (c) Methyl chloride
   (d) Carbon tetrachloride **(MNR, 1981)**

   (v) Haloforms are trihalogen derivatives of
   (a) Ethane *(b)* Methane
   (c) Propane *(d)* Benzene **(PMT, 1985)**

   (vi) Only two isomeric monochloro derivatives are possible from:
   (a) *n*-Butene *(b)* 2,4-Dimethyl pentane
   (c) Benzene *(d)* 2-Methyl propane **(IIT, 1986)**

   (vii) Dipole moment is shown by
   (a) 1,4-Dichlorobenzene
   (b) *cis*-1,2-Dichloroethene
   (c) *trans*-1,2-Dichloroethene
   (d) *trans*-1,2-Dichloro-2-Pentane **(IIT, 1986)**

   (viii) Fire extinguisher is:
   (a) CO_2
   (b) CCl_4
   (c) CS_2
   (d) $CHCl_3$ **(Delhi PMT, 1985)**

(ix) Chloroform reacts with conc. HNO_3 to give

(a) Chloropicrin *(b)* Nitromethene

(c) Picric acid *(d)* Acetylene **(CPMT, 1986)**

(x) Ethyl alcohol is obtained when ethyl chloride is boiled with:

(a) Alcoholic KOH

(b) Aqueous KOH

(c) Water

(d) H_2O_2 **(MNR, 1982)**

3. What happens when:

(i) Chloral is heated with aq. NaOH

(ii) CCl_4 is treated with steam at 500 °C

(iii) CCl_4 is hydrolyzed with aq. KOH. **(IIT, 1984)**

4. Complete the following reactions:

(i) $(CH_3)_2-\underset{\displaystyle Cl}{\underset{|}{C}}-CH_2CH_3 \xrightarrow{\text{alc. KOH}}$ **(IIT, 1992)**

(ii) $CH_3CH_2CHCl \xrightarrow{\text{boil/alkali}}$ **(IIT, 1992)**

(iii) $CCl_4 + SbF_3 \xrightarrow{SbCl_5}$

(iv) $CCl_4 + HF \xrightarrow{C/FeCl_3}$

(v) $CHCl_3 \xrightarrow{SbF_3} ? \xrightarrow{850^\circ C} > CF_2 = CF_2 \xrightarrow[\text{Polymerization}]{\text{Free radical}} ?$

(vi) $\begin{matrix} CHCl_2 \\ | \\ CHCl_2 \end{matrix} \xrightarrow{Ca(OH)_2} ?$

(vii)  + $CHBr_3$ + t-Buok ⟶ H ?

(IIT, 1997)

5. For the following subquestions, four choices are given of which one or more may be chosen as the correct answers

(i) The compound, which reacts fastest with Lucas reagent at room temperature.

(a) Butane-1-ol

(b) Butane-2-ol

(c) 2-Methylpropane-1-ol

(d) 2-Methyl propane-2-ol **(IIT, 1981; CPMT, 1989)**

(ii) Which of the following compounds will give a yellow precipitate with I_2 and alkali.

(a) 2-Hydroxy propane *(b)* Methyl acetate

(c) Acetophenone *(d)* Acetamide **(IIT, 1984)**

(iii) An industrial method of preparation of methanol is

(a) Catalytic reduction of carbon monoxide in presence of Zno-Cr_2O_3

(b) By reacting methane with steam at 900 °C with a Ni-catalyst

(c) By reducing HCHO with $LiAlH_4$

(d) By reacting HCHO with aq. NaOH **(IIT, 1984)**

(iv) The compound with the highest boiling point is

(a) CH_4

(b) CH_3OH

(c) CH_3Cl

(d) CH_3Br **(MNR, 1985)**

(v) HBr reacts fastest with

(a) 2-Methyl propane-2-ol

(b) Propan-1-ol

(c) Propan-2-ol

(d) 2-Methyl propan-1-ol **(IIT, 1986)**

(vi) Which one of the following compounds is oxidized to prepare dimethyl ketone

(a) 2-Propanol

(b) 1-Butanol

(c) 2-Butanol

(d) *t*-Butyl alcohol **(IIT, 1987)**

(vii) Lucas reagent is used to distinguish between primary, secondary and tertiary:

(a) Alkyl halides

(b) Alcohols

(c) Aliphatic amines

(d) Aromatic amines **(IIT, Screening 1987, 1990; AISB, 1990; Roorkee, 1988)**

(viii) The reagent anhy. $ZnCl_2$ and conc. HCl is used to identify the type of

(a) Nitroalkanes *(b)* Alkyl amines

(c) Alkyl halides *(d)* Alcohols **(IIT, Screening, 1990)**

(ix) 23 gm of sodium reacts with 23 gm of ethyl alcohol. The volume of H_2 evolved at NTP is:

(a) 22.4 lit *(b)* 11.2 lit

(c) 5.6 lit *(d)* 2.8 lit

(x) Rectified spirit is a mixture of:

(a) 95% Ethyl alcohol + 5% water

(b) 94.4% Ethyl alcohol + 5.43% water

(c) 94% Ethyl alcohol + 4.53% water

(d) 95.57% Ethyl alcohol + 4.43% water **(Delhi PMT, 1982)**

(xi) Primary, secondary and tertiary alcohols may be distinguished by employing:

(a) Fehling solution test

(b) Victor-Meyer test
(c) Hoffmann test
(d) Beilstein test (CPMT, 1980)

(xii) Crown compounds are:
(a) Simple ether (b) Mixed ethers
(c) Cyclic ethers (d) Primary alcohols

(xiii) What is 'A' in the following reaction:

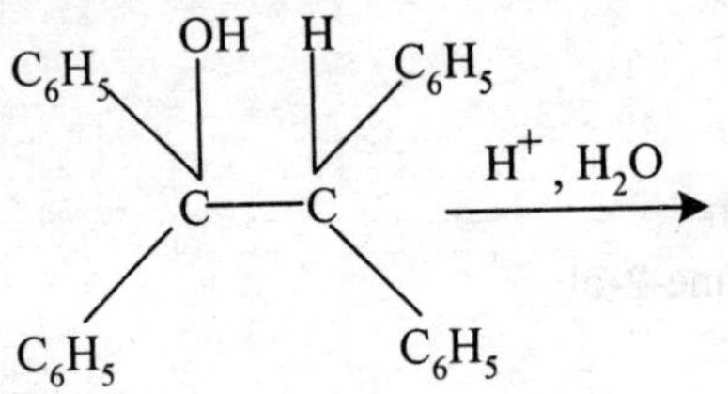

(a) $(C_6H_5)_3C-\overset{O}{\overset{\|}{C}}-C_6H_5$

(b) $(C_6H_5)_3-\overset{O}{\overset{\|}{C}}-\overset{O}{\overset{\|}{C}}-C_6H_5$

(c) C_6H_5COOH
(d) None of the above.

(xiv) Acetyl chloride does not react with:
(a) Diethyl ether (b) Aniline
(c) Phenol (d) Ethanol (MNR, 1995)

(xv) Denatured spirit is mainly used as a
(a) Good fuel
(b) Drug
(c) Solvent in preparing varnishes
(d) Material in the preparation of oil (MNR, 1995)

(xvi) Tonics in general contain:
(a) Ether
(b) Methanol
(c) Ethanol
(d) Rectified spirit (MNR, 1995)

(xvii) Which of the following gives +ve iodoform test:
(a) Formic acid
(b) Propionic acid
(c) Pyruvic acid
(d) Benzoic acid

(xviii) The enzyme which can catalyze the conversion of glucose to ethanol is:

(a) Zymase *(b)* Maltase

(c) Diastase *(d)* Lipase **(CPMT, 1983, 1984, 1986)**

(xix) Hydrogen combines with CO in the presence of catalyst consisting of the oxides of Zn and Cr at 30 °C to give:

(a) Ethyl alcohol

(b) Methyl alcohol

(c) Formaldehyde

(d) Formic acid

(xx) Ethyl alcohol is industrially prepared from ethylene by

(a) Permanganate oxidation

(b) Catalytic reduction

(c) Absorbing of H_2SO_4, followed by hydrolysis

(d) Fermentation **(CPMT 1985)**

(xxi) Ethylene glycol on oxidation with HIO_4 gives

(a) Oxalic acid *(b)* Glycol

(c) Formaldehyde *(d)* Glycollic acid

(xxii) Which of the following is not a characteristic of alcohols:

(a) They are lighter than water.

(b) Their boiling points rise fairly uniformly with rising molecular weight.

(c) Lower members are insoluble in water with increasing molecular weight.

(d) Lower members have a pleasant smell and burning taste, higher members are odourless and tasteless.

(xxiii) 'Grain alcohol' is the common name of:

(a) Amyl alcohol *(b)* Ethyl alcohol

(c) Methyl alcohol *(d)* None of these

(xxiv) What is the decreasing order of strength of the bases: OH^-, NH_2^-, $HC\equiv C^-$ and $CH_3\ CH_2^-$?

(a) $CH_3CH_2^- > NH_2^- > H-C\equiv C^- > OH^-$

(b) $H-C\equiv C^- > CH_3-CH_2^- NH_2^- > OH^-$

(c) $OH^- > NH_2^- > HC\equiv C^- > CH_3-CH_2^-$

(d) $NH_2^- > HC\equiv C^- > OH^- > CH_3-CH_2^-$

(xxv) Identify the set of reagents/reaction conditions 'x' & 'y' in the following transformations

$$CH_3-CH_2-CH_2-Br \xrightarrow{X} \text{Product} \xrightarrow{Y} CH_3-\underset{\underset{Br}{|}}{CH}-CH_3$$

(a) X = dilute aq. NaOH; 20°C, Y = HBr/Acetic acid, 20°C

(b) X = Concentrated alcohlic NaOH; 80°C, Y = HBr/Acetic acid, 20°C

(c) X = dilute aqueous NaOH; 20°C, Y = $Br_2/CHCl_3$, O°C

(d) X = Concentrated alcoholic NaOH; 80°C, Y = $Br_2/CHCl_3$, 0°C

(xxvi) Compound A (molecular formula C_3H_8O) is treated with acidified potassium dichromate to form a product B (molecular formula C_3H_6O). 'B' forms a shining silver mirror on warming with ammonical $AgNO_3$. 'B' when treated with an aqueous solution of $H_2NCONHNH_2{\cdot}HCl$ and sodium acetate gives a product 'C'. Identify the structure of 'C':

(a) $CH_3CH_2CH{=}N.NHCONH_2$

(b) $CH_3-\underset{\underset{CH_3}{|}}{C}=NNHCONH_2$

(c) $CH_3-\underset{\underset{CH_3}{|}}{C}=NCONHNH_2$

(d) $CH_3CH_2CH{=}NCONHNH_2$

(xxvii)

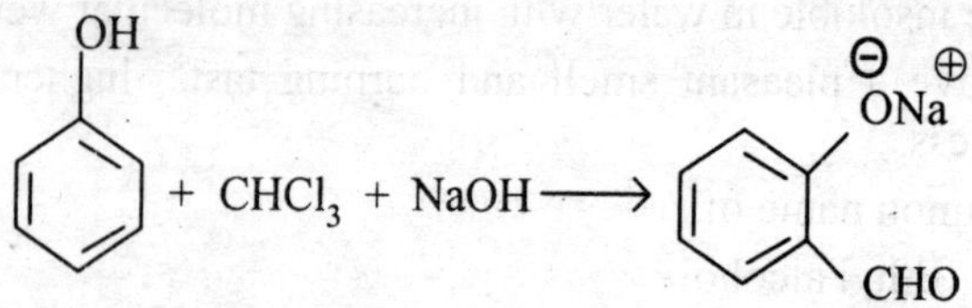

The electrophile involved in the above reaction is **(AIEEE 2006)**

(a) formyl cation ($\overset{\oplus}{C}HO$)

(b) Dichloromethyl cation $\overset{\ominus}{C}HCl_2$

(c) Dichloro Carbone (: CCl_2)

(d) Trichloromethyl anion $\left(\overset{\ominus}{C}Cl_3\right)$

(xxviii) The increasing order of the rate of HCN addition in compound A — D is

(AIEEE 2006)

(A) HCHO (B) CH_3COCH_3

(C) $PhCOCH_3$ (D) PhCOPh

(a) C < D < B < A *(b)* A < B < C < D

(c) D < B < C < A *(d)* D < C < B < A

6. State whether *True* or *False*. If *False*, give correct statement:
   (i) When 3-methyl-2 butanol reacts with HBr, the product is 3-bromo-3-methyl butane
   (ii) All alkanol-2 are optically active and give +ve iodoform test
   (iii) Isobutanol on dehydration with conc. H_2SO_4 gives 2-butene
   (iv) Trimethyl acetaldehyde undergoes Cannizzaro reaction **(MNR, 1990)**
7. Fill in the blanks:
   (i) When 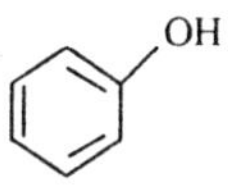 reacts with $SOCl_2$, the product is ....
   (ii) Aliphatic ethers are purified by shaking with a solution of ferrous salt to remove .... which are formed on prolonged standing in contact with air. **(IIT, 1992)**
   (iii) .... is used for preparing alkyl resins, acid-proof cement and inks.
   (iv) ....is used for preparing terylene.
   (v) Cyclic ethers are called ....
8. Assuming the choice to be limited to alkane, alkene, alkyl halide, secondary, identify A, B, C, D and E on the basis of the following information.

Compound	*Elemental analysis*	*H_2SO_4*	*Br_2/CCl_4*	*KMno$_4$*	*CrO_3*
A	—	Insoluble	—	—	—
B	—	Soluble	—	—	+
C	Cl	Insoluble	—	—	—
D	—	Soluble	+	+	—
E	—	Soluble	—	—	—

9. An alkaloid ($C_{20}H_{21}O_4N$) was heated with HI followed by treatment with $AgNO_3$, when yellow ppt. was obtained, thereby showing presence of methoxy group, 4.34 mg of compound $C_{20}H_{21}O_4N$ gave 11.62 mg of AgI. Calculate the number of methoxy groups/molecule.
10. Convert ethyl alcohol into vinyl acetate. **(MNR, 1995)**
11. Complete the following reactions:

(i)
$$\begin{array}{l} CH_2OH \\ | \\ CHOH + KHSO_4 \rightarrow ? \\ | \\ CH_2OH \end{array}$$

(ii)
$$\begin{array}{l} CH_2OH \\ | \\ CHOH + HI \rightarrow ? \\ | \\ CH_2OH \end{array}$$

(iii) $\begin{array}{l} CH_2OH \\ | \\ CHOH \\ | \\ CH_2OH \end{array} + \begin{array}{l} COOH \\ | \\ COOH \end{array} \rightarrow ?$

(iv) $\begin{array}{l} CH_2OH \\ | \\ CHOH \\ | \\ CH_2OH \end{array} + HNO_3 \rightarrow ?$

12. An organic compound forms different compounds under different conditions'. Justify the statement, taking oxidation of glycerol as suitable example.

13. Choose the correct answer from the alternatives given below (one or more may be correct):

(i) A compound that gives +ve iodoform test is:
(a) 1-Pentanol *(b)* 2-Pentanone
(c) 3-Pentanone *(d)* Pentanol-2 **(IIT, 1982)**

(ii) The compound that will not give iodoform on treatment with alkali and iodine is.
(a) Acetone *(b)* Ethanol
(c) Diethyl ketone *(d)* Isopropyl alcohol **(IIT, 1985)**

(iii) Which one of the following compounds is oxidized to prepare ethylmethyl ketone
(a) 2-Propanol *(b)* 1-Butanal
(c) 2-Butanol *(d)* *t*-Butyl alcohol **(IIT, 1987)**

(iv) When calcium acetate and calcium formate are heated, they form
(a) Acetone *(b)* Formic acid
(c) Acetic acid *(d)* Acetaldehyde **(MNR, 1982)**

(v) Of the following, the functional group reduced by Wolif-Kishner reduction is
(a) 
(b) >C=C <
(c) —C≡N
(d) >C=O **(IIT Screening, 1990; AISB, 1989)**

(vi) Cannizzaro's reaction is related to:
(a) Formaldehyde
(b) Acetaldehyde
(c) Acetone
(d) All of the above **(Roorkee, 1989)**

(vii) Aldehydes may be distinguished from ketones by the use of
(a) Fehling's solution
(b) Conc. H_2SO_4
(c) Pyrogallol

(d) Grignard reagent **(Delhi PMT, 1982)**

(viii) Which compound is obtained when calcium acetate is distilled dry

(a) HCHO

(b) CH_3CHO

(c) $CH_3-\overset{\overset{O}{||}}{C}-CH_3$

(d) CH_3COOH **(MNO, 1985; Delhi PMT, 1981)**

(ix) When acetaldehyde is heated with Fehling's solution it gives a ppt. of

(a) Cu + CuO

(b) Cu_2O

(c) Cu + Cu_2O + CuO

(d) $CuSO_4$ **(IIT, 1983)**

(x) The Cannizzaro's reaction is not given by

(a) Trimethyl acetaldehyde *(b)* Acetaldehyde

(c) Benzaldehyde *(d)* Formaldehyde **(BHU, 1980)**

(xi) Aldehydes are produced in atmosphere by

(a) Oxidation of secondary alcohols

(b) Reduction of alkenes

(c) Reaction of oxygen atoms with hydrocarbons

(d) Reaction of oxygen atoms with alkanes **(NCERT, 1982)**

(xii) When 0.66 gm of a mixture of benzaldehyde and acetaldehyde is heated with Fehling solution, 1.43 gm of brown ppt. is obtained. The percentage of benzaldehyde in the mixture is:

(a) 25.5% *(b)* 33.3%

(c) 66.6% *(d)* It cannot be calculated

(xiii) Which of the following compounds would give yellow ppt. with I_2 and NaOH.

(a) 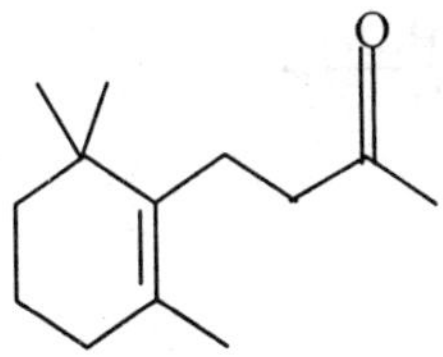

β – ionone

(b) 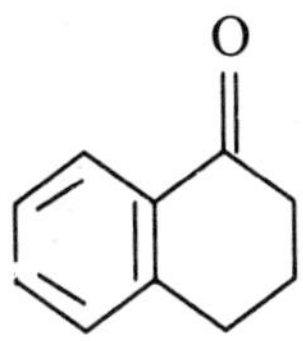

Tetralone

(c) 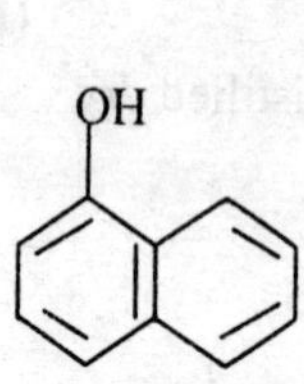

α – napthol

(d) 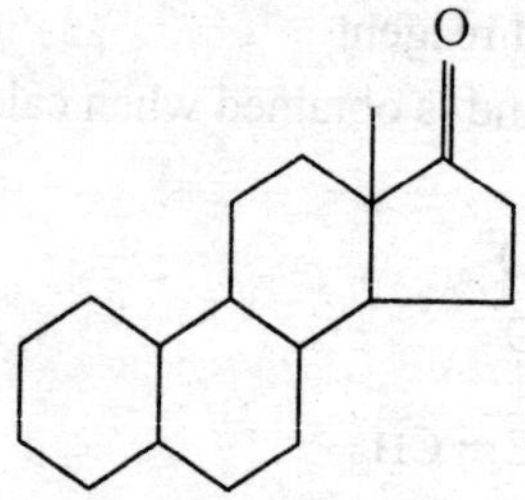

Progesterone

(xiv) The carbon in the carbonyl group is normally most susceptible to attack by

(a) Nucleophiles

(b) Free radicals

(c) Electrophiles

(d) Acids

14. Fill in the blanks:

(i) Fehling's solution 'a' contains .... and 'b' contains............. **(IIT, 1990)**

(ii) Enolic form of acetylacetone showing H-bonding is .... **(IIT, 1993)**

(iii) Reduction of ketone with $NaBH_4$ generates a ....alcohol. **(GATE, 1994)**

(iv) Structure of polymer of formaldehyde is.................

(v) Structure of polymer of acetaldehyde is .... **(MNR, 1994)**

(vi) Formalin is ....

15. Arrange the following according to the instruction:

(i) $HCHO$, CH_3CHO, $(CH_3)_2CO$ (in order of reactivity)

(ii) A ketal, acetal and formal (ease of hydrolysis).

(iii) CH_3CHO, CH_3COCH_3, $HCHO$, $C_2H_5COCH_3$ (increasing reactivity towards HCN). **(IIT, 1985)**

16. 0.532 gm of a mixture of ethanol and acetaldehyde when heated with Fehling's solution produced 1.2 gm of brown ppt. What is the percentage of ethanol? **(Roorkee, 1984)**

17. Complete following reactions:

(i) $CH_3CH{=}CHCHO \xrightarrow{NaBH_4} ? \xrightarrow[Zncl_2]{HCl} ? \xrightarrow[H^+]{HCN} ?$ **(IIT, 1991)**

(ii) 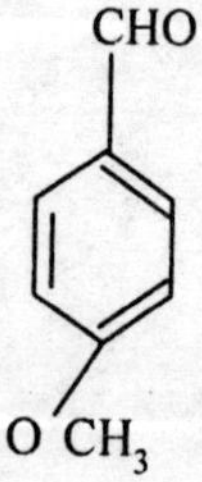 $+ CH_2O \xrightarrow{NaOH} ?$

(iii) $C_6H_5CHO + CH_3COCH_2CH_3 \xrightarrow{NaOH/H_2O} ?$

(iv) $C_6H_5CHO + CH_3COCH_2CH_3 \xrightarrow[CH_3COOH]{H_2SO_4} ?$

(v) $2CH_3COCH_3 \xrightarrow{\text{Base catalyst}} ? \xrightarrow{\text{Acid catalyst}} ?$ **(IIT, 1998)**

18. How will you distinguish between acetaldehyde and acetone? **(IIT, 1987)**

19. Carry out the following conversion: Ethanal to 2-hydroxy-3-butenoic acid. **(IIT, 1990)**

20. Amongst the four answers given below, one or more than one, is correct. Write the correct answer(s):

(i) The acids which do not contain –COOH group is:
- *(a)* Ethanoic acid
- *(b)* Picric acid
- *(c)* Lactic acid
- *(d)* Palmitic acid **(IIT, 1990)**

(ii) When formic acid reacts with PCl_3, it forms
- *(a)* Formyl chloride *(b)* Acetyl chloride
- *(c)* Methyl chloride *(d)* Propionyl chloride **(MNR, 1982)**

(iii) Which acid reduces Tollen's reagent
- *(a)* Formic acid *(b)* Acetic acid
- *(c)* Oxalic acid *(d)* Propionic acid

(iv) Which of the following gives +ve iodoform test
- *(a)* Acetic acid *(b)* Pyruvic acid
- *(c)* Propanol-1 *(d)* Pentanone-3

(v) Formic acid and acetic acid may be distinguished by reaction with:
- *(a)* Sodium
- *(b)* Dilute acid permanganate
- *(c)* 2,4-Dinitrophenyl hydrazine
- *(d)* Sodium ethoxide **(AIMS, 1983)**

(vi) Which of the following is not a fatty acid
- *(a)* Stearic acid *(b)* Palmitic acid
- *(c)* Oleic acid *(d)* Phenyl acetic acid **(CPMT, 1998)**

(vii) When ammonium acetate is strongly heated, the product is:
- *(a)* Acetamide *(b)* Methyl cyanide
- *(c)* Urea *(d)* Nitromethane **(MLNR, 1995)**

(viii) The product when $CH_3CH{=}CHCOOH$ is reduced by $LiAlH_4$ is
- *(a)* $CH_3CH_2CH_2COOH$
- *(b)* $CH_3CH{=}CHCHO$
- *(c)* $CH_3CH_2CH_2OH$
- *(d)* $CH_3{-}CH{=}CHCH_2OH$

(ix) The strongest acid amongst the following is:
- *(a)* Formic acid *(b)* Acetic acid

(c) Benzoic acid (d) Trichloroacetic acid

(x) The order of nucleophilic attack at the carbonyl carbon of the acid derivative:
(a) Acid chloride > Ester > Acid amide
(b) Acid amide > Ester > Acid chloride
(c) Acid amide > Acid chloride > Ester
(d) Acid chloride > Acid amide > Ester

(xi) Acetamide is treated separately with the following reagents. Which one of these would give methyl amine?
(a) PCl_5 (b) $NaOH + Br_2$
(c) Sodalime (d) Hot conc. H_2SO_4

(xii) Reaction of $RCONH_2$ with a mixture of Br_2 and KOH gives RNH_2 as the main product. The intermediates involved in this reaction are:
(a) RCONHBr
(b) RNHBr
(c) $RCON\begin{smallmatrix}Br\\Br\end{smallmatrix}$
(d) R—N=C=O **(IIT, 1994)**

(xiii) On strong heating, ammonium acetate gives:
(a) Acetamide (b) Methyl cyanide
(c) Urea (d) Fromamide **(MNR, 1995)**

(xiv) Diazomethane reacts with carboxylic acid to yield
(a) Amine
(b) Esters
(c) Alcohol
(d) Imines

(xv) In the Cannizzaro reaction given below,. $2PhCHO \xrightarrow{OH^-} Ph—CH_2OH + PhC\overline{O}_2$. The slowest step is:
(a) The attack of $OH^{\ominus}$ at the carbonyl group
(b) The transfer of hydride to the carbonyl group
(c) The abstraction of proton from the carboxylic acid
(d) The deprotonation of Ph—CH_2OH.

21. In the following subquestions, identify the compounds/reaction conditions represented by the alphabets A, B, C, D, E, F, G and H:

(i) $CH_3COONH_4 \xrightarrow{\Delta\, P_2O_5} B \xrightarrow{\Delta\, P_2O_5} C$

(ii) $CH_3COONH_4 \xrightarrow{PCl_5} E \xrightarrow{NH_3} F \xrightarrow{\Delta\, P_2O_5} G \xrightarrow{H_2/Ni} H$

(iii) $$\begin{array}{l} CH_2OH \\ | \\ CHOH \\ | \\ CH_2OH \end{array} + \begin{array}{l} COOH \\ | \\ COOH \end{array} \xrightarrow{110\,^{\circ}C} A + B$$

(iv) $SOCl_2 + CH_3COOH \rightarrow A + SO_2\ HCl$

(v) $HCOONa \xrightarrow{360\ ^{\circ}C} A + B$

(vi) $$\begin{array}{l} CH_3CO \diagdown \\ \qquad\qquad O \\ CH_3CO \diagup \end{array} \xrightarrow[\text{ether}]{LiAlH_4\ \text{in}} A$$

(vii) $CH_3COOH + A \rightarrow ClCH_2COOH + \text{Excess } NH_3 \rightarrow B$ **(IIT, 1988)**

22. State whether *True* or *False*. If false, give correct statement with proper reasoning:
Amongst HCOOH, CH_3COOH, $CH_2ClCOOH$, $CHCl_2COOH$, CCl_3COOH and C_6H_5COOH, only CH_3COOH will react with P and Cl_2.

23. Give reasons in one or two sentences:
Acetic acid reacts with P and Cl_2 but formic acid does not **(IIT, 1981)**

24. Fill in the blanks:

(i) Acetyl chloride on hydrogenation in presence of coarsely-divided Pd supported on $BaSO_4$ gives ....

(ii) Strongest aliphatic carboxylic acid is.............. **(GATE, 1990)**

(iii) Formic acid when heated with conc. H_2SO_4 is ....

25. Complete the following with appropriate structures:

$$(CH_3CO)_2O \xrightarrow{C_2H_5OH} CH_3COOH^+\ ?$$ **(IIT, 1986)**

26. Carry out the following conversions:

(i) Ethanoic acid into a mixture of methanoic acid and diphenyl ketone. **(IIT, 1990)**

(ii) Formic acid into oxalic acid

(iii) Acetaldehyde into acetoacetic ester

(iv) Acetone into methyl methacrylate

27. Four alternate answers are provided for each subquestion. One or more is correct. Write correct answer(s)

(i) Ethylamine reacts with nitrous acid to form

(a) Methyl alcohol *(b)* Ethyl alcohol

(c) Ethane *(d)* Ethylnitrate **(MNR, 1982)**

(ii) The compound which on reaction with aq. nitrous acid at low temperature produces an oily nitrosoamine, is

(a) Methyl amine *(b)* Ethyl amine

(c) Diethyl amine (d) Triethyl amine **(IIT, 1981; CPMT, 1989)**

(iii) $CH_3-\overset{\overset{\large CH_3}{|}}{\underset{\underset{\large CH_3}{|}}{C}}-NH_2$ is a

(a) Primary amine
(b) Secondary amine
(c) Tertiary amine
(d) Quaternary salt.

(iv) The reaction between primary amine, chloroform and few drops of alc. KOH is known as:

(a) Cannizzaro reaction
(b) Frankland reaction
(c) Wurtz reaction
(d) Carbylamine reaction **(MNR, 1987)**

(v) Which compound is obtained at the end of the following reactions: Ethyl amine $\xrightarrow{HNO_2}$ A $\xrightarrow{PCl_5}$ B $\xrightarrow{NH_3}$ C

(a) Ethyl cyanide (b) Ethylamine
(c) Methylamine (d) Acetamide **(CPMT, 1989)**

(vi) Which one of the following will give primary amine on hydrolysis:

(a) Nitroparaffin (b) Alkyl cyanide
(c) Aldoxime (d) Alkyl isocyanide **(BHU, 1982)**

(vii) Gabriel's phthalimide synthesis is used in the preparation of

(a) Primary aromatic amine
(b) Secondary amine
(c) Primary aliphatic amine
(d) Tertiary amine **(Delhi, PMT, 1983; CPMT, 1982)**

(viii) When ethyl amine is heated with chloroform and alc. KOH, a bad odour compound is formed. This compound is

(a) A secondary amine
(b) An isocyanide
(c) A cyanide
(d) An acid **(CPMT, 1983, 1984)**

(ix) Which of the following reactions does not yield an amine

(a) $R-X + NH_3 \rightarrow$
(b) $RCH{=}NOH + [H] \xrightarrow{Na/C_2H_5OH}$ **(CPMT, 1989)**
(c) $RC{\equiv}N + H_2O \longrightarrow H^+$
(d) $R-CONH_2 + 4[H] \xrightarrow{LiAlH_4}$

(x) Carbylamine test is performed in alcoholic KOH by heating a mixture of
- (a) Chloroform and silver powder
- (b) Trihalogenated methane and a primary amine
- (c) An alkyl halide and a primary amine
- (d) An alkyl cyanide and a primary amine **(IIT, 1984)**

(xi) How many primary amines are possible for the formula $C_4H_{11}N$.
(a) 1 (b) 2 (c) 3 (d) 4

28. Fill in the blanks:
- (a) Reaction of acetaldoxime produces...............
- (b) If the alkyl groups of a tertiary amine are all different, the amine will be ....
- (c) The structure of ammonium ion is ....
- (d) When acyl chloride is treated with sodium azide, acyl azide is formed. When heated and hydrolyzed it produces primary amine. The rearrangement is known as.........

29. State whether '*True*' or '*False*'. If false, write correct statement.
- (i) Acetamide is more basic than ethylamine.
- (ii) 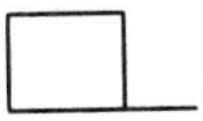 NH_2 is a primary amine.
- (iii) Amides and amines both dissolve in dilute aqueous HCl, HBr, HI or H_2SO_4

30. Complete the following reactions:
- (i) $C_2H_5NH_5 + CS_2 + HgCl_2 \rightarrow$
- (ii) $RNH_2 + R'CHO \rightarrow$
- (iii) 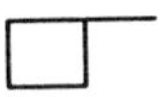 $CH_2\overset{\oplus}{N}(CH_3)_3 \xrightarrow{OH^-/\text{Heat}}$
- (iv) $RMgBr + CO_2 \rightarrow A \xrightarrow{H_2} \xrightarrow{SnCl/CH_2N_2} B \xrightarrow{Ag_2O} C \xrightarrow{H_2O} D$
- (v) $RCH_2NH_2 + 2CH_2O + 2HCOOH \rightarrow$
- (vi) $EtNH_2 + KCN + Br_2 \rightarrow$

31. Explain the following:
- (i) Benzylmethyl amine cannot be resolved into enantiomeric forms under ordinary conditions.
- (ii) In a group of isomeric amines, the tertiary amine has the lowest boiling point.

32. How will you distinguish between:
- (i) Benzyl amine and benzamide
- (ii) Allyl amine and propyl amine
- (iii) Cyclohexyl amine and aniline
- (iv) Ethyl amine and diethyl amine **(IIT, 1988)**

33. Amine (A) $C_6H_{15}N$ on treatment with CH_3I and then KOH, gives (B) $C_8H_{20}NOH$. This on heating produces isobutene and amine (C) $C_4H_{11}N$. What is A?

34. 'A' (C_3H_9N) reacts with benzene sulphonyl chloride to give a solid insoluble in alkali. What is 'A'? **(IIT, 1993)**

35. Write the structure of the foul-smelling compound obtained when aniline is treated with chloroform in the presence of KOH. **(IIT, 1996)**

ANSWERS

1. $$CH_3CHCl_2 \xrightarrow{\text{aq. KOH}} CH_3CH\begin{matrix} OH \\ OH \end{matrix} \longrightarrow CH_3CHO$$

Ethylidene chloride

$$\begin{matrix} CH_2Cl \\ | \\ CH_2Cl \end{matrix} + \text{aq. KOH} \longrightarrow \begin{matrix} CH_2OH \\ | \\ CH_2OH \end{matrix}$$

Ethylidene chloride — Glycol

2. *(i)* **(b)**, with ethanolic KOH undergoes elimination reaction
(ii) **(d)**.
(iii) **(b)** and **(d)**
(iv) **(a)**
(v) **(b)**.
(vi) **(a)** and **(b)** (see alkanes, both (a) and (b) contain two types of H atom.
(vii) **(b)** and **(d)**.
(viii) **(b)**.
(ix) **(a)**.
(x) **(b)**.

3. *(i)* See Cannizzaro reaction.

(ii) $CCl_4 + H_2O \xrightarrow{500°C} COCl_2 + 2HCl.$

(iii) $CCl_4 + 6\ \text{aq.} KOH \rightarrow 3H_2O + K_2CO_3 + 4KCl$

4. *(i)* $$CH_3-\overset{\overset{CH_3}{|}}{C}=CH-CH_3$$ See Saytzeff rule

2 – methyl Butene – 2

(ii) CH_3CH_2CHO

(iii) $3CCl_4 + 2SbF_3 \xrightarrow{SbCl_5} 2HCl + 3CCl_2F_2$ (Freon)

(iv) $CCl_4 + 2HF \xrightarrow{C/FeCl_3} CCl_2F_2 + 2HCl$

(v) $CHCl_3 \xrightarrow{SbF_3} CHF_2Cl \xrightarrow{850°C} CF_2{=}CF_2 + 2HCl \quad (CF_2—CF_2)_n$

Free radical polymerization ↑ Teflon

(vi)

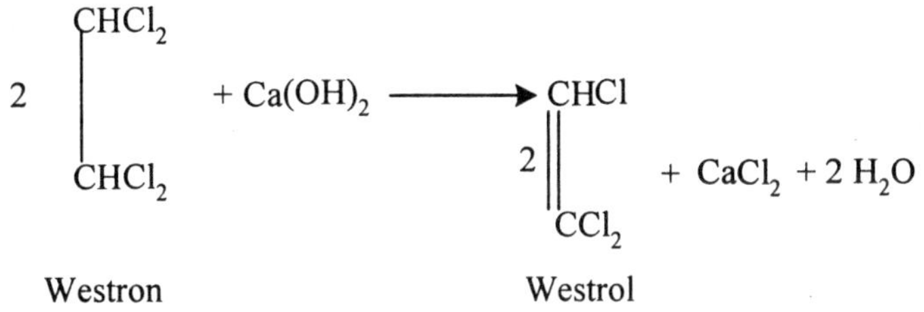

(vii)

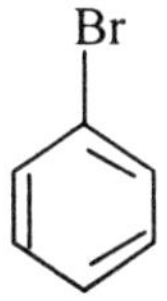

(i) (**d**), (**b**) and (**c**) are primary alcohol (**a**) is secondary alcohol

(ii) (**a**) and (**b**), see Iodoform test, Acetamide (CH_3CONH_2) and methylacetate (CH_3COOCH_3) do not contain

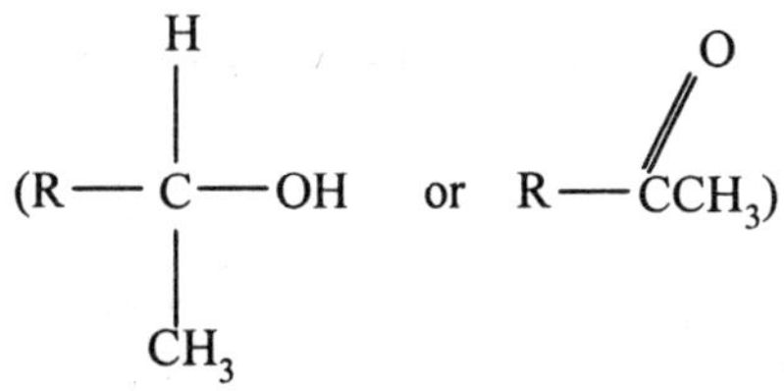

(iii) (**a**)

(iv) (**b**), due to hydrogen bonding

(v) (**a**), alcohol reactivity tertiary > secondary > primary

(vi) (**a**), secondary alcohol on oxidation gives ketone

(vii) (**b**) is primary alcohol and (d) is tertiary alcohol

(viii) (**c**) butanol-2 on oxidation gives methylethyl ketone

(ix) (**c**),

Na	+	C_2H_5OH	→	C_2H_5OHa	+ $\frac{1}{2}H_2$
23		46			11.2 lit at NTP
23		23			5.6 lit at NTP

(x) (**d**).

(xi) (**b**).

(xii) (**c**).

(xiii) (**a**), see rearrangement of carbonium ion.

(xiv) (**a**), It is because acetyl chloride reacts with NH_2 and –OH group.

(xv) (**c**).

(xvi) (**c**).

(xvii) $CH_3-\overset{\displaystyle O}{\overset{\|}{C}}-COOH$ is pyruvic acid. (*Note*: contains CH_3CO group.)

(xviii) (**a**).

(xix) (**b**).

(xx) (**c**), see Chapter on Carbonhydrates.

(xxi) (**a**).

(xxii) (**c**).

(xxiii) (**b**).

(xxiv) (**a**).

(xxv) (**b**).

(xxvi) (**d**).

6. *(i)* *False*. Product is 2-bromo-2-methyl butane.

(ii) *True*, Except propanol-2 which is optically inactive

False,

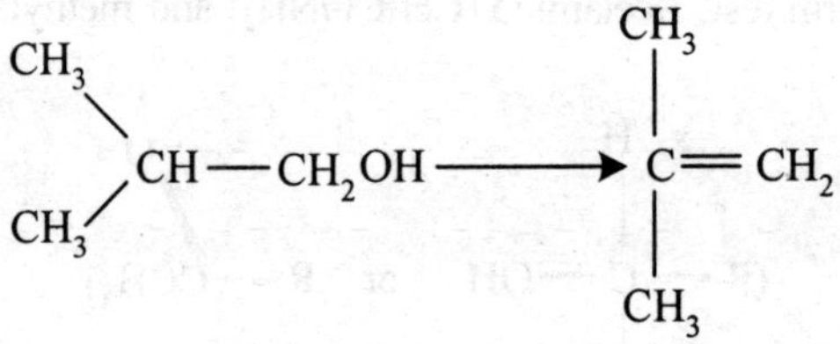

2-methyl propene

(iii)

(iv) *True*.

7. *(i)*

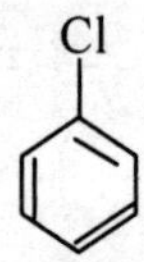

(ii) Peroxides.

(iii) Glycerol.

(iv) Glycol.

(v) Crown ether.

8. A—Alkane, because it is insoluble and is not affected by any reagent.
B—Secondary alcohol, because response is +ve with CrO_3
C—Alkyl, because it contains Cl.
D—Alkene, because it reacts with bromine water, $KMnO_4$
E—Tertiary alcohol, because response is –ve with CrO_3

9. 3 methoxy groups, moles of AgI obtained from 1 mole of compound $= \dfrac{11.62\times10^{-3}\times 339}{4.24\times10^{-3}\times 235} = 3.$

10.

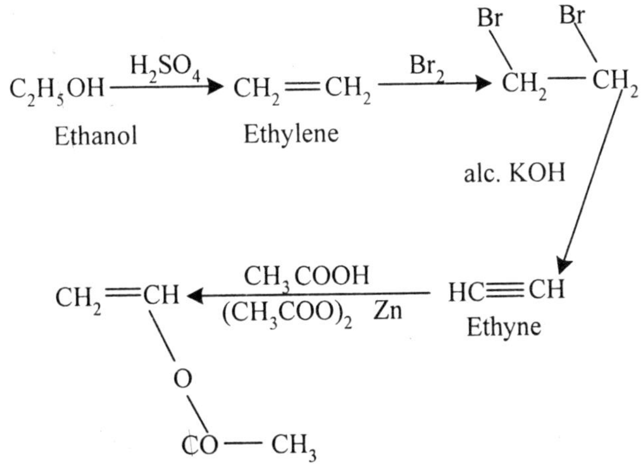

Vinyl acetate

11. When glycerol is treated with a dehydrating agent like P_2O_5 or $KHSO_4$, acrolein is formed.

(i)

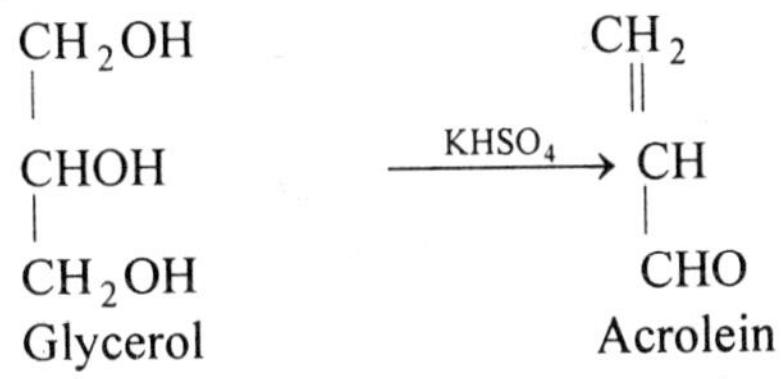

($CH_2{=}C{=}CHOH$ is the reaction intermediate)

(ii) [9]

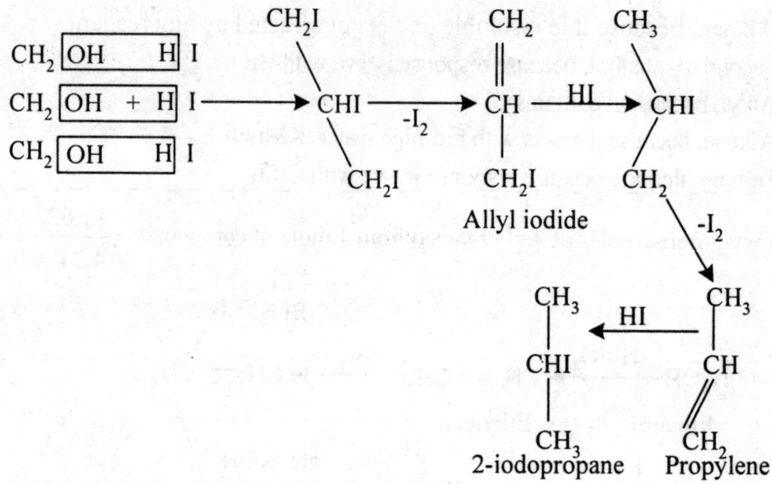

(iii) *(a)*

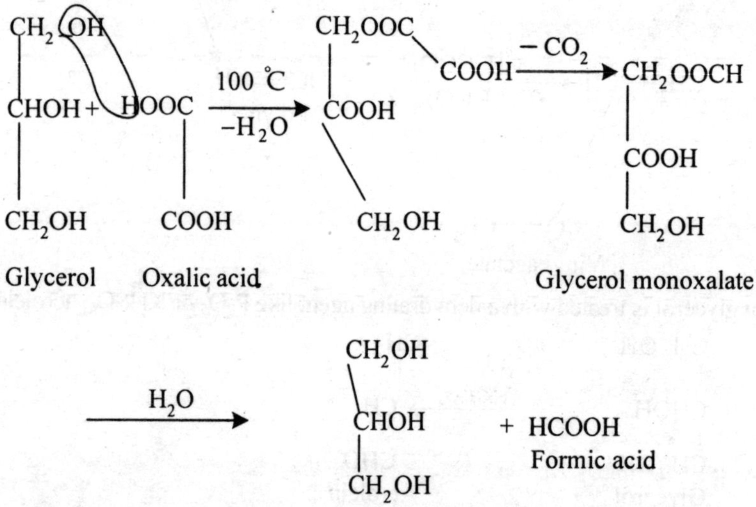

(b)

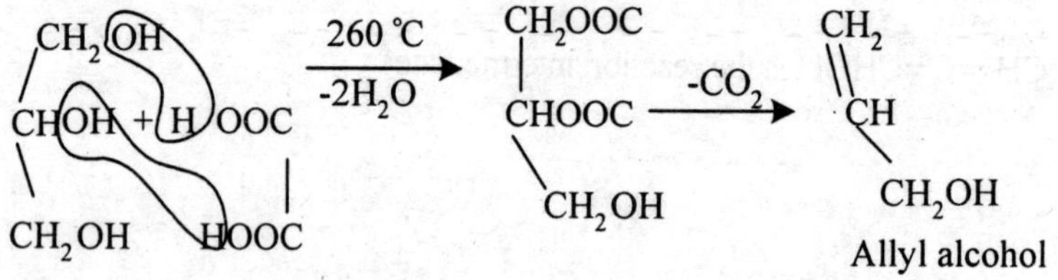

---

[9] Addition takes place according to Markonivkov's rule.

(c) 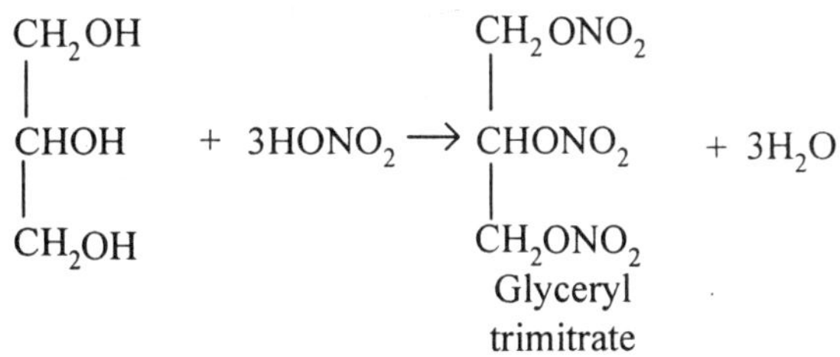

12. Oxidation of glycerol using different oxidizing agent yields different products, as illustrated in the following table:

	Oxidizing agent	*Product*
1.	Dil HNO_3	Glyceric acid
2.	Conc. HNO_3	Glyceric acid and tartonic acid
3.	Bismuth and $NaNo_2$	Mesoxalic acid
4.	Fenton's reagent ($H_2O_2 + FeSO_4$)	Glyceraldehyde and dihydroxy acetone
5.	Sodium hypobromite	Glyceraldehyde + dihydroxy acetone
6.	Bromine water in presence of sodium carbonate	Glyceraldehyde + dihydroxy acetone
7.	Solid $KMnO_4$	Oxalic acid and carbon dioxide.

The oxidation route is as follows:

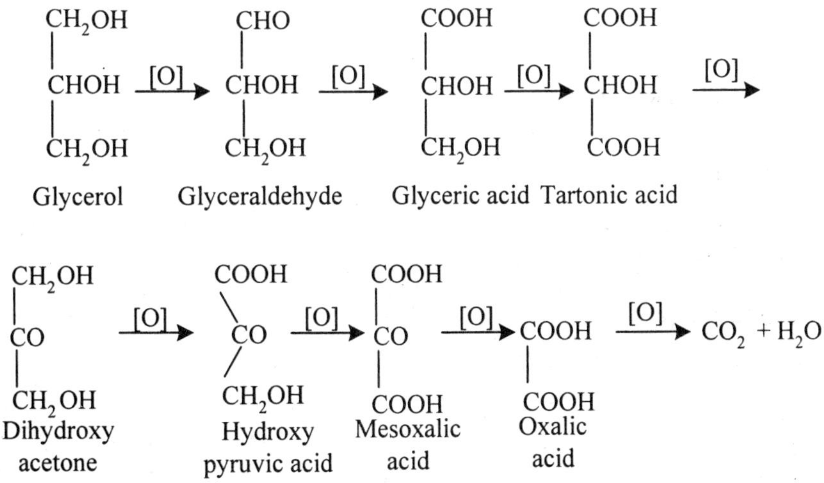

13. *(i)* (**b**), Rest do not contain —CH_3 group. Requirement for iodoform test is:

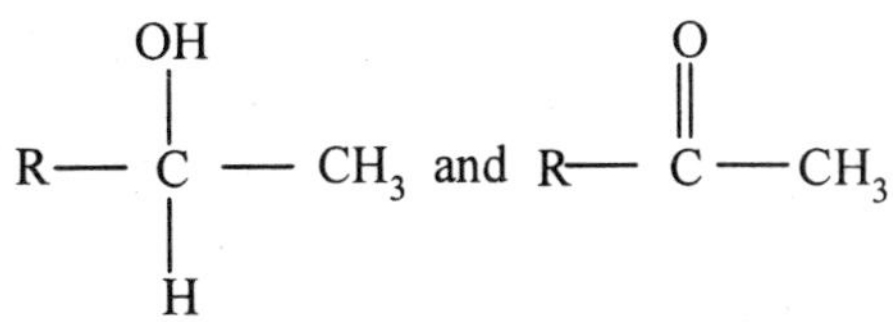

(ii) (**c**), It does not contain methyl group.

(iii) (**c**), Propanol-2 yields propanone, Butanol-1 yields butanol, tertiary butyl alcohol does not oxidize easily.

(iv) (**d**), see preparation of aldehydes and ketones.

(v) (**d**).

(vi) (**a**), acetaldehyde contains three α-H atoms, ketones do not undergo Cannizzaro reaction.

(vii) (**a**).

(viii) (**c**), see preparation of aldehydes and ketones.

(ix) (**c**), see properties of aldehydes.

(x) (**b**), Rest contain no α-H atoms

(xi) (**c**).

(xii) (**b**), Benzaldeyde does not reduce Fehling solution.

$$\underset{44}{CH_3CHO} + \underset{\text{Fehling solution}}{2CuO} \rightarrow CH_3COOH + \underset{\substack{\text{Brown ppt.} \\ 143}}{Cu_2O}$$

143 gm of brown ppt. is obtained from 44 gm CH_3CHO

1.43 gm of brown ppt. is obtained = 44 × 1.43/143 = 0.44 gm.

Therefore, amount of benaldehyde in mixture = 0.66 – 0.44 = 0.22 gm of benzaldehyde

=0.22 × 100/0.66 = 33.3%.

(xiii) (**a**), as it is methyl ketone. Rest have no $(-\overset{\overset{\Large O}{\|}}{C}-CH_3)$, grouping.

(xiv) (**a**).

14. *(i)* $CuSO_4 \cdot 5H_2O$ and NaOH

(ii)

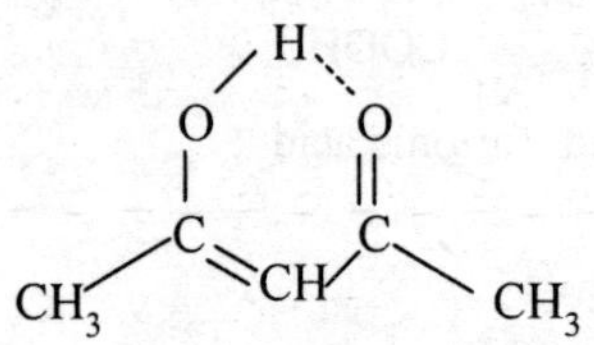

(iii) Secondary alcohol

(iv)

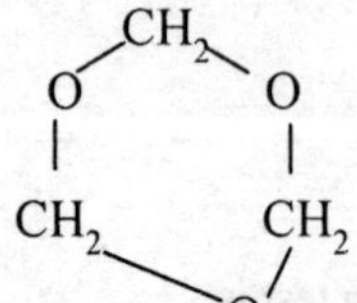

Trioxane

(v) 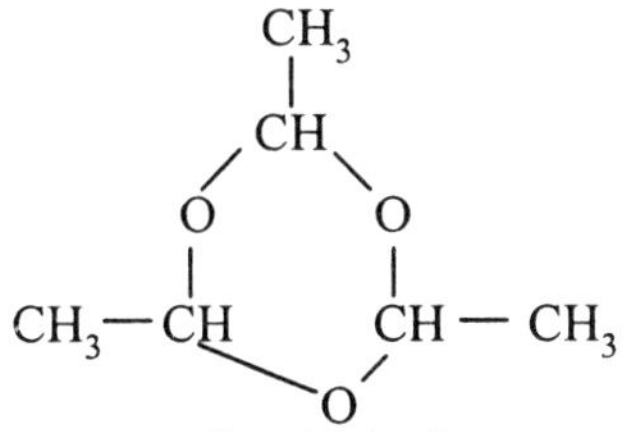

Paraldehyde

(vi) 40% solution of formaldehyde

15. (i) Reactivity of alhehydes and ketones is in the following order:

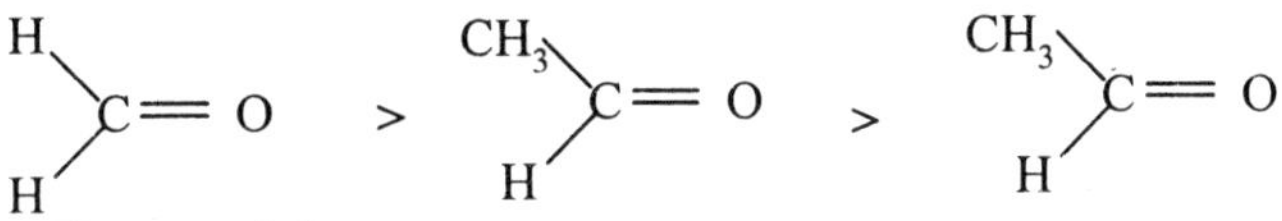

The reactivity of aldehydes or ketones $\propto$ **1/steric hindrance**

(ii) Ketal < formal < acetal.

(iii) $C_2H_5COCH_3 < CH_3, COCH_3 < CH_3CHO < HCHO$

The reason is same as above.

16. Ethanol does not reduce Fehling solution but acetaldehyde does so.

$$\underset{44}{CH_3CHO} + \underset{\text{Fehling solution}}{2CuO} \xrightarrow{\Delta} CH_3COOH + \underset{\substack{\text{Brown ppt.}\\143}}{Cu_2O}$$

weight of acetaldehyde which yields 1.2 gm brown ppt. = 44 × 1.2/143 gm.

Percentage of acetaldehyde = 44 × 1.2 × 100/143 × .532 = 87%.

17. (i) $CH_3CH{=}CHCH_2OH$, $CH_3CH{=}CHCH_2Cl$, $CH_3CH{=}CHCH_2COOH$

(ii) 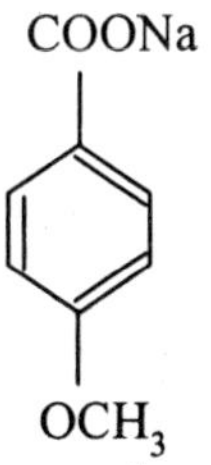

p-methoxy sodium benzoate

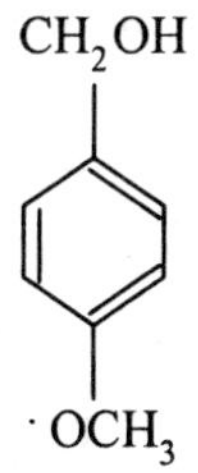

p-methoxy benzyl alcohol

(iii) $C_6H_5CH{=}CHCOCH_2CH_3$

(iv) $C_6H_5CH{=}\underset{\substack{|\\CH_3}}{C}{-}COCH_3$

(v) 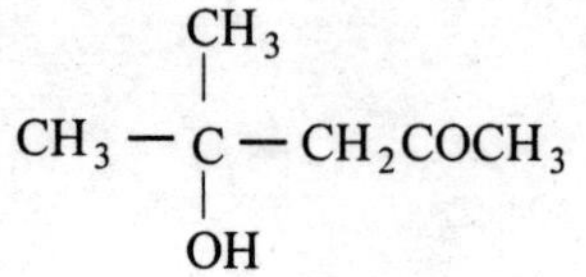

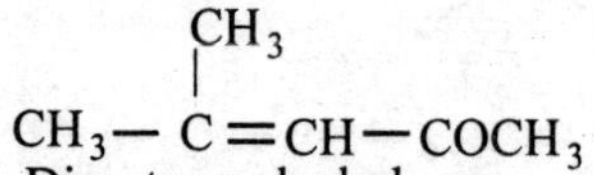

Diacetone alcohol

18. Acetaldehyde reduces Fehling solution and Tollen's reagent but acetone does not.

19. 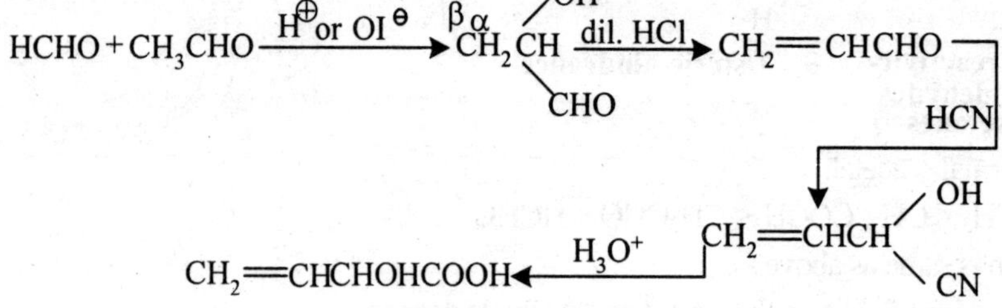

2-hydroxy-3-butenoic acid

20 *(i)* **(b)** 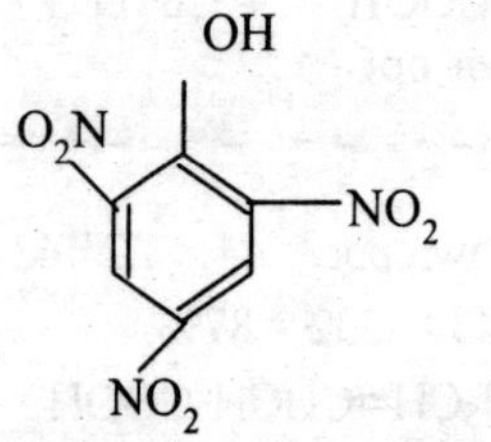

2,4,6-Trinitrophenol or picric acid

(ii) **(a)**, HCOCl Formyl chloride.

(iii) **(a)**, Because it has the structure

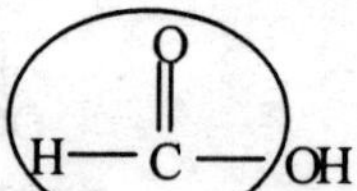

(iv) **(b)** $CH_3-\overset{\overset{\large O}{\|}}{C}-COOH$, it contains $CH_3-\overset{\overset{\large O}{\|}}{C}-COOH$

Pyruvic acid

(v) **(b)**.

(vi) (**d**), 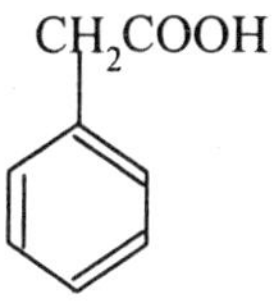 It is because rest of the acids are saturated.

(vii) (**b**), On simple heating acetamide is obtained but on strong heating with P_2O_5 further, a molecule of water is lost and methyl cyanide is obtained. The reaction is as follows:

$$CH_3COONH_4 \xrightarrow[-H_2O]{\Delta} \underset{\text{Acetamide}}{CH_3-\overset{O}{\overset{\|}{C}}-NH_2} \xrightarrow[-H_2O]{\Delta} CH_3CN$$

(viii) (**d**), It is because $LiAlH_4$ reduces —COOH group to $—CH_2OH$ group without affecting unsaturation.

(ix) (**d**), because of —IE effect of three chlorine atoms (for details, see Chapter 3).

(x) (**a**), It is because minimum is delocalization of electrons, greater is its reactivity. Hence, the order of reactivity is

$$RCOCl > (RCO)_2O > RCOOR' > RCN > RCONH_2$$

(xi) (**b**), Hoffmann bromide reaction.

(xii) (**a**) and (**b**).

(xiii) (**b**)

(xiv) (**b**)

(xv) (**b**)

21. *(i)* B: CH_3CONH_2, C: CH_3CN.

(ii) E, CH_3COCl, F: $CH_3CO\ NH_2$G, CH_3CH; H: $CH_3CH_2NH_2$.

(iii) A: HCOOH and B: Glycerol.

(iv) A: CH_3COCl.

(v) A: $\begin{matrix}COONa\\|\\COONa\end{matrix}$ B: $\begin{matrix}COOH\\|\\COOH\end{matrix}$

The reaction is

$$2HCOONa \xrightarrow{360\,°C} \underset{\text{Sodium oxalate}}{\begin{matrix}COONa\\|\\COONa\end{matrix}} + H_2$$

$$\downarrow$$

$$\begin{matrix}COOH\\|\\COOH\end{matrix} + NaCl$$

(vi) A: C_2H_5OH

(vii) A: Cl_2 and P, B: CH_2NH_2COOH.

22. *False. Acids* having α-Hydrogen atom would undergo Hell-Volhard-Zelinsky reaction, i.e., react with P and Cl_2 forming α-chloro substituted carboxylic acids with P and Cl_2. Hence, only CH_3COOH, $CHI_2C\ COOH$, $CHCl_2COOH$ will be with P and Cl_2 due to the presence of α-hydrogen (s).
23. It is because formic acid has no α-hydrogen atom.
24. *(i)* CH_3CHO Rosenmund Reduction.
    (ii) CH_3COOH.
    (iii) CO and H_2.
25. C_2H_5OH
26. *(i)*

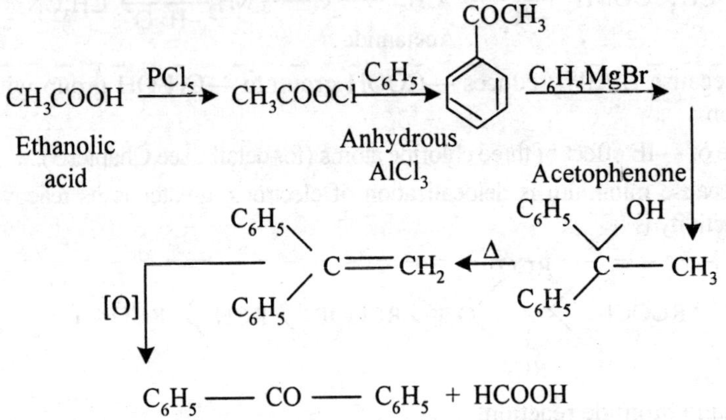

(ii)

$$CH_3CHO \xrightarrow{[O]} CH_3COOH \xrightarrow{C_2H_5OH/H^+} CH_3COOC_2H_5$$

$$\xrightarrow{NaOC_2H_5} \underset{\text{Acetoacetic ester}}{CH_3COCH_2COOC_2H_5}$$

Note: The reaction of ethyl acetate with sodium ethoxide is called Claisen condensation.

(iii)

$$HCOOH \xrightarrow{NaOH} HCOONa \xrightarrow[-H_2]{360^\circ C} \begin{array}{c} COONa \\ | \\ COONa \end{array}$$

$$\xrightarrow{H_3O^+} \begin{array}{c} COOH \\ | \\ COOH \end{array} + NaCl$$

(iv) 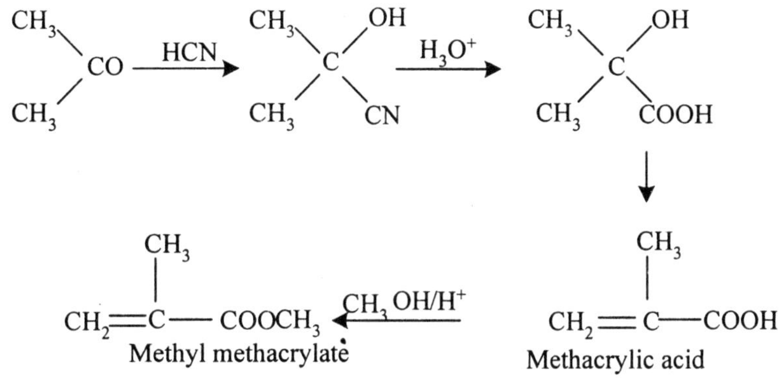

27. *(i)* (**b**).

(ii) (**c**).

(iii) (**a**), because primary amine is RNH_2. In the problem R = $(NH_3)_3$—C.

(iv) (**d**)

(v) (**b**) $C_2H_5NH_2 \xrightarrow{HNO_2} C_2H_5OH \xrightarrow{PCl_5} C_2H_5Cl \xrightarrow{HNH_2} C_2H_5NH_2$

(vi) (**c**).

(vii) (**c**).

(viii) (**b**) see carbylamine reaction.

(ix) (**c**).

(x) (**b**).

(xi) 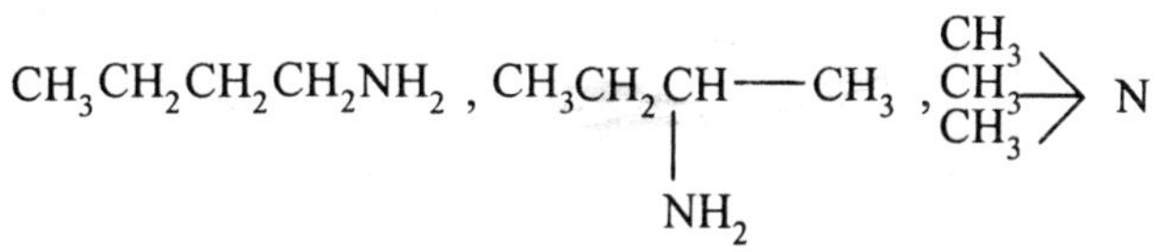

28. *(i)* Ethylamine *(ii)* Chiral *(iii)* 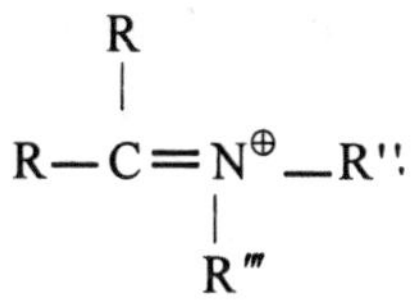

(iv) Curtius rearrangement.

29. *(i)* False. Acetamide is less basic than ethylamine due to resonance

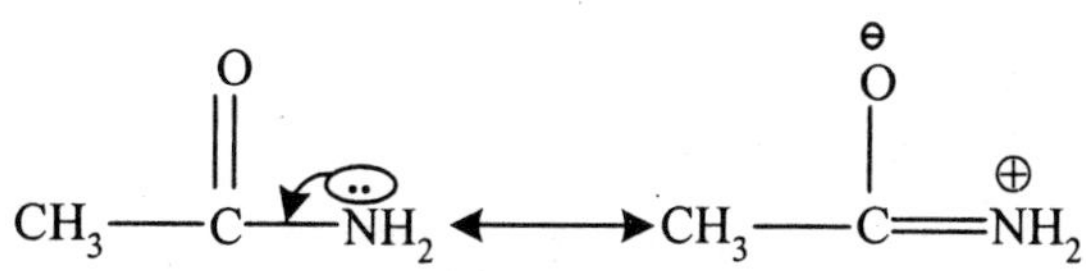

(ii) False. It is secondary amine.

(iii) False. Amines dissolves, but amides do not dissolve.

30. (i) $C_2H_5NH_2 + CS_2 + HgCl_2 \rightarrow C_2H_5NCS + HgS + 2HCl$
Ethyl isothiocyanate

(ii) $RNH_2 + R'CH{=}O \rightarrow RN{=}CHR' + H_2O$

(iii) 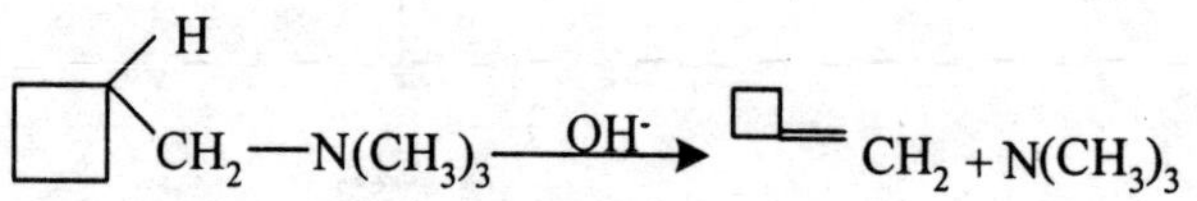

(iv) $R{-}\overset{O}{\overset{\|}{C}}{-}OMgBr$, $RCOOH$, $R{-}\overset{O}{\overset{|}{C}}CH\ N{\equiv}N$, $O{=}C{=}CHR$, $RCHCOOH$

(v) $RCH_2NH_2 + 2CH_2O + 2HCOOH \rightarrow$
$RCH_2NMe_2 + 2CO_2 + 2H_2O$

(vi) $KCN + Br_2 \rightarrow BrCN + KBr$
$EtNH_2 + BrCN \xrightarrow{KOH} EtNHCN + KBr$

31. (i) Inversion about asymmetric nitrogen is too rapid to permit separation of enantiomers
(ii) Because tertiary Amines have no hydrogen bonding

32. (i) Benzamide when heated with NaOH gives off ammonia whereas benzyl amine does not
(ii) Allyl amine decolourizes bromine water but propyl amine does not
(iii) Aniline forms dye but cyclohexyl amine does not
(iv) Ethylamine when treated with benzene sulphonyl chloride forms alkali-soluble product whereas diethyl amine, being secondary amine water insoluble product with benzene sulphonyl chloride

33. A is $(CH_3)_2CHCH_2NHC_2H_5$ (ethylisobutyl amine) or $Me_3CNHC_2H_5$ (t – butylethyl amine)

34. Since the solid is insoluble in alkali, it should be a secondary amine. Hence, A is ethylmethyl amine.

$$C_2H_5{-}\underset{\underset{H}{|}}{N}{-}CH_3$$

35. $C_6H_5N \equiv C$
Phenyl isocyanide

Chapter 11

Aromatic Hydrocarbons And Their Derivatives

This chapter presents an account of aromaticity and aromatic hydrocarbons and their derivatives. The chapter includes following:

(1) Aromaticity; (2) Aromatic hydrocarbons; (3) Aryl halides; (4) Phenols and aromatic ethers; (5) Aromatic aldehyde, ketones and quinines; (6) Aromatic carboxylic acids; (7) Benzene sulphonic acids; and (8) Aromatic nitro, amino compounds and diazonium salt.

11.1 AROMATICITY

The term 'Aromaticity' may be defined as the ability of an organic compound, like benzene, to undergo substitution rather than addition reaction, and to resist oxidation. A compound is identified as 'Aromatic' if it fulfils the following conditions:

(i) The organic compound should have conjugated; planar[1], cyclic ring.

(ii) Although unsaturated, it must undergo substitution rather than addition reaction.

(iii) It must resist oxidation.

(iv) It must obey Hückel's Rule. It states, "If in a conjugated planar cyclic polyene, the number of delocalized π electrons is $(4n + 2)$, where n is an integer, the compound would have aromatic stability".

(a) If $n = 0$, the aromatic compound is cyclopropenium cation which is obtained as cyclopropenium perchlorate.

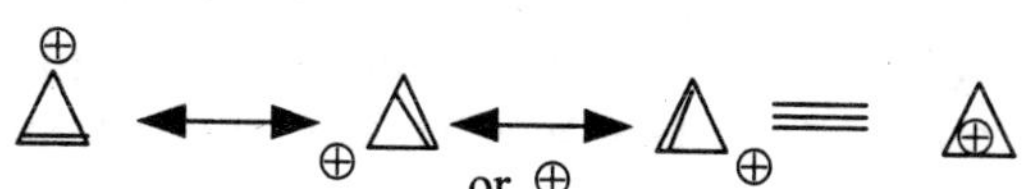

or $\overset{\oplus}{C_3H_3}$

---

[1] Planar means hybridization state of all the carbons must be sp^2

(b) If $n = 1$, the aromatic compounds are:

;

or $\overset{\ominus}{C}_5H_5$;
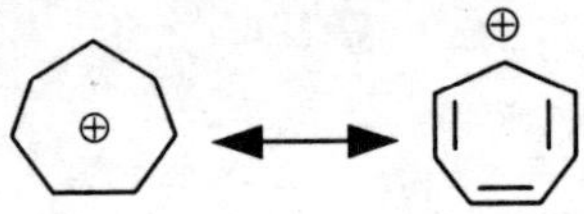
or $\overset{\oplus}{C}_7H_7$

Benzene, Cyclopentdienylanion, Tropylium cation or Cycloheptatrienyl cation

(c) If $n = 2$, i.e., 10 delocalized π-electrons

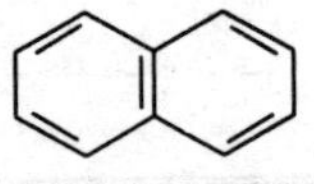

Naphthalene

(d) If $n = 3$, i.e., 14 delocalized π-electrons

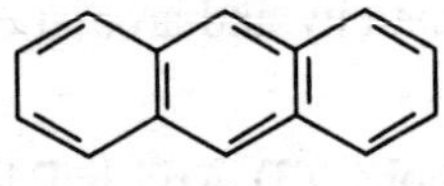

Anthracene

Tricks:

(i) *Number of delocalized π electrons = No. of Carbon atoms in a Compound*

e.g., in benzene = 6

It is in accordance with Hückel rule, hence aromatic

(ii) *Number of delocalized π electrons in cation = No. of carbon atom — No. of +ve charge*

e.g., $C_7H_7^+ = 7 - 1$

Hence aromatic = 6

(iii) *Number of delocalized π electrons in anion = No of carbon atoms + No. of –ve charge*

e.g., $C_5H_5^{\ominus} = 5 + 1 = 6$

11.1.1. Directive Influence of Substituents

Benzene contains 6 delocalized π-electrons and therefore, acts as a nucleophile in organic reaction and undergoes electrophilic substitution reactions, i.e., one or more H-atom(s) of benzene can be replaced by electophile(s). Since electron density on each carbon atom is uniform, when hydrogen atom of benzene is replaced by any group or atom, it can attach to any carbon atom of the ring.

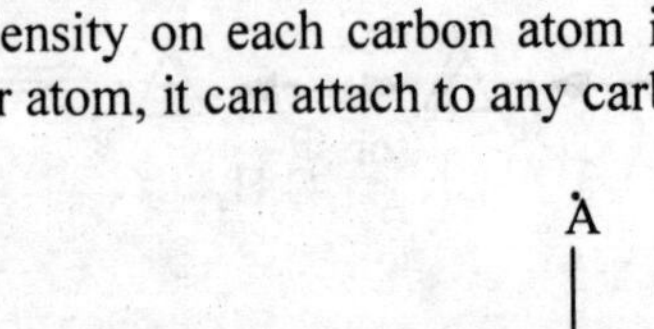

However, when monosubstituted benzene is converted into disubstituted benzene, the electron density is no more uniform on carbon atoms of ring. It is either increased or decreased on ortho or para position due to previous group/atom present in the ring. We can divide substituent/groups into two classes according to the way they influence the orientation of attack by the incoming electrophile:

(a) Ortho-para directing groups

(b) Meta-directing group.

(a) ***Ortho-para directing groups (o-p)***: The characteristics of these groups are as follows:

(i) *o-p* direction groups *increase electron density* at *o-p* position.

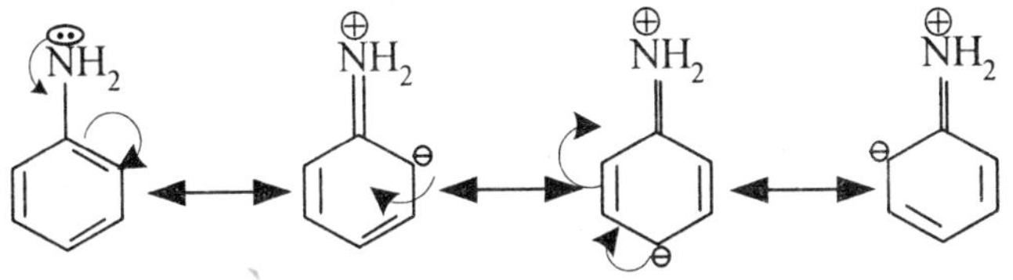

(ii) All *o, p*-directing groups have *either lone pair* of electrons, or *negative charge* (except alkyl groups like methyl group), on the atom directly attached to the ring.

$$-\ddot{\underset{\cdot\cdot}{O}}H,\quad -\ddot{N}H_2,\quad -\ddot{N}H-\overset{O}{\overset{\|}{C}}-R,\quad -\ddot{\underset{\cdot\cdot}{C}}l:$$

Lone pair

The *o, p* directing nature of methyl group can be explained on the basis of *hyperconjugation* (for details, see Chapter 3).

Trick: Electronegativity of atom attached to ring is higher than second atom e.g.,

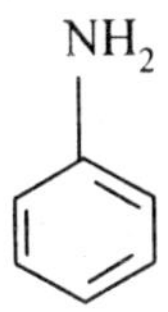

i.e., electronegativity of N > electronegativity of H in aniline.

(iii) All *o, p*-directing groups increase electron density at ortho and para position *hence activate the ring except the halogens.* Halogens, because of high electronegativity, deactivate the ring. Following is the order of activation of ring.

$$-O^- > -NR_2\ -NHR > -NH_2 > -OH >$$

$$-CH_3 > -C_2H_5 > (CH_3)_2\ CH^-$$

$$\text{Activation of the ring} \propto \frac{1}{\text{electronegativity of atom attached to the ring}}$$

(b) ***Meta-directing groups***: The characteristics are as follows:

(i) All *meta*-directing groups *decrease* electron density at *ortho* and *para* positions. There is no change of electron density at *meta* position, e.g.,

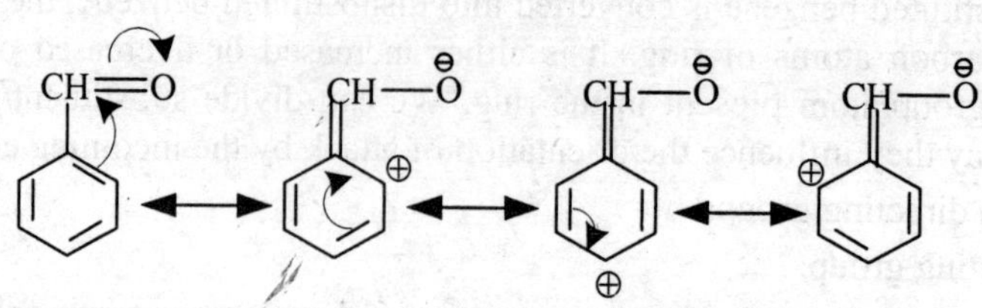

(ii) All meta directing groups have π *electrons* or positive charge on the atom attached to benzene ring, e.g.,

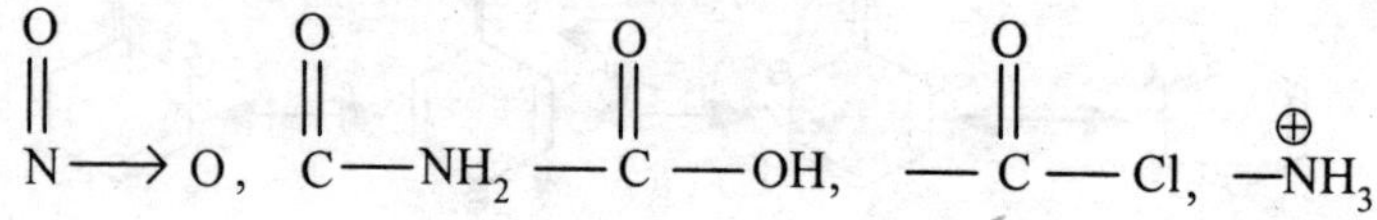

Trick: electronegativity of atom attached to ring e.g. < electronegativity of second atom.

$N < O$ in NO_2

$C < F$ in CF_3

(iii) All meta-directing groups decrease electron density at *ortho* and *para* positions, hence deactivate the ring. The order is as follows:

$$-NO_2 > -CF_3 > -CN > -CHO > -COOH > -COOR > -CCl_3 > CHCl_2$$

Deactivation of the ring ∝ electronegativity of atom attached to the ring.

11.2 AROMATIC HYDROCARBONS

The aromatic hydrocarbons have general formula C_nH_{2n-6}. Some of the important aromatic hydrocarbons are:

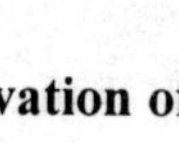

Benzene

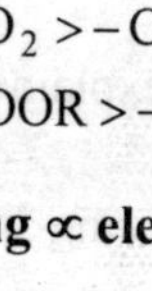

Toluene

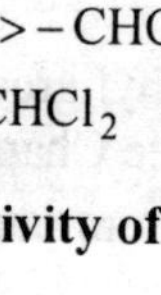

Ethyl benzene

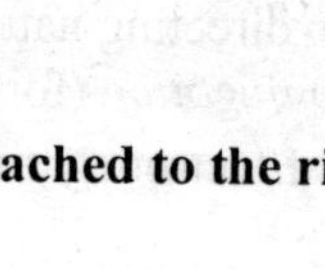

Xylene (*o, m, p*)

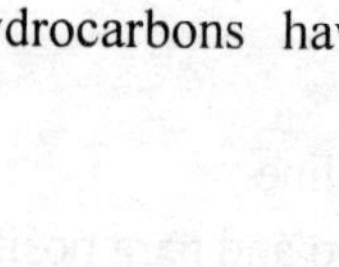

o-xylene

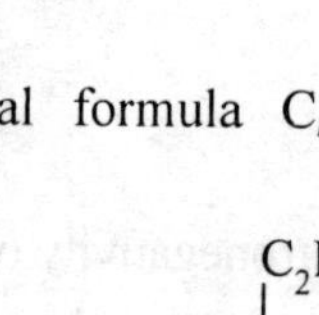

Styrene

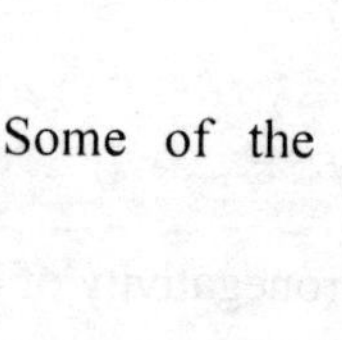

Isopropyl benzene

or

Cumene

(a) Preparation

(i) By decarboxylation of aromatic acid, e.g.,

(a)

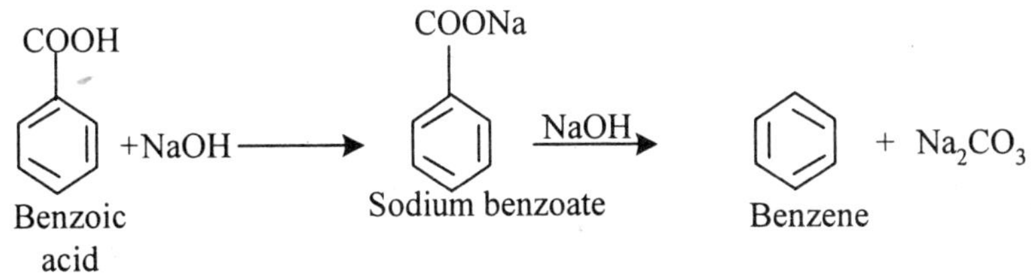

(b)

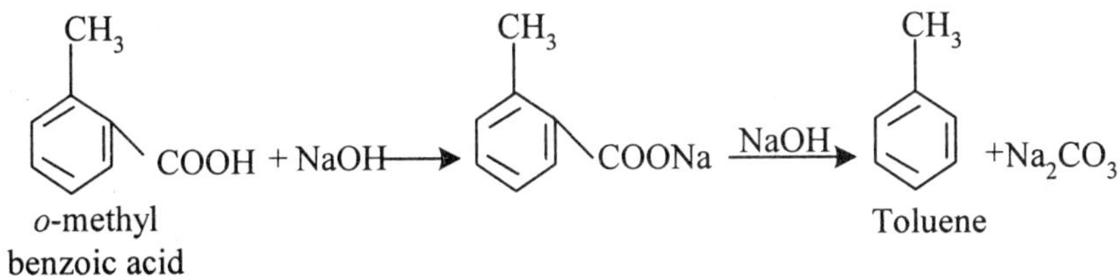

(ii) By passing vapours of phenols over zinc dust

(a)

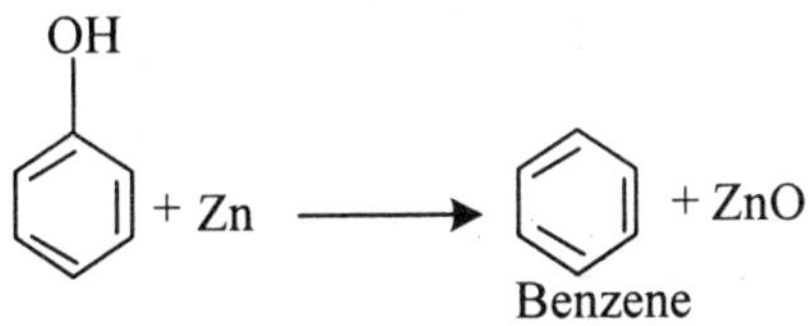

(b)

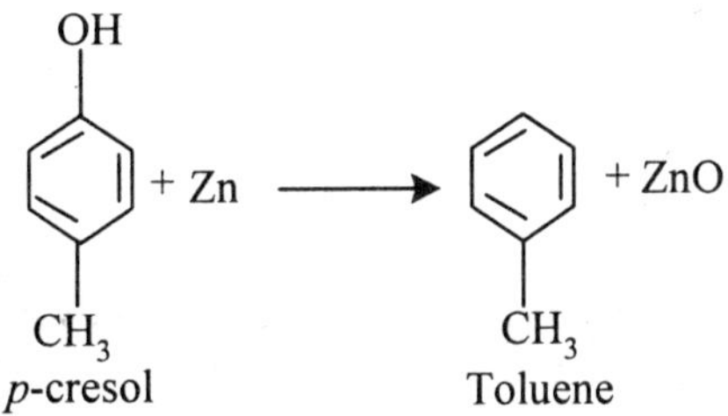

(iii) By boiling benzene sulphonic acid with dil. HCl at 160°C under pressure:

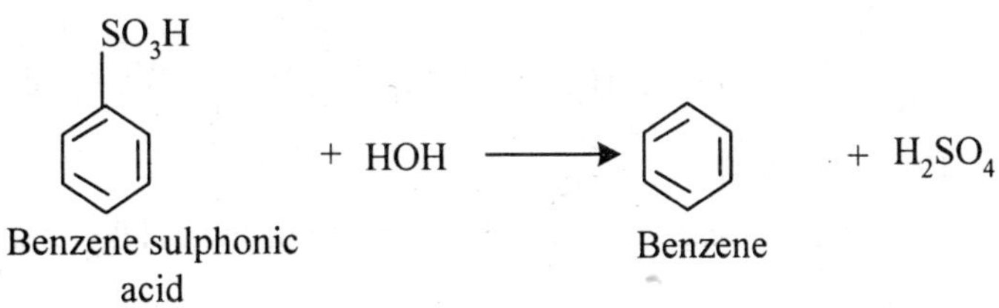

(iv) By reduction of benzene diazonium chloride with H_3PO_2

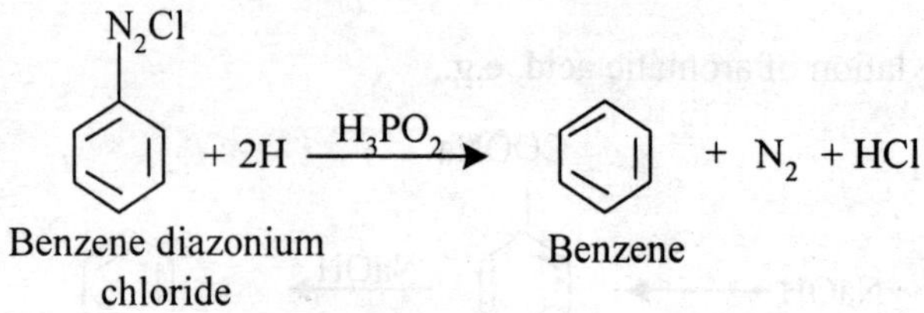

(v) By Friedel-Craft alkylation reaction

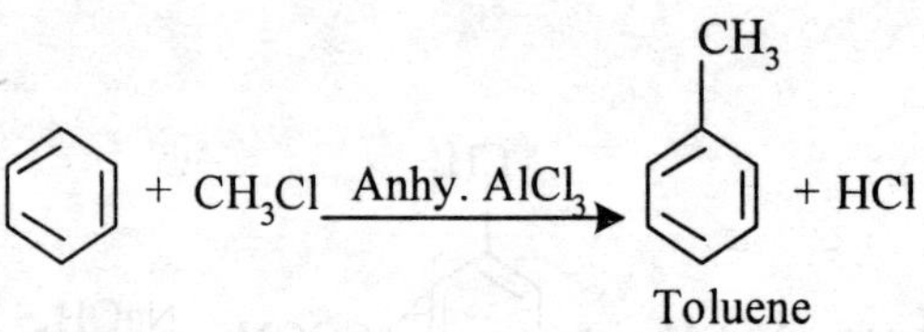

(vi) By Friedel-Craft acylation reaction[2]

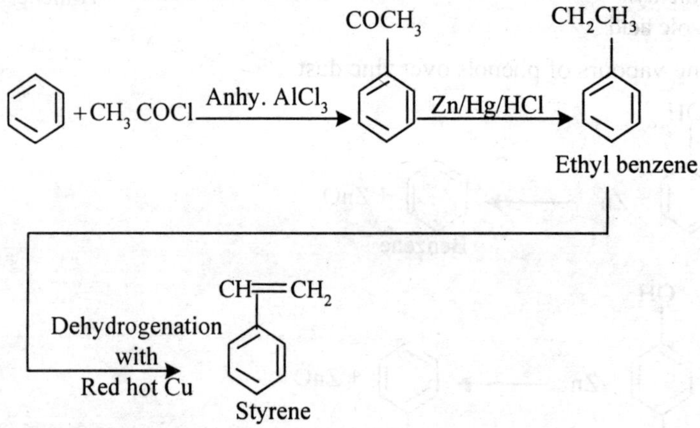

(vii) *Wurtz-Fittig reaction:* When an aryl halide is heated with alkyl halide in the presence of metallic sodium in ether, benzene homologues are obtained, e.g.,

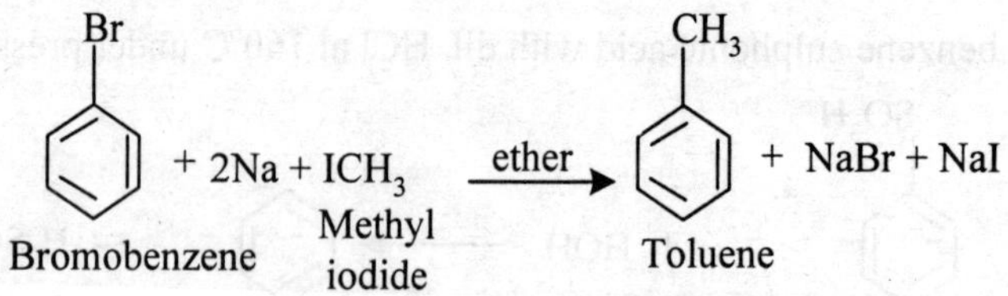

(viii) By reaction of Grignard reagent (where R = aryl group) with alkyl halide

---

[2] See Chapter 5 for mechanism.

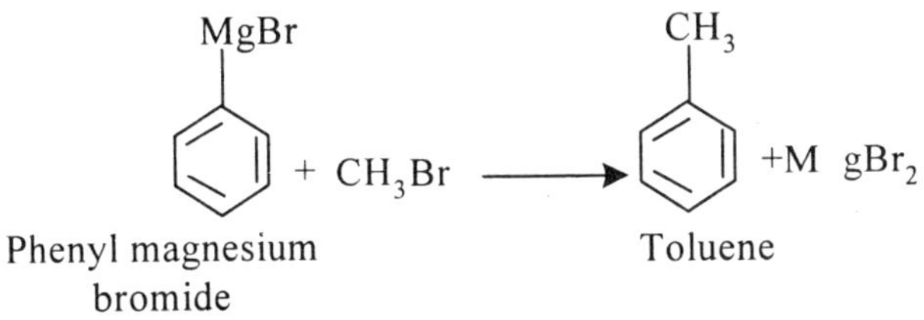

(ix) From coal tar

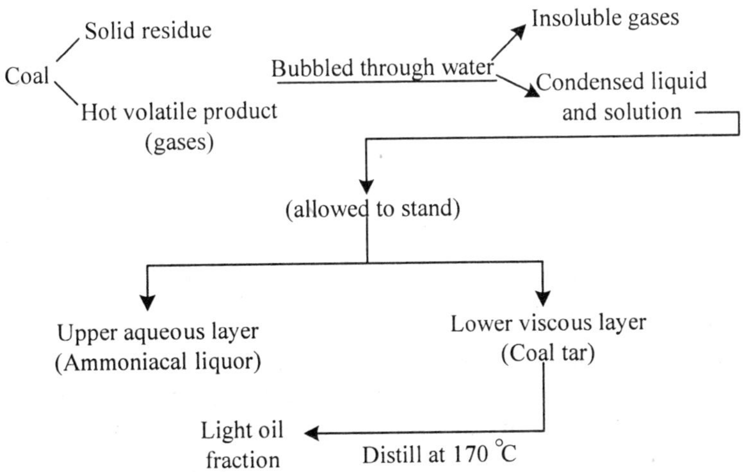

The light oil fraction contains about 65% benzene, 15% toluene with naphthalene, 8-10% phenol and 1-3% basic substances like pyridine.

(x) *From petroleum products:* Aliphatic hydrocarbons, obtained from petroleum, are subjected to cyclization, isomerization and dehydrogenation, e.g.,

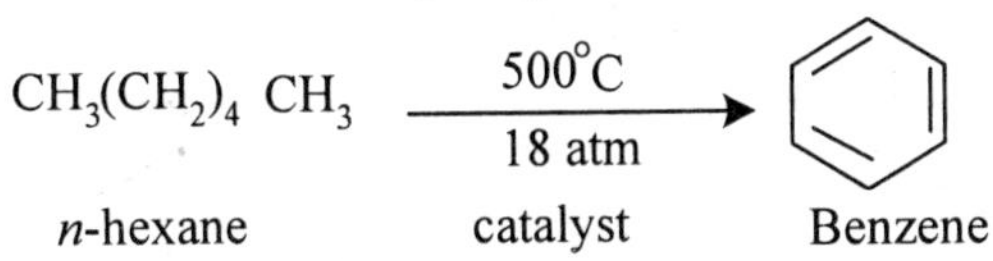

Catalyst consists of oxides of chromium, molybdenum and vanadium deposited on alumina.

(b) Properties

(a) *Physical:* Colorless, volatile liquid having characteristic odour.

(b) *Chemical:*

(i) Due to benzene ring

(ii) Due to side chain

(i) Due to benzene ring: These include halogenation, nitration, sulphonation and Friedel-Craft reaction. These may be briefly mentioned as below (For details, see Chapter 5).

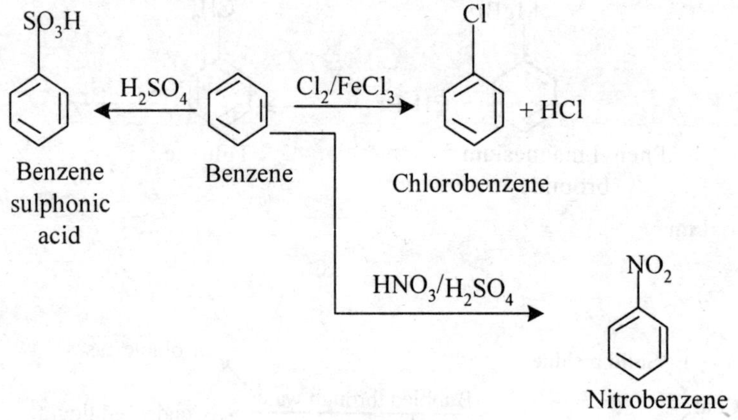

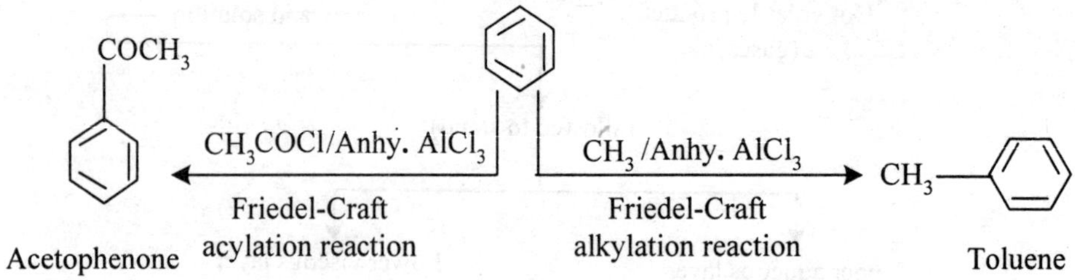

(ii) Due to side chain: An alkyl group attached to benzene ring is termed as 'side chain'. The important reaction is the oxidation of a 'side chain'. On oxidation, the alkyl group, irrespective of the number of carbon atoms, is converted into —COOH group, e.g.,

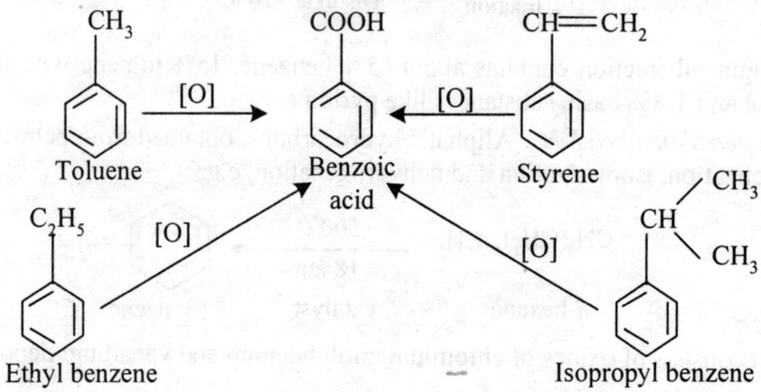

11.3 ARYL HALIDES

Organic compounds in which the halogen atom is directly attached to benzene ring are called aryl halides.

(a) Preparation

(1) By direct halogenation of aromatic hydrocarbons like benzene, in the presence of a halogen carrier ($FeCl_3$ $AlCl_3$, $SbCl_3$, etc.), at low temperature, e.g.,

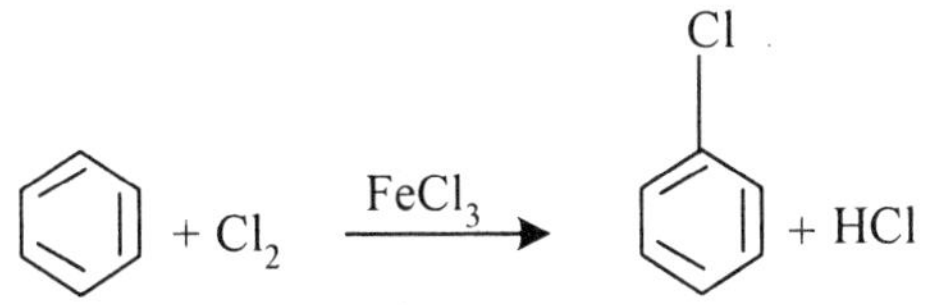

1 molecule 1 molecule

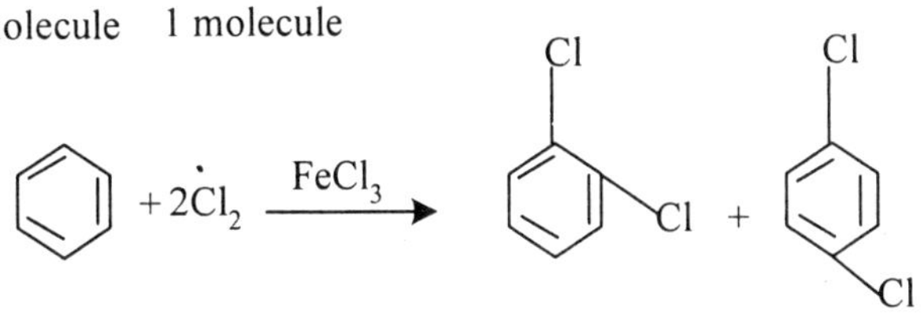

o-dichloro benzene *p*-dichloro benzene

(2) By reaction of phenol with PCl_5

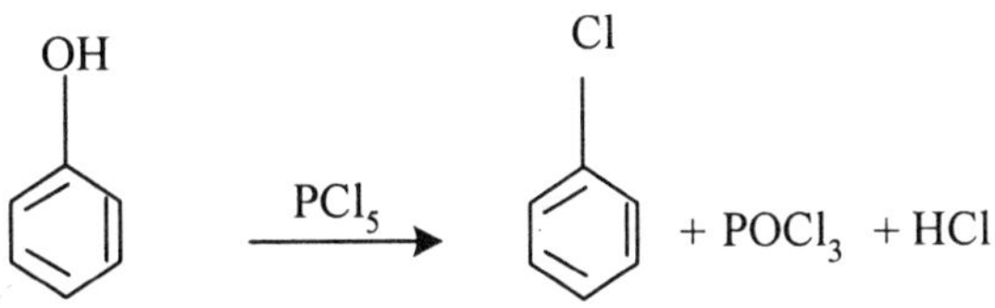

Phenol

(b) Properties

Physical: Aryl halides are generally colorless liquid or crystalline solids.

Chemical: Chemical properties can be discussed as below:

(i) Properties due to benzene ring

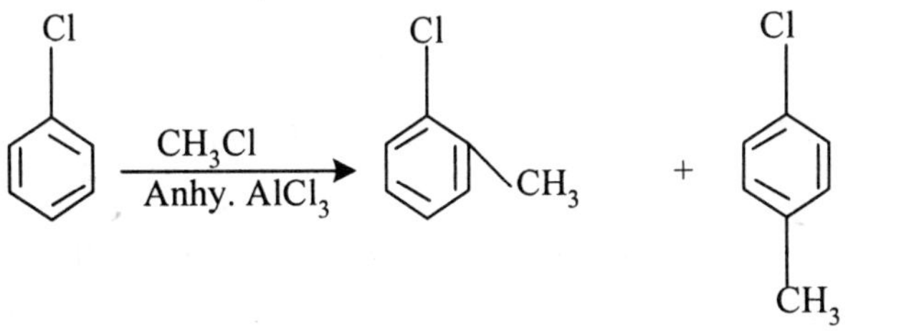

o-chloro toluene *p*-chloro toluene

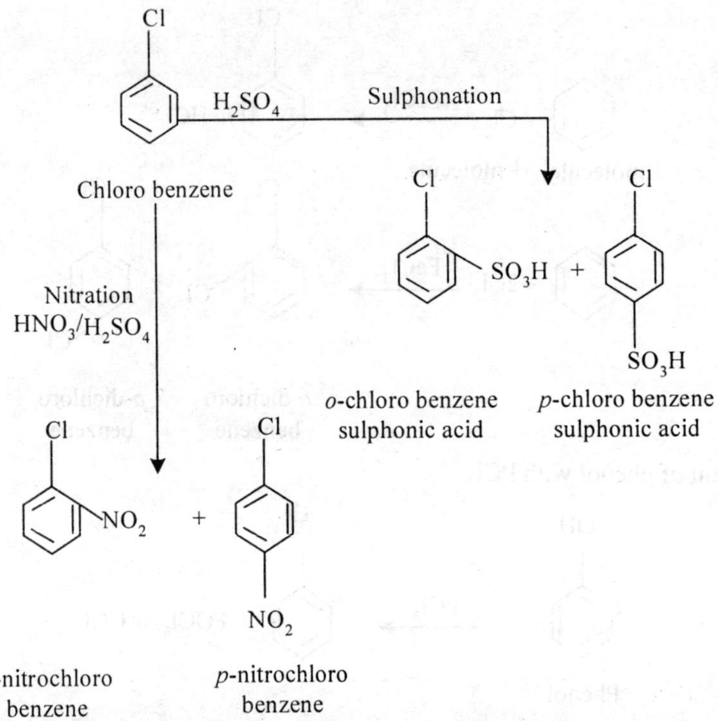

(ii) Properties due to –Cl atom: Aryl halides undergo nucleophilic substitution with difficulty because carbon-chlorine bond acquires double bond character (see Resonance, Chapter 3 for details). The nucleophilic substitution reactions are as follows:

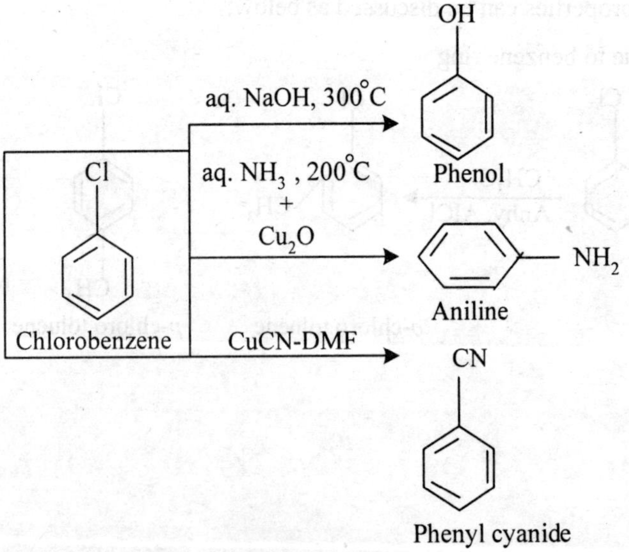

If a powerful electron-withdrawing group is present in *ortho* or *para* position to the halogen atom, the substitution becomes easy, e.g.,

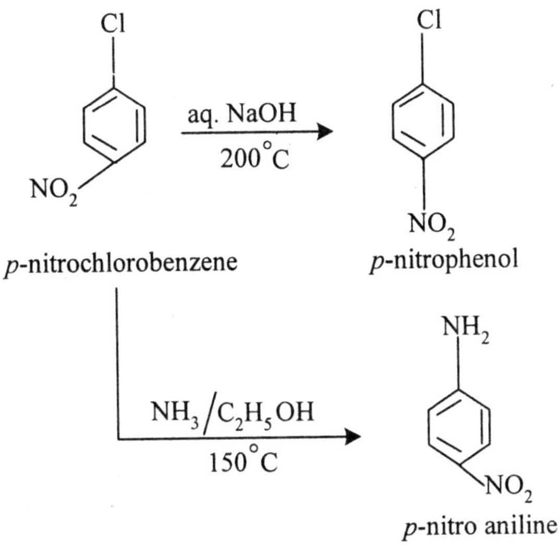

(iii) *Reaction with chloral*

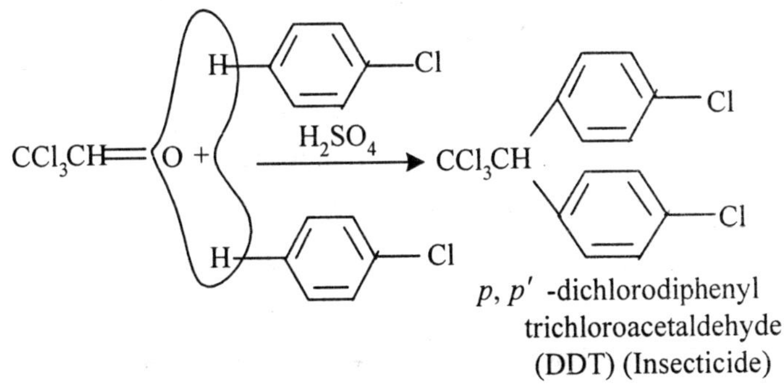

11.4 PHENOLS AND AROMATIC ETHERS

11.4.1 Phenols

Phenols are aromatic compound in which hydroxyl group is directly attached to benzene ring. Structures of some phenols are as follows:

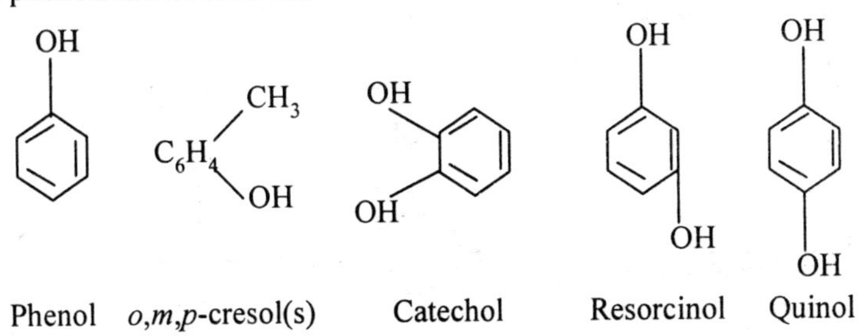

(a) ***Preparation***

(i) By boiling benzene diazonium chloride with water.

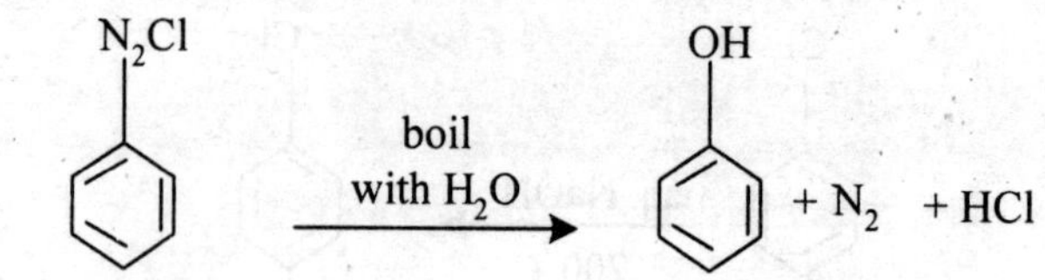

Benzene diazonium chloride

(ii) By fusion of sodium salt of benzene sulphonic acid with NaOH

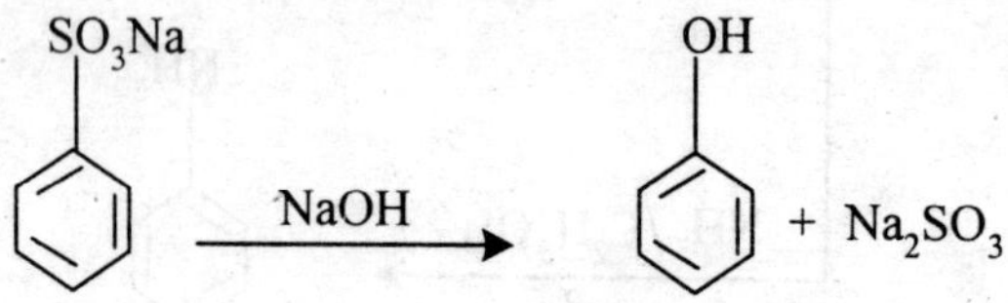

(iii) By reaction of sodium hydroxide with aryl halide containing activated halogen atom

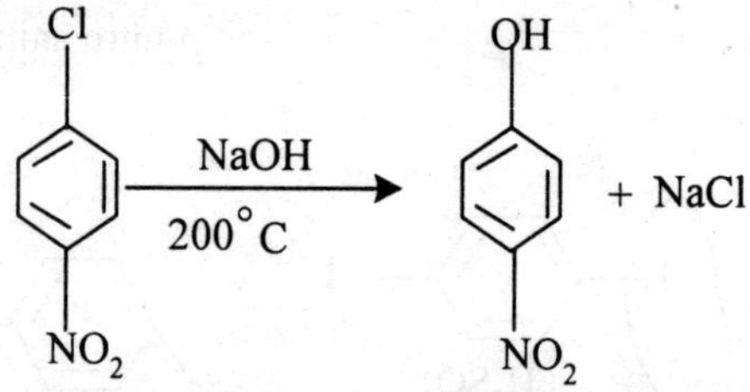

p-nitrophenol

In case of aryl halides, drastic conditions are needed, e.g.,

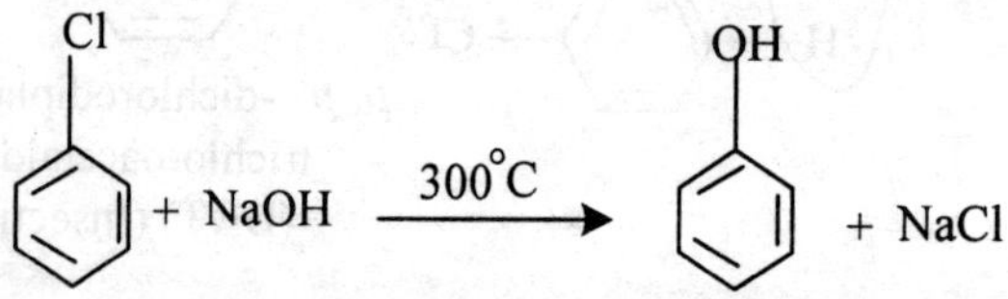

(iv) By oxidation of toluene by air in the presence of manganous and cupric salts as catalyst.

$$\underset{\text{Toluene}}{C_6H_5CH_3} + O_2 \xrightarrow{\text{Catalyst}} C_6H_5OH + CO_2$$

(v)

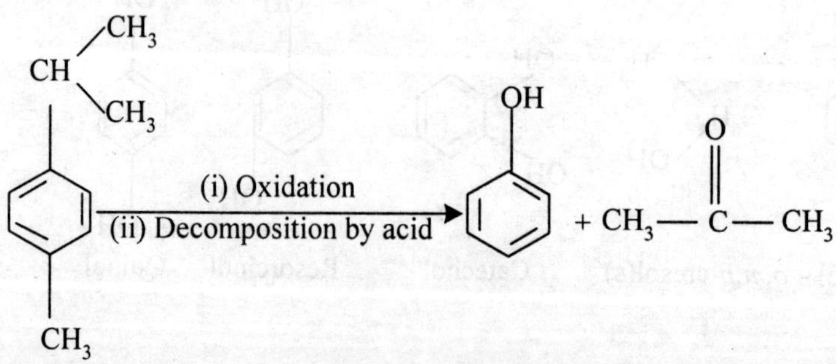

Cumene

(b) *Properties*

Physical: Colorless crystalline solid, which turns pink, one exposure to air and light due to oxidation, less soluble in water, but readily soluble in ethanol and ether.

Chemical: Properties of phenol can be discussed under the following two categories:

(1) *Reaction due to —OH group:*
   (a) Reactions different from alcohols.
   (b) Reactions similar to alcohols.

(2) Reaction due to benzene ring.

(1) *Reaction due to –OH group* (1) Acidity of phenols: Phenol is acidic in nature (For details, see Resonance in Chapter 3).

(2) *Phenols give different colours with $FeCl_3$.* Although exact reason is not known, it is proposed that phenol forms complex with $FeCl_3$ in which iron is hexavalent. The different colors are as follows:

Phenol—violet
o-cresol—green
m-cresol—Blue or Blue-violet
Quinol—green
p-cresol—Blue or Blue violet
Resorcinol—violet
Catechol—green

(3) ***Phthalein reaction*:** Phenol reacts with phthalic anhydride in the presence of H_2SO_4 to form phenolphthalein, the indicator.

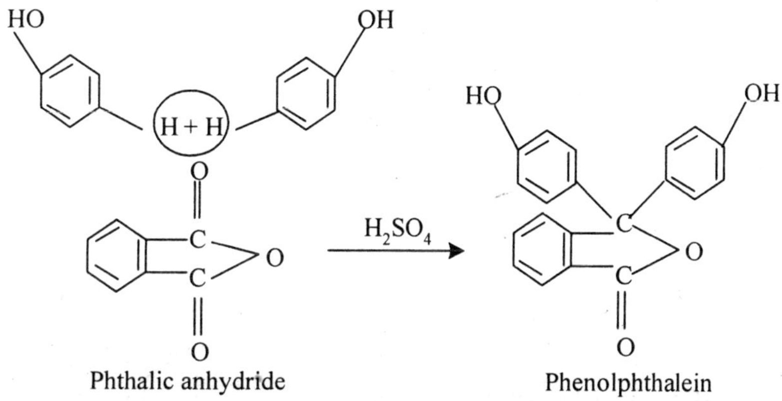

Phenols in which *p*-position to —OH group like *p*-cresol or quinol is filled do not undergo phthalein reaction

(4) ***Kolbe-Schmidt reaction*:** When sodium or potassium phenoxide is treated with CO_2 under 6–7 atmospheric pressure and at 120–130°C, salicylic acid is obtained.

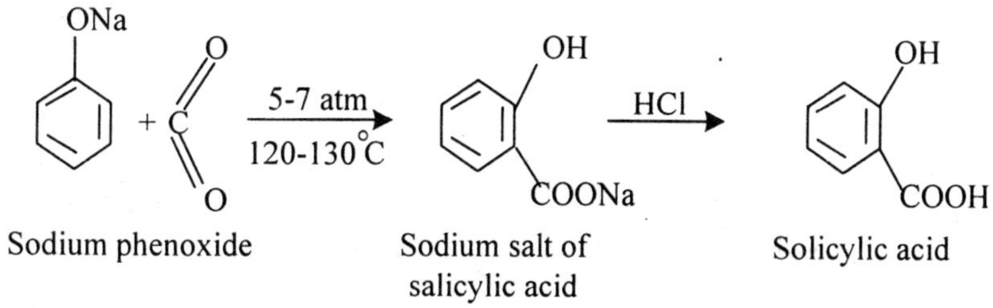

Reaction intermediates:

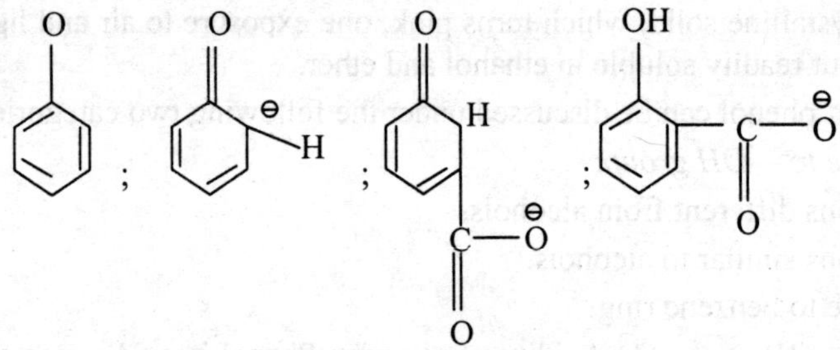

(5) ***Reimer–Tiemann Reaction***: When phenol is treated with $CHCl_3$ and KOH, salicylic aldehyde is obtained. However, if CCl_4 and KOH is used salicylic acid is obtained.

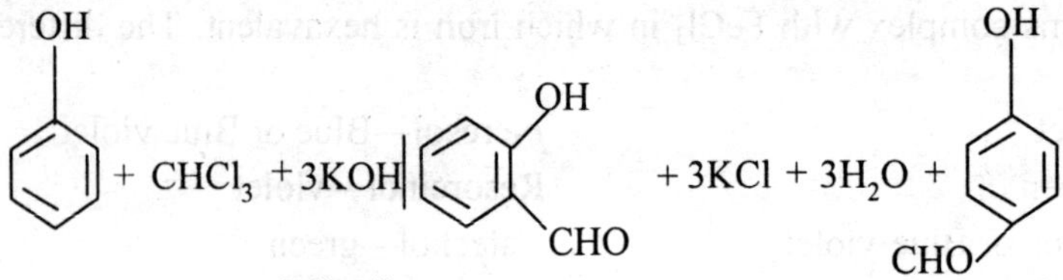

Salicylic aldehyde (Major product, 40%)

p-hydroxy benzaldehyde (Minor product, 10%)

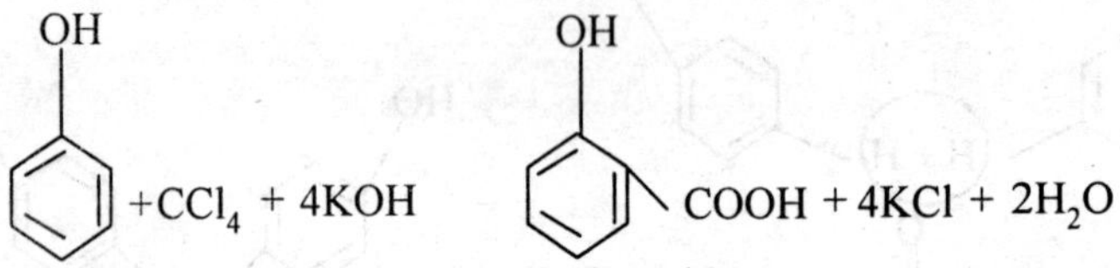

Salicylic acid

(Reaction intermediate): CCl_2;

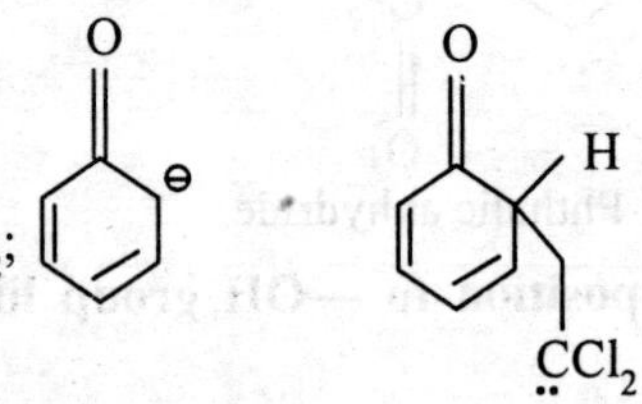

(6) ***Ledrer–Mannase Reaction***: When phenol reacts with CH_2O in the presence of H^+/OH^- phenol-formaldehyde resin (bakelite) is obtained.

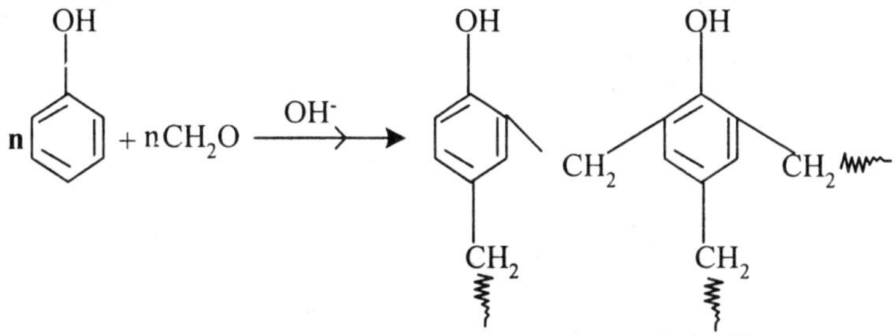

Bakellite

(7) ***Liebermann Nitrosation*** *Reaction:* It is used to identify phenol. In brief,

Phenol $\xrightarrow{NaNO_2 \text{ and } H_2SO_4}$ Blue or Green colour $\xrightarrow{H_2O}$ Red colour $\xrightarrow{NaOH}$ Blue or Green colour

The reactions are as follows:

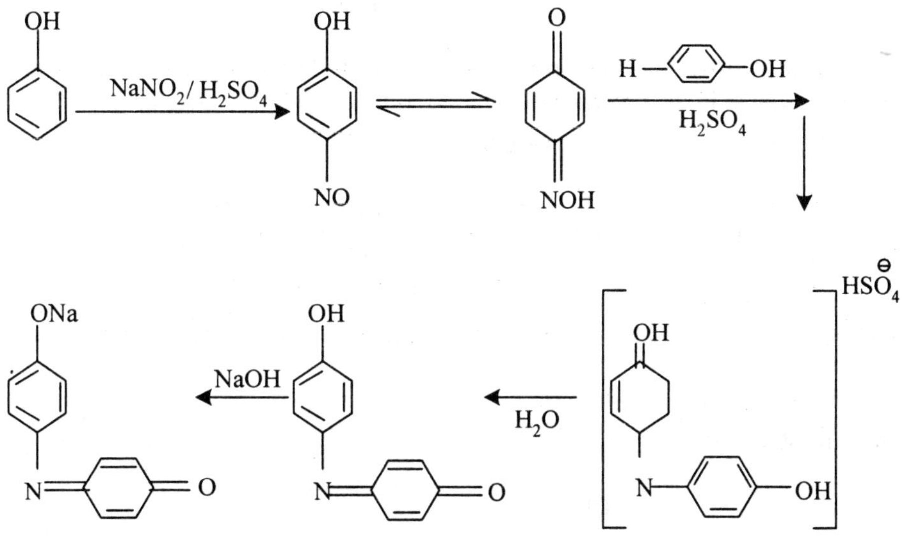

Sodium salt of phenol indophenol (Blue or Green)

Phenol indophenol (Red)

Phenol indophenol hydrogen sulphate (Blue)

(8) *Oxidation:* Phenol on oxidation gives quinone

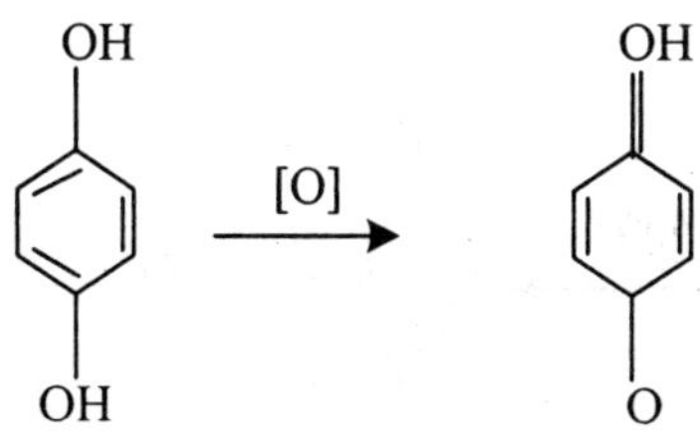

Quinol

Benzoquinone

Phenols also undergo Elbs persulphate oxidation by reaction with alkaline potassium persulphate to dihydric phenol. The —OH group comes to *para* position and if this position is blocked, it comes to *ortho* position, but never to *meta* position.

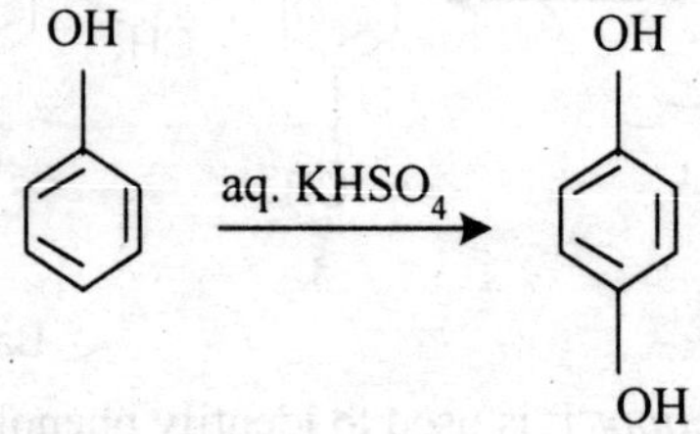

(9) When distilled with Zn, it forms benzene

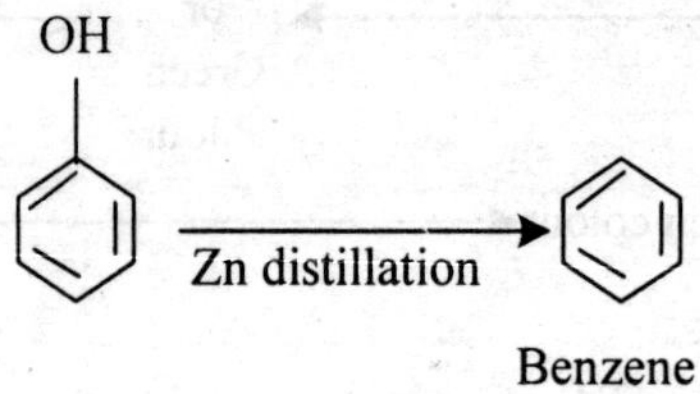

(10) Phenol when treated with acid chloride or acetic anhydride forms phenol esters which undergo Fries Rearrangement in the presence of $AlCl_3$ to form *o* or *p*-hydroxy ketone.

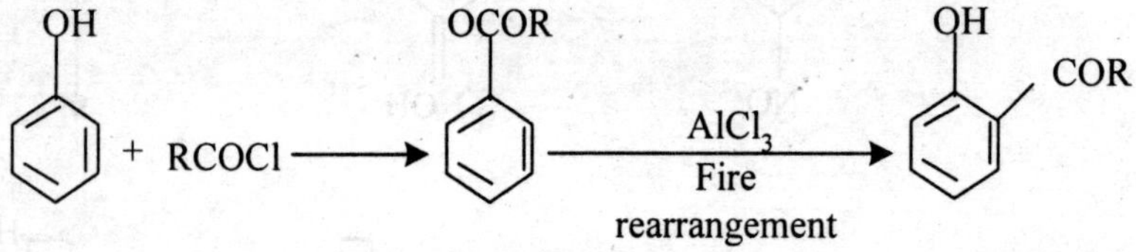

If *ortho* position is not vacant, the —COR group migrates to *para* position; but never to *meta* position.

(a) *Reactions similar to alcohols*

(1) With PCl_5, $SOCl_2$, PCl_3

(i) 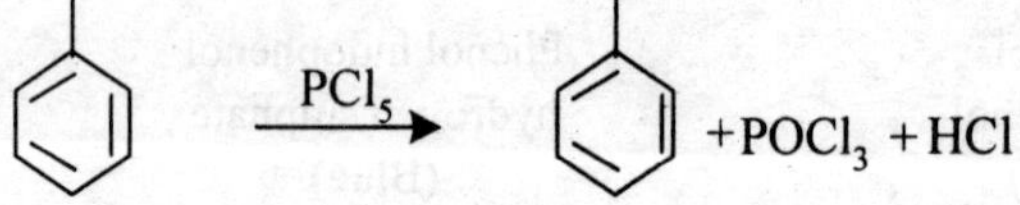

(ii) 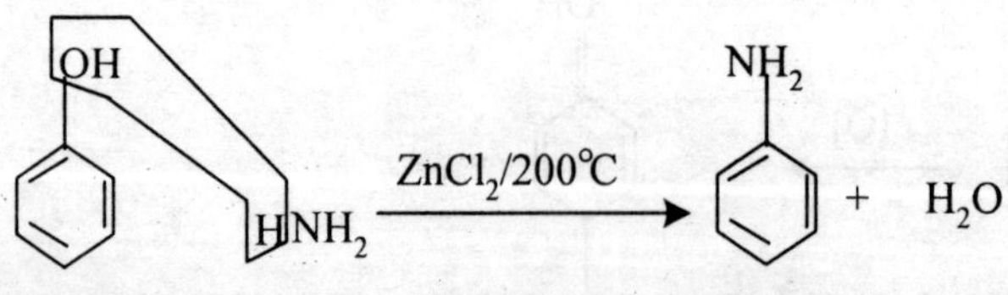

(ii) 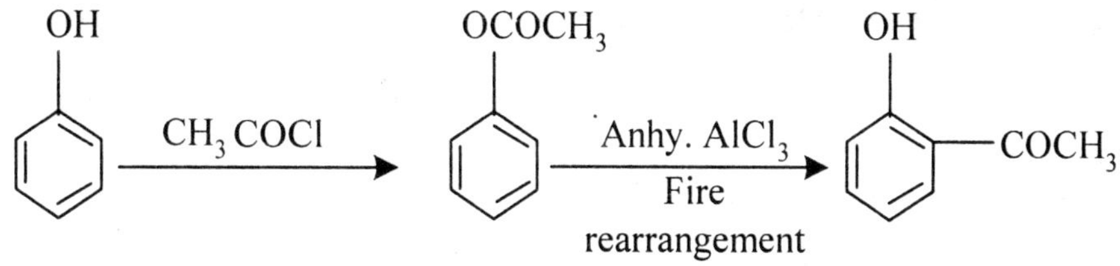

(2) *Reactions due to benzene ring*

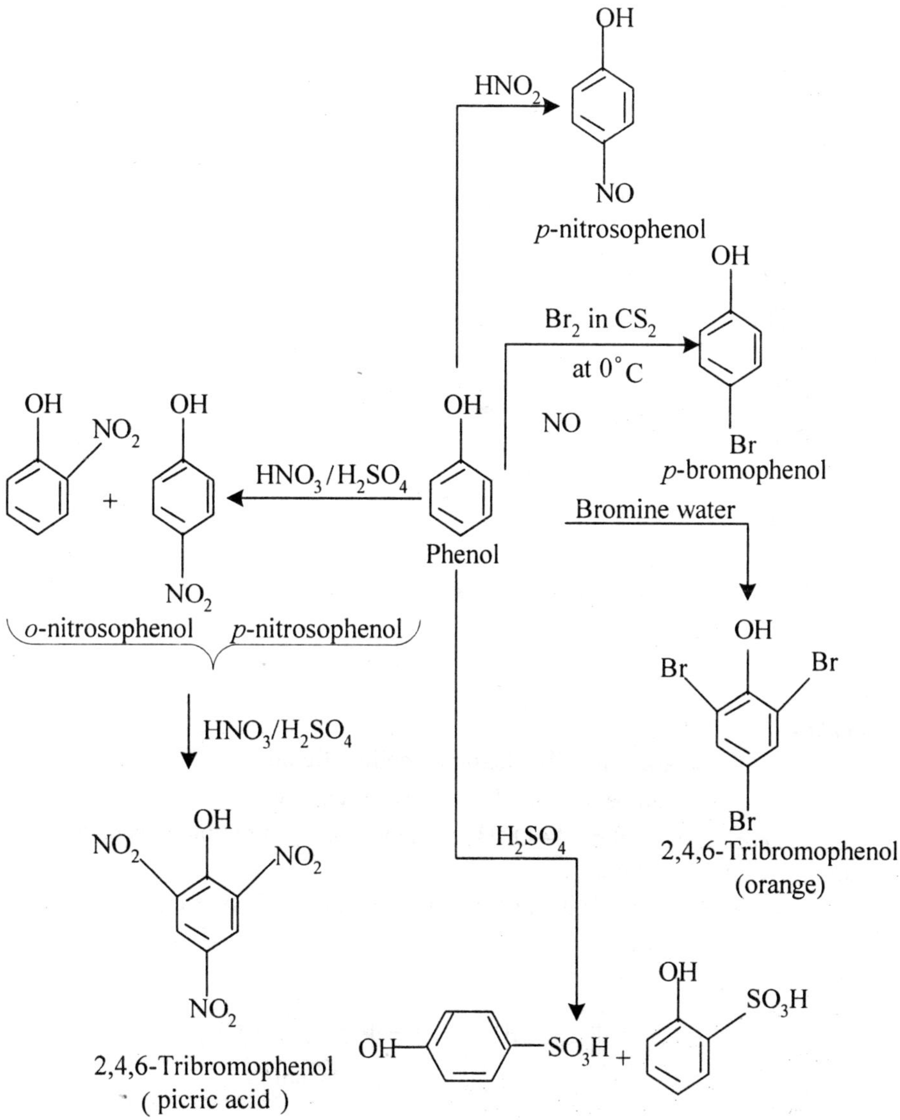

11.4.2 AROMATIC ETHERS

Aromatic ethers are organic compounds having general formula ArOR or ArOAr (where R=alkyl group, Ar = aryl group), e.g.,

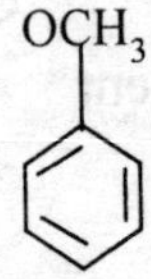

Phenylmethyl ether or anisole

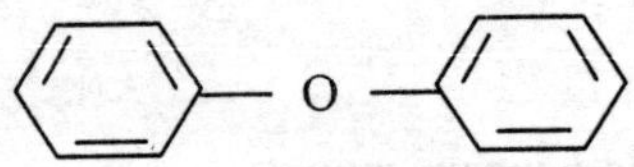

Diphenyl ether

(a) *Preparation*

(i)

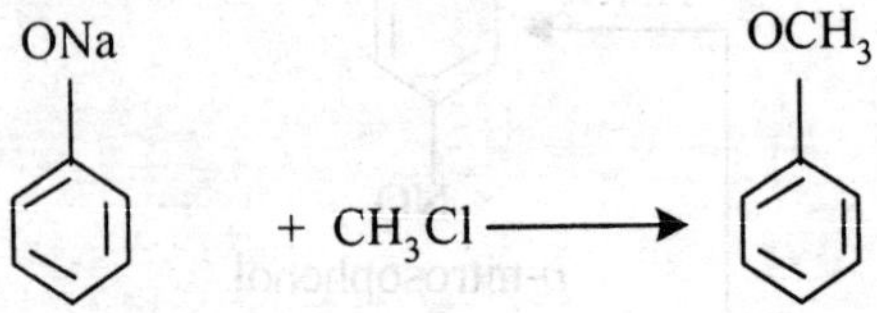

Sodium phenoxide — Anisole

(ii)

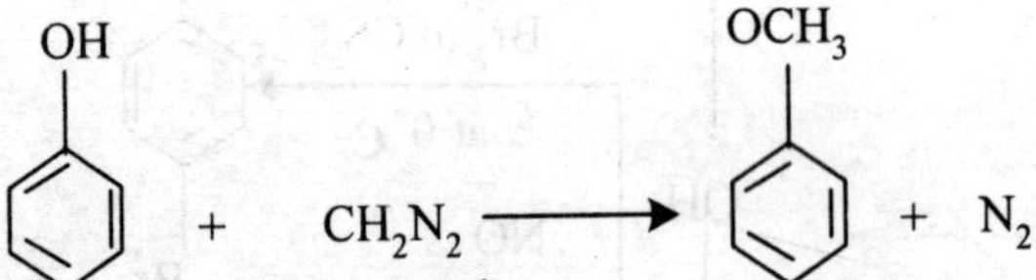

Diazomethane

(iii)

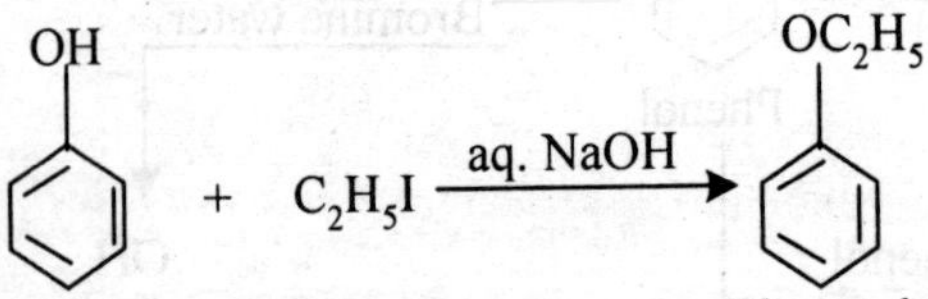

Phenetole

(b) *Properties*

(a) *Physical:* These are generally pleasant-smelling liquids.

(b) *Chemical:* These can be discussed under two categories:

(i) *Due to benzene ring:* Since —OCH_3 group is *o* and *p*-directing hence, *o*- and *p*-substituted derivatives are obtained.

(ii) *Due to —OCH_3 group*: These are similar to aliphatic ethers, e.g.,

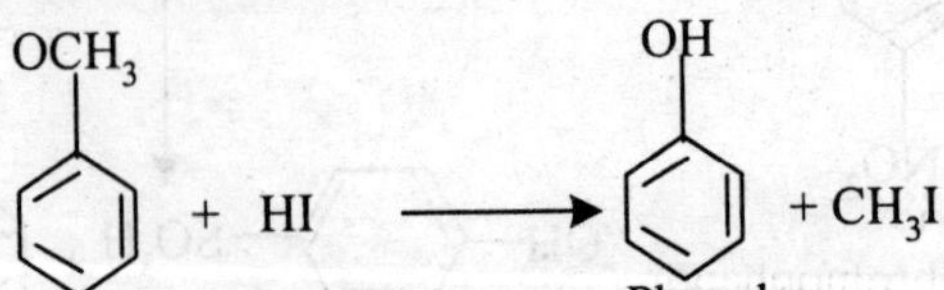

Cold — Phenol

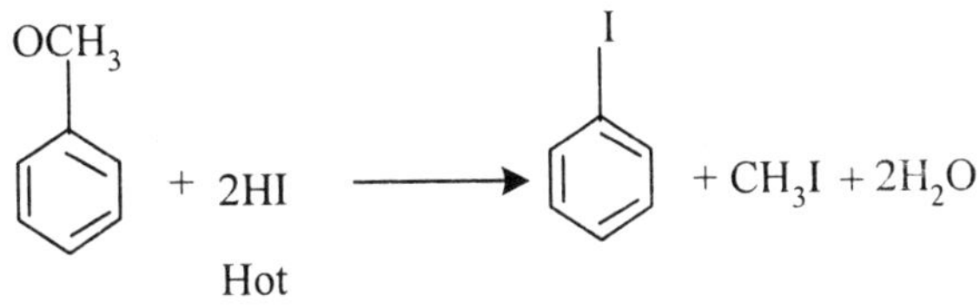

11.5 AROMATIC ALDEHYDES, KETONES AND QUINONES

Aromatic aldehydes are the organic compounds in which aldehydes group is directly attached to benzene ring, e.g,

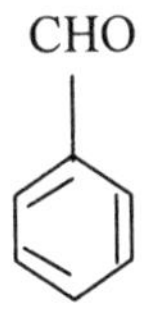

Benzaldehyde or oil of bitter almonds

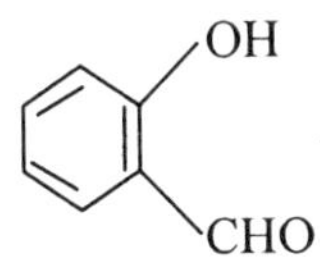

Salicylic aldehyde

(a) Preparation of benzaldehyde

The general methods are same as those for aliphatic aldehydes, e.g.,

(i) By heating calcium benzoate with calcium formate

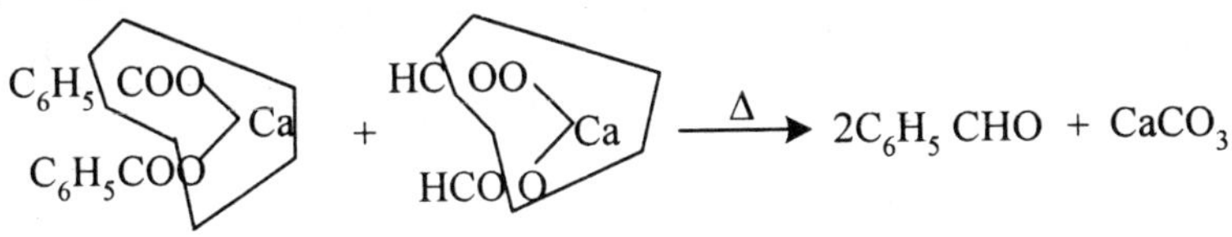

(ii) By oxidation of benzyl alcohol

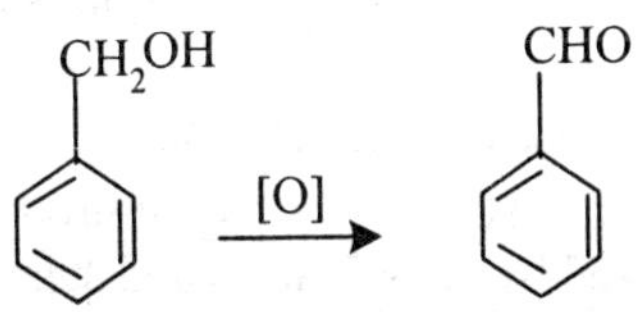

Some other important methods of preparation are as follows:

(iii) By boiling benzalchloride with aqueous copper.

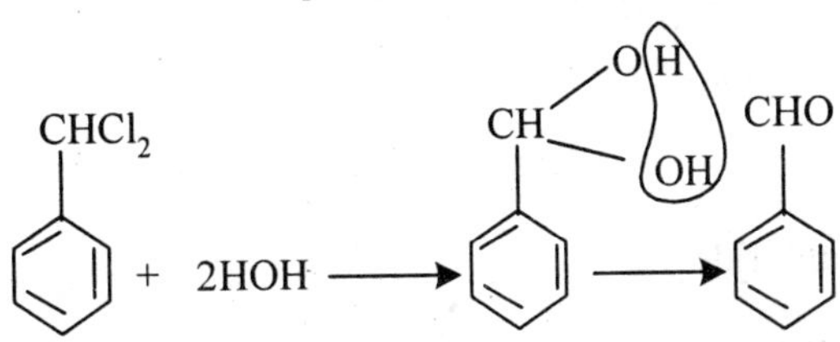

(iv) By oxidation of Toluene with chromyl chloride (*Etard reaction)*

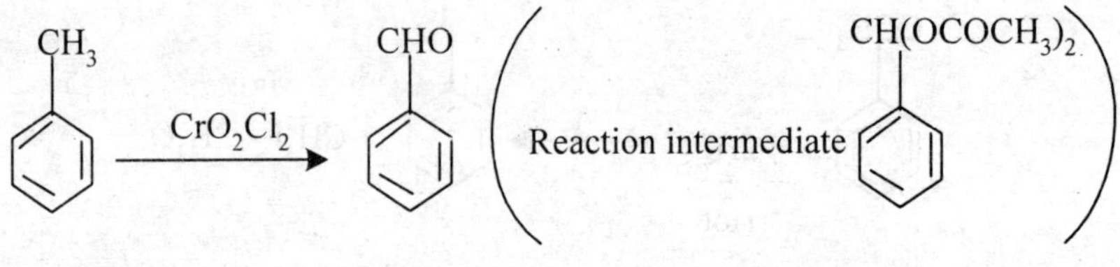

(v) ***Gattermann-Koch aldehyde synthesis***: When a mixture of carbon monoxide and hydrogen chloride is bubbled through a solution of benzene in ether in the presence of a catalyst (aluminium chloride and small amount of cuprous chloride) the benzaldehyde is obtained.

$$C_6H_6 + CO + HCl \xrightarrow{AlCl_3} C_6H_5CHO + HCl$$

$$\left(\text{Reaction intermediate} \quad HC{\equiv}O \longleftrightarrow H\overset{\oplus}{C}{=}\overset{\ominus}{O}\right)$$

This reaction fails in case of phenols and their ethers or when strongly deactivating group is present, e.g., nitrobenzene.

(vi) ***Gattermann aldehyde synthesis***: It consists of treatment of benzene with a mixture of hydrogen cyanide and hydrogen chloride in the presence of aluminium chloride followed by decomposition with water.

$$C_6H_6 + HCN + HCl \xrightarrow{ZnCl_2} C_6H_5CH{=}\overset{\oplus}{N}H_2Cl \xrightarrow{H_2O} C_6H_5CHO + NH_4Cl$$

(Intermediate reaction: $HN{=}\overset{\oplus}{C}HCl$. This reaction is applicable in case of phenols and phenolic ethers but not in case of nitrobenzene.

(vii) ***Sommelet's reaction***: It consists of refluxing of urotropin with benzyl chloride in aqueous ethanolic solution, followed by acidification and steam distillation.

$$C_6H_5CH_2Cl + (CH_2)_6N_4 \rightarrow C_6H_5CHO$$

(Reaction intermediate: $C_6H_5CH{=}\overset{\oplus}{N}H_2$)

(viii) ***Rosenmund Reduction***

$$\underset{\text{Benzoylchloride}}{C_6H_5COCl} + H_2 \xrightarrow{\text{Pd suspended in } BaSO_4} C_6H_5CHO + HCl$$

(ix) *Stephen's method*

$$C_6H_5CN \xrightarrow[\text{(ii)}H_2O]{\text{(i)}SnCl_2/HCl \text{ or moist Raney Ni}} C_6H_5CHO$$

(x) By reaction of Grignard reagent with ethyl formate

$$C_6H_5MgBr + HCOOC_2H_5 \rightarrow C_6H_5CHO + MgBr(OC_2H_5)$$

(xi) By reaction of benzene diazonium chloride with formladoxime

$$C_6H_5N_2Cl \xrightarrow{CH_2=NOH} C_6H_5CH=NOH \xrightarrow{H_2O} C_6H_5CHO$$

(b) Properties

Physical: Colorless liquids with a smell of almonds.

Chemical:

(i) Due to benzene ring

(ii) Due to aldehyde group

(a) Similar to aliphatic aldehydes

(b) Different from aliphatic aldehydes

(i) Due to benzene ring: It does not undergo Friedel-Crafts alkylation and acylation reaction because —CHO group deactivates the ring. However, it undergoes nitration and sulphonation to yield meta-substituted product.

(ii) Due to aldehyde group

(a) **Similar to aliphatic aldehydes**

(1) Addition Reactions

(2) Removal of 'Oxygen' atom in the form of water

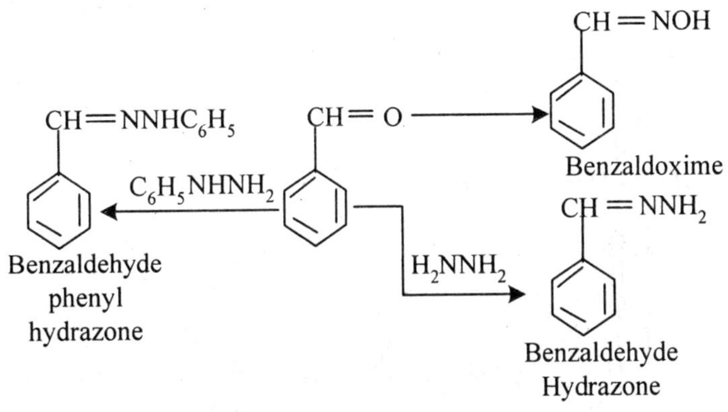

(3) It stores pink color with the Schiff's reaction reagent.

(b) **Reaction different from aliphatic aldehydes**

(1) It does not reduce Fehling's solution

(2) It undergoes Cannizz aro reaction, unlike acetaldehyde.

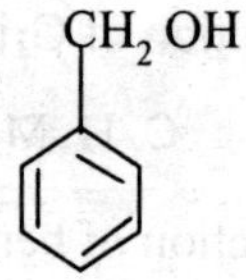

$C_6H_5CHO + NaOH \longrightarrow C_6H_5COONa + C_6H_5CH_2OH$

Sodium benzoate   Benzyl alcohol

(IIT, 1992)

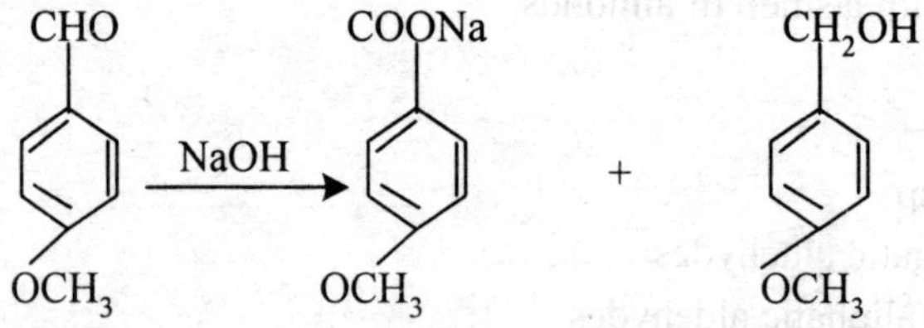

p-methoxy sodium benzoate   *p*-methoxy benzyl alcohol

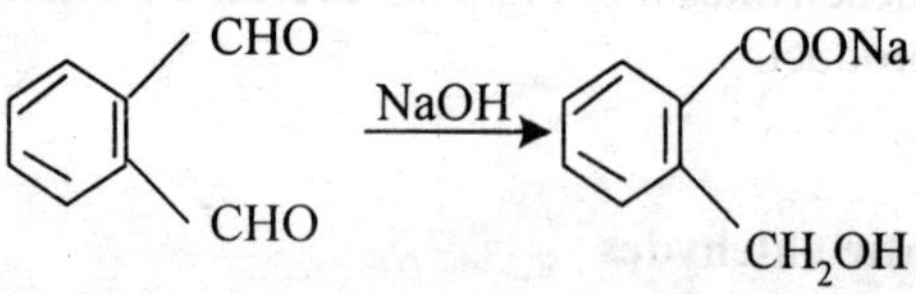

o-methylol sodium benzoate

(GATE, 1993)

The mechanism involves hydride ion transfer. It is as follows:

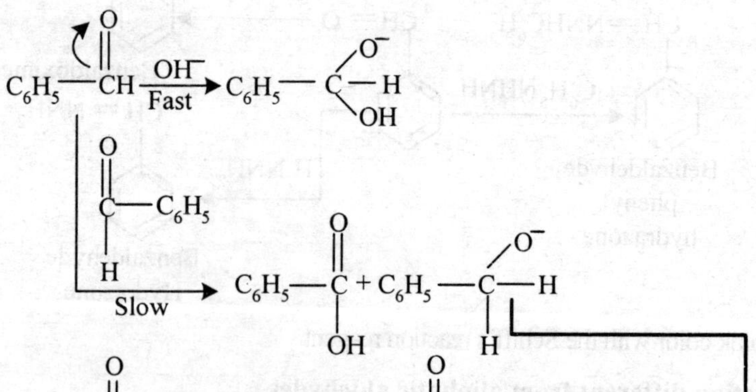

Note the reaction intermediates involved in the mechanism.

(3) It reacts with ammonia to form hydrobenzamide

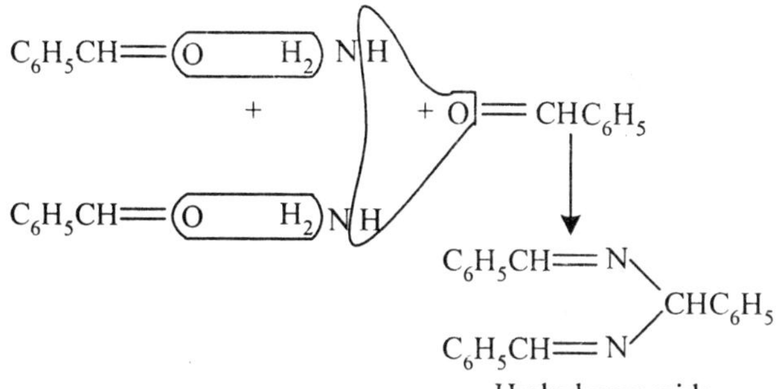

Hydrobenzamide

(4) It al so reacts with primary amines to form anils or Schiff's base.

$$C_6H_5CH{=}(O + H_2)NC_2H_5 \longrightarrow C_6H_5CH{=}NC_2H_5$$

$$C_6H_5CH{=}(O + H_2)NC_6H_5 \longrightarrow C_6H_5CH{=}NC_6H_5$$

(5) ***Perkin Reaction***: When benzaldehyde is treated with an acid anydride in the presence of sodium salt of the same fatty acid, α-β-unsaturated acid is obtained. It is important to mention here that the α-hydrogen atoms of the anhydride are involved in the condensation, e.g.,

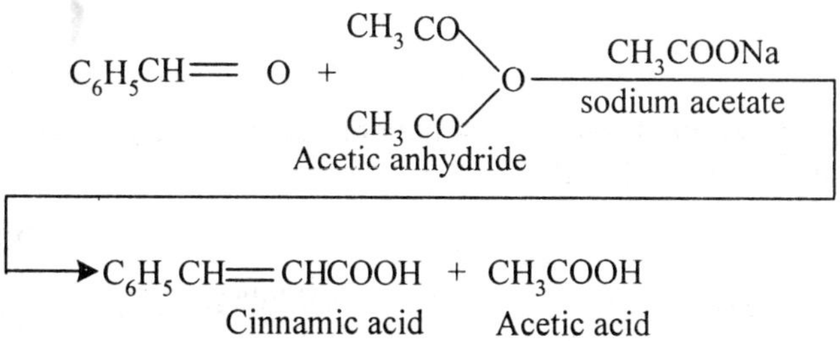

It can be written below to understand the method of obtaining products:

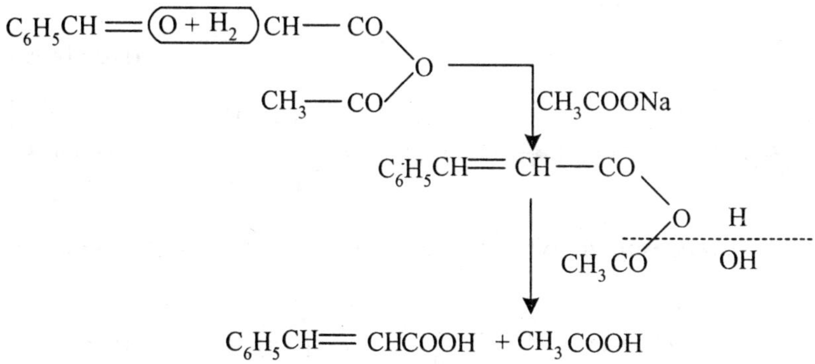

The mechanism is as follows:

(i)

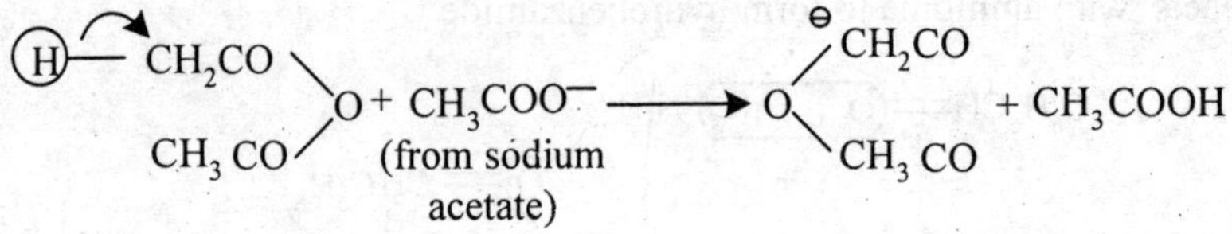

(ii)

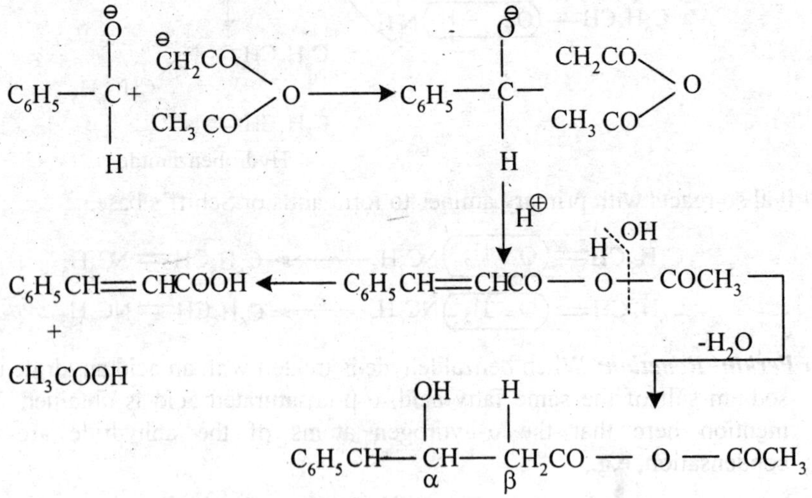

(6) Knov engel Reaction: It reacts with malonic ester or malonic acid in ethanolic ammonia to form α, β-unsaturated acid.

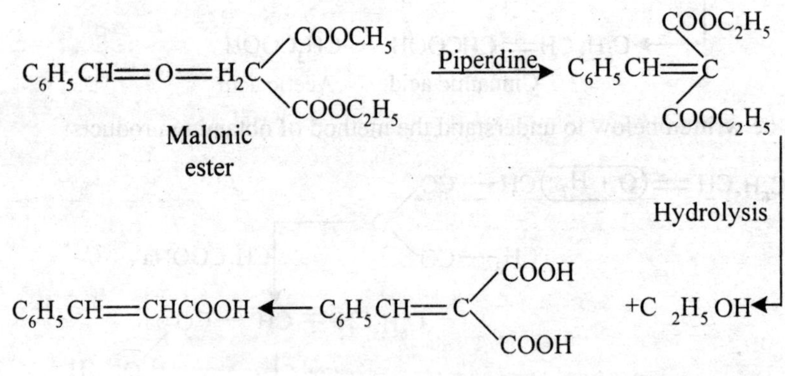

(7) Benzoin condensation: When benzaldehyde if refluxed with alcoholic KCN, benzoin is obtained.

$$2C_6H_5CHO \xrightarrow{\text{alcoholic KCN}} C_6H_6CHOHCO\ C_6H_5$$

(Reaction intermediate $\overset{\ominus}{CN}$) Benzoin

(8) Tishenk o Reaction: When benzaldehyde is treated with aluminium isopropoxide, it forms ester.

$$C_6H_5CHO \xrightarrow{\text{Aluminium isopropoxide}} \underset{\text{Benzyl benzoate}}{C_6H_5CH_2OOC—C_6H_5}$$

(9) ***Claisen -Schmidt Reaction of Claisen Reaction*:**

$$C_6H_5CH = O + H_2CHCHO \xrightarrow{NaOH} \underset{\text{Cinnamic aldehyde}}{C_6H_5CH = CHCHO} + H_2O$$

(10) It also undergoes condensation reaction with acetone, acetophenone and nitromethane. The reactions are as follows:

$$C_6H_5CH = O + H_2CHCOCH_3 \xrightarrow{NaOH} \underset{\text{Benzylidene acetone}}{C_6H_5CH = CHCOC_6H_5}$$

$$C_6H_5CH = O + C_6H_5COCHH_2 \xrightarrow{NaOH} \underset{\text{Phenyl styryl ketone or Chalcone or Benzylidene acetophenone}}{C_6H_5CH = CHCOC_6H_5}$$

$$C_6H_5CH = O + H_2CHNO_2 \xrightarrow{NaOH} \underset{\text{Co-nitro styrene}}{C_6H_5CH = CHNO_2} + H_2O$$

11.5.1 Aromatic Ketones

Aromatic ketones are organic compounds in which carbonyl group is linked to either aryl and alkyl group or two aryl groups, e.g.,

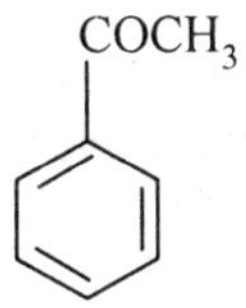

Acetophenone

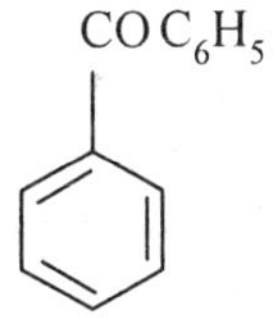

Benzophenone

(a) ***Preparation***

The general methods are same as those for aliphatic ketones, e.g.,

(i) By distilling a mixture of calcium benzoate and calcium acetate.

$$\begin{matrix} C_6H_5COO \\ C_6H_5COO \end{matrix} \!\!> Ca \xrightarrow{\text{distill}} \underset{\text{Benzophenone}}{C_6H_5COC_6H_5} + CaCO_3$$

$$(C_6H_5COO)_2Ca + \begin{matrix} CH_3COO \\ CH_3COO \end{matrix} \!\!> Ca \longrightarrow \underset{\text{Benzophenone}}{C_6H_5COC_6H_5} + CaCO_3$$

(ii) By Friedel-Craft acylation reaction

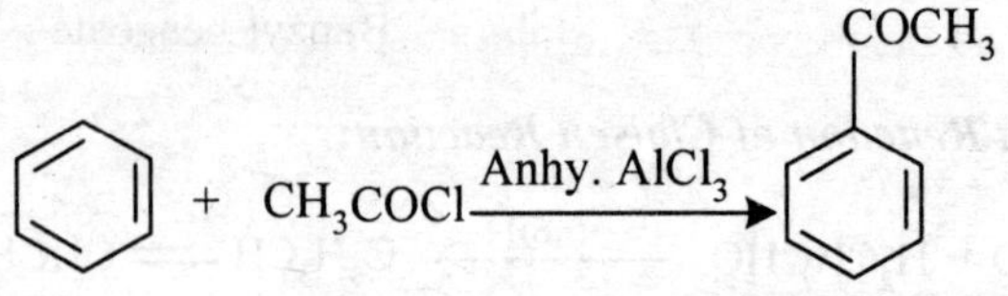

Acetophenone

(b) *Properties*

Physical properties: Acetophenone: Solid (m.p.20 °C), Benzophenone solid (m. p. 48 °C).

Chemical properties: They are mostly same as alphatic ketones. Some important properties are as follows:

(i) $$C_6H_5-\overset{\overset{\displaystyle O}{\|}}{C}-CH_3 + CH_3-\overset{\overset{\displaystyle O}{\|}}{C}-C_6H_5 \longrightarrow C_6H_5-\overset{\overset{\displaystyle CH_3}{|}}{C}=CHCOC_6H_5$$

Dyphone

(ii)

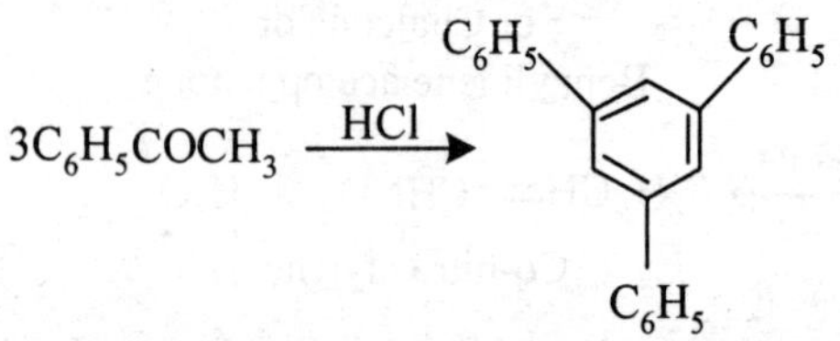

1,3,5-Triphenyl benzene

(iii) It undergoes Munnich reaction

$$C_6H_5COCH_3 + CH_2O + Me_2NHHCl \rightarrow C_6H_5COCH_2NMe_2{\cdot}HCl$$

(iv) *Willgerodt reaction*: It is consists of conversion of a carbonyl compound into an amide with the same number of carbon atoms, e.g., when acetophenone reacts with aq. solution of yellow ammonium polysulphide, an amide of phenyl acetic is obtained.

$$C_6H_5COCH_3 \xrightarrow{(NH_4)_2\ SX} C_6H_5CH.CONH_2 + C_6H_5CH_2COONH_4$$

11.5.2 Quinones

The cyclic unsaturated diketones in which keto groups are conjugated with the double bond are known as quinones, e.g.,

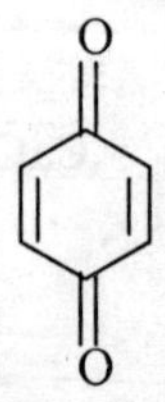

p-benzoquinone

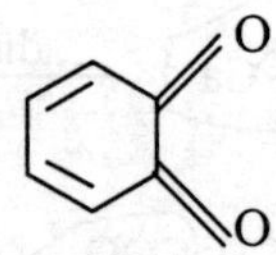

o-benzoquinone

(a) ***Preparation***

(1) *p*-benzoquinone can be prepared by mild oxidation of hydroquinone with ferric chloride

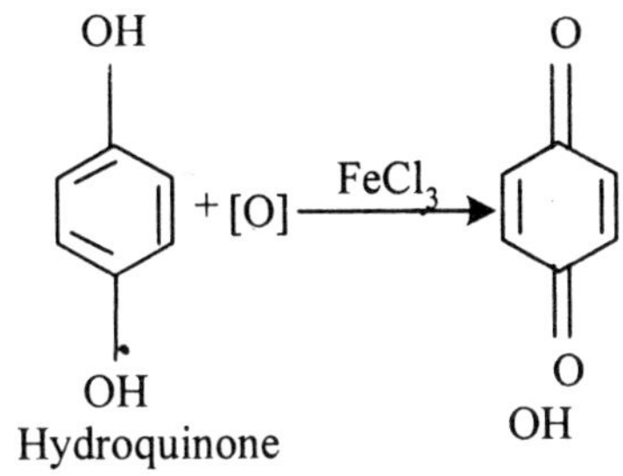

(2) *p*-benzoquinone can also be prepared by oxidation of aniline with manganese dioxide and sulphuric acid

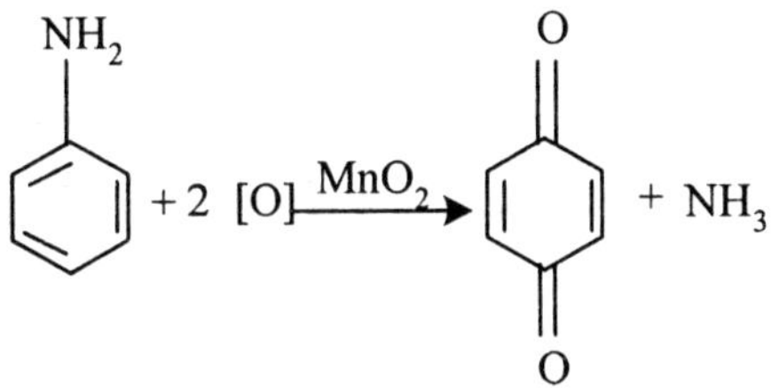

(3) *P*-benzoquinone can be prepared by mild oxidation of hydroquinone with ferric chloride.

(b) **Properties**

Physical: It is steam-volatile and undergoes sublimation.

Chemical: (1) It reacts with hydroxylamine

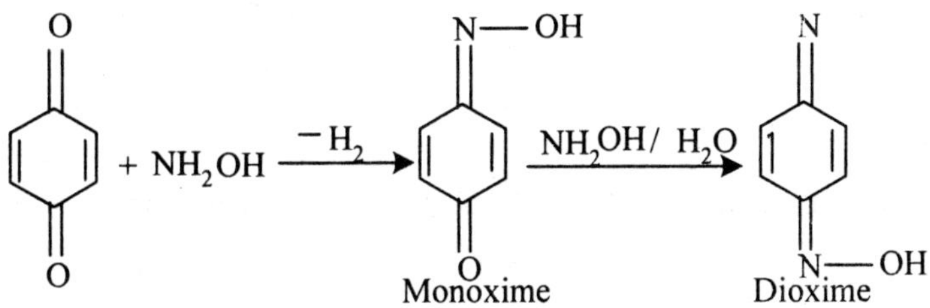

(2) It is reduced by sulphurous acid to form hydroquinone

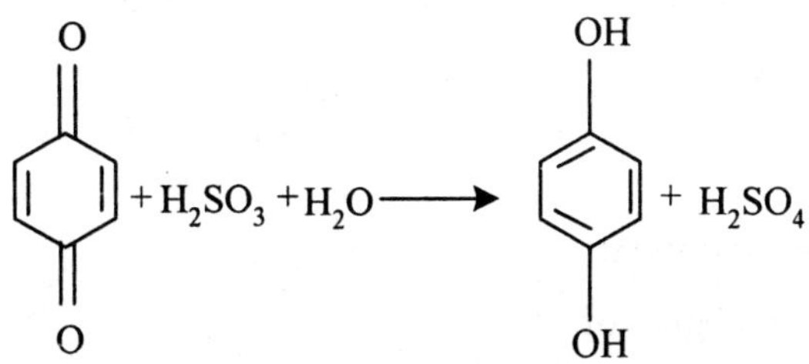

(3) It undergoes reaction with 1,3-butadiene to form cycloaddition product. The reaction is called Diels-Alder reaction.

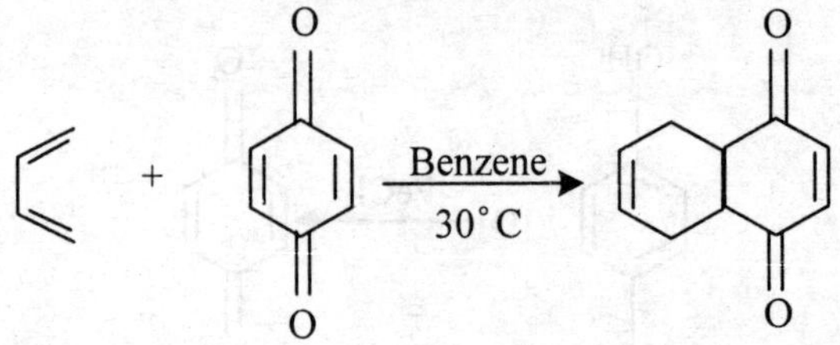

1,3-butadiene Tetrahydro-1,4 naphthoquinone

(4) It also reacts with HCl to form 1,4-addition products.

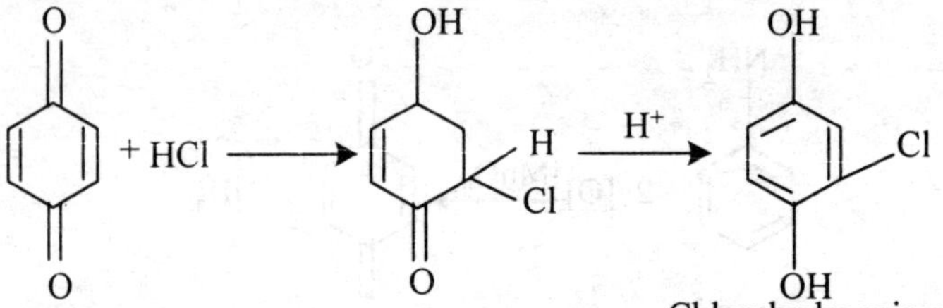

Chlorohydroquinone

11.6 AROMATIC CARBOXYLIC ACIDS

These are compounds in which one or more carboxyl groups are linked directly to benzene ring. Some examples are:

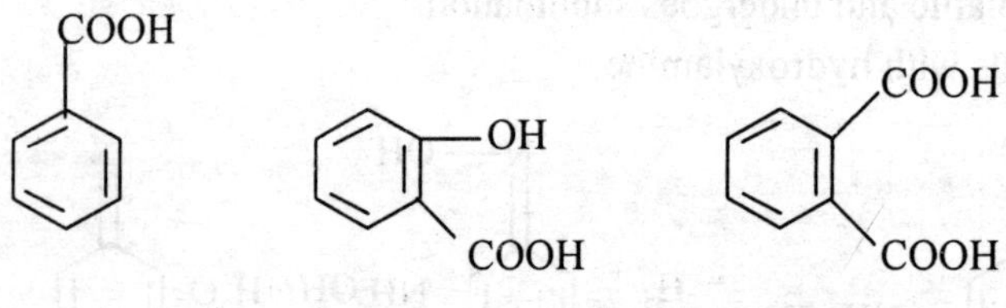

Benzoic acid Salicyclic acid Phthalic acid

(a) ***Preparation***

The methods are almost same as those for aliphatic carboxylic acids, e.g.,

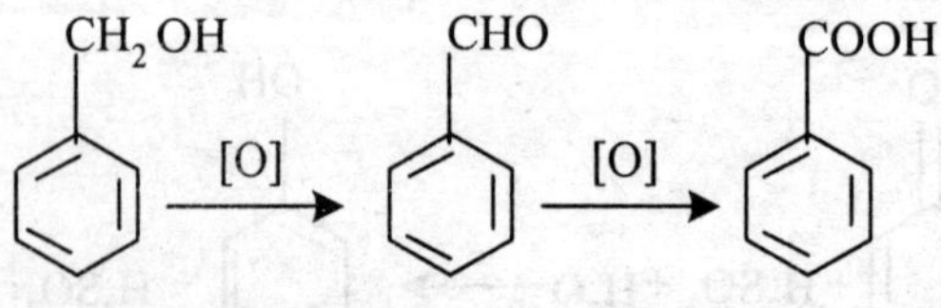

A few additional methods are as follows:

(i) By oxidation of homologues of benzene

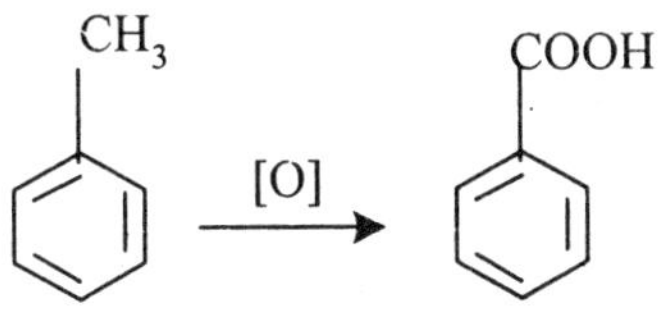

(ii) By hydrolysis of benzotrichloride

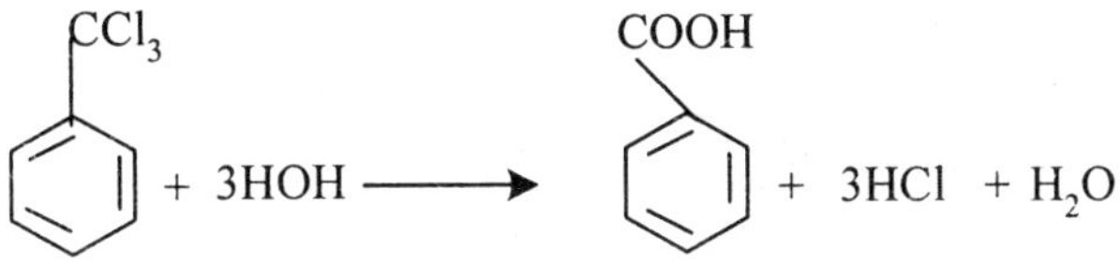

(iii) Salicylic acid can be prepared either by Reimer-Tiemann reaction or Kolbe-Schmidt reaction (See Phenols)

(iv) Phthalic acid can be obtained by oxidation of *o*-xylene.

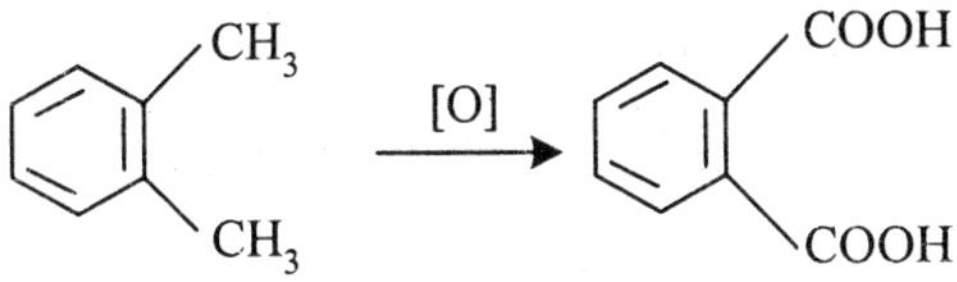

(b) *Properties*

Physical: These are colorless solids and generally soluble in hot water, and steam volatile.

Chemical: Can be studied under two categories.

(i) Due to —COOH group: These are similar to those of aliphatic acids.

(ii) Due to benzene ring: See electrophilic substitution reactions.

Miscellaneous

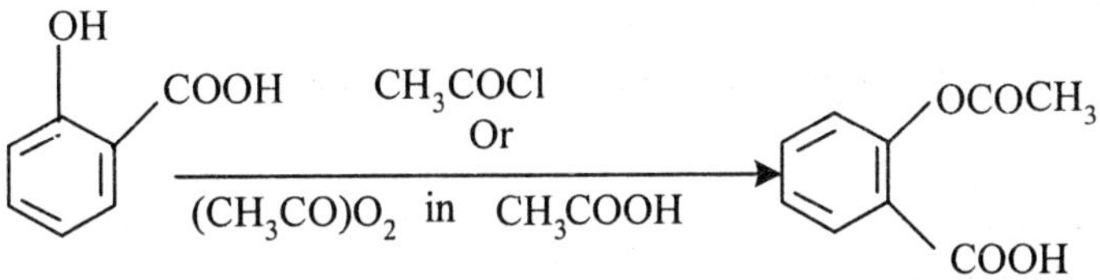

Acetyl salicylic acid
or aspirin
(used as analgesic)

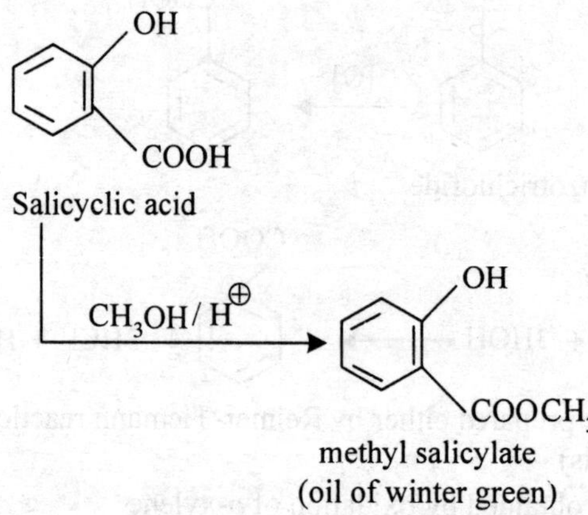

11.7 BENZENE SULPHONIC ACIDS

Reaction of aromatic hydrocarbons with sulphuric acid results in the formation of sulphonic acids.

(a) ***Preparation***

By the sulphonation of aromatic hydrocarbons

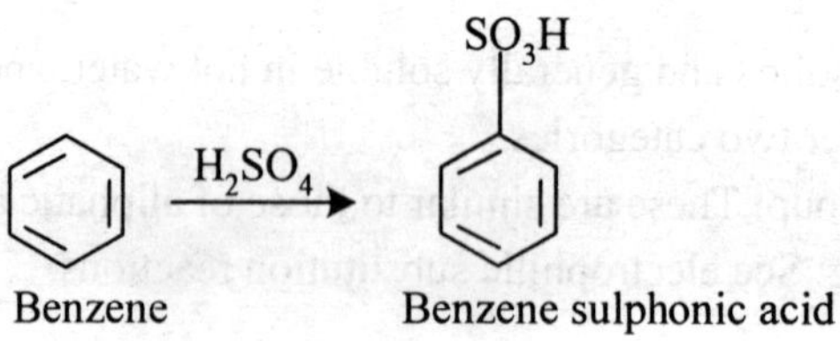

11.7.1 (b) *Properties*

Physical: Colorless, crystalline, deliquescent solid, soluble in water.

Chemical:

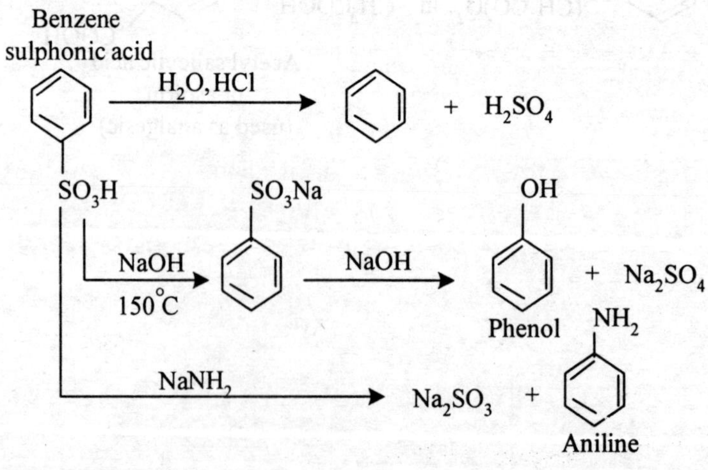

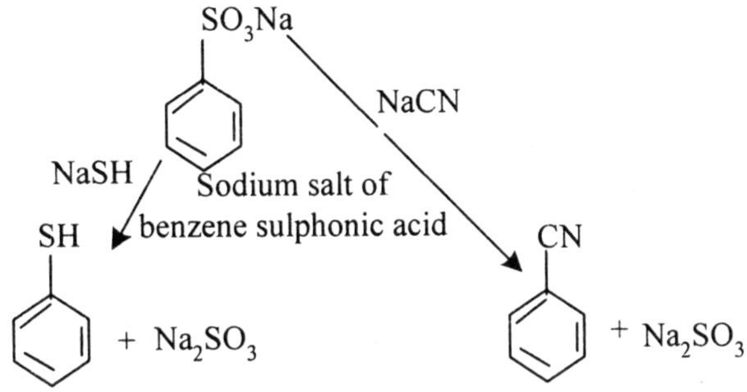

Thiophenol Phenyl cyanide

11.8 AROMATIC NITRO, AMINO COMPOUNDS AND DIAZONIUM SALT

11.8.1 Nitrobenzene

(a) *Preparation*

By nitration of benzene

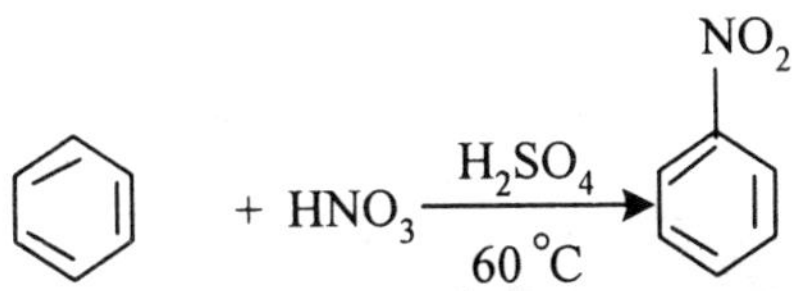

Nitrobenzene

(b) *Properties*

Physical: Yellow crystalline solid (nitro-benzene is yellow liquid) whereas *m*-dinitro benzene is insoluble in water.

Chemical: Chemical properties can be discussed in following categories:

(i) *Properties due to benzene ring*: Aromatic nitrocompounds do not undergo Friedel-Craft reactions.

(ii) *Properties due to nitro group*: Nitrobenzene can be reduced by six methods.

(a) *Reduction in acidic medium:* In acidic medium, nitrobenzene is reduced by metal-acid combination like Sn+HCl, Zn+HCl or Fe+CH_3COOH. The iron + acetic acid combination is used for industrial purposes because it is economical.

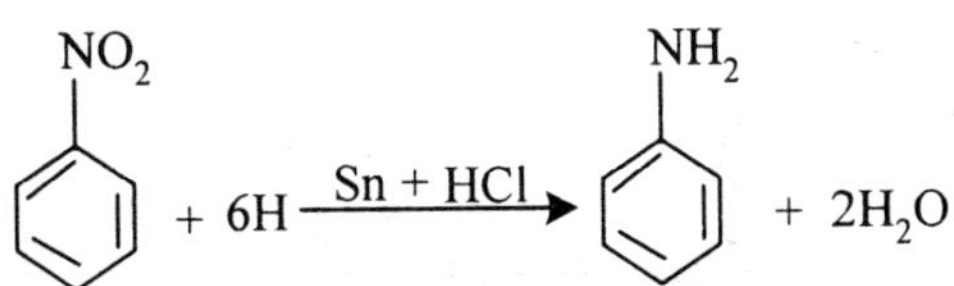

Aniline

(b) *Reduction in neutral medium:* The end product depends upon the reagent used:

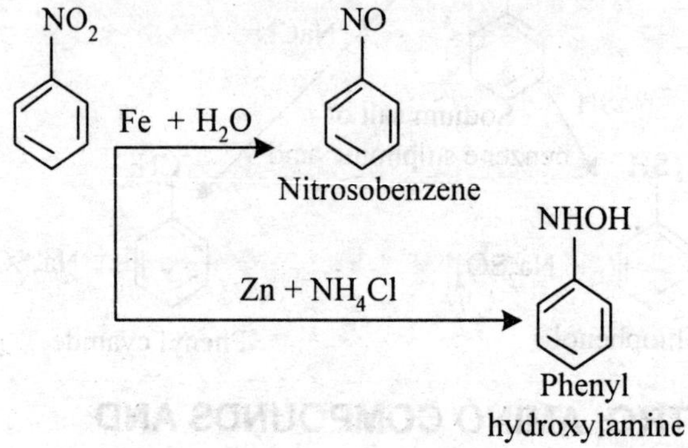

(c) *Reduction in alkaline medium:* The end product depends upon the reagent used. The general scheme is as follows:

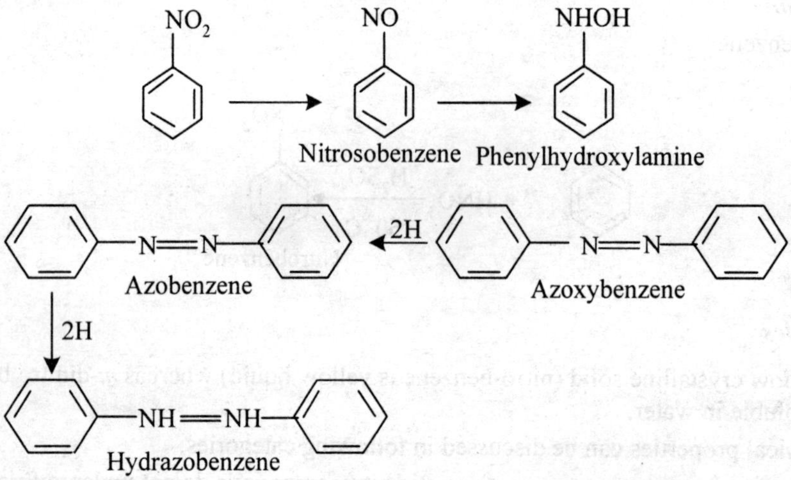

	Reducing agent	Product
1.	CH_3ONa/CH_3OH	Azoxybenzene
2.	SnCl/NaOH	Azobenzene
3.	Zn/NaOH	Hydrazobenzene

(d) *Selective reduction*

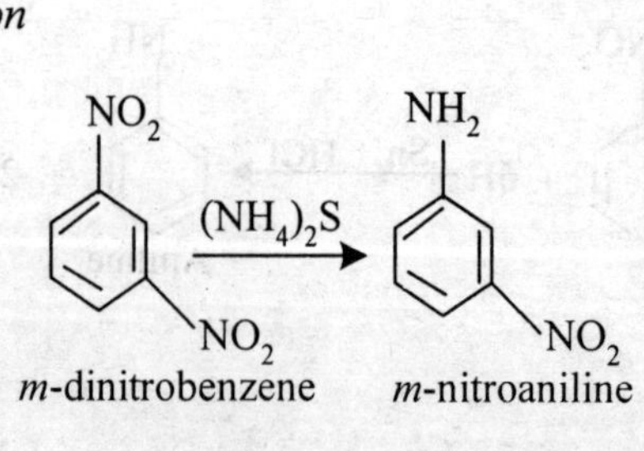

(e) *Catalytic reduction* N′

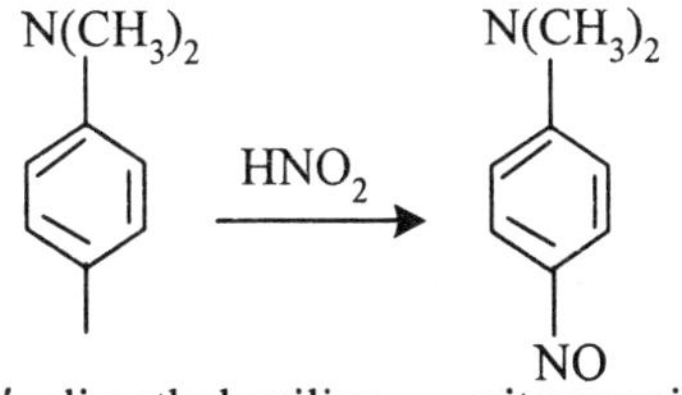

N, N′ - dimethyl aniline *p*-nitrosoaniline

(f) *Electrolytic reduction*

(i) Weak acidic medium

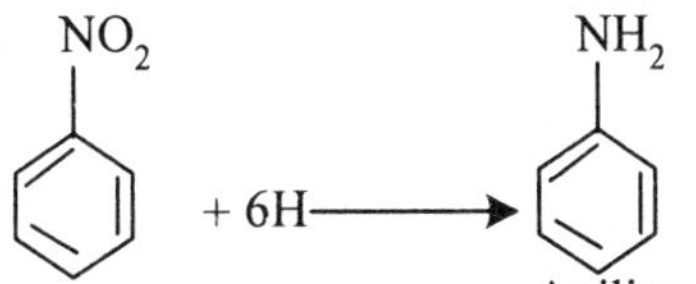

Aniline

(ii) Strong acidic medium

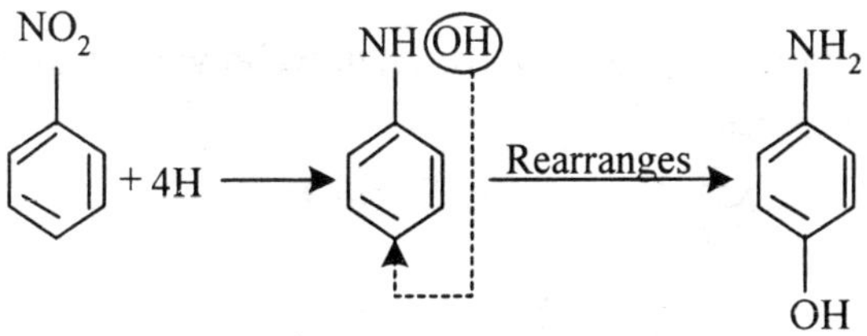

p-hydroxyaniline

(iii) Alkaline medium

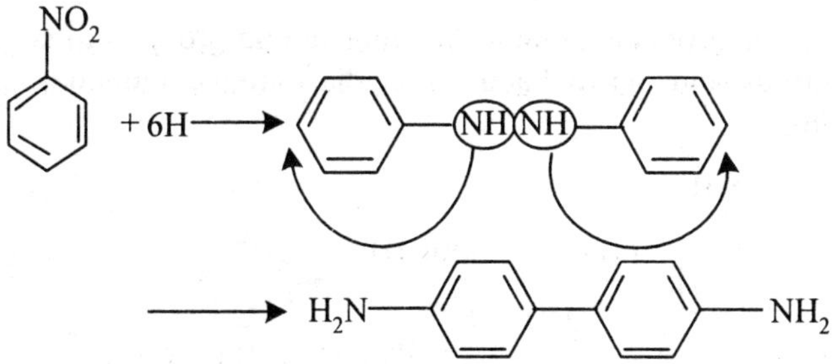

Benzidine

11.8.2 Aryl amines

Organic compounds in which amino group is directly linked to benzene ring are called aryl amines. Some examples are:

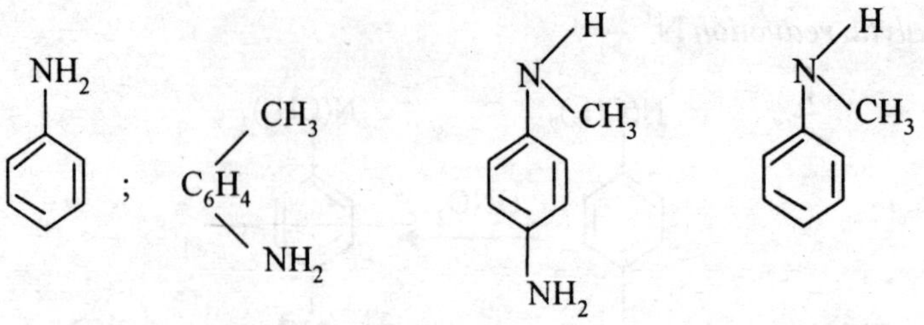

Aniline ; *o,m,p*-toluidine N-methyl aniline N,N'-dimethyl aniline

(a) ***Preparation of aniline***

1. By reaction of nitrobenzene under different conditions (Sec. 1).
2. By reaction of benzamide with Br_2 + KOH

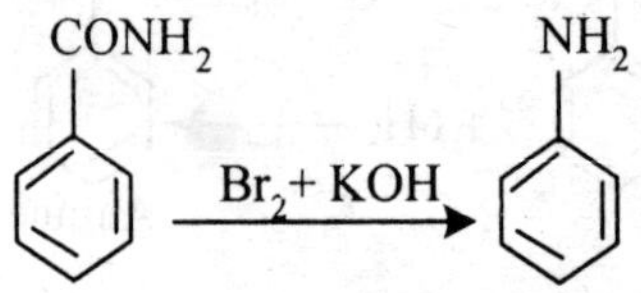

Aniline

(b) ***Properties of aniline***

Physical: Pure aniline is colorless liquid. However, it turns brown due to oxidation, when stored.

Chemical: 1. Reaction with nitrous acid

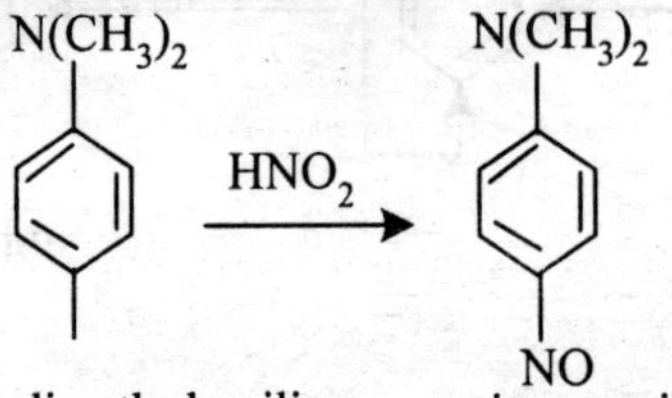

N, N'-dimethyl aniline *p*-nitrosoaniline

(2) Diazotization: When aromatic amines, in which amino group is directly attached to benzene ring, are treated with nitrous acid and HCl at 0 – 5 °C, diazonium compounds are formed, the reaction is known as diazotization.

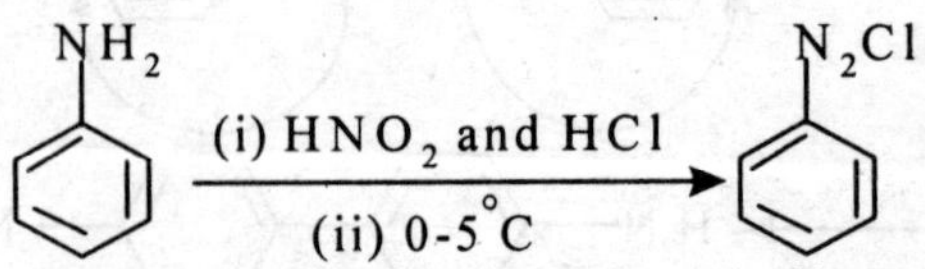

Benzene diazonium chloride

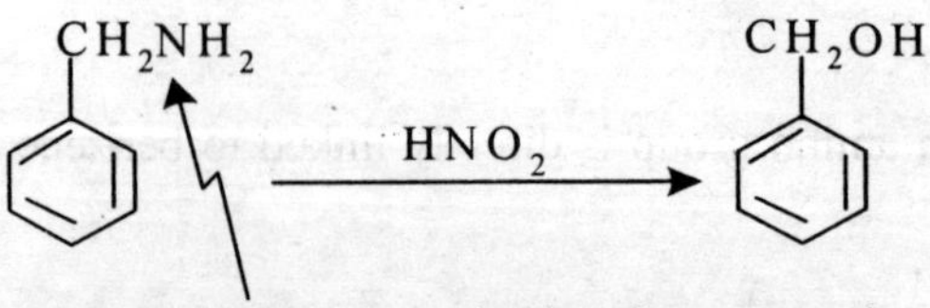

Not directly attached to benzene ring.

11.8.3 Diazonium Salt

Organic compounds, having formula R—N=N—X, is called a diazonium salt.

—N=N—	R—N=N—R	R—N=N—X
Azo group	Azo compounds	Diazo compounds

(a) *Preparation*

The diazotization results in formation of diazonium salt. The reaction is as follows:

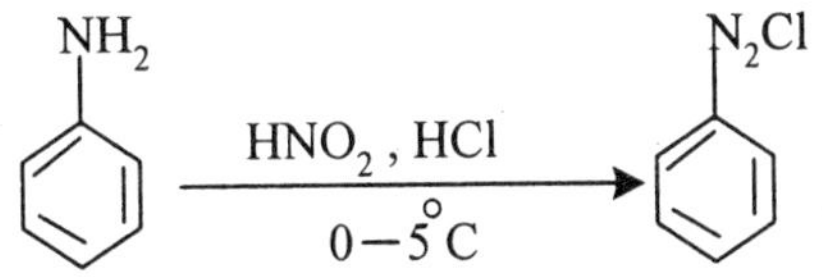

Benzene diazonium chloride

The mechanism is as follows:

(i) $HO—NO + H^+ \underset{}{\overset{\text{Fast}}{\rightleftharpoons}} H_2O^+—NO \xrightarrow{\text{Fast}} H_2O + NO^+$

(ii)

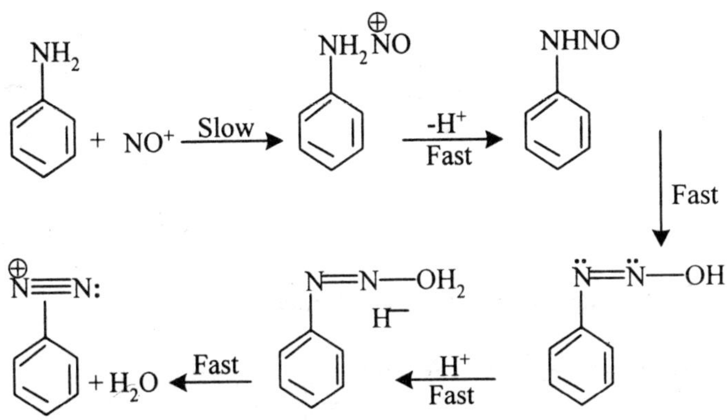

(b) *Properties*

(a) *Physical:* Diazonium salts are colorless crystalline solids which turns brown, soluble in water.

(b) *Chemical:* Properties may be discussed under two categories:

- *(a)* Reactions involving liberation of nitrogen gas
- *(b)* Reaction in which two nitrogen atoms are retained.

(a) Reactions involving liberation of nitrogen gas:

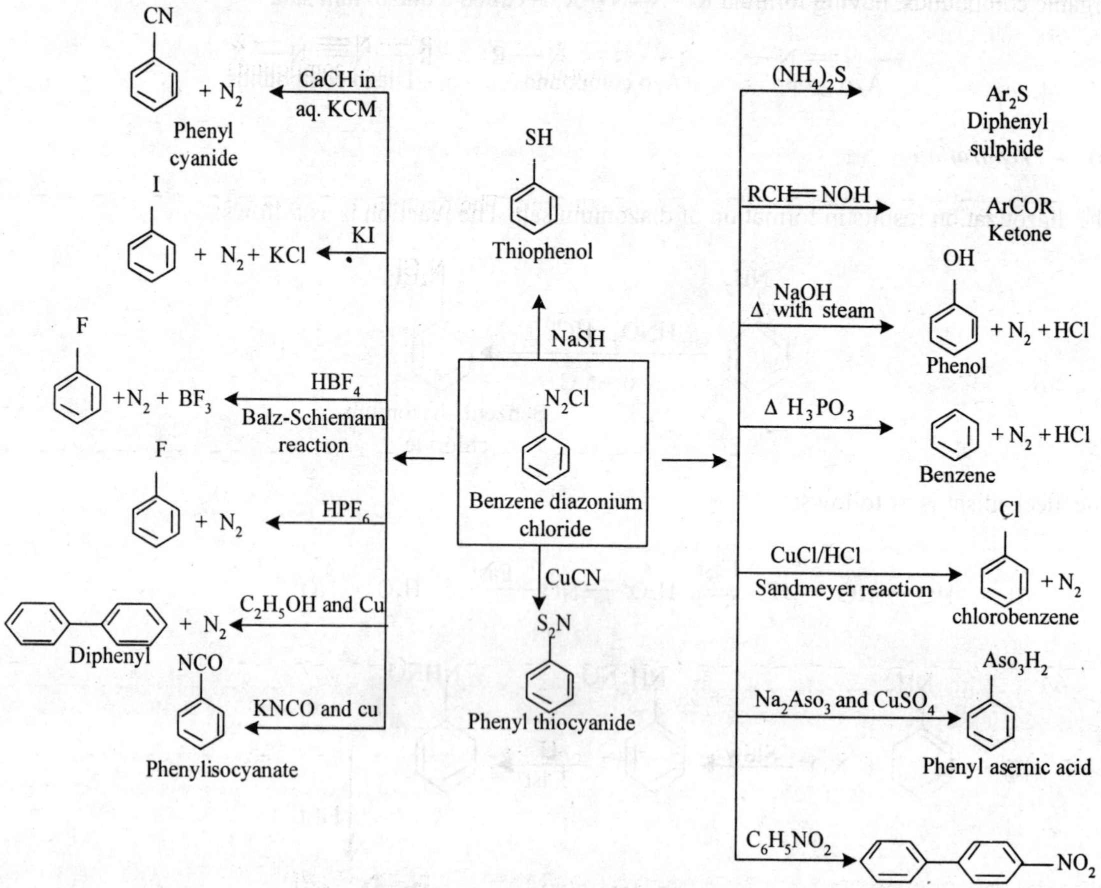

(b) Reaction in which nitrogen atoms are retained

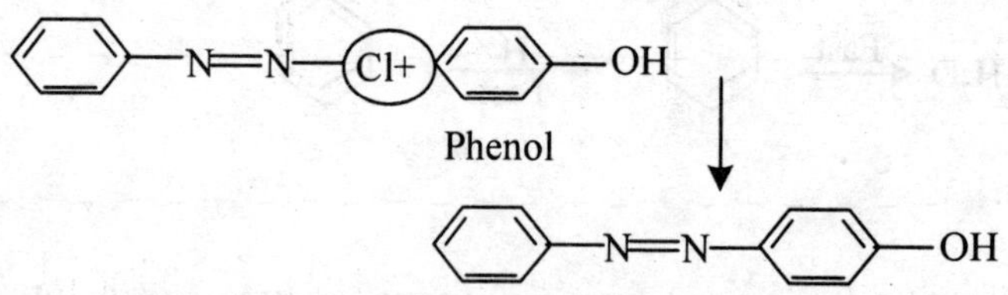

p-Hydroxyazobenzene

Reaction of diazonium salts with phenols, naphthols and aromatic amines to form highly-colored azocompounds is known as coupling reaction.

PROBLEMS

1. Choose the correct answers from the following answers:

   (i) Aniline in acidic medium on nitration gives

(a) *o*-nitroaniline *(b)* *p-nitroaniline*
(c) *m*-nitroaniline (d) mixture of *o*- and *p*-nitroaniline

(ii) Which of the following would undergo chlorination fast

(a)  *(b)* 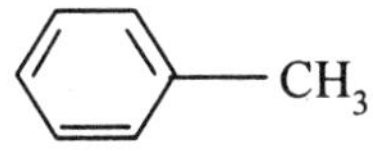

(c) 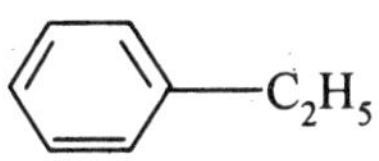 *(d)* 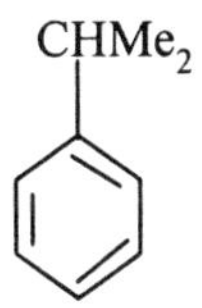

(iii) Among the following compounds one towards sulphonation is:
(a) Toluene *(b)* Chlorobenzene
(c) Nitrobenzne *(d)* *m*-xylene **(GATE, 1992)**

(iv) The compound that is most reactive towards electrophilic nitration is:
(a) Toluene
(b) Benzene
(c) Benzoic acid
(d) Nitrobenzene

(v) When nitrobenzene is treated with Br_2 in presence of $FeBr_3$, the major product formed is *m*-bromonitrobenzene. Statements which are related to obtain the *m*-isomer are:
(a) The electron density of *meta* carbon is more than that on *ortho* and *para* positions
(b) The intermediate carbonium ion formed after initial attack of BR^+ at the meta position is least destabilized
(c) Loss of aromaticity when Br^+ attacks at the ortho and para positions and not at meta position
(d) Easier loss of H^+ to regain aromaticity from the meta position than from ortho and para position **(IIT, 1992)**

(vi) The most stable carbocation is:

(a)  *(b)* 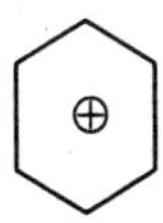

(c)  *(d)* 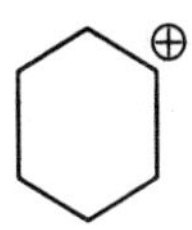

(vii) Which of the following molecules/species are aromatic in nature.
(a) $C_{20}H_{20}$ *(b)* $C_{20}H_{20}^{\oplus}$
(c) $C_{20}H_{20}^{++}$ *(d)* $C_{20}H_{20}^{--}$.

(viii) Among the following compounds the one that exhibits aromaticity is
   (a) Cyclopentadenyl cation
   (b) Cyclopentadienyl radical
   (c) Cycloheptatrienyl cation
   (d) Cycloheptatrienyl radical

(ix) Identify the correct order of reactivity in electrophilic substitution reactions of the following compounds:

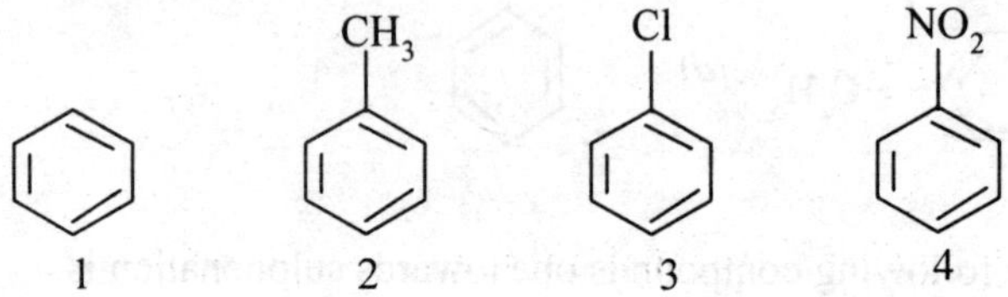

1 2 3 4

   (a) 1 > 2 > 3 > 4
   (b) 4 > 3 > 2 > 1
   (c) 2 > 1 > 3 > 4
   (d) 2 > 3 > 1 > 4

(x) Fluorobenzene (C_6H_5F) can be synthesized in the laboratory **(AIEEE 2006)**
   (a) by reacting bromobenzene with NaF solution
   (b) by heating phenol with HF & KF
   (c) from aniline by diazotisation followed by heating the diazonium salt with HBF_4
   (d) by direct fluorination of benzene with F_2 gas

2. Give reasons in two or three sentences:
   (i) Phenyl group is known to exert negative inductive effect. But each phenyl ring in biphenyl is more reactive than benzene towards electrophilic substitution. **(IIT, 1992)**
   (ii) Pyridine is aromatic.
3. Outline a synthesis of *p*-bromonitrobenzene from benzene in two steps. **(IIT, 1993)**
4. Although benzene is highly unsaturated, it does not undergo addition reactions. **(MNR, 1995)**
5. What happens when:
   (i) *p*-xylene is reacted with conc. H_2SO_4 and the resultant product is fused with KOH. **(IIT, 1984)**
   (ii) Excess chlorine is passed through boiling toluene in the presence of sunlight. **(IIT, 1987)**
6. *(i)* How can you prepare benzene from lime? **(IIT, 1987)**
   (ii) How can you convert toluene to *m*-nitrobenzene? **(IIT, 1997)**
   (iii) How can you convert toluene to *p*-nitrobenzylethyl ether. **(MNR, 1995)**
7. Complete following equations:
   (i) $C_6H_5C_2H_5 \xrightarrow{Br_2,\ \text{Heat, light?}} ?$
   (ii) $C_6H_5C_2H_5 \xrightarrow{NaCN} ?$

8. Write all possible aromatic hydrocarbons for the compounds having molecular formula C_8H_{10}. How will you identify them?

9. For each subquestion given below four answers are provided of which only one is correct. Write down the correct answer:

(i) The reaction of toluene with chlorine in presence of $FeCl_3$ gives predominantly:

(a) Benzoylchloride *(b)* *m*-chlorotoluene

(c) benzylchloride *(d)* *o*- and *p*-chloro toluene **(IIT, 1986)**

(ii) Chlorination of toluene in the presence of light and heat followed by treatment with aq. NaOH gives:

(a) *o*-cresol *(b)* *p*-cresol

(c) *2,4*-Dihydroxytoluene *(d)* Benzoic acid

10. Predict the structures of the intermediate products in the following reaction sequence:

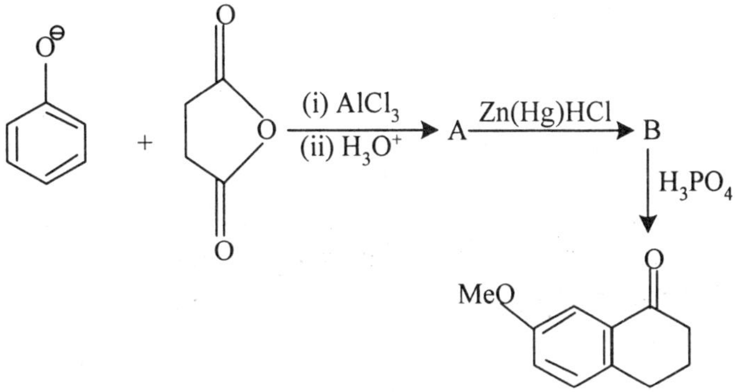

(IIT, 1996)

11. Toluene reacts with bromine in the presence of light to give benyzl bromide while in presence of $FeBr_3$,. *p*-promotoluene is formed. Explain. **(IIT, 1996)**

12. Complete the following equations:

(i)

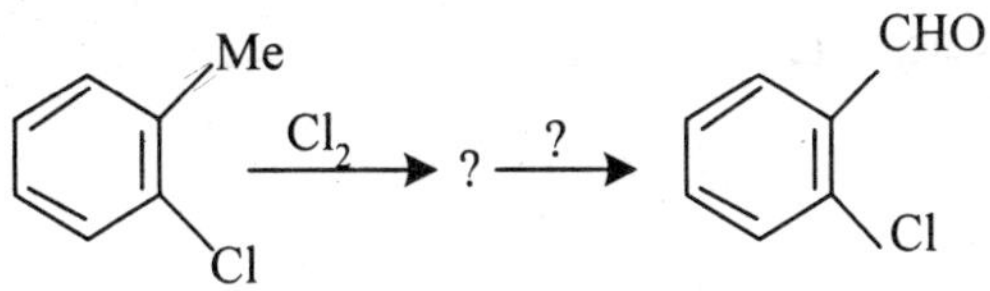

(ii) $2PH_2CCl_2 \xrightarrow{Cu/PhH} ?$

(iii)

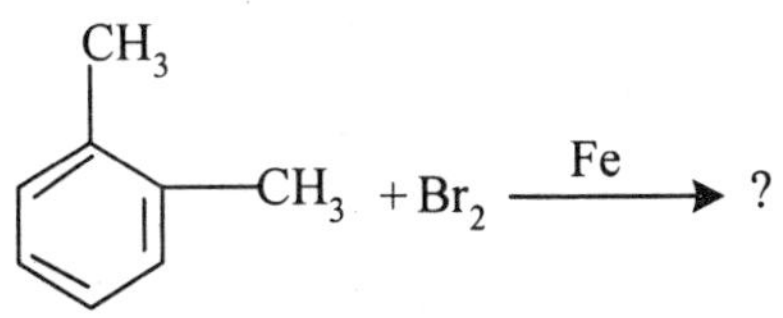

(iv) $C_6H_5CH_3 + C.\ BrCl_3 \rightarrow$(No ring substitution)

(v) $C_6H_5Br \xrightarrow{NaNH_2/liq.NH_3} ? \xrightarrow{[EtCHCHMe]^- Na^+} ?$

(vi) 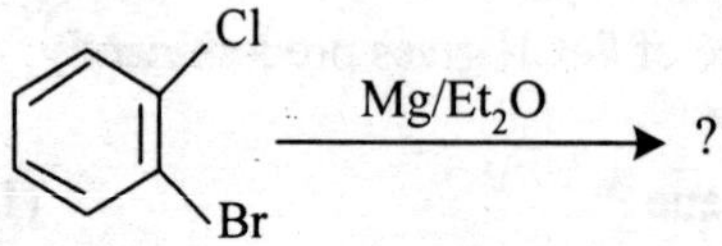

(vii) 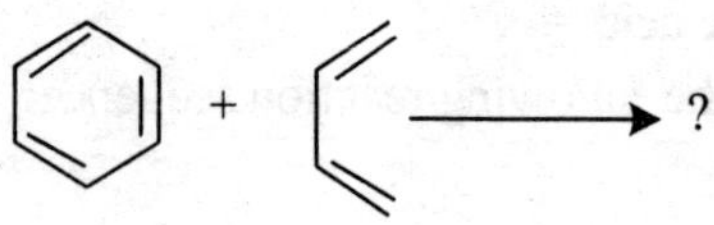

13. Fill in the blanks:

(i) Preparation of chlorobenzene from benzenediazonium chloride with cuprous chloride and aq. HCl is known as .... Reaction. **(GATE, 1994)**

(ii) The structure of the intermediate product, formed by oxidation of toluene with CrO_3 and acetic anhydride whose hydrolysis gives benzaldehyde is.......... **(IIT, 1992)**

14. Complete following reactions:

(i) 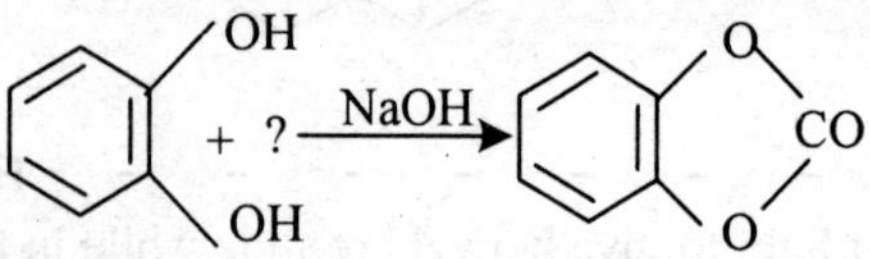

(ii) 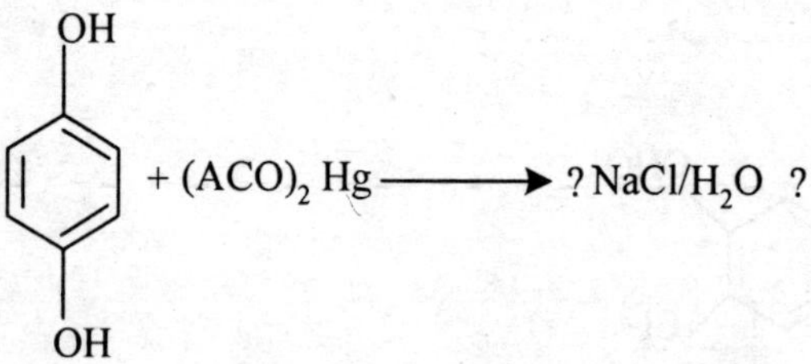

(iii) 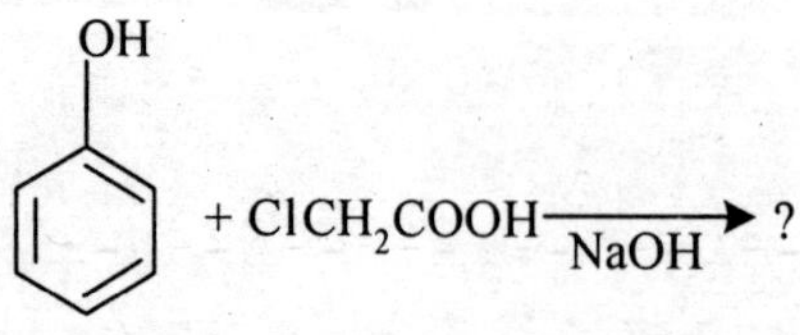

(iv) 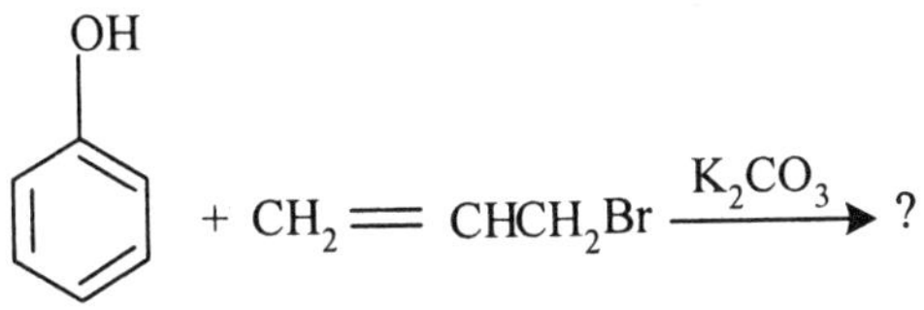

(v) 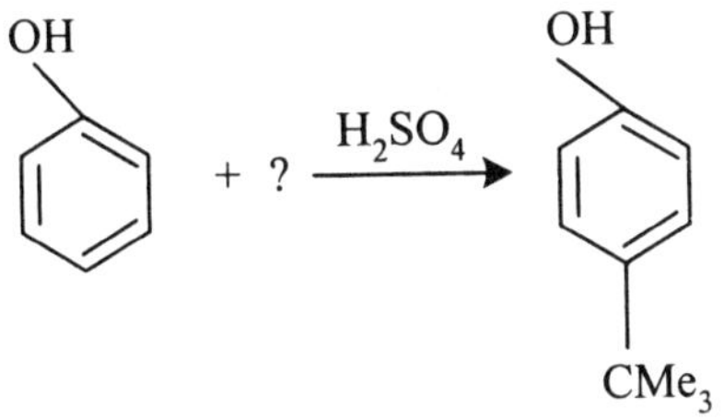

(vi) 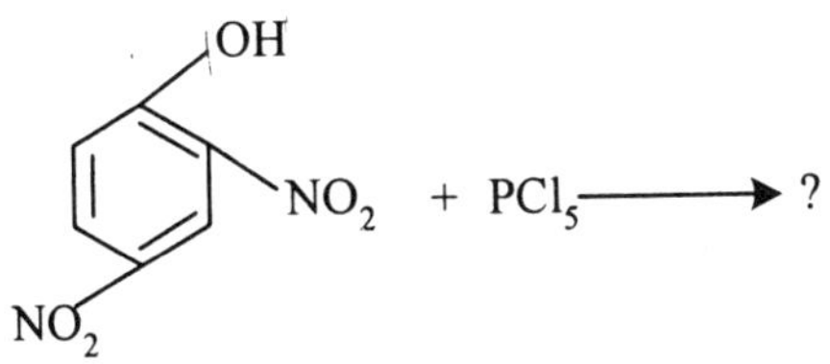

(vii) 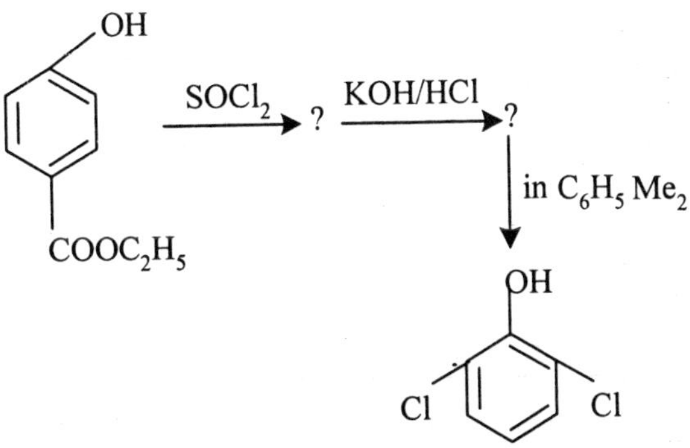

(viii) 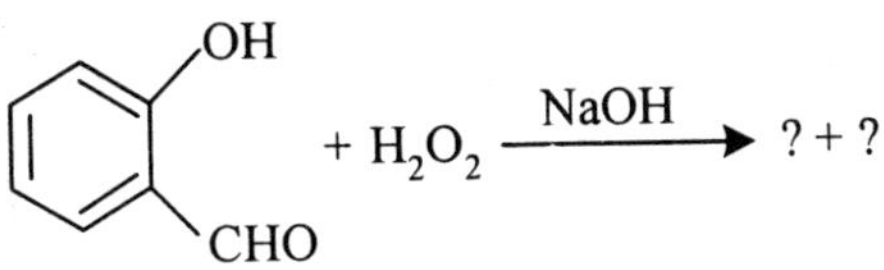

(ix) 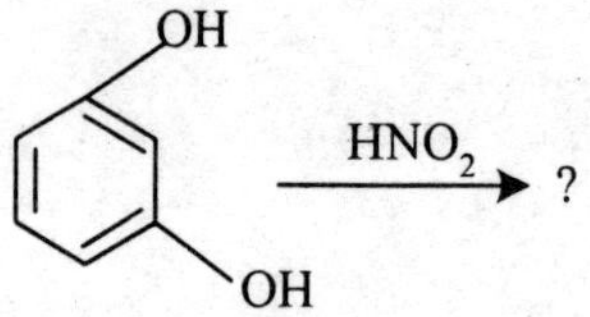

(x) 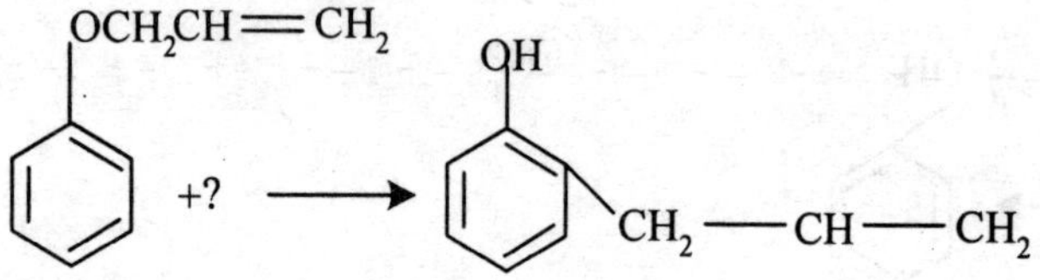

(xi) 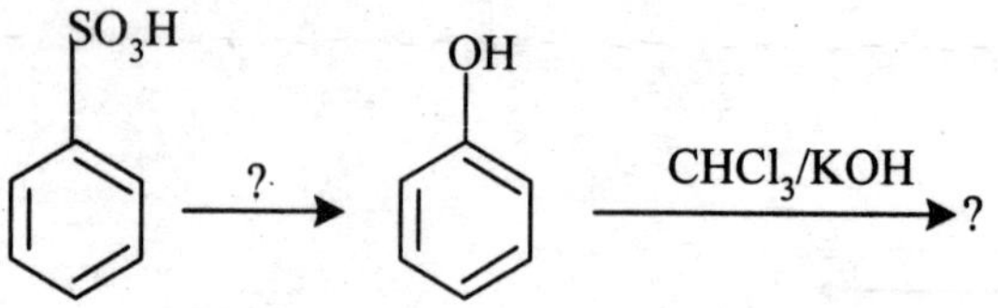

(IIT, 1988)

(xii) 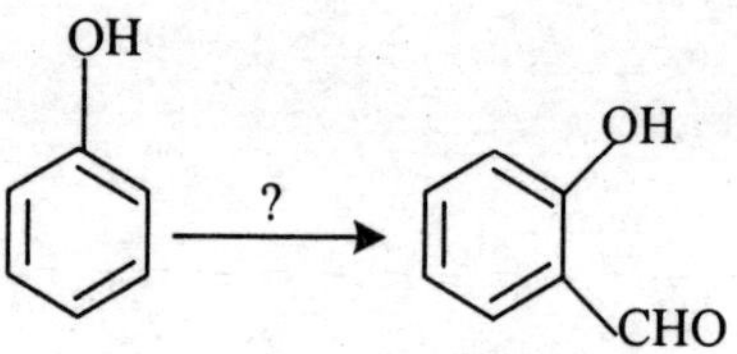

(IIT, 1986)

15. Carry out the following conversions:

(i) 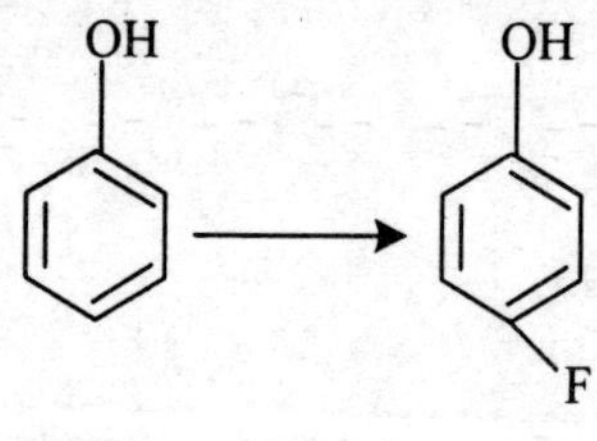

(ii) 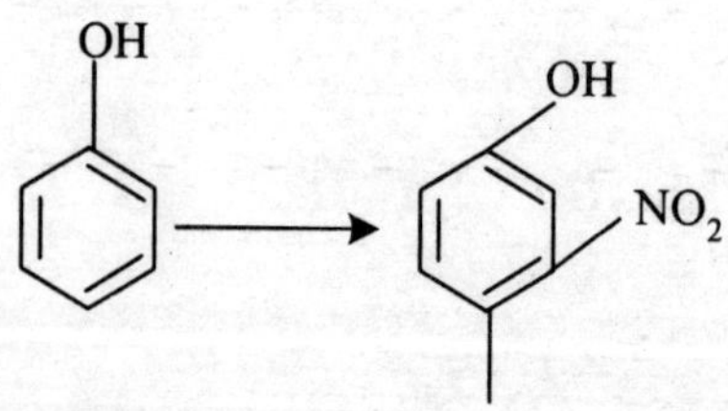

(iii) 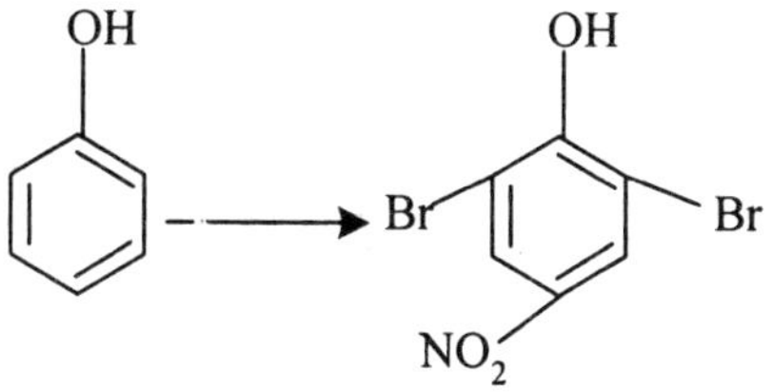

(iv) 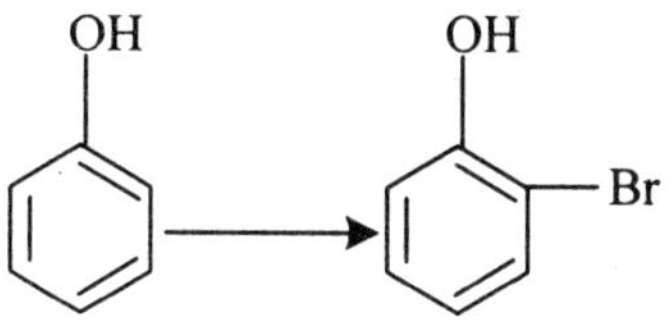

(v) 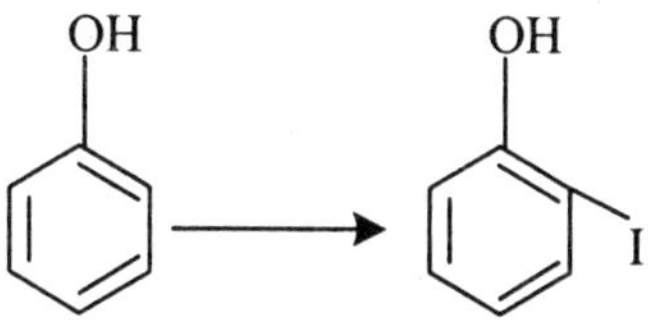

(vi) 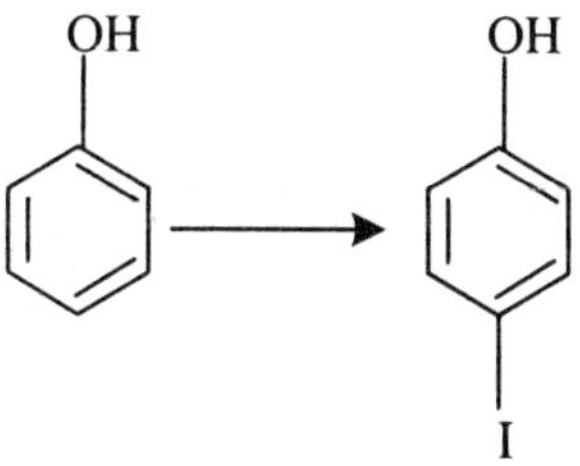

(vii) 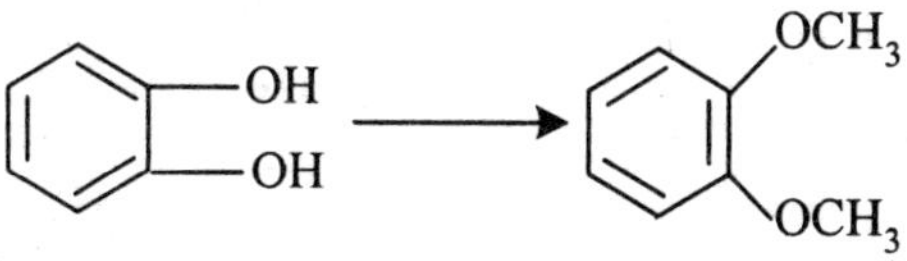

(viii) 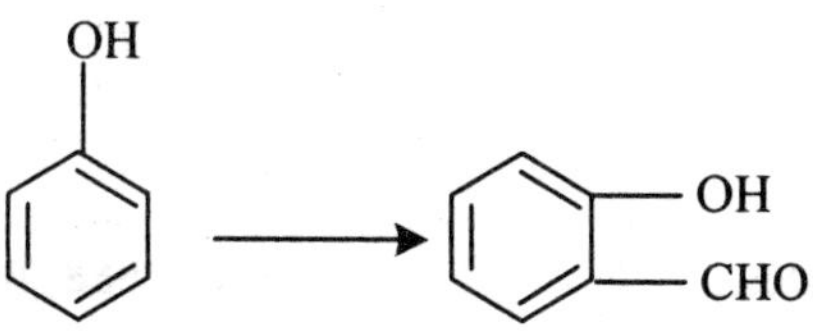

(IIT, 1986)

16. Write Tautomeric structure of *p*-nitrosophenol.

17. State 'True' or 'False'. If 'False' give correct statement.

(i) Catechol, Resorcinol and quinol reduce Fehling's solution and Tollen's reagent.

(ii) o- & p-nitrophenols, both are steam volatile.

18. Fill in the blanks:

(i) The acidity of phenol is due to the ....of its anion. **(IIT, 1984)**

(ii) Amongst the three isomers of nitrophenol, the one that is least soluble in water is .... **(IIT, 1992)**

19. Amongst the following answers one or more are correct. Write correct answers:

(i) When phenol is treated with excess of bromine water, it gives:

(a) m-Bromophenol (b) o- and p-bromophenol

(c) 2,4-Dibromophenol (d) 2,4,6-Tribromophenol **(IIT, 1984)**

(ii) Phenol reacts with bromine in CS_2 at low temperature to give:

(a) m-bromophenol (b) p-bromophenol

(c) o- and p-bromophenol (d) 2,4,6-tribromophenol **(IIT, 1988)**

(iii) Phenol is heated with CCl_4 and alk. KOH when salicylic acid is produced. The reaction is known as

(a) Friedel-Craft reaction

(b) Reimer-Tiemann reaction

(c) Rosenmund's reaction

(d) Sommlet reaction **(Delhi PMT, 1983)**

(iv) Which of the following is formed when sodium phenoxide is reacted with C_2H_5I.

(a) Phenetole (b) Phenol (c) Ethylphenyl alcohol

(d) None of these **(Delhi PMT, 1989)**

(v) Which of the following are reaction intermediates in Reimer-Tiemann reaction?

(a) 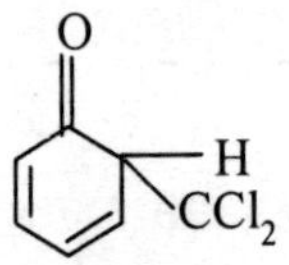

(b) 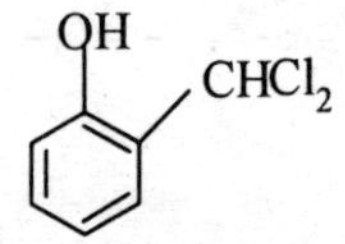

(c) 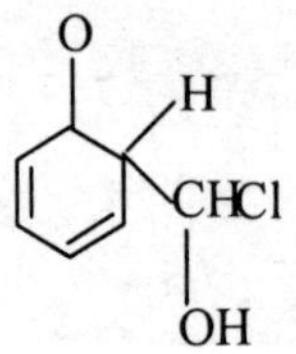

(d) 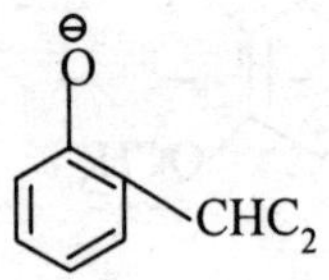

(IIT, 1995)

(vi) Phenol is less acidic than:

(a) Ethanol (b) Methanol

(c) o-nitrophenol (d) p-methylphenol **(IIT, 1988; MNR, 1995)**

20. What happens when phenol is treated with chloroform and conc. NaOH and the product is hydrolyzed. **(R, 1982, 1985)**

21. How will you distinguish between: Phenol and Ethyl alcohol. **(MNR, 1990)**

22. Amongst the four answers given for questions below, choose the correct answer:

(i) Acetophenone and Benzophenone can be distinguished by

(a) Fehling's solution (b) Tollen's reagent
(c) Iodoform test (d) Lucas reagent

(ii) Which of the following compounds will react with ethanolic KCN?
(a) Ethyl chloride
(b) Acetyl chloride
(c) Chlorobenzene
(d) Benzaldehyde **(IIT, 1984)**

(iii) m-chlorobenzaldehyde on reaction with conc. KOH at room temperature gives:
(a) potassium m-chlorobenzoate and m-hydroxybenzaldehyde
(b) m-hydroxybenazldehyde and m-chlorobenzyl alcohol
(c) m-chlorobenzyl alcohol and m-hydroxybenzyl alcohol
(d) potassium m-chlorobenzoate and m-hydroxybenzyl alcohol **(IIT, 1991)**

(iv) Which one of the following aldehydes gives Cannizzaro reaction when heated with strong alkali:
(a) Benzaldehyde (b) Acetaldehyde
(c) Propanaldehyde (d) All of them **(MNR, 1995)**

23. Complete the following reactions:

(i) $C_6H_5COC_6H_5 \xrightarrow{Zn/NaOH} ?$

(ii) 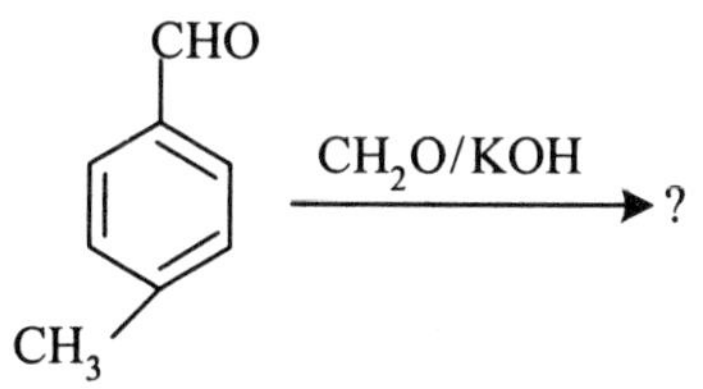

(iii) 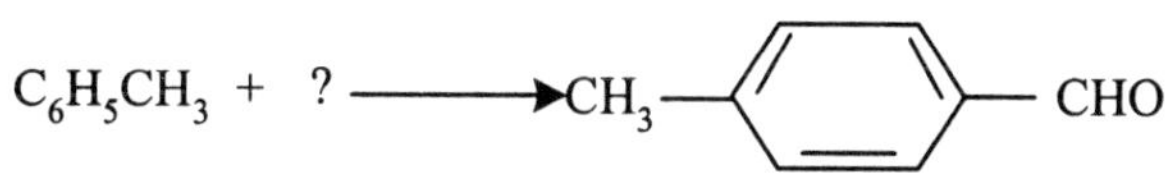

(iv) $C_6H_5COCH_3 \xrightarrow{?} C_6H_5COCHO$

(v) 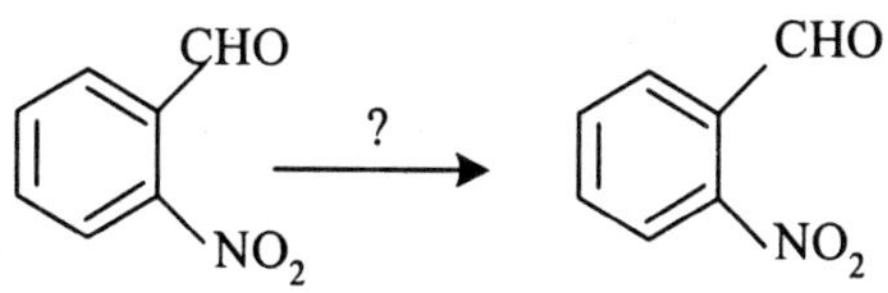

(vi) 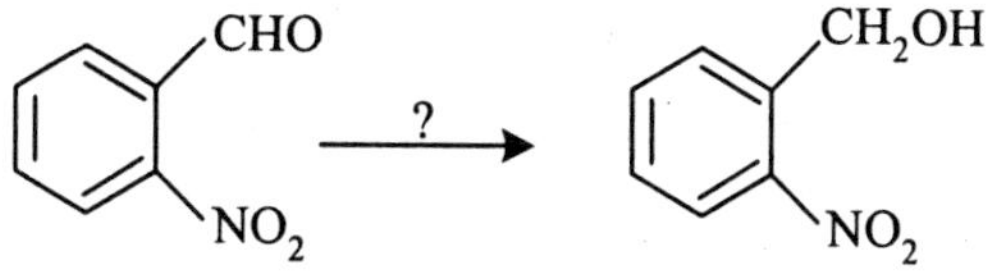

(vii) ? + ? NaOH→ $C_6H_5CH{=}CHCHO$ **(IIT, 1986)**

(viii) 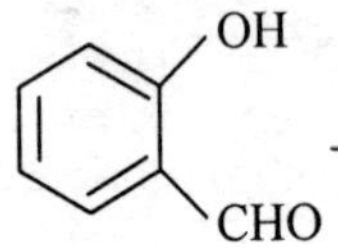 $\xrightarrow{AC_2O/CH_3COONa}$? **(R, 1993)**

(ix) $C_6H_5CHO + CH_3COOEt \xrightarrow[\text{ethanol and Heat}]{NaOC_2H_5 \text{ in absolute}}$? **(IIT, 1995)**

24. 4.4 gm of a mixture of acetaldehyde and benzaldehyde, when boiled with Fehling solution, yielded 1.43 gm of brown precipitate. Find the composition of the mixture.

25. Fill in the blanks:

(i) The structure of the intermediate product, formed by the oxidation of toluene with CrO_3 and acetic anhydride whose hydrolysis gives benzaldehyde is .... **(IIT, 1992)**

(ii) Between acetophenone and anisole, iodoform test is given by .... **(GATE, 1994)**

26. Convert benzene into 4-nitrobenzaldehyde.

27. Identify alphabets, B, C, D, E in the following sequence

(i) $C_6H_5COOH \xrightarrow{PCl_5} B \xrightarrow{NH_3} C \xrightarrow{P_2O_5} D \xrightarrow{H_2/Ni} E$ **(IIT, 1991)**

28. Complete following reactions:

(i) 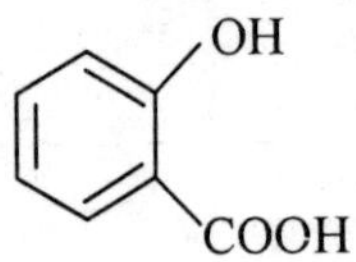 + 2ICl ⟶ ?

(ii) 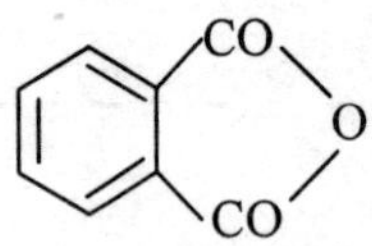 + PCl_5 ⟶ ?

(iii)  Mg/Et_2O ⟶ ? $\xrightarrow{CO_2/H}$?

(iv) 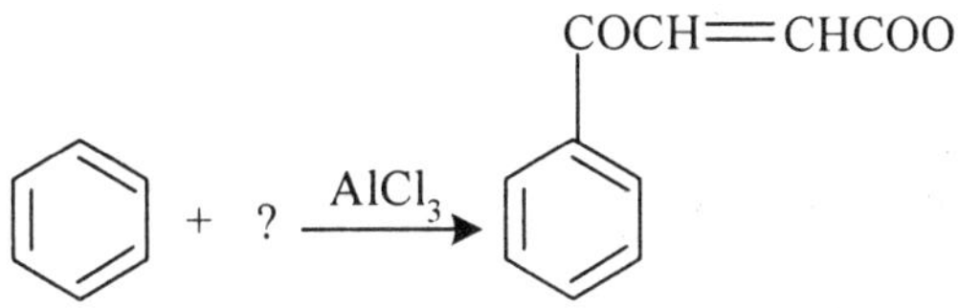

29. Acetyl chloride is readily decomposed by water, but benzoyl chloride is decomposed very slowly, Explain.

30. Amongst following answers only one is correct. Write the correct answer.

(i) Hydrogenation of benzoyl chloride in the presence of Pd on $BaSO_4$ gives:

(a) Benzyl alcohol *(b)* Benzaldehyde

(c) Benzoic acid *(d)* Phenol

(ii) Aspirin is

(i) Antipyretic *(ii)* Analgesic *(iii)* Antimalarial

(iv) Insecticide

31. Fill in the blanks

(i) The high melting point and insolubility in organic solvents of suphanilic acid are due to its .... structure. **(IIT, 1994)**

(ii) Salicylic acid is ....in cold water but soluble in........

32. Convert: Toluene into *p*-nitrobenzoic acid. **(GATE, 1994)**

33. Write the correct answer (one or more is correct):

(i) Which of the following carboxylic acids undergoes decarboxylation easily? Explain briefly.

(a) $C_6H_5COCH_2COOH$ *(b)* $C_6H_5COCOOH$

(c) $C_6H_5CH—COOH$ *(d)* $C_6H_5—\underset{NH_2}{\overset{OH}{CH}}—COOH$ **(IIT, 1995)**

34. Complete the following equations:

(i) 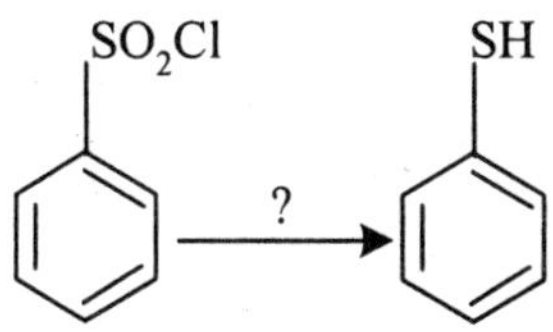

(ii) 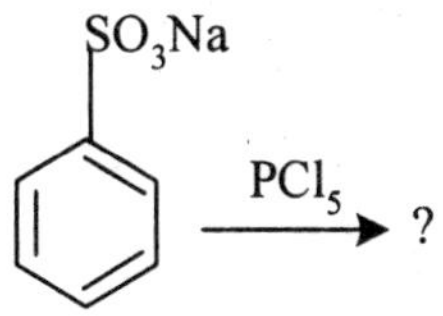

(iii) 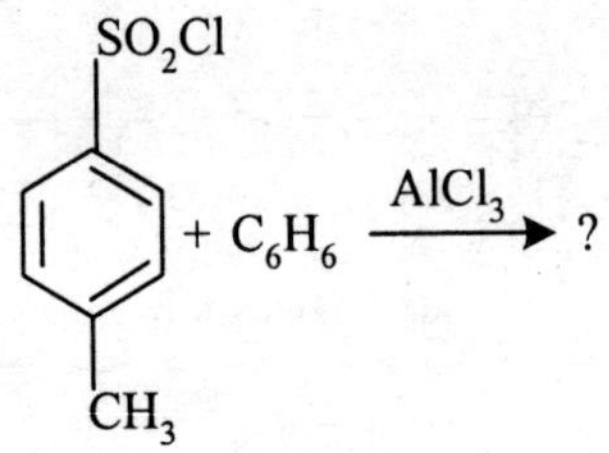

(iv) 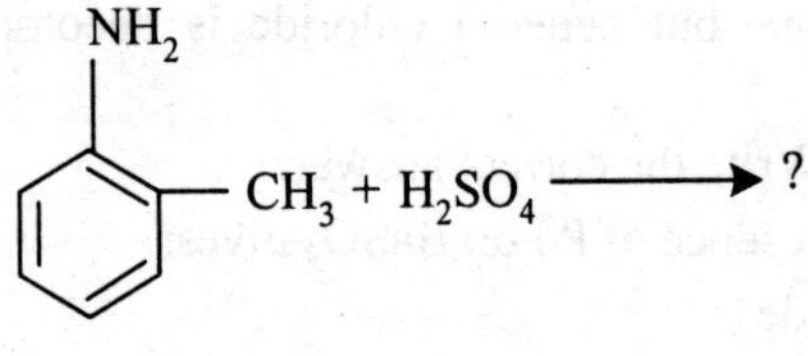

(v) 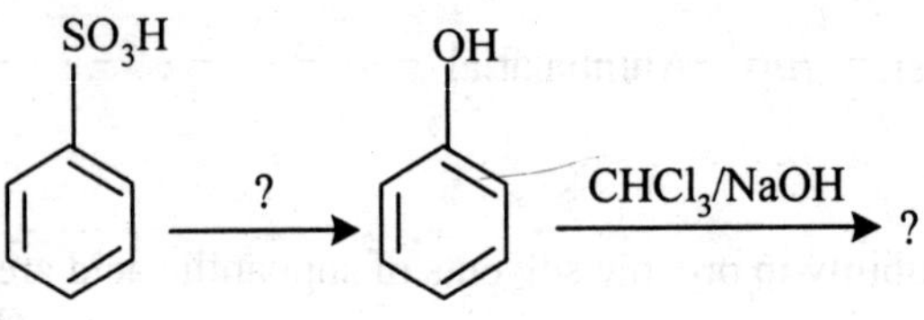 **(IIT, 1988)**

(vi) 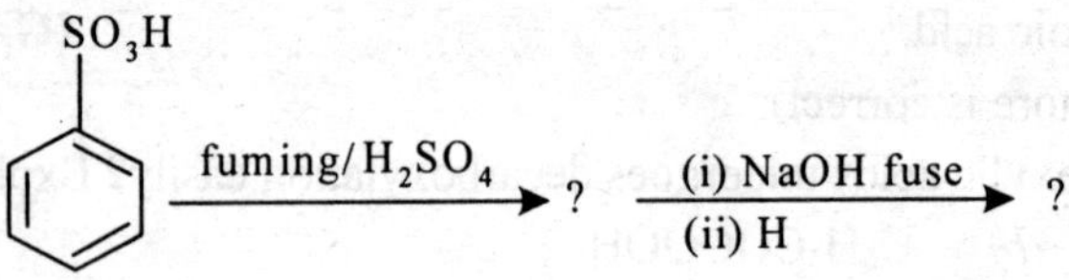 **(IIT, 1992)**

35. Complete the equations:

(i) $C_6H_5NH_2 + ? \rightarrow C_6H_5NHCOCH_3$

(ii) 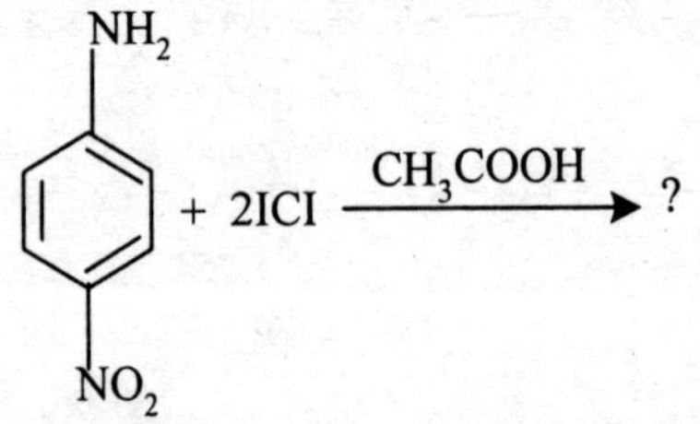

(iii) 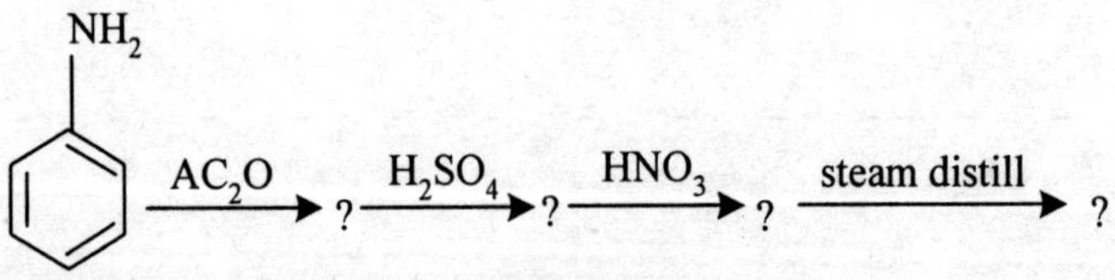

(iv) 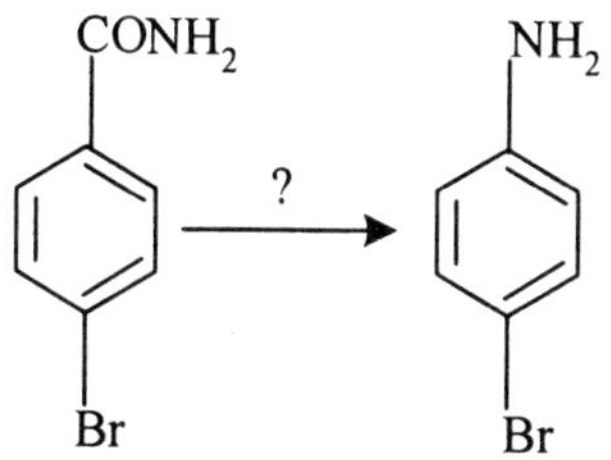

(v) 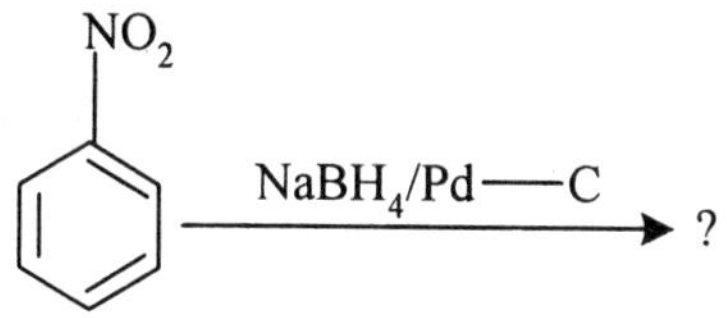

(vi) 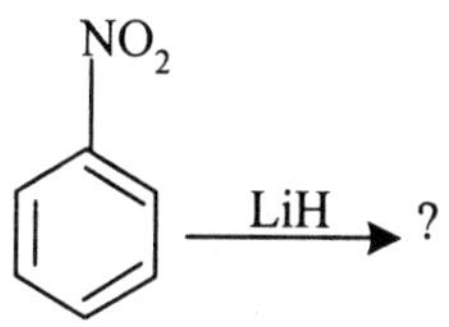

(vii) 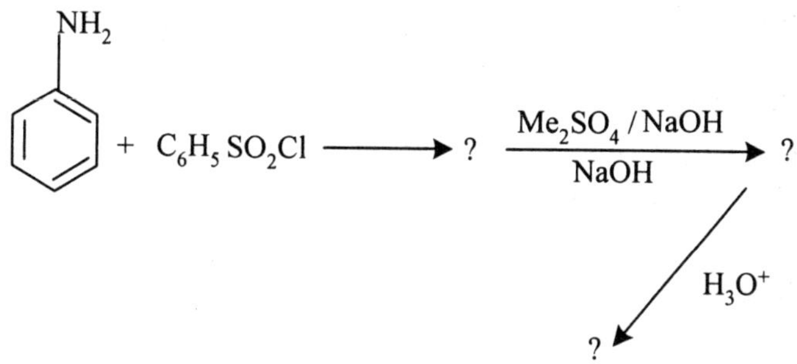

(viii) 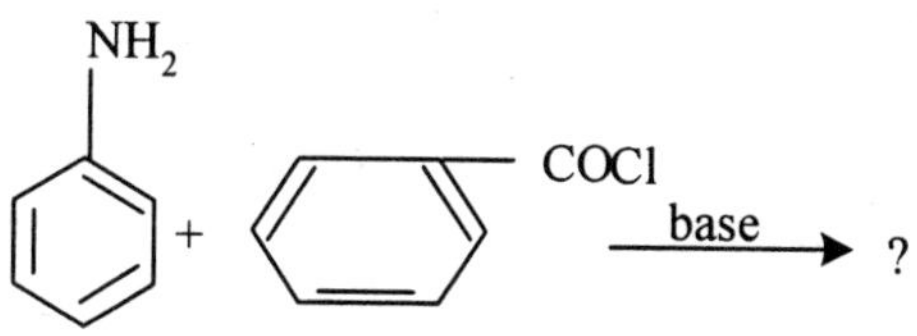 **(IIT, 1986)**

(ix) 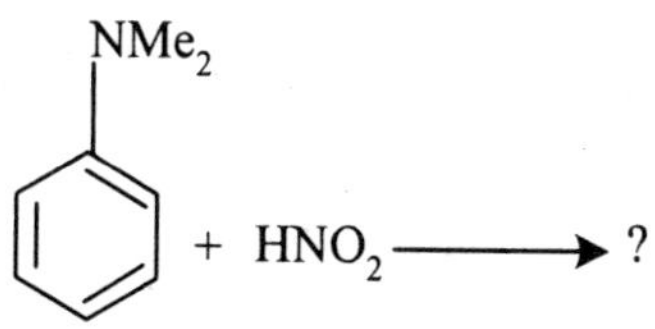 **(IIT, 1992)**

(x) 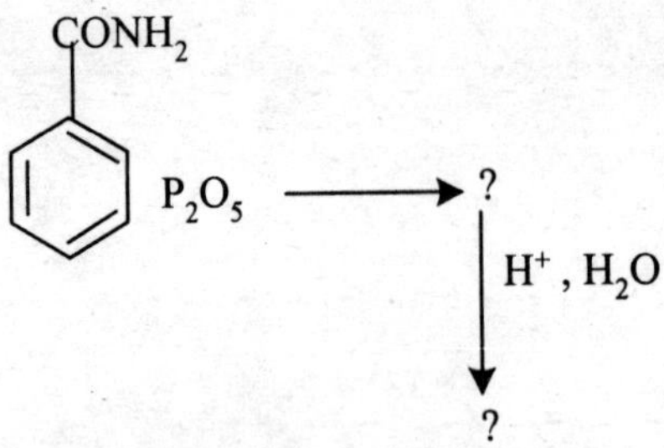 (IIT, 1992)

36. Carry out following conversions:

(i) 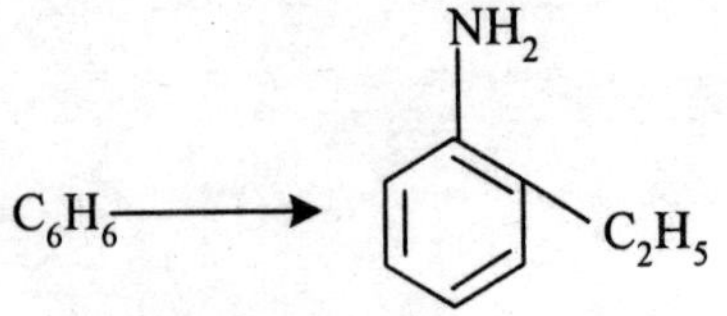

(ii) 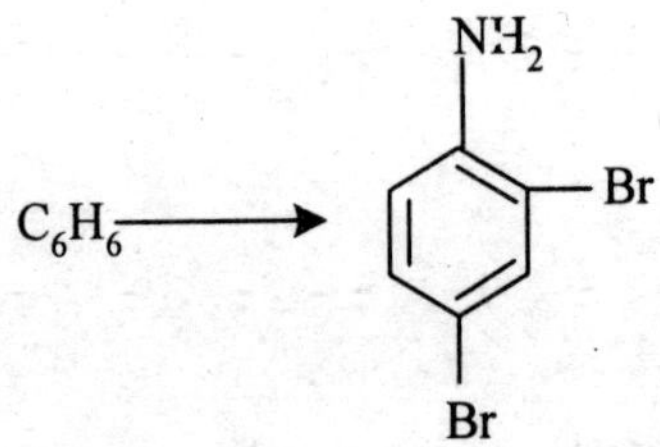

37. Which of the following is most basic?

(a) Aniline
(b) N-methyl aniline
(c) N,N-dimethyl aniline
(d) *o*-methyl aniline

38. Complete following reactions:

(i) 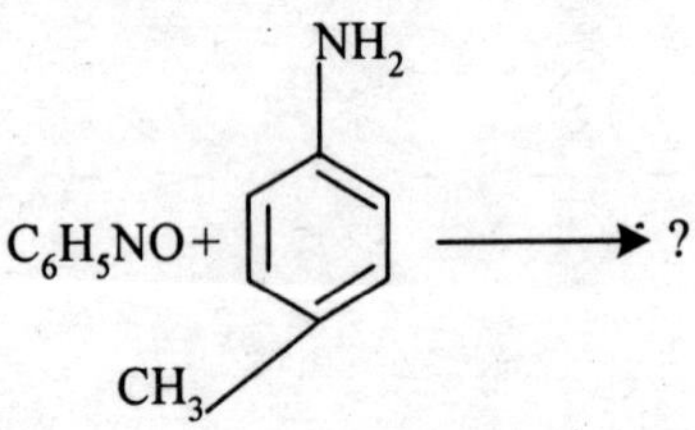

(ii) 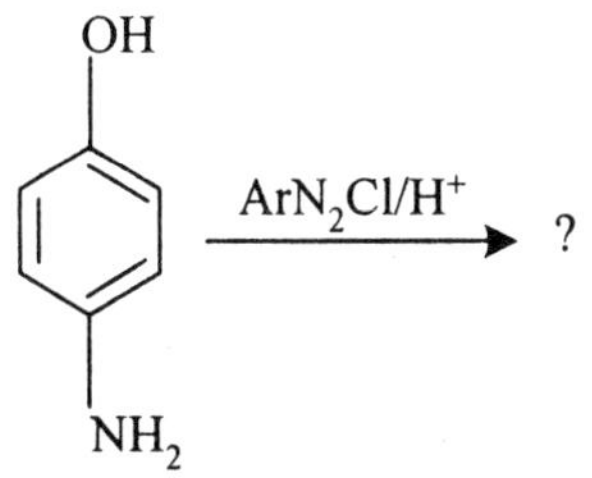

(iii) 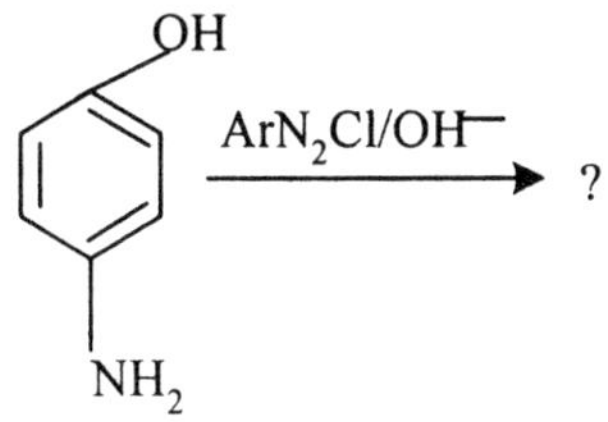

(iv) 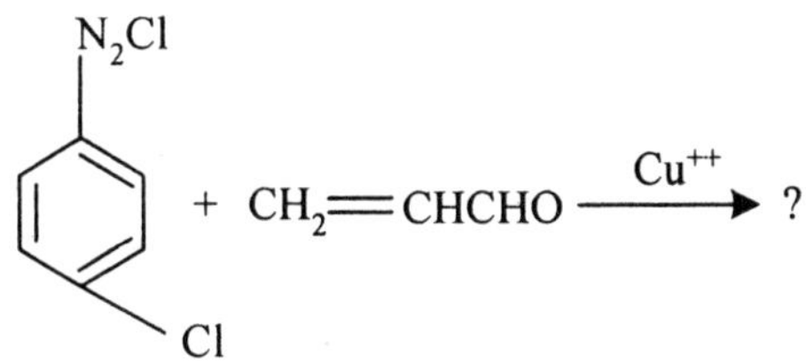

(v) 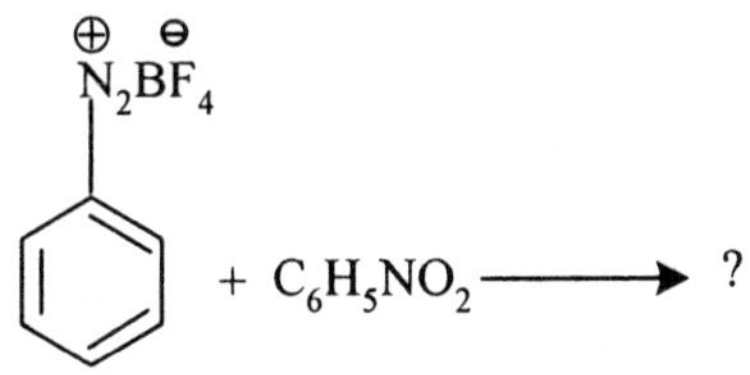

(vi) 2,4 – dinitroaniline $\xrightarrow[\text{(ii) Anisole}]{\text{(i) NaNO}_2\text{ and HCl at 5 °C}}$? **(IIT, 1995)**

39. Carry out following conversions:

(i) 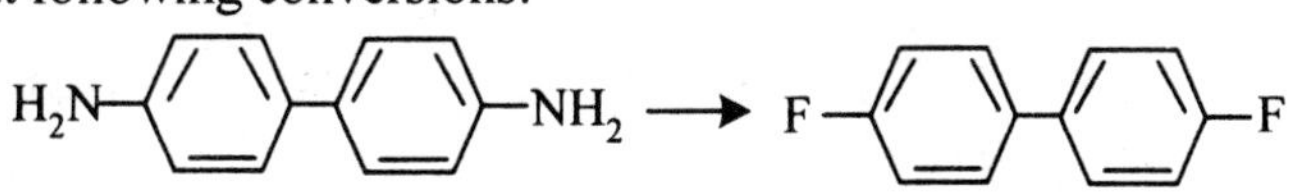

(ii) 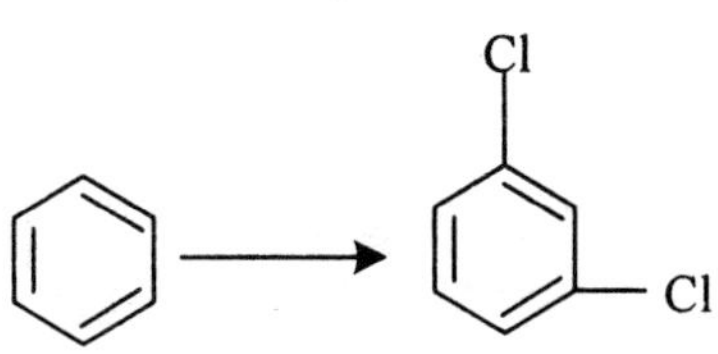

(iii) 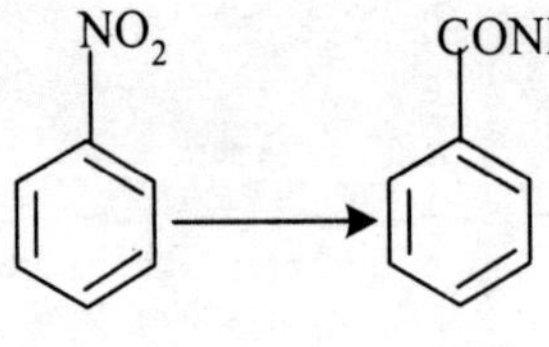 **(IIT, 1994)**

(iv) 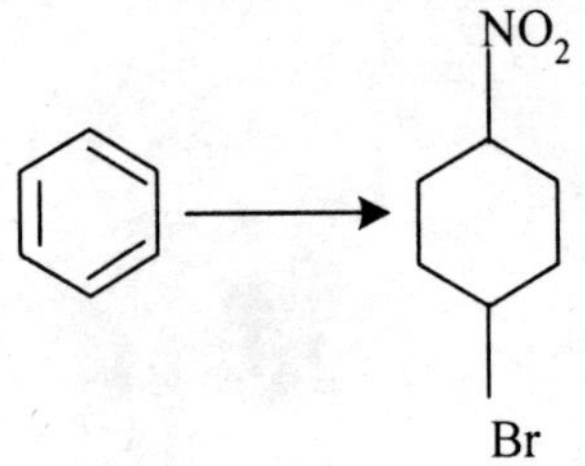

(v) 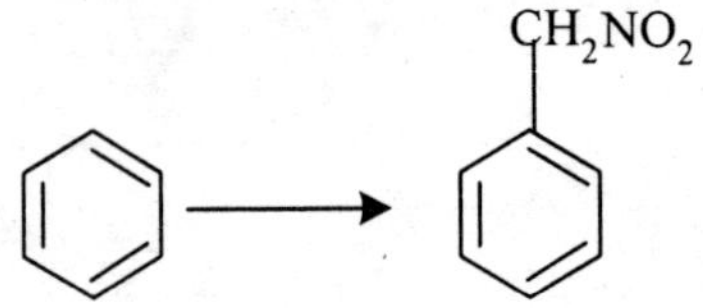

(vi) 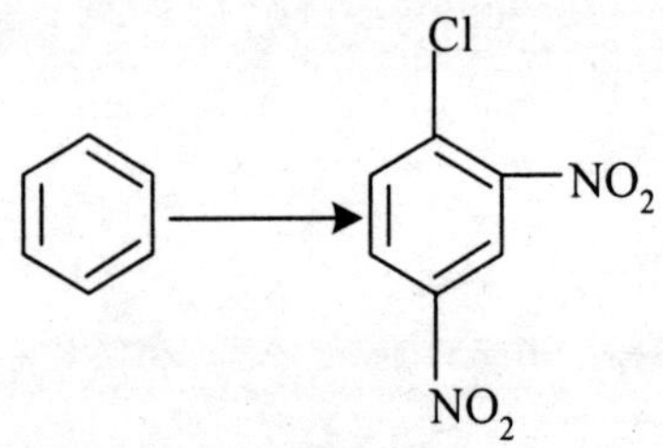

40. Complete following reactions:

(i) 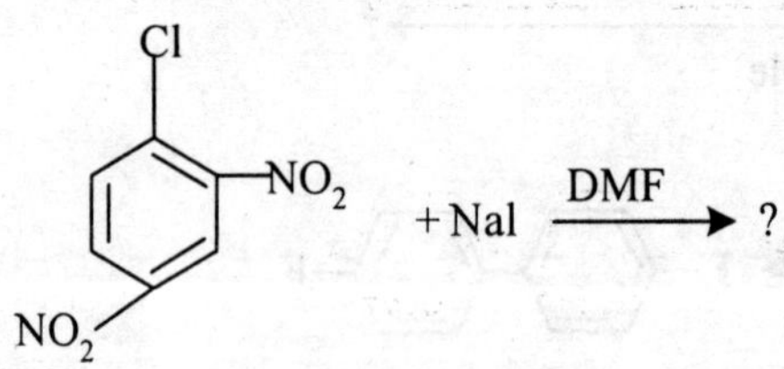

(ii) 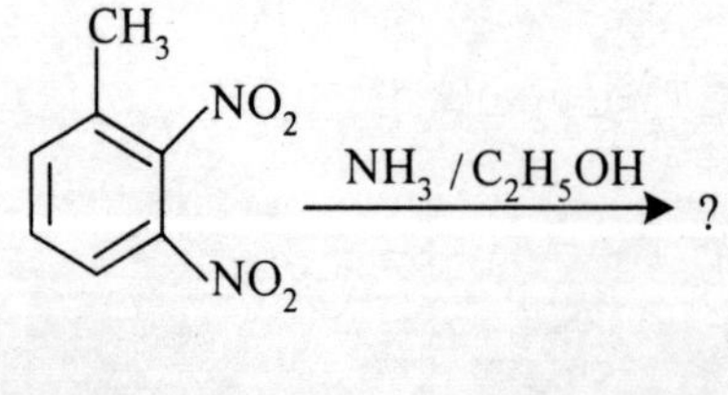

(iii) 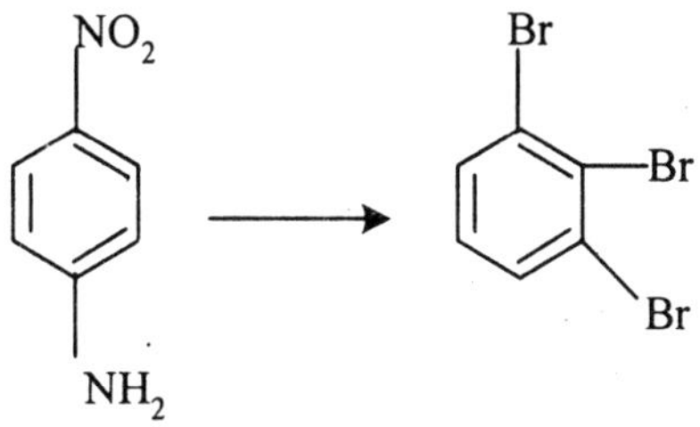

41. Carry out following conversions:

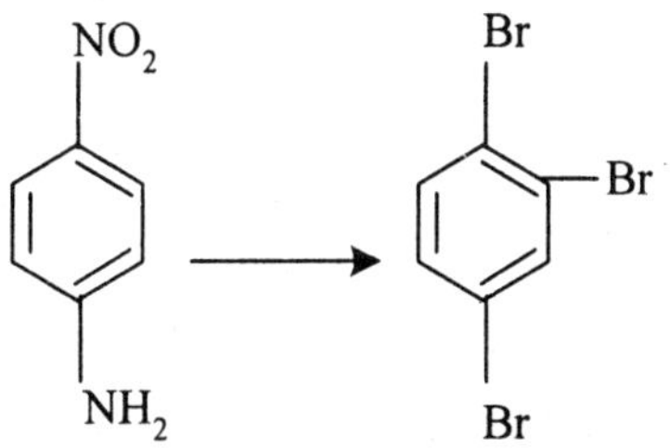

(IIT, 1990)

42. The aromatic compound (A), C_8H_9Br, on treatment with DEM/EtoNA, followed by reflux with H_2SO_4, gave (B), $C_{10}H_{12}O_2$. (B) on vigorous oxidations gave (C), $C_8H_6O_4$ which on mononitration gave only one mononitroderivative, (d) Identify A to D.

43. A positive carbylamine test is given by

(a) N,N-dimethyl aniline *(b)* 2,4-dimethyl aniline
(c) N-methyl-*o*-methyl aniline *(d)* *p*-methylbenzyl aniline **(IIT, 1999)**

ANSWERS

1. *(i)* (**c**), *m*-nitroaniline. It is because —NO_2 group in acidic medium forms $-\overset{\oplus}{N}H_3$, which decreases electron density at *o*, *p*-position.

(ii) (**b**), It is because there are 3, 2 and 1 Hyperconjugating structure (s) 3 for —CH_3, 2 for —CH_2CH_3 and for $\begin{matrix} CH_3 \\ CH_3 \end{matrix}\!\!>\!CH$ group.

(iii) (**d**), two methyl groups activate the ring, chlorobenzene and nitrobenzene would be deactivated. Toluene is less activated due to one methyl group.

(iv) (**a**), because methyl group will activate the ring, nitro and —COOH group will deactivate the ring.

(v) (**a**) and (**d**), see notes on meta-directing groups.

(vi) (**b**) & (**c**), see Hückel's rule.

(vii) (c) and, see Hückel's rule (you may count number of delocalized π-electrons as equal to the number of carbon atoms. In case of cation, substract the charge and in case of anion, add the charge to the number of carbon atoms).

(viii) (c), see Hückel's rule

(ix) (c)

(x) (c)

2. *(i)* One phenyl group behaves as electron donor and the other electron acceptor. As a result, phenyl ring is more reactive.

(ii) It is because it undergoes substitution reactions.

3. 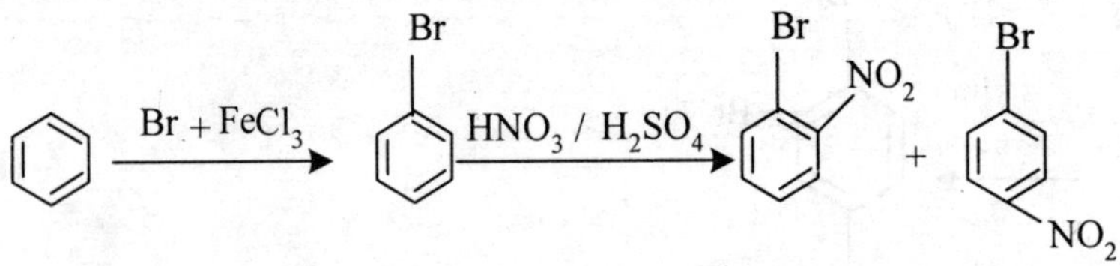

carry out bromination first because Br is *o*, *p*-directing whereas —NO_2 group is meta-directing.

4. It is because it is aromatic due to (i) presence of conjugated cyclic ring, (ii) resistant towards oxidation, and (iii) undergoes substitution reaction and obey Hückel's rule.

5. *(i)* 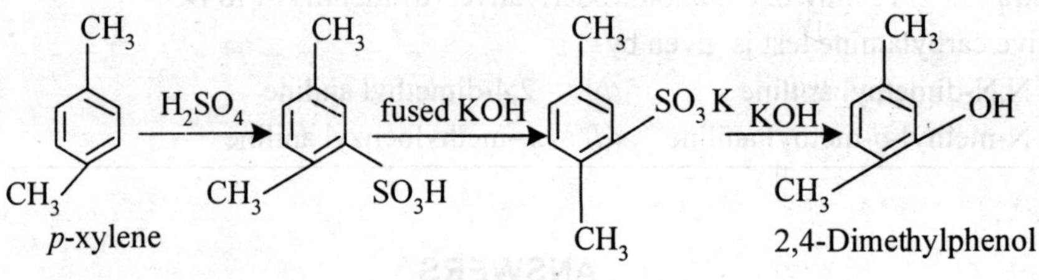

(ii) 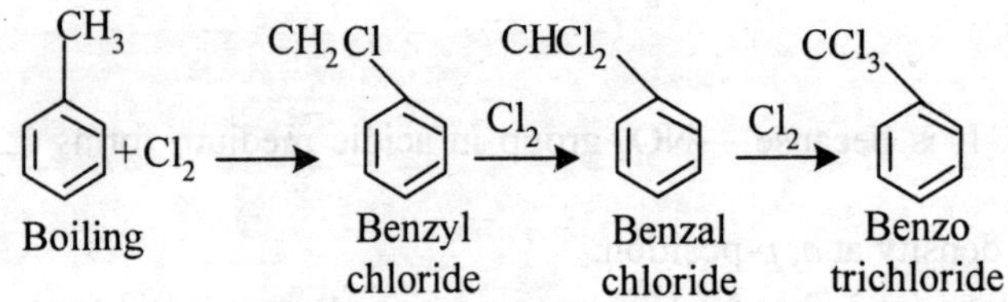

6. *(i)*

$$3C + CaO \xrightarrow{2000^\circ C} CaC_2 + CO$$

$$CaC_2 \xrightarrow{2H_2O} C_2H_2 + Ca(OH)_2$$

$$\underset{\text{Benzene}}{C_6H_6} \xleftarrow{450^\circ C} \underset{\text{Acetylene}}{C_2H_2} + Ca(OH)_2$$

(ii)

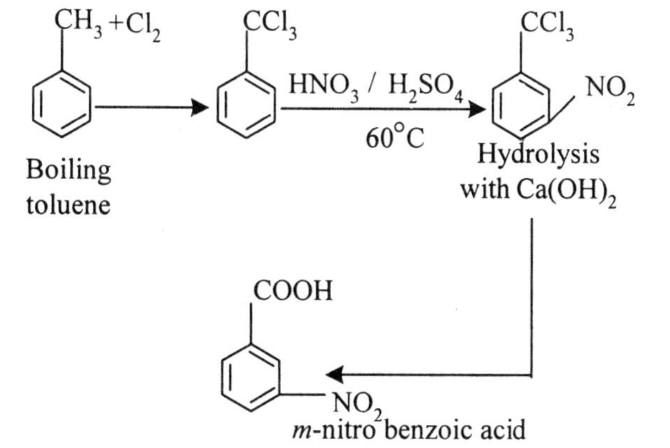

(iii)

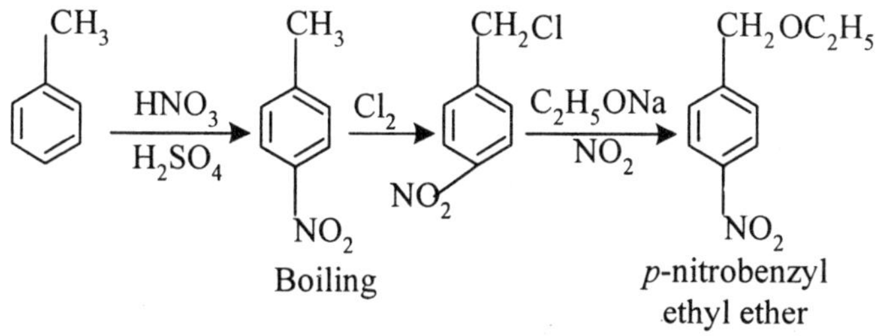

7. (i) $C_6H_5—\underset{\displaystyle Br}{\underset{|}{CH}}—CH_3$

(ii) $C_6H_5—\underset{\displaystyle CN}{\underset{|}{CH}}—CH_3$

8. The possible isomers and their identification is as follows:

(i)

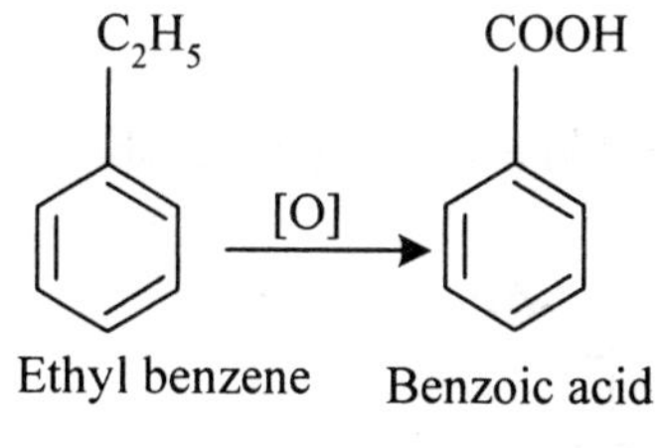

(ii)

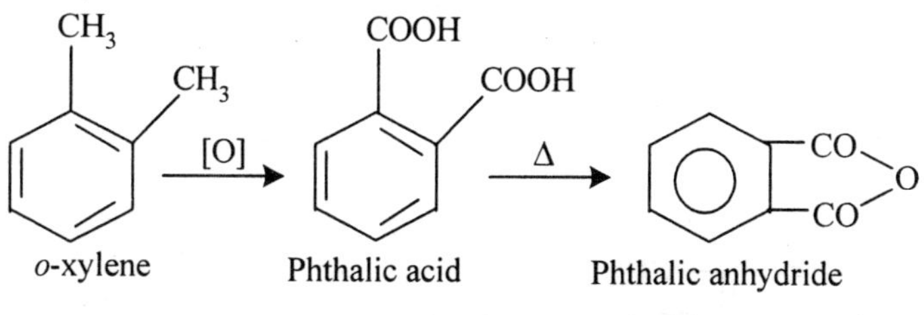

(iii) 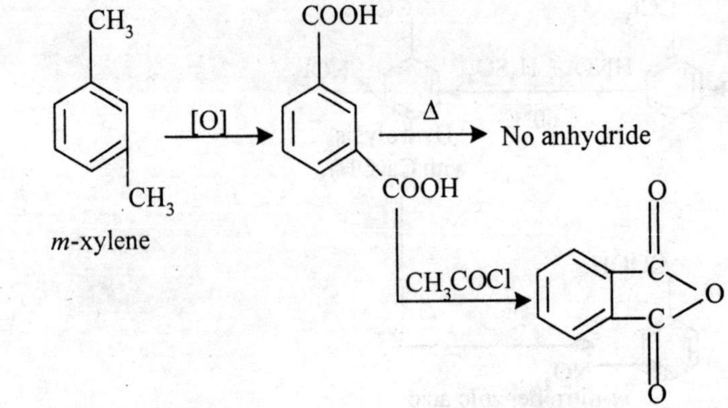

m-xylene

Phthalic anhydride

(iv) 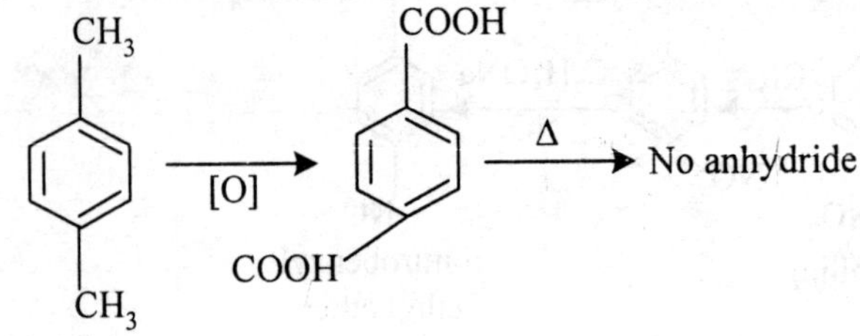

p-xylene

9. (i) **(d)** (ii) **(d).**

10. 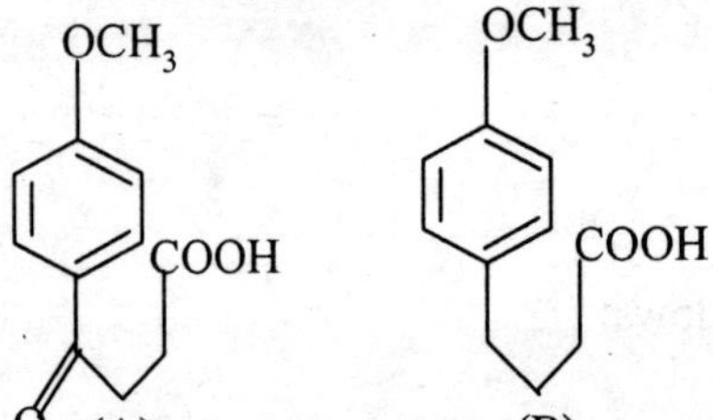

(A) (B)

11. In presence of light, free radicals are intermediate to yield benzyl bromide while in case of $FeBr_2$ the electrophilic substitution takes place.

12. (i) 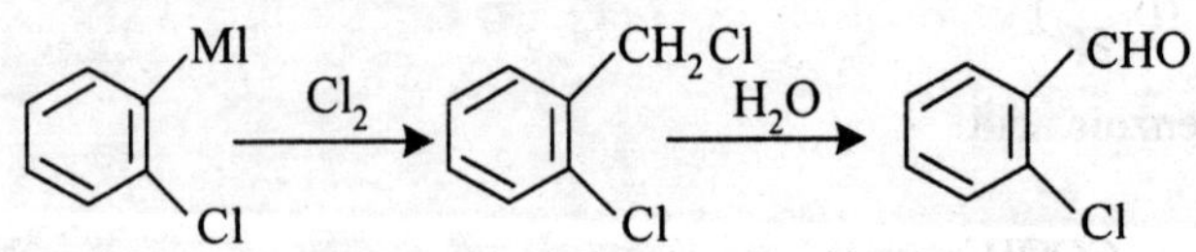

(ii) $Ph_2C{=}CPh_2$

(iii) 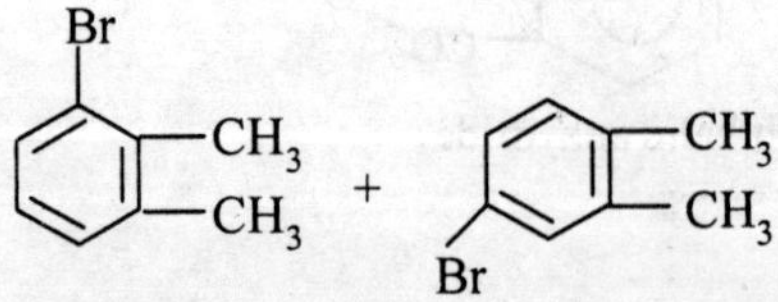

(iv) $CBrCl_3 \xrightarrow{h\upsilon} Br + CCl_3$

$PhCH_3 \xrightarrow{CCl_3} CHCl_3 + PhCH_2$

$PhCH_2 \xrightarrow{CBrCl_3} PhCH_2Br + CCl_3$

(v) 

Benzyne

, 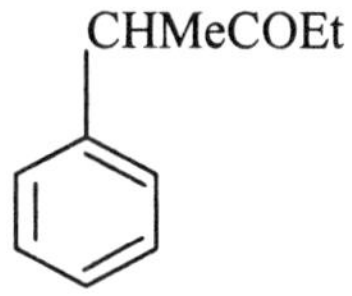

(vi) 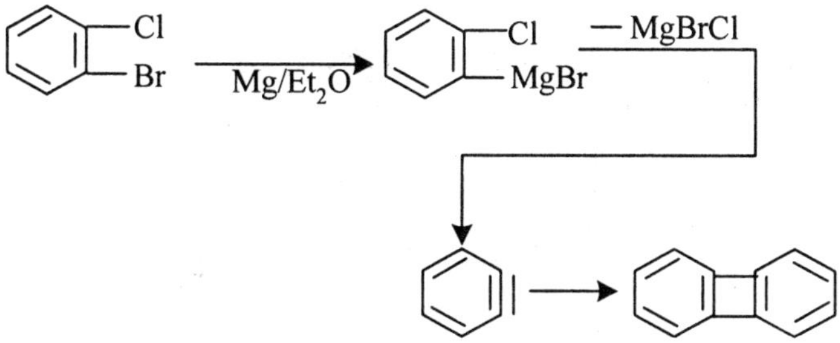

(vii) 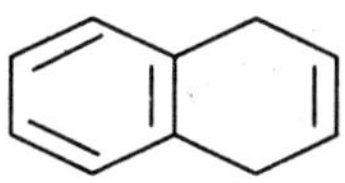

Bisphenylene

13. *(i)* Sandmeyer reaction

(ii) CH $(OCOCH_3)_2$.

14. *(i)* $COCl_2$

(ii) 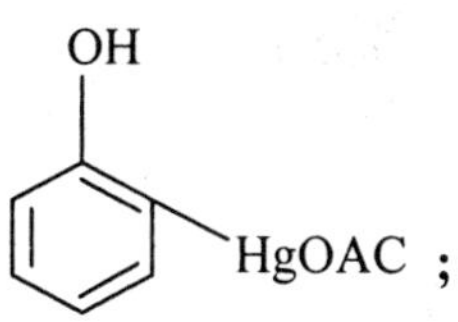

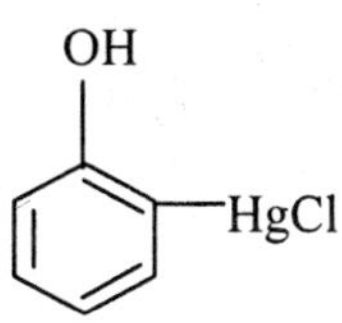

(iii) 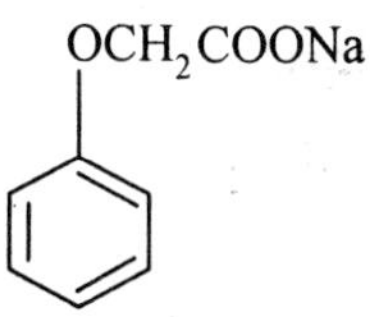

(iv) 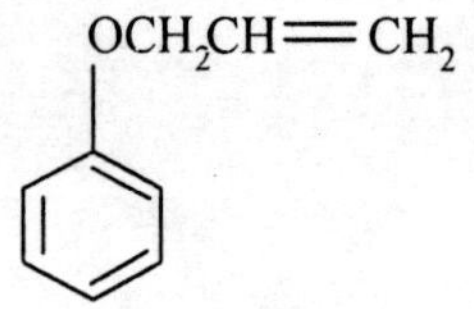

(v) $Me_3C—OH$

(vi) 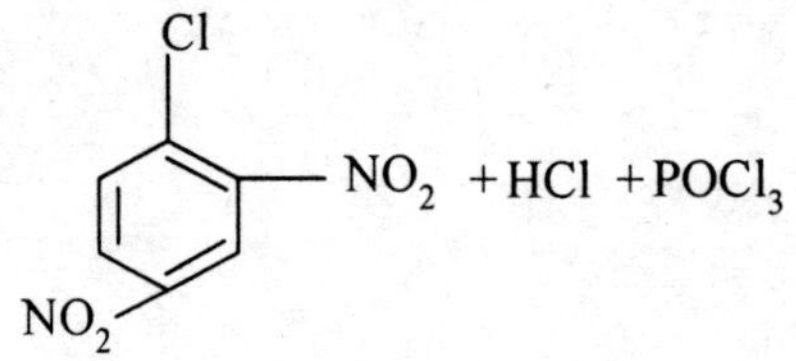

(vii) 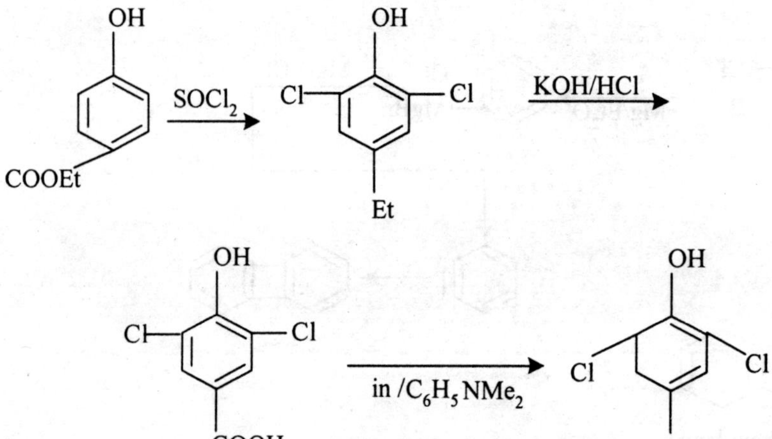

(viii) 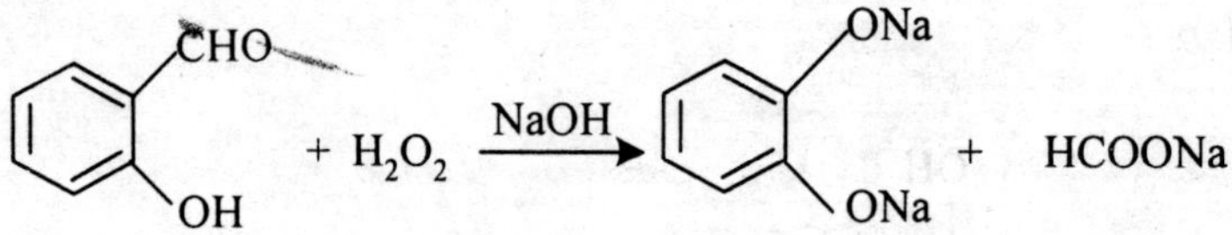

This is known as the Dakin reaction.

(ix) 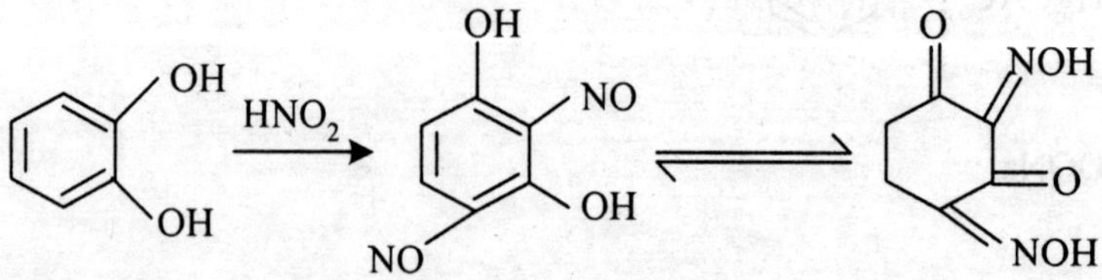

(x) 200 °C, it is an example of Claisen condensation.

(xi) $NaOH_2$ 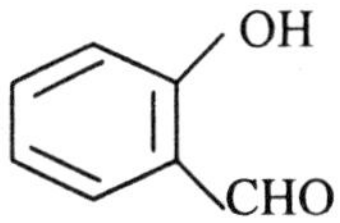

(xii) $CHCl_3$ and KOH.

15. *(i)*

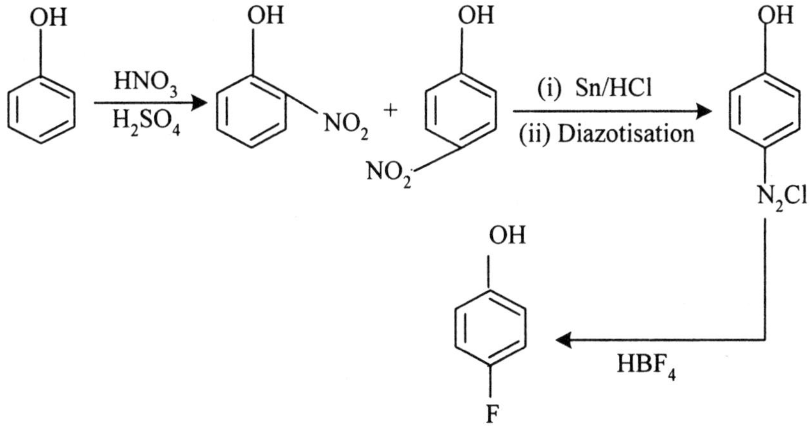

(ii)

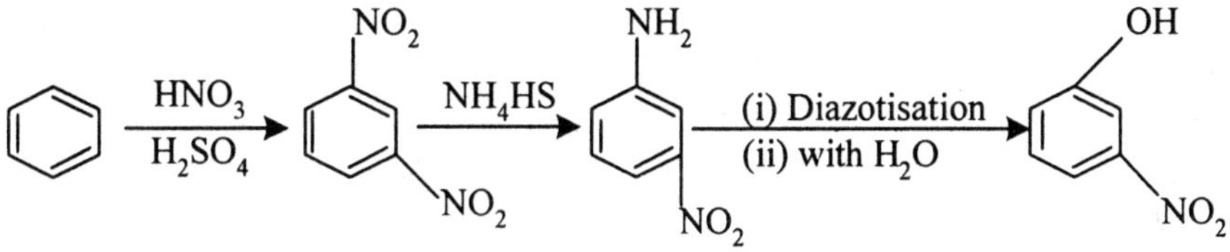

(iii)

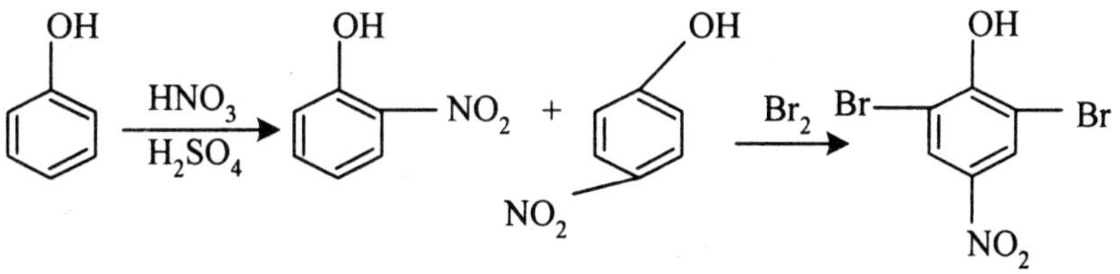

(iv)

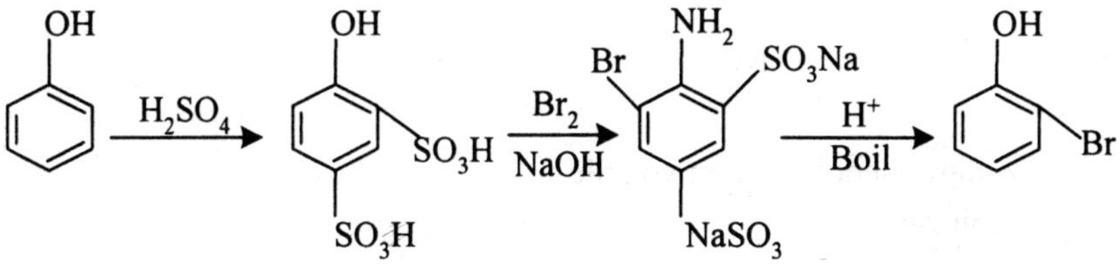

(v)

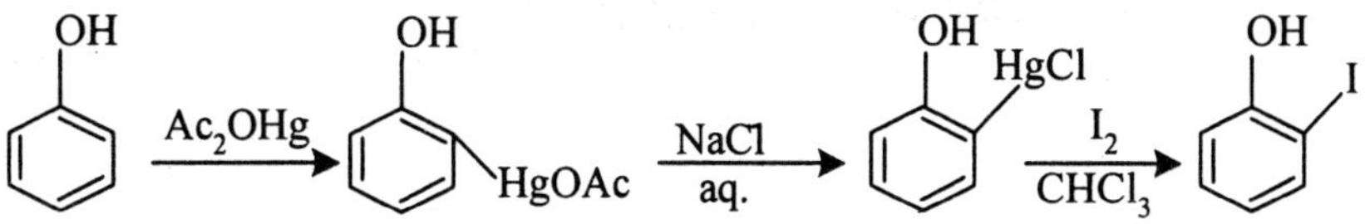

(vi) 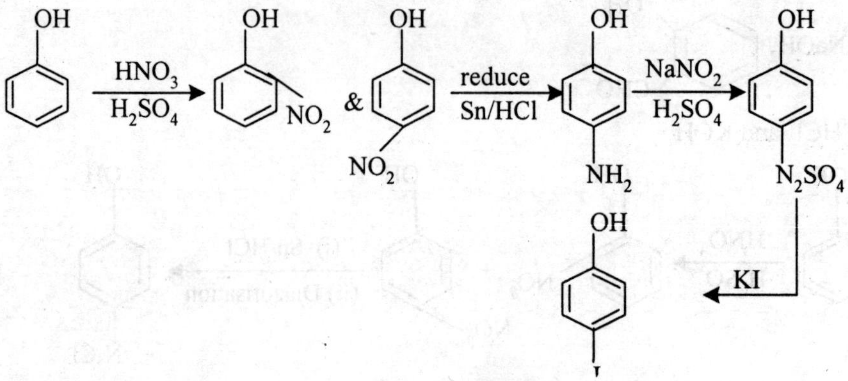

(vii) 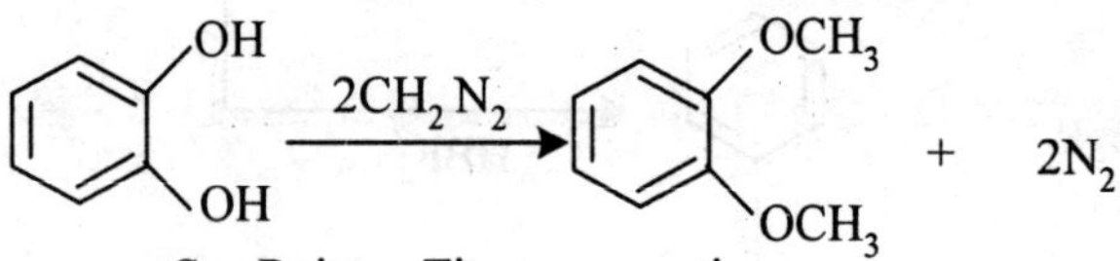

See Reimer-Tiemann reaction

(viii) 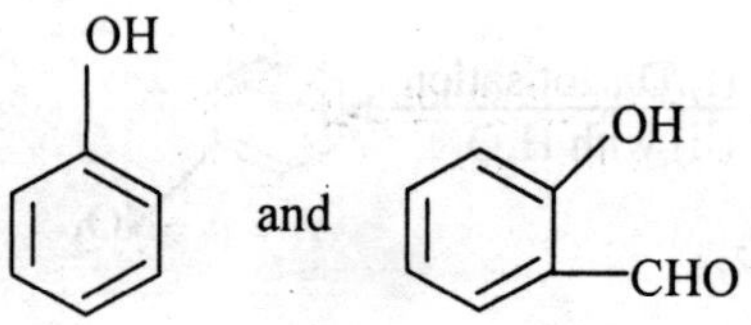

16. 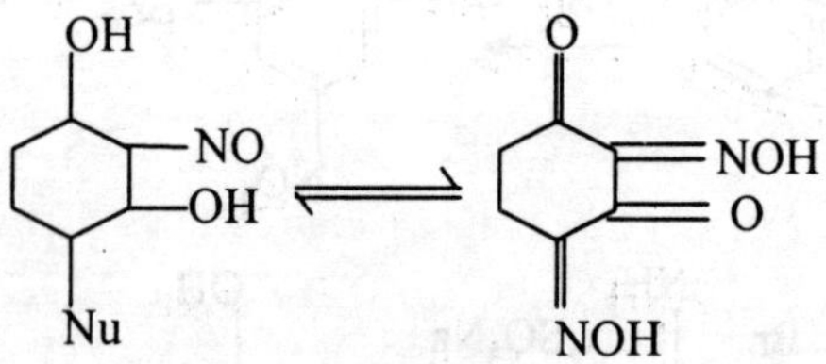

17. *(i)* *True.*

(ii) *False.* Only *o*-nitrophenol is steam volatile.

18. *(i)* Resonance stabilization.

(ii) *o*-nitrophenol due to intramolecular hydrogen bonding.

19. *(i)* **(d)**.

(ii) **(b)**.

(iii) **(b)**.

(iv) **(a)**, See notes on phenol.

(v) **(c)** Because of more resonance stabilization of substituted phenoxide anion (See Resonance).

(vi) **(c)**.

20. See Reimer-Tiemann reaction.

21. Phenol gives violet color with $FeCl_3$ whereas ethanol does not. Moreover ethanol gives the iodoform test whereas phenol does not.

22. *(i)* (**c**).

(ii) (**a**) and (**d**).

(iii) (**d**), the *m*-chlorobenzaldehyde will undergo Cannizzaro reaction. Moreover, chlorine is tightly attached to the ring; hence it will not be substituted by –OH group (See Resonance).

(iv) (**a**), rest contain α-H.

23. *(i)*

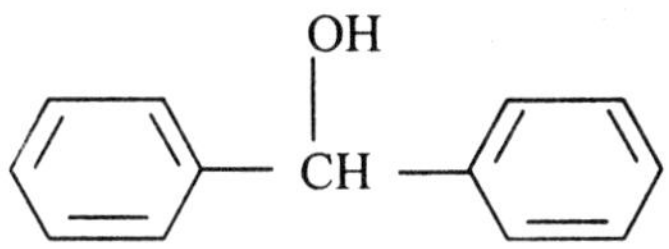

(ii)

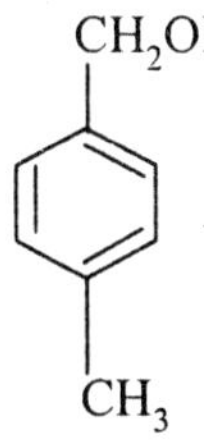

+ HCOOK (An example of crossed Cannizzaro reaction)

(iii)

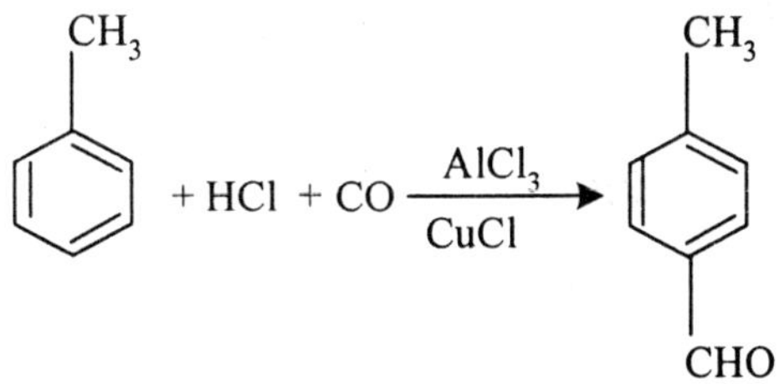

(iv) SeO_2.

(v) $FeSO_4/NH_4OH$ or $SnCl_2/HCl$.

(vi) $NaBH_4$or LiH + $AlCl_3$.

(vii) C_6H_5CHO and CH_3CHO (See Claisen-Schmidt reaction).

(viii)

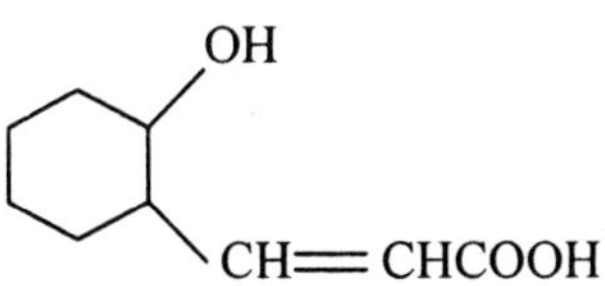

(See Perkin reaction)

(ix) $C_6H_5CH{=}CHCOOCH_2CH_3$.

24. The Fehling solution would be reduced by acetaldehyde only. So

$$\underset{44\ \text{gm}}{CH_3CHO + 2CuO} \rightarrow \underset{\text{Brown ppt. 143 g}}{CH_3COOH + Cu_2O}$$

143 g brown ppt. is obtained from 44 g of acetaldehyde.

$1.43 \rightarrow 44 \times 1.43/143 = 0.44$ gm.

So, percentage of acetaldehyde $= 0.44 \times 100/4.4 = 10\%$

$CH_3CHO = 10\%$

$C_6H_5CHO = 90\%$.

25. *(i)* $C_6H_5CH(OCOCH_3)_2$ *(ii)* Acetophenone

26.

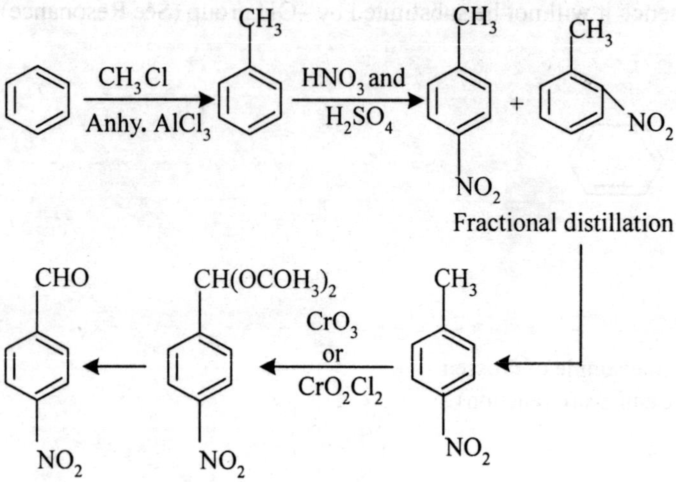

27. C_6H_5COCl, $C_6H_5CONH_2$, C_6H_5CN, $C_6H_5CH_2NH_2$.

28. *(i)*

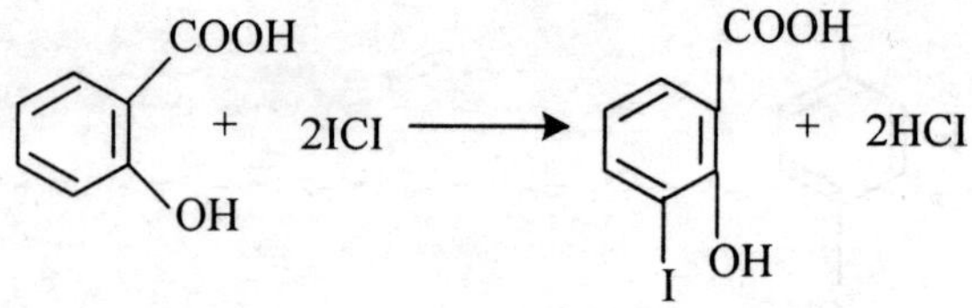

(ii)

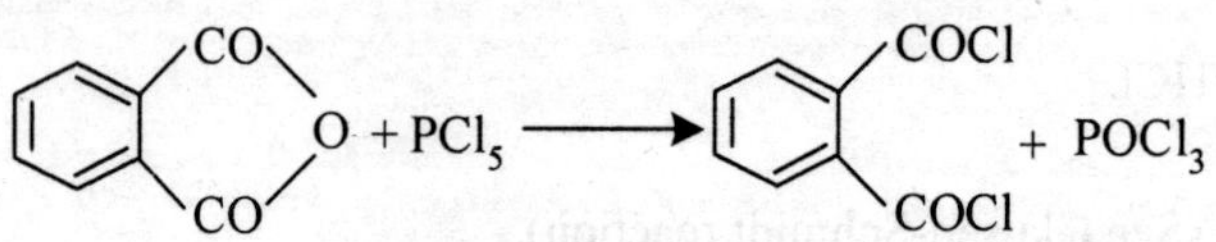

(iii)

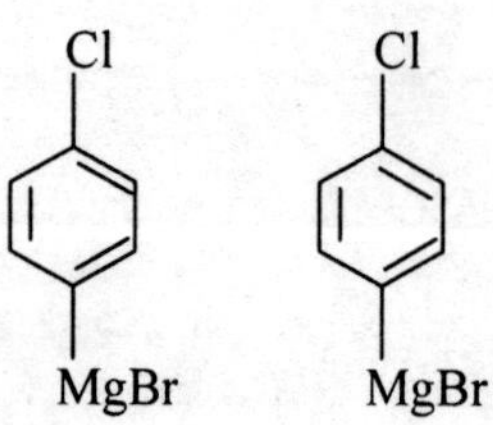

(iv)

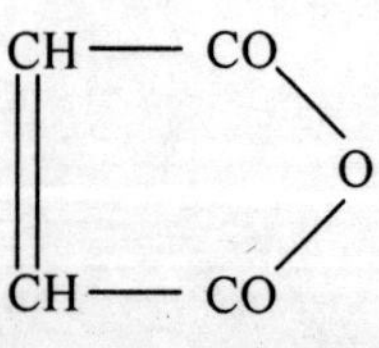

29. Benzoyl chloride is reasonably stabilized.

30. *(i)* (**b**), see Rosenmund reduction.

(ii) Analgesic.

31. *(i)* Zwitter ion 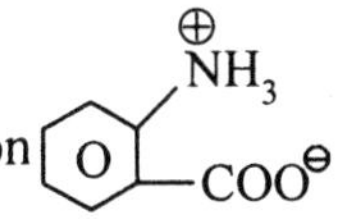 (See Chapter 14, on Amino Acids).

(ii) insoluble, hot water.

32.

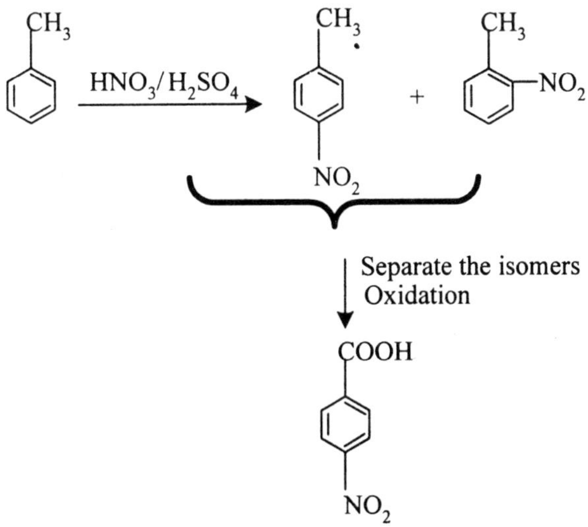

33. *(i)* (**a**), it is because it is unstable. The mechanism involves a cyclic transition state.

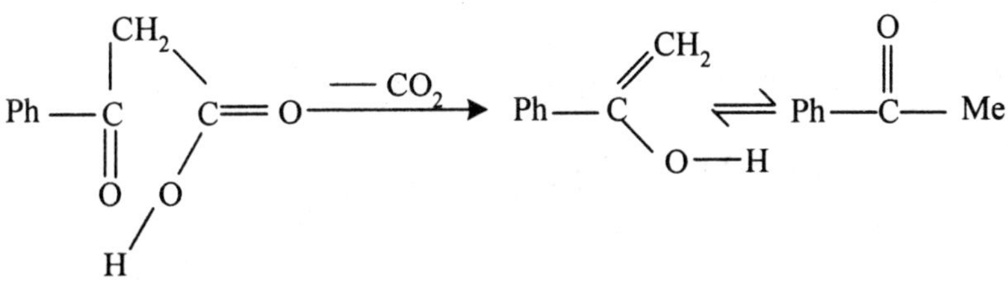

34. *(i)* Zn/H_2SO_4 or $LiAlH_4$.

(ii) $PhSO_2Cl + NaCl + POCl_3$.

(iii)

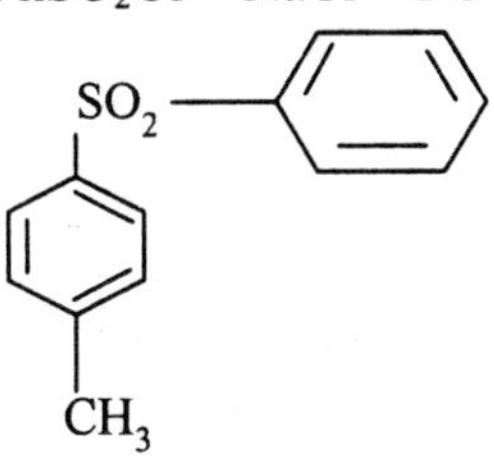

(iv) 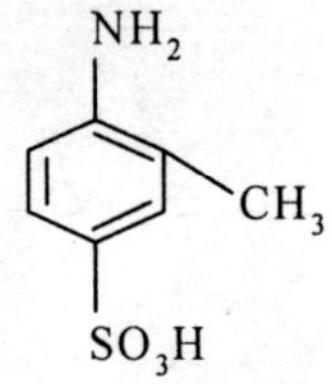

(v) NaOH, 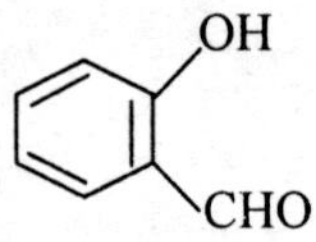

(vi) 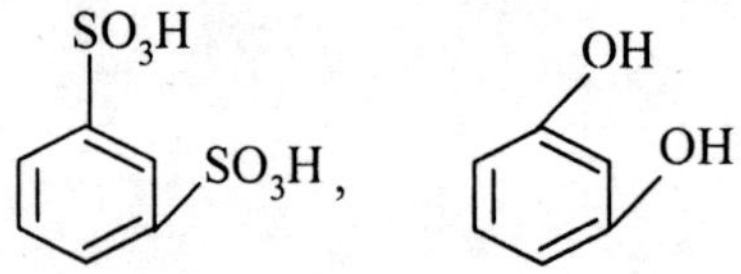

35. (i) $C_6H_5NH—COCH═CHCOOH$

(ii) 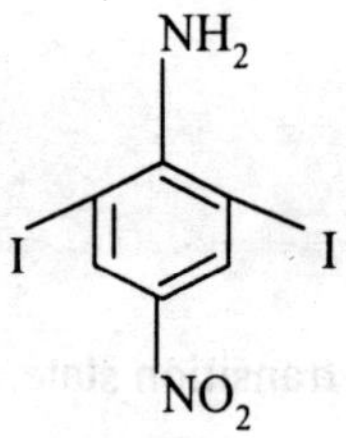

(iii) 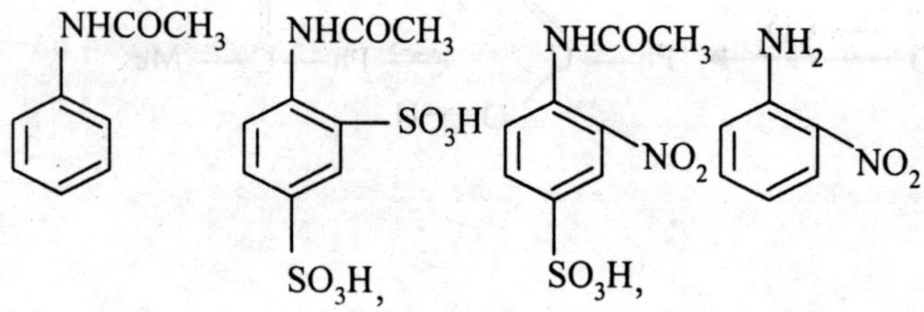

(iv) Br_2/KOH

(v) 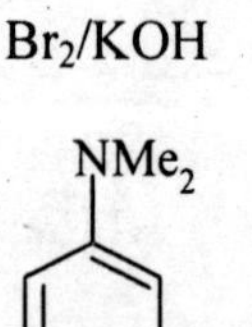

(vi) $C_6H_5—N═N—C_6H_5$
Azobenzene

(vii) $C_6H_5NHSO_2Ph$, $C_6H_5NCH_3SO_2Ph$,
$C_6H_5NHCH_3$, $C_6H_5SO_3H$

(viii) 

(ix) 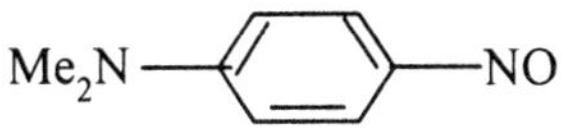

(x) 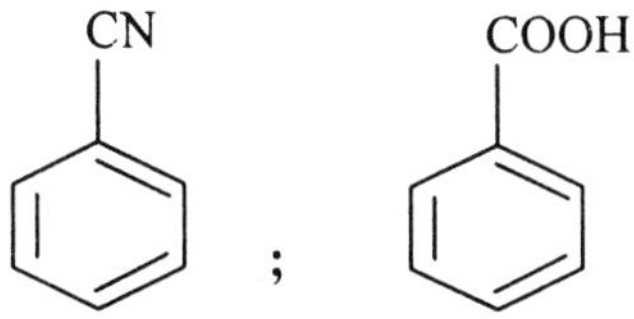

36. Conversions:

(i) 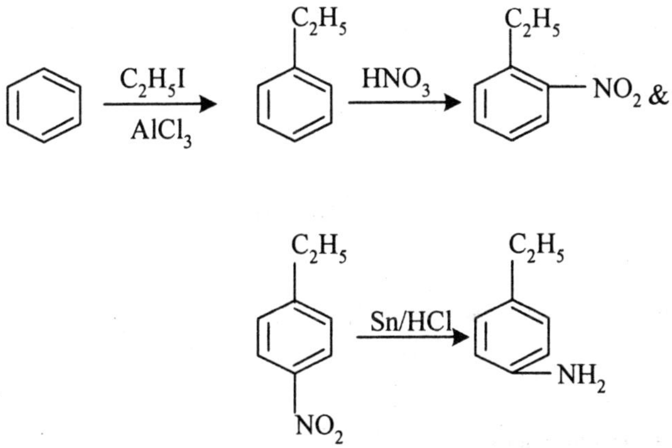

(ii) 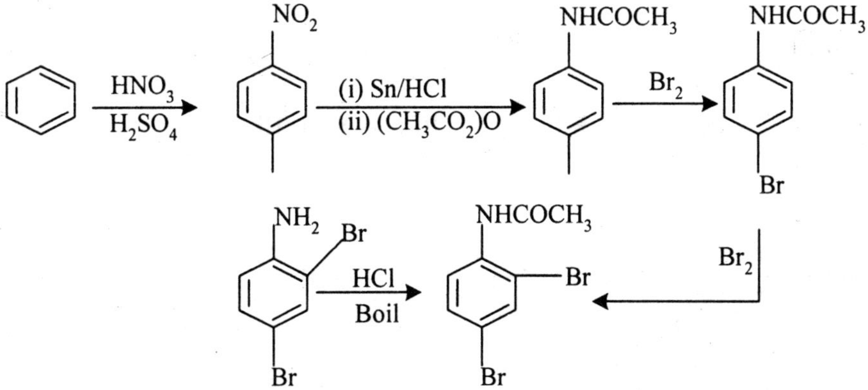

37. **(c)**

38. *(i)*

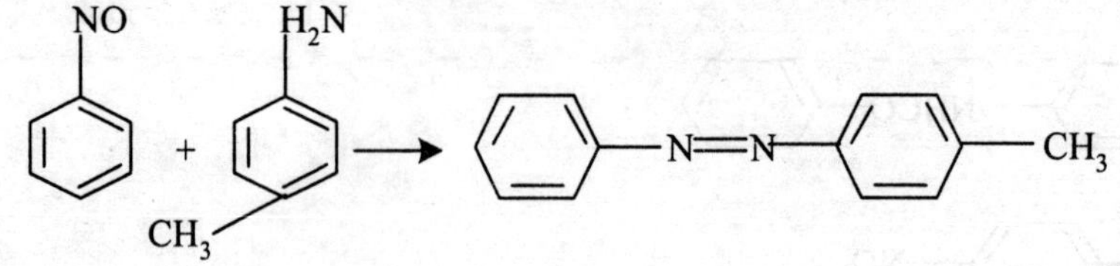

(ii)

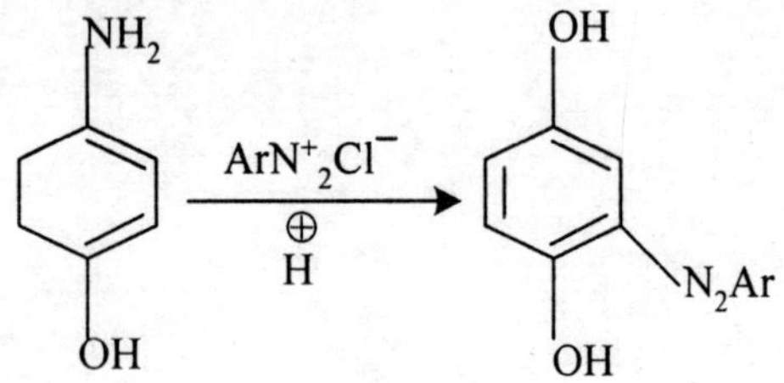

(iii)

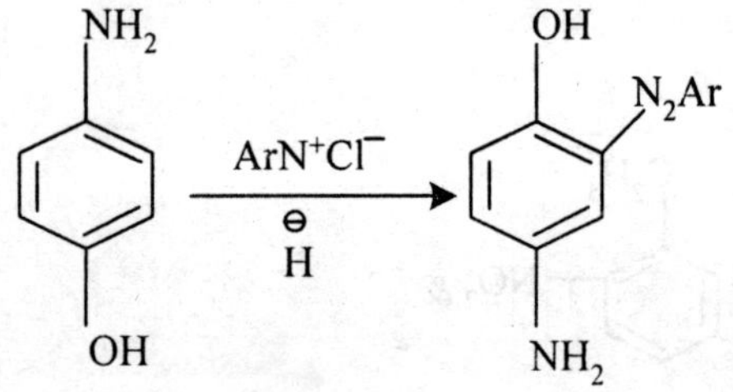

(iv)

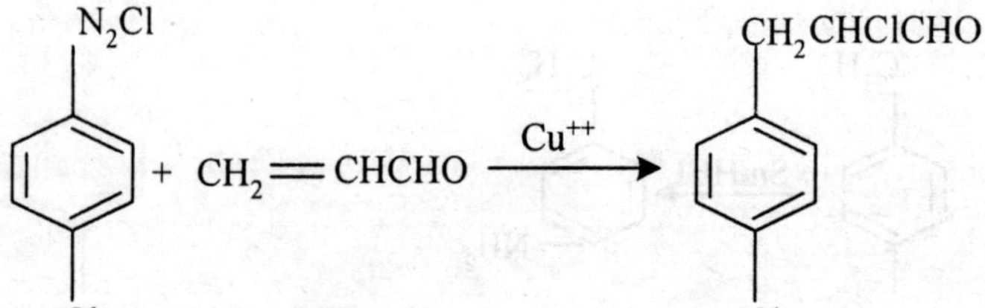

(Note: See Meerwein reaction)

(v)

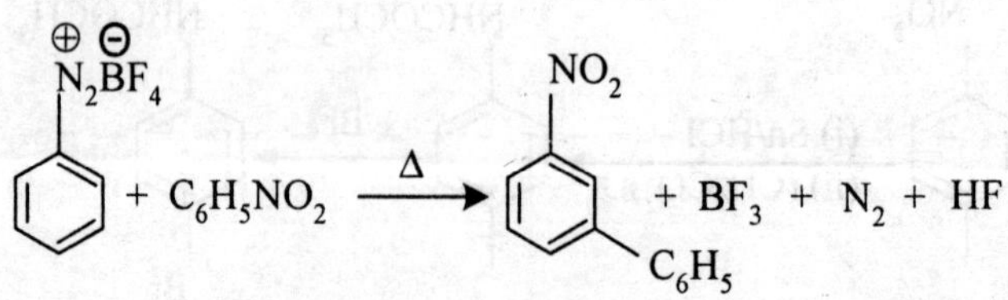

(vi)

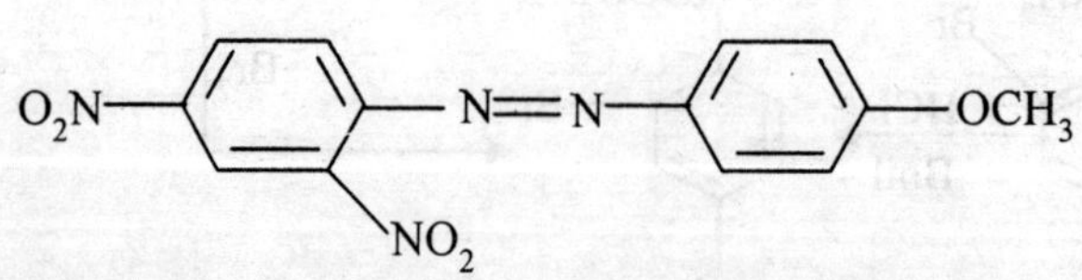

39. *(i)*

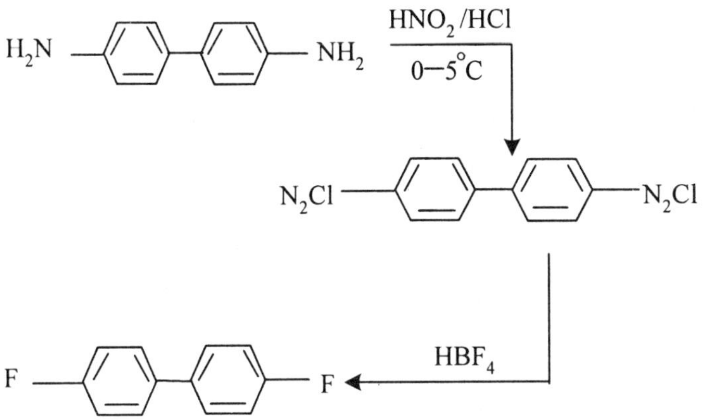

(ii)

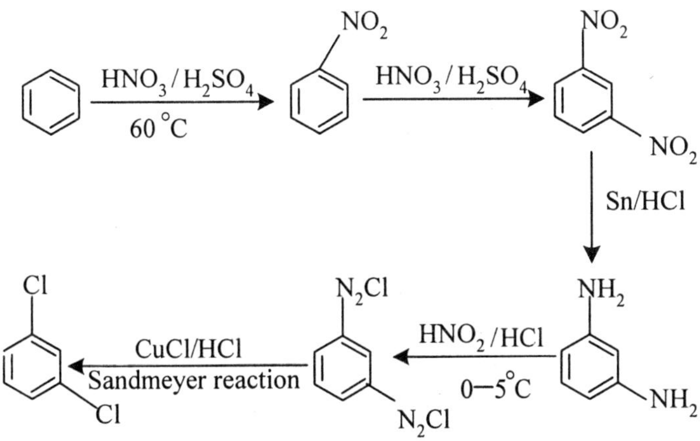

(iii)

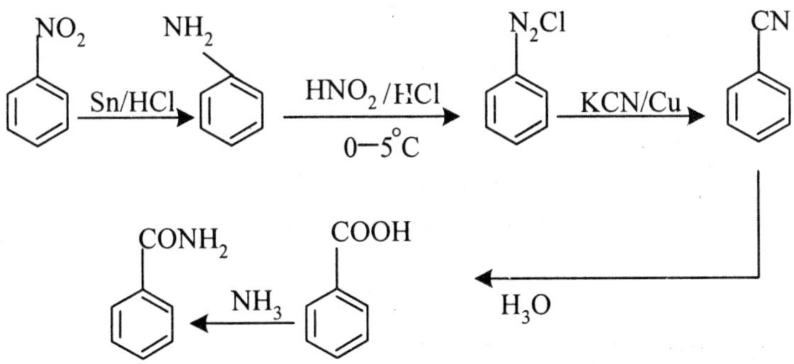

Benzamide

(iv) 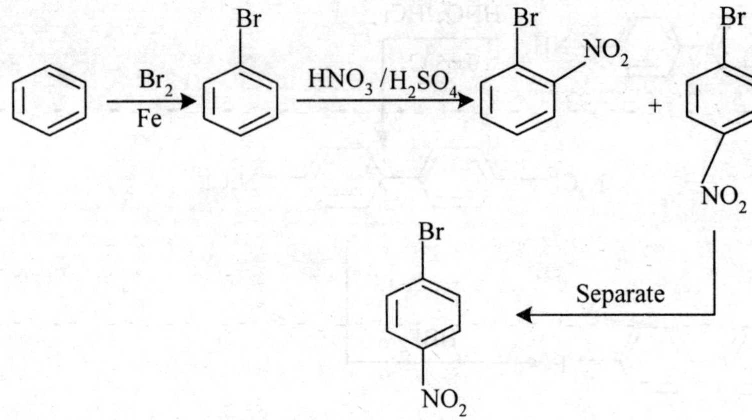

(v) 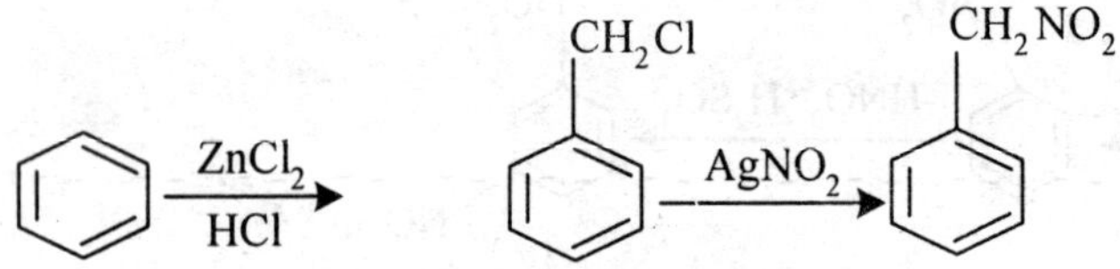

(vi) 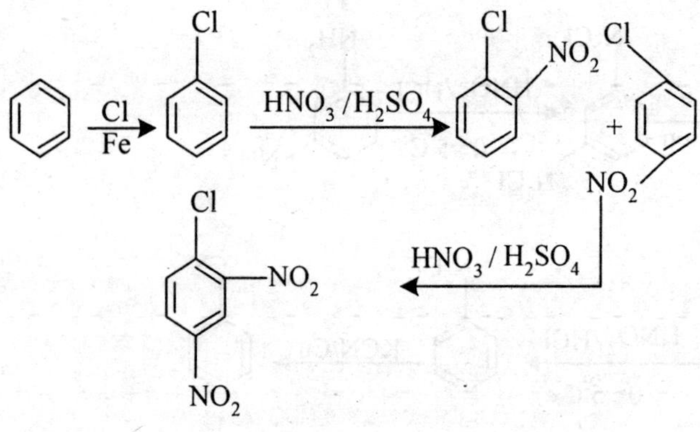

40. Examples of activated nucleophilic substitution reactions:

(i) 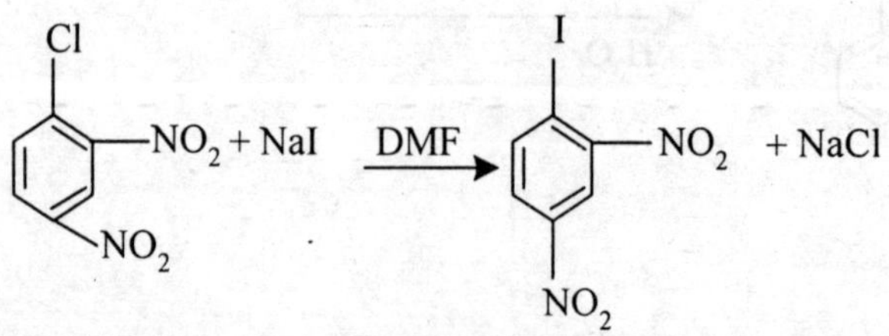

(ii) 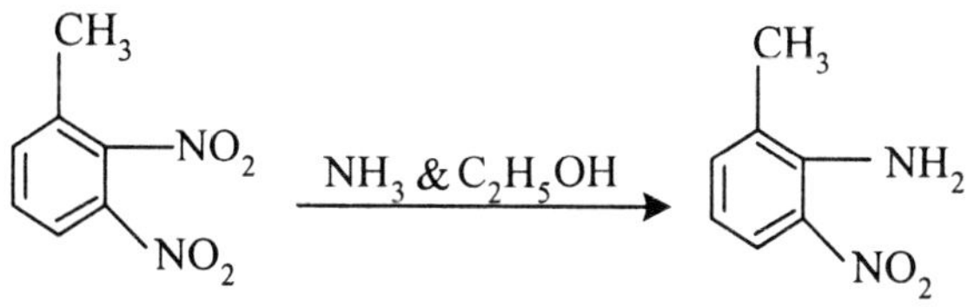

(iii) 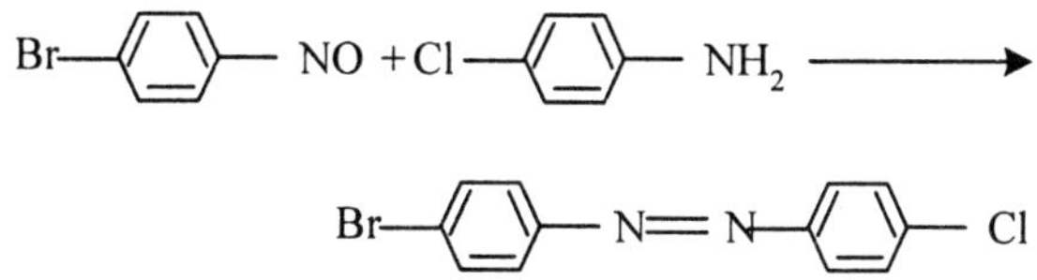

41 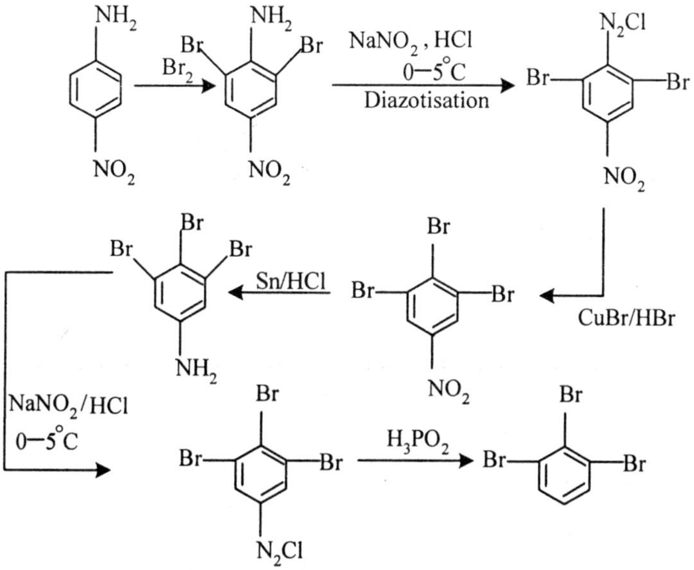

42. 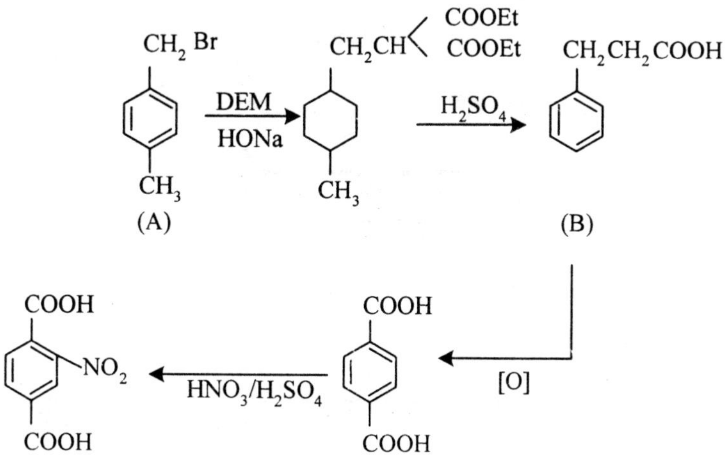

43. (**A**) & (**D**).

Chapter 12

Polymers

12.1 INTRODUCTION

Polymerization is a chemical reaction in which two or more than two molecules of one or more substances combine together with or without evolution of anything like water, ammonia, heat, etc., to form a molecule of high molecular weight. The product is called a polymer and the starting material is called a monomer.

Monomer may be any substance which can act as a bifunctional compound. The chemical unit which repeats itself throughout the structure of the polymer is called 'repeating unit' or 'structural unit' of the polymer. The degree of polymerization ($\overline{D}p$) refers to the number of repeating units in the polymer chain. The average molecular weight of a polymer is equal to the product of weight of one repeat unit and the average degree of polymerization.

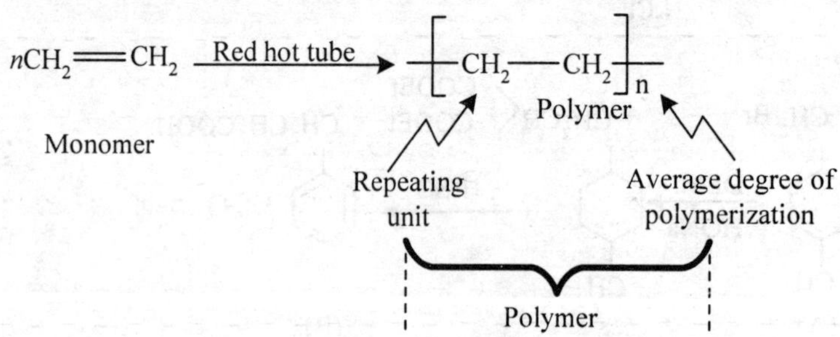

12.2 KINDS OF MONOMERS

Monomers may be classified in two categories:

(i) *Non-vinyl monomer*—These do not contain vinyl group ($CH_2{=}CH$—). For example: Thiophene, furan.

(ii) *Vinyl monomer*—They contain vinyl group. For example styrene 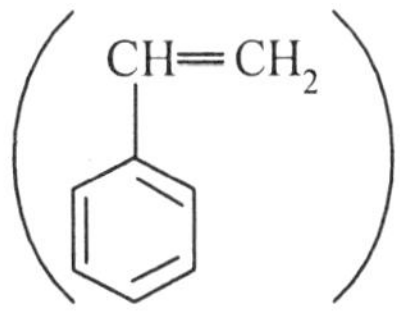,

acrylonitrile (CH_2=CHCN).

12.3 CHARACTERISTICS OF POLYMERS

1. They have a chain of molecules in which a large number of atoms combine consecutively.
2. They have two kinds of bonds—chemical and intermolecular; the bonds are mainly covalent bonds.
3. They have average molecular weight. It is because polymer chain length varies and there are two modes of termination. Average molecular weight is of the following types:

(i) *Weight average molecular weight* $(\overline{M}_w)$: This may be expressed mathematically:

$$\overline{M}_w = \frac{\sum M_i^2 N_i}{\sum N_i M_i}$$

Where N_i = number of molecules of weight. M_i. It may be determined by light-scattering or by ultra-centrifugation methods

(ii) *Number average, molecular weight* $(\overline{M}_n)$ This may be expressed mathematically as below:

$$\overline{M}_n = \frac{\sum M_i N_i}{\sum N_i}$$

It may be determined by the methods depending on colligative properties. End group analysis method.

(iii) *Viscosity average molecular weight* $(\overline{M}_v)$: It can also be calculated by using Mark–Houwink equation:

$$[\eta_{\text{int}}] = K[M_v]^{\alpha}$$

where K and α are constants;

$(\overline{M}_v)$ is viscosity average molecular weight; and

$[\eta_{\text{int}}]$ = Intrinsic viscosity.

(iv) *Z average molecular weight* $(\overline{Mz})$: (z is a sedimentation constant); it may be expressed mathematically as below:

$$(\overline{M}_z) = \frac{\sum M_i^3 N_i}{\sum M_i^2 N_i}$$

The ratio $\frac{\overline{M}_w}{\overline{M}_n}$ is termed as polydispersity index.

(v) $\overline{M}_{(z+1)}$: It is expressed mathematically as below:

$$\overline{M}_{(z+1)} = \frac{\sum Ni\, M_i^4}{\sum Ni\, M_i^3}$$

The inequality is as follows:

$$\overline{M}_w > \overline{M}_v > \overline{M}_n > \overline{M}_{(z+1)} > \overline{M}_z$$

4. Polymers have no sharp melting point, but a softening temperature range.
5. Molecular weight distribution is another important property of polymers.
6. A polymer exists in crystalline form if it has a high stereoregular structure with little or no chain branching or when it contains highly polar groups.
7. Each polymer has a characteristic glass transition temperature (T_g), i.e., the temperature above which it behaves as a glass, and below which, as a polymer.
8. Polymers dissolve slowly in the solvent and there is no saturation point. The viscosity of the solution is higher than that of the solvent.

12.4 CLASSIFICATION OF POLYMERS

Polymers can be broadly classified into following categories:

(i) On the basis of chemical nature

A. Inorganic: (a) Natural —clays and sands; and (b) Synthetic—fibers like glass fiber.

B. Organic: (a) natural—(i) Polysaccharides; (ii) Proteins; and (iii) Polyisoprene (rubber).

(b) Synthetic—(i) Adhesives; (ii) Fibers; (iii) Paints; (iv) Plastics; and (v) Rubber.

(ii) On the basis of stereochemistry of polymer

(a) *Atactic*—A polymer in which there is no systematic configuration on carbon atoms is called atactic polymer.

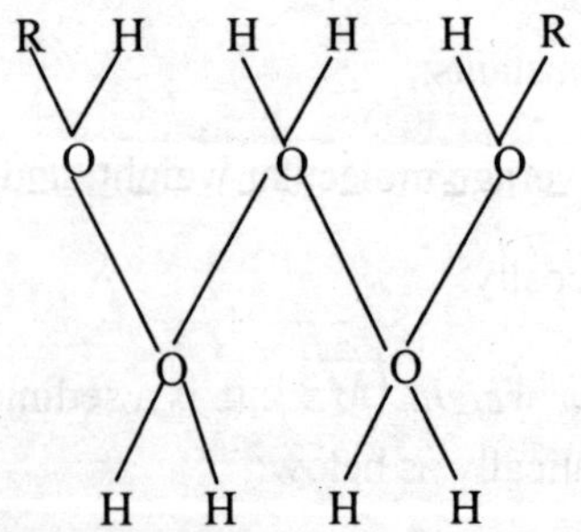

(b) *Syndiotactic*—A polymer in which alternate carbon atoms have same configuration is called syndiotactic.

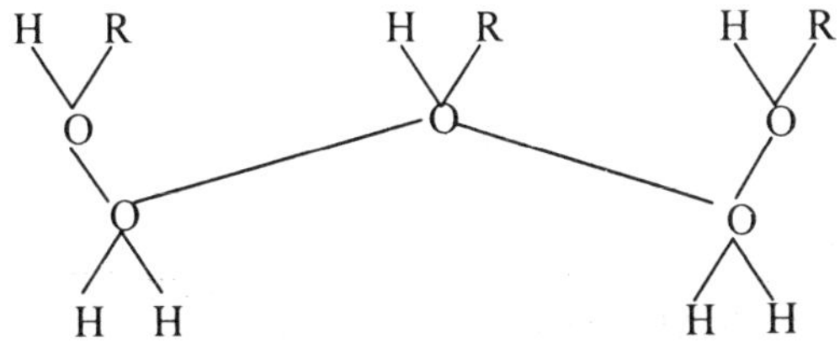

(c) *Isotactic*—A polymer in which each carbon atom has same configuration on carbon atoms is called isotactic polymer.

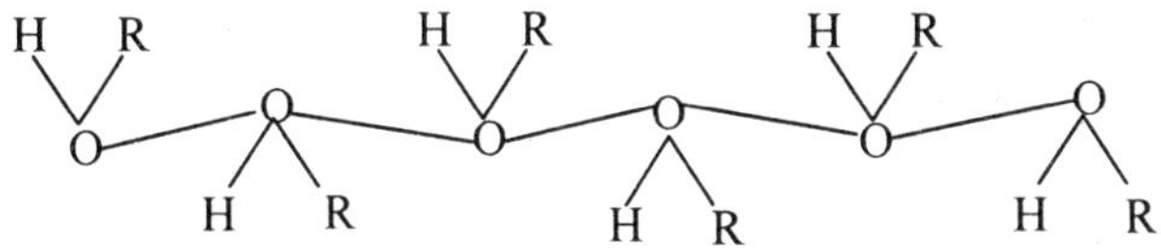

The isotactic polymers can be prepared in the presence of Zeigler-Natta catalyst, which consists of two components

(a) *Catalyst*: It consists of halides or oxyhalides of element of 4th to 8th groups of periodic table like Ti, ZnR, Mb—

(b) *Co*-catalyst: It consists of hybrids alkyl of elements of 1st, 2nd & 13th groups of periodic table. e.g., $TcCl_4 + (C_2H_5)_4Al$

(iii) On the basis of number of the kinds of monomers used

(a) If a polymer is obtained from one monomer only, that it is known as *homopolymer* and the reaction is called *homopolymerization*, e.g.,

$$n\mathrm{CH_2{=}CH_2} \xrightarrow{\text{Red hot tube}} \left[\!-\mathrm{CH_2}-\mathrm{CH_2}-\!\right]_n$$

(b) If a polymer is obtained by the polymerization of two different monomers then it is said to be a *copolymer* and the reaction is called *copolymerization*. Copolymers are of following types:

(i) *Random copolymer:* A copolymer in which the two monomer units are placed in a random manner is termed as random copolymer:

$$(\text{—A—B—A—A—A—B—A—B—A—B—B—B})_n\text{—}$$

(ii) *Alternate copolymer:* A copolymer in which the two monomer units are placed in an alternate manner is called alternate copolymer:

$$(\text{—A—B—A—B—A—B—A—B—A—B—})_n\text{—}$$

(iii) *Block copolymer:* It is a linear copolymers in which each monomer unit forms a block:

$$(\text{—A—A—A—B—B—B—A—A—A—A})_n\text{—}$$

(iv) *Graft copolymer:* It is a branched copolymer in which one monomer unit forms one block and other forms another block.

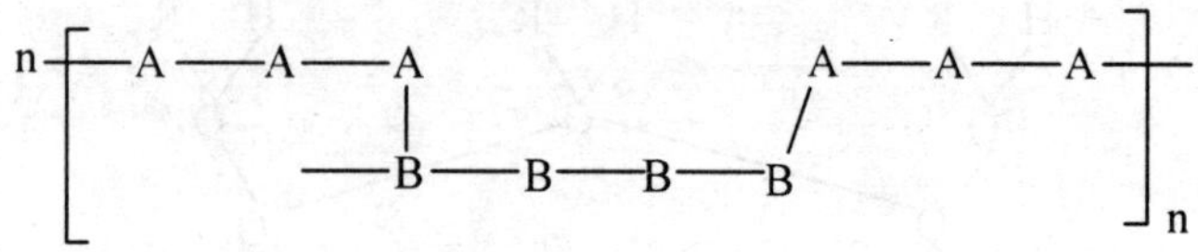

(iv) On the basis of engineering properties

The polymers can be classified in two categories:

(i) *Thermoplastics*: These polymers can be remoulded, e.g., Addition polymers like polystyrene, polyacrylonitrile.

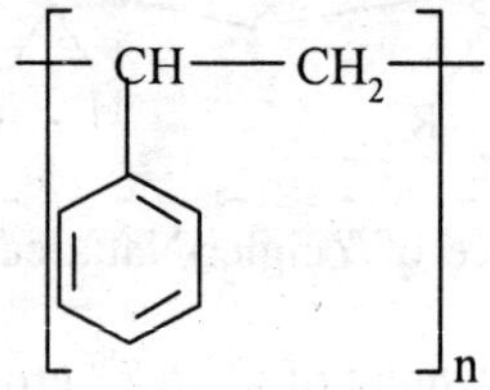

Polystyrene

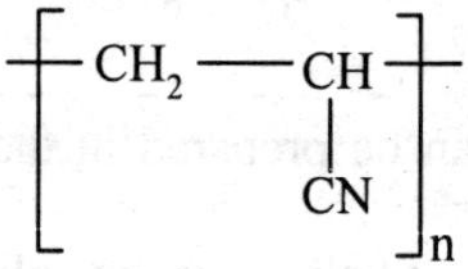

Polyacrylonitrile

(ii) *Thermosets*: These polymers cannot be remoulded due to cross-linking, e.g., condensation polymers like bakelite, nylons.

12.5 NOMENCLATURE OF POLYMERS

Generally polymers prepared from vinyl monomers, are named simply by placing the prefix 'poly' before the name of monomer:

e.g., Styrene→Polystyrene.

12.6 TYPES OF POLYMERIZATION

W.H. Carothers, in 1929, classified polymerization into two categories:

12.6.1 Addition Polymerization

It is the polymerization in which *nothing is eliminated* like water, ammonia, etc., during the course of reaction; so the molecular weight of the polymer is *exact multiple* of that of the monomer.

12.6.2 Condensation Polymerization

It is the polymerization in which something like water, ammonia is eliminated during the course of reaction, so that the molecular weight of the polymer is not be exact multiple of that of the monomer.

The differences between addition and condensation polymerization have been listed in Table 12.1.

Table 12.1 Differences between addition and condensation polymerization

Addition polymerization	*Condensation polymerization*
1. Nothi ng is eliminated during polymerization.	1. So mething like water or ammonia is eliminated.
2. It is a reversible reaction.	2. It is an irreversible reaction.
3. It needs a subs tance called initiator to start the reaction.	3. It does not need an init iator to start the reaction.
4. A pol ymer of high molecular weight even at low percent conversion is obtained.	4. A pol ymer of low molecular weight even at high percent conversion is obtained.
5. Three step s namely, initation, propagation, termination are identified.	5. No such steps can be identified.

But this classification leads to confusion because certain polymers can be obtained from both the methods, e.g.,

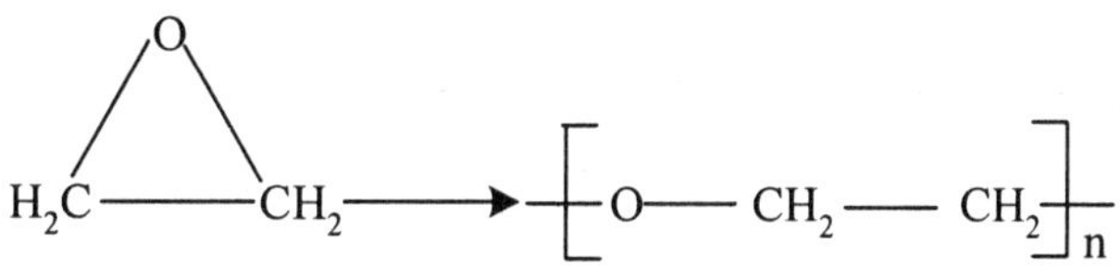

Addition polymerization

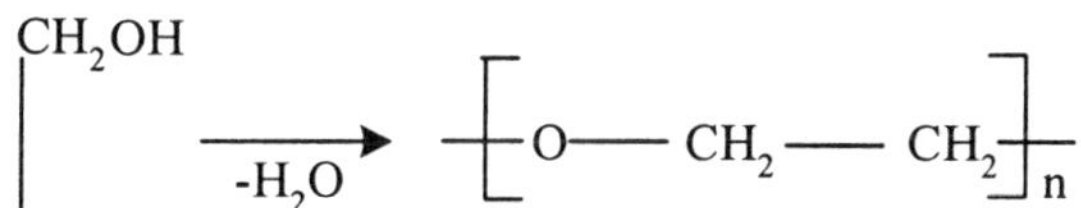

Polyether condensation polymerization

Therefore in more recent years, P.J. Flory has classified polymerization in two categories:

(1) Step-growth polymerization

(2) Chain-growth polymerization

12.6.3 Step-growth Polymerization

In this type of polymerization reaction occurs in a stepwise manner. This can happen in two ways:

(a) Both reaction functional groups are present in one molecule.

A—B $\quad$ $-[$A—B$]-$

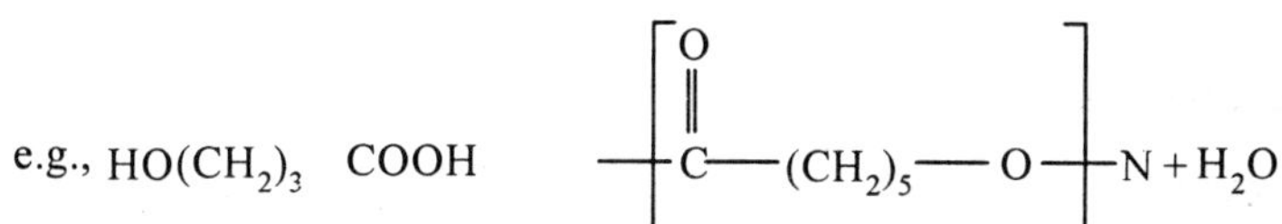

Polyester

(b) Two bifunctional monomers undergo polymerization

A—A + B—B ⟶ $-[$A—A—B—B$]-$

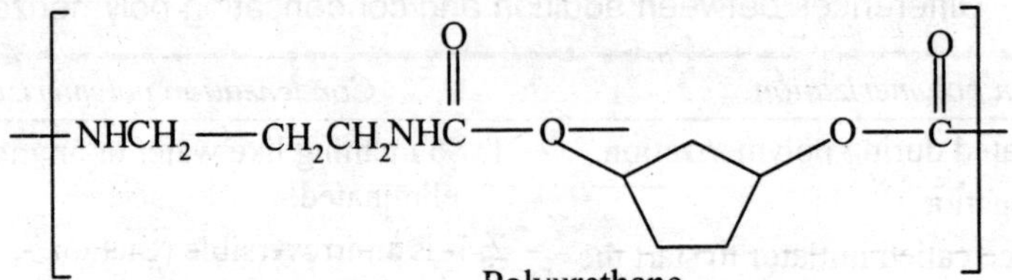

Polyurethane

This type of polymerization has following characteristics:

(i) Monomer molecules are consumed rapidly without any large increase in molecular weight.

(ii) No initiator is needed for the reaction.

(iii) Average degree of polymerization ($\overline{\text{Dp}}$) can be related to percentage conversion of the monomer.

Let us suppose that there are N_0 molecules present initially and N molecules present after a given reaction period. Then the amount reacted is $N_0 - N$. The per cent reaction conversions is given by the expression

$$P = \frac{N_0 - N}{N_0} \times 100 \qquad (1)$$

$$N = N_0\,(1 - p) \qquad (2)$$

The average degree of polymer (D_p) is N_0/N

$$D_p = \frac{1}{1-p}$$

12.6.4 Chain-Growth Polymerization

This type of polymerization has following characteristics:

(i) It needs an initiator to start the reaction.

(ii) The molecular weight increases rapidly even though large amount of monomers remains unreacted.

(iii) This type of polymerization consists of three steps:

(a) Ini tiation; *(b)* Propagation; and *(c)* Termination.

Chain-growth polymerization can further be classified into two categories depending upon the nature of initiator used.

1. Free-radical polymerization
2. Ionic polymerization

12.6.4.1 Free-Radical Polymerization

In this type of polymerization, the initiator produces free radicals and then polymerization proceeds. The following initiators may be used:

(a) Azobisisobutyronitrile (AIBN)

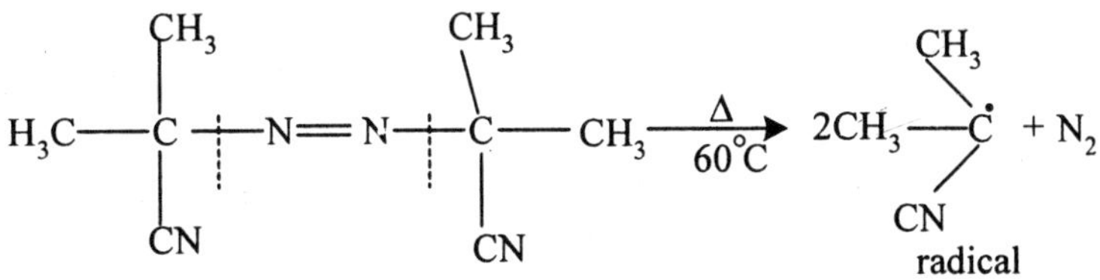

(b) Benzoyl peroxide

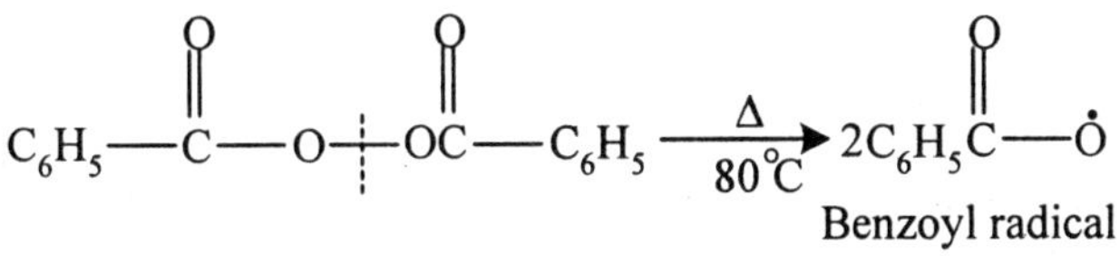

(c) Redox initiators:

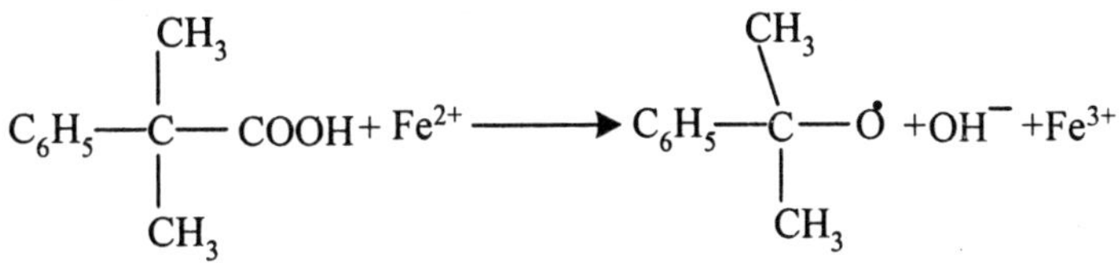

$$H_2O_2 + Fe^{2+} \longrightarrow {}^{\bullet}OH + OH^- + Fe^{3+}$$

(d) Ylides: These initiators may be broken photolytically and thermally.

12.6.4.2 Ionic Polymerization

It is of two types:

(a) *Cationic polymerization:* In this polymerization, initiating species is the carbonium ion ($—C^{\oplus}$). Among the cationic initiators are included the mineral acids, Friedel-Craft catalyst like anhydrous $AlCl_3$.

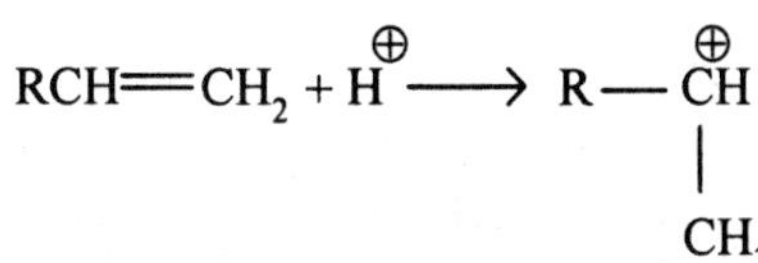

(b) *Anionic polymerization:* In this polymerization, the initiating species is a carbon ion ($—C^{\ominus}$). The anionic initiators include butyl lithium, KNH_2 in ammonia.

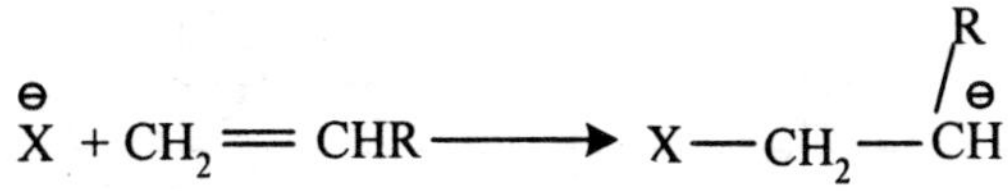

12.7 COPOLYMERIZATION

It is a useful technique to modify properties of a polymer, e.g., polystyrene has good mechanical strength but, poor transparent property. On the contrary, polymethylmethacrylate has good transparent property, but poor mechanical strength. But a copolymerization of styrene and methylmethacrylate has good mechanical strength as well as transparent property.

Copolymerization may be defined as a reaction in which two or more than two molecules of two monomers combine together to form a copolymer. Copolymerization can be carried out by using free-radicals, cationic and anionic initiators. Copolymers are of four types: Random, alternate, block and graft copolymers.

12.8 CHAIN-TRANSFER AGENTS

A substance which transfers activity of a polymer chain to a new chain is called a chain-transfer agent and the reaction is called a chain-transfer reaction. They affect the average molecular weight of the polymer adversely, e.g., CCl_4 is a chain-transfer agent.

12.9 CHEMICAL STRUCTURE OF SOME COMMON POLYMERS

1. *Natural Rubber*

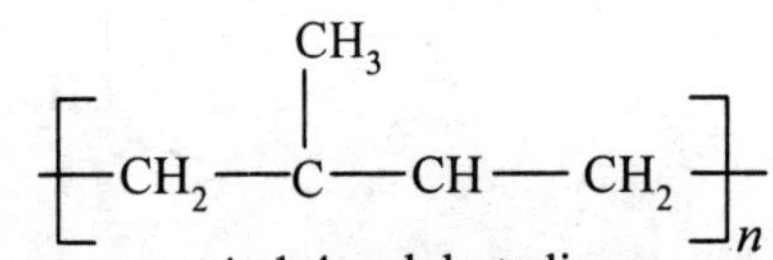

cis-1,4-polybutadiene

2. *Buna-s or SBR rubber:* It is a copolymer of butadiene and styrene.

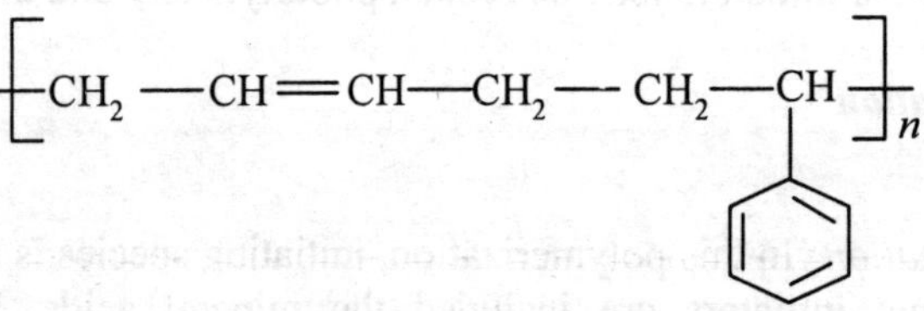

3. *Buna-N*: It is a copolymer of butadiene and acrylonitrile. It has high resistance to solvents, oil and chemicals.

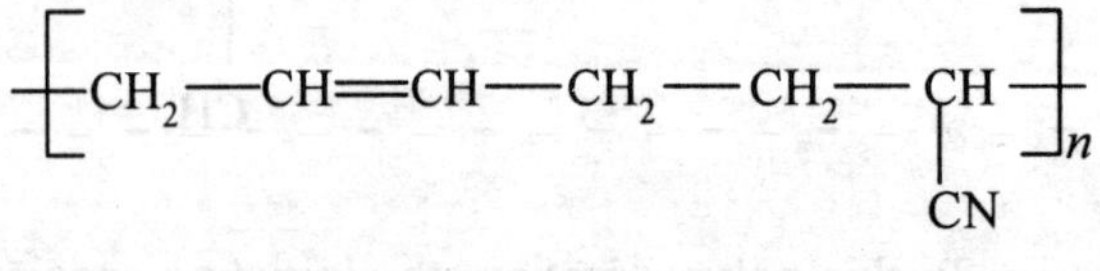

4. *Neoprene*: It is a homopolymer of chloroprene and is used for house wire insulation.

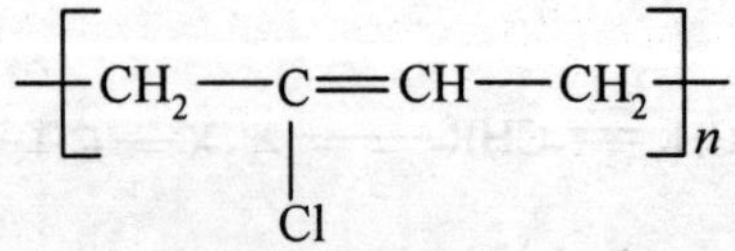

5. *Thiokol*: Prepared by copolymerization of ethylene dichloride with inorganic polysulphides.
6. *Polystyrene*: It is of two types:
(i) High impact; and *(ii)* Low impact.
7. Polymethylmethacrylate.
8. *Silicones*: Have Si—O chain to which various organic groups are attached.
9. *Urea formaldehyde resin*: Can be used as adhesive laminates.
10. *Phenol formaldehyde resin or bakelite resin*:

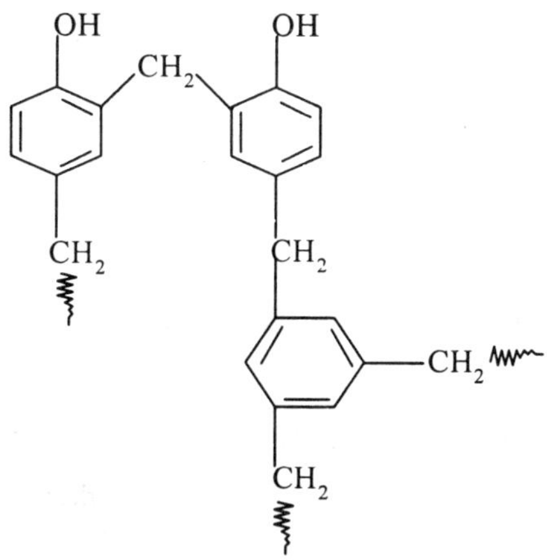

11. *Polyester*: Prepared by condensation of diamine with diacid.
12. *Polyethylene*: It is of two types:
    (i) Low-density polythene (LDPE)—It is branched and prepared under high pressure.
    (ii) High-density polythene—It is linear and prepared under low pressure.
13. *Nylons*: It is a generic term used to designate any long-chain synthetic polymeric amide with amide group as an integral part of the main polymer chain. The nylons are of following types:

(i) Nylon-1: 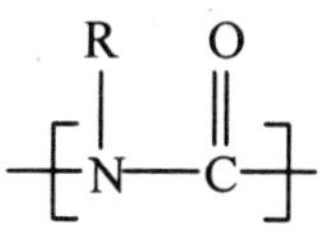

(ii) Nylon-2: 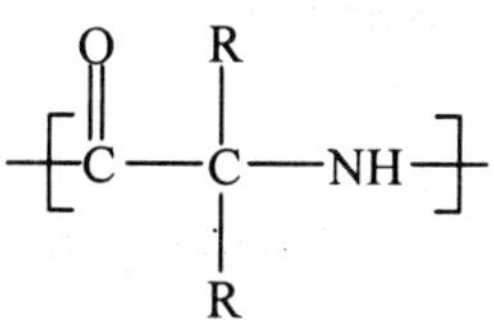

(iii) Nylon-3: 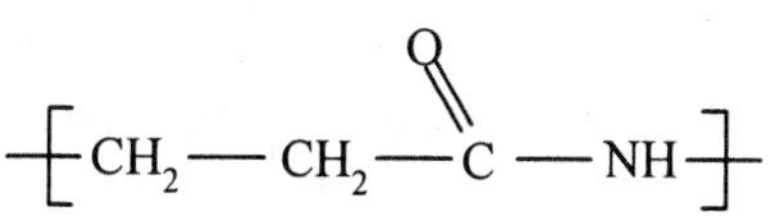

(iv) Nylon-6: $\left[\!-\overset{O}{\overset{\|}{C}}-(CH_2)_5-NH-\!\right]$

(v) Nylon-8: $\left[\!-\overset{O}{\overset{\|}{C}}-(CH_2)_7-NH-\!\right]$

(vi) Nylon-11: $\left[\!-\overset{O}{\overset{\|}{C}}-(CH_2)_{10}-NH-\!\right]$

(vii) Nylon-12: $\left[\!-\overset{O}{\overset{\|}{C}}-(CH_2)_{11}-NH-\!\right]$

(viii) Nylon-66: Number of carbon atoms contributed by diamine and dicarboxylic acid is six each.

$$\underset{\text{Adipic acid}}{HOOC\,(CH_2)_4\,COOH} + \underset{\text{Hexamethylene diamine}}{H_2N(CH_2)_6\,NH_2}$$
$$\downarrow$$
$$H\left[\!-\overset{O}{\overset{\|}{C}}-(CH_2)_4-\overset{O}{\overset{\|}{C}}-NH-(CH_2)_6-NH-\!\right]$$

(ix) Nylon 6:10: $\left[\!-\overset{O}{\overset{\|}{C}}-(CH_2)_8-\overset{O}{\overset{\|}{C}}-NH-(CH_2)_6-NH-\!\right]$

(x) Nylon 6:12: $\left[\!-\overset{O}{\overset{\|}{C}}-(CH_2)_{10}-\overset{O}{\overset{\|}{C}}-NH-(CH_2)_6-NH-\!\right]$

(xi) Dacron:

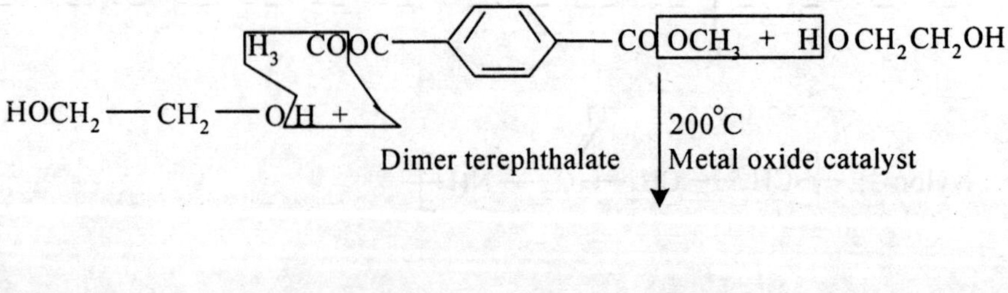

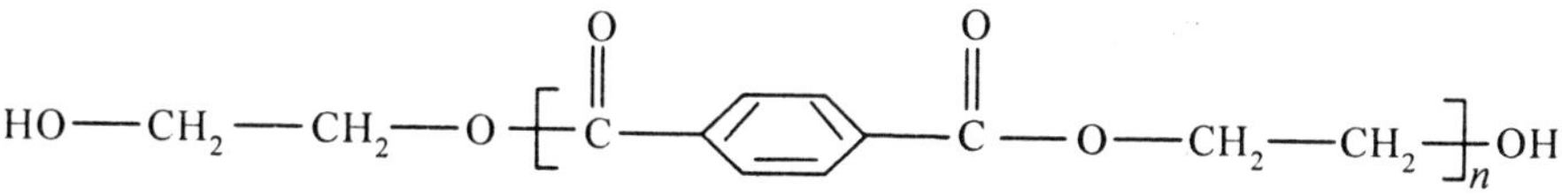

Terylene or Dacron

Polyethylene glycol terphthalate (Dacron)

14. Lexan:

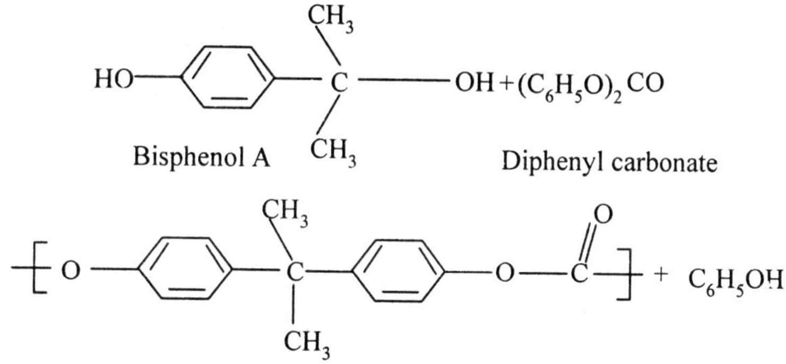

Lexan

15. Epoxy Resin-1:

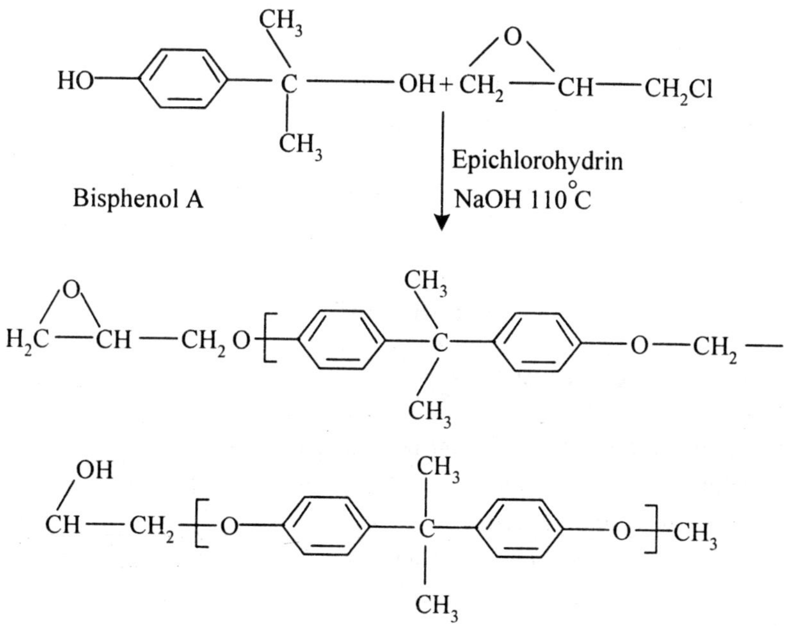

16. Polycarbonate: These are of two types:

(i) Aromatic polycarbonate

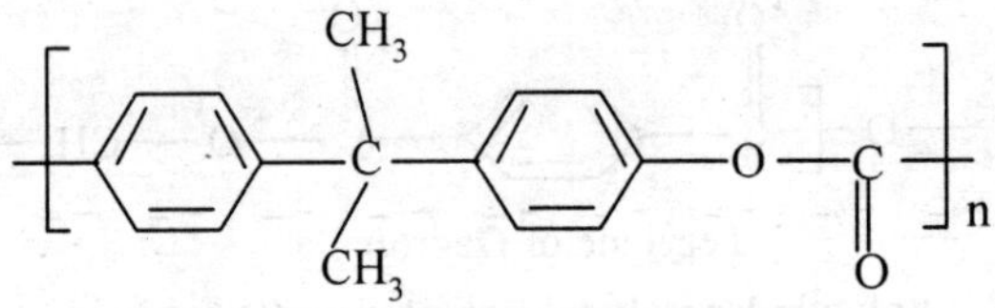

(ii) Aliphatic polycarbonate

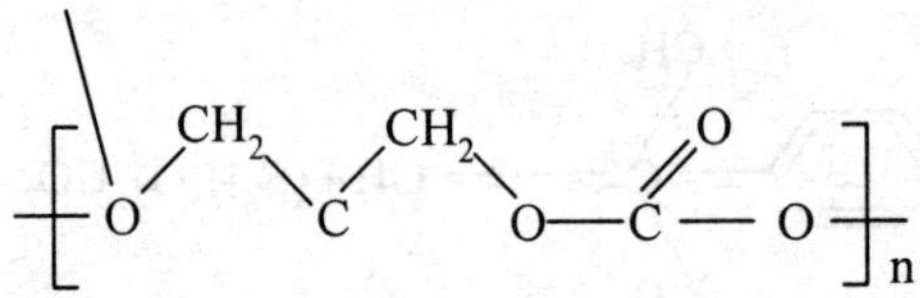

17. *Polyimide*: $-\!\!\left[CO-NR-CO \right]_n\!\!-$

18. *Addition Polymers*:

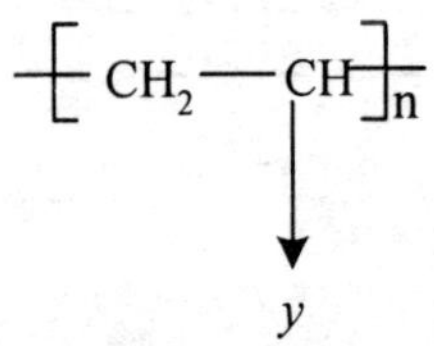

(a) If *y* is H, the polymer is polyethene.
(b) If *y* is CN, the polymer is polyacrylonitrile.
(c) If *y* is C_6H_5, the polymer is polystyrene.
(d) If *y* is Cl, the polymer is polyvinyl chloride.
(e) It *y* is $COOCH_3$, the polymer is polymethylacrylate.

PROBLEMS

1. Among the four answers for the questions given below, only one is correct. Tick out the correct answer:

(i) Which of the following monomer-polymer pair does not exist?
- *(a)* Ethylene and polyethylene
- *(b)* Styrene and polystyrene
- *(c)* Vinyl alcohol and polyvinyl alcohol
- *(d)* Vinyl chloride and polyvinyl chloride

(ii) The monomer of starch (natural polymer) is
- *(a)* α-D (+) glucose *(b)* β-D (+) glucose
- *(c)* amylase *(d)* amylopectin

(iii) Which of the following is suitable for preparation of acrylonitrile?
- *(a)* $CH_2{=}CHCl + KCN \rightarrow$
- *(b)* $CH_2{=}CH_2 + HCN \rightarrow$

(c) 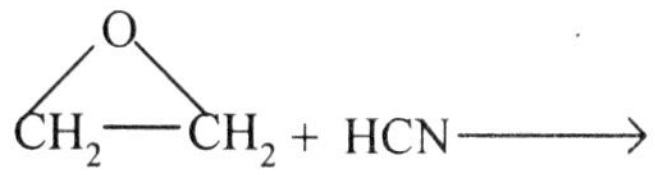

(d) $CH_3CHO + HCN \rightarrow$

(iv) Which of the following can be used as starting material for the synthesis of a few vinyl monomers?

(a) $CH_2{=}CH_2$ *(b)* $CH{\equiv}CH$ *(c)* C_6H_6 *(d)* CH_3CHO

(v) Buna-S is a

(a) Homopolymer of butadiene

(b) Copolymer of butadiene and acrylonitrile

(c) Natural rubber

(d) None of the above

(vi) Which of the following polymers would have the highest resistance to combustion?

(a) 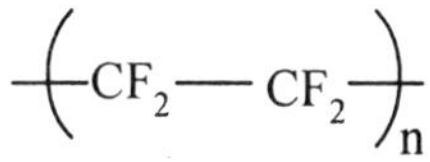

(b) 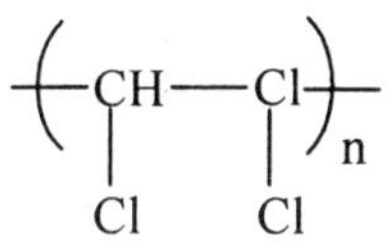

(c) 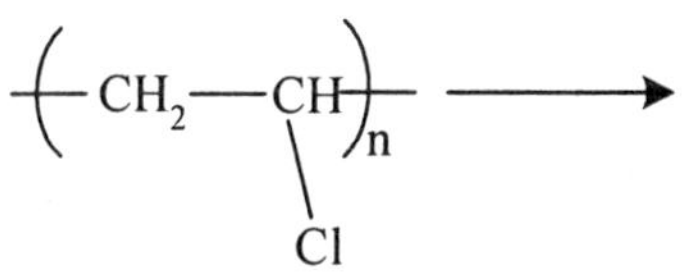

(d) 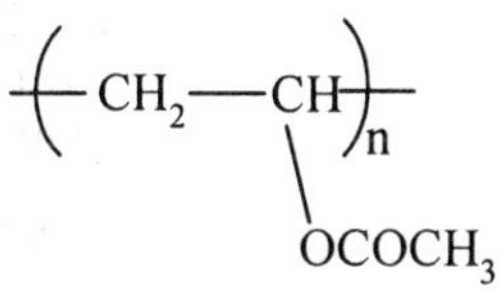

(vii) Which of the following would yield a polymer of high molecular weight:

(a) 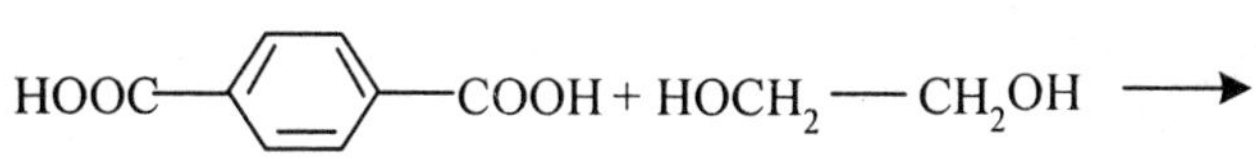

(b) $H_2N\,(CH_2)_6\,NH_2 + HOOC{-}(CH_2)_6{-}COOH \rightarrow$

(c) 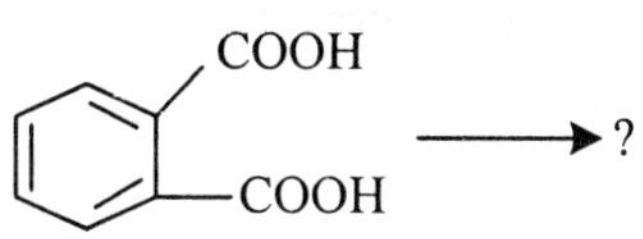

(d) $CH_2{=}CH_2 \rightarrow ?$

(viii) A sample of polystyrene has 10^{20} chains. If average degree of polymerization of each chain is 10^{30}, the average molecular weight of the sample is:

(a) 104×10^{20} *(b)* 10^3 *(c)* 104×10^{17} *(d)* 104×10^{23} **(GATE, 1992)**

(ix) If average degree of polymerization of Teflon $\left[CF_2 - CF_2 \right]_n$ is 10^3, its average molecular weight is:

(a) 6×10^4 *(b)* 10^3 *(c)* 10^5 *(d)* 4.8×10^4

(x) Which of these is a polyamide? **(AFMC, 2001)**

(a) Nylon *(b)* rayon *(c)* Orlon *(d)* None of these

2. Examine the structures of the following synthetic polymers. Identify the class of compounds they belong to:

(a)

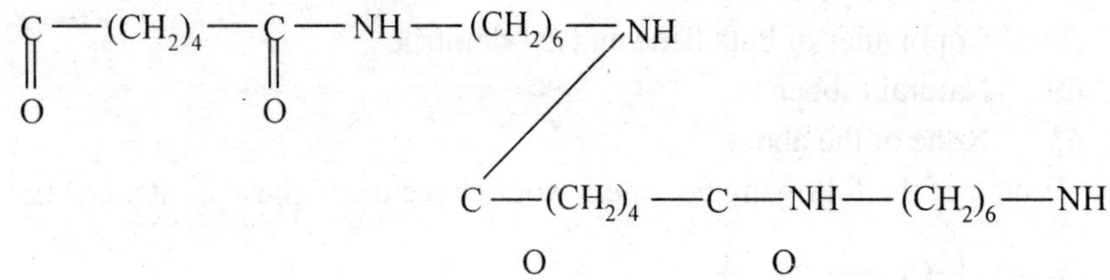

(b)

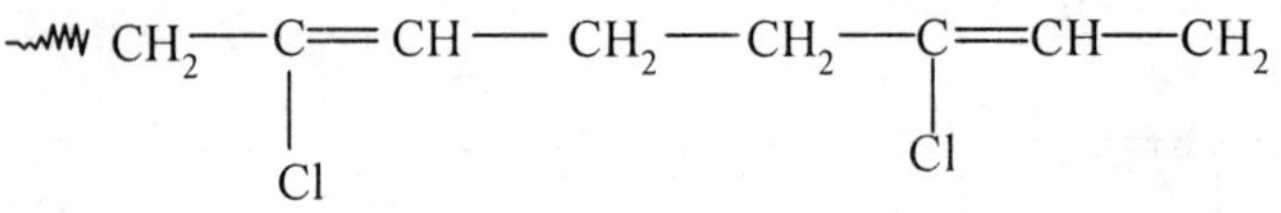

(c) ~~~ $CH_2CCl_2 - CH_2\ CCl_2$ ~~~

3. What is the molecular weight of polytertrafluoroethylene with the degree of polymerization 10^6. What is the ratio of its molecular weight to that of polyethylene with the same degree of polymerization.

4. A polymer has an average molecular weight of 48000 and a degree of polymermization 460.

(i) What is the molecular weight of the monomer?

(ii) Which of the following molecular structure is this monomer?

(a) $CH_2{=}CHNH_2$

(b) $CH_2{=}CHF$

(c) $C_6H_5CH{=}CH_2$

(d) $CF_2{=}CF_2$ **(GATE, 1991)**

5. State whether *True* or *False:*

(i) If average degree of polymerization of polymethylmethacrylate 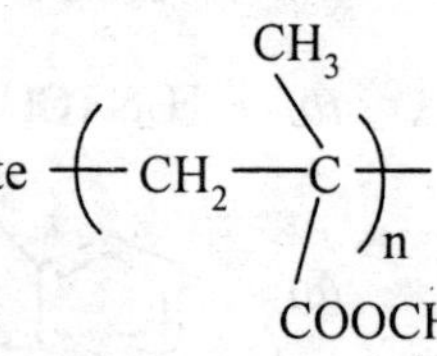 and Teflon $(CF_2—CF_2)_n$ is 10^3, the ratio of average molecular weight of both is 1.

(ii) Polyvinyl alcohol is obtained by the hydrolysis of polyvinyl acetate.

6. Write short note on Copolymerization. **(R, 1984)**

ANSWERS

1. *(i)* (**c**), vinyl alcohol is tautomeric form of acetaldehyde.

(ii) (**a**)

(iii) (**a**)

(iv) (**b**)

(v) (**b**)

(vi) (**a**), F has highest resistance to combustion

(vii) (**d**), it is an example of addition polymerization whereas rest are examples of condensation polymers.

(viii) (**d**), Average molecular weight = weight of monomer unit × average degree of polymerization. $= 104 \times 10^3 \times 10^{20} = 104 \times 10^{23}$.

(ix) (**c**), Average molecular weight $= 10^2 \times 10^3 = 10^5$ (remember the atomic weight of fluorine is 19 and atomic number is 9).

(x) (**a**).

2. *(a)* Condensation polymers.

(b) Addition polymers.

(c) Addition polymers.

3. Average molecular weight of Teflon $= 100 \times 10^6 = 10^8$.

Average molecular weight of polyethylene $= 28 \times 10^6$.

Ratio of molecular wt. of Teflon to polyethylene $= 100 \times 10^6 / 28 \times 10^6 = 3.57$.

4. $$\text{Weight of monomer} = \frac{\text{Average molecular weight of polymer}}{\text{Average degree of polymerization}}$$

$$= \frac{48000}{760}$$

$$= 104.$$

Hence monomer is styrene ($C_6H_5CH{=}CH_2$).

5. *(i)* True, because weight of both monomers, i.e., tetrafluoroethylene and methylmethacrylate is 100

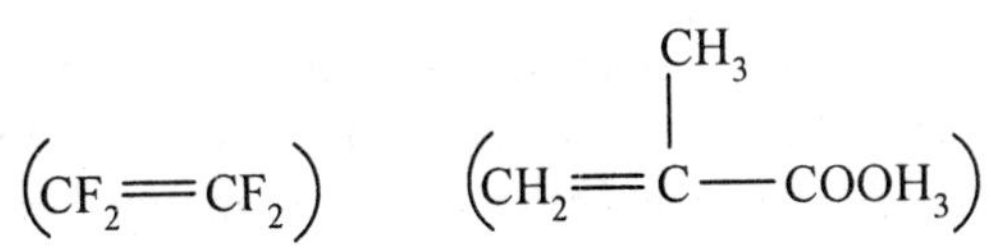

(ii) True.

6. See Page 337

Chapter 13

Carbohydrates

13.1 INTRODUCTION

Carbohydrates are compounds which are chemically polyhydroxyl aldehydes or ketones or substances which yield them upon hydrolysis, e.g.,

Glucose	Fructose
CHO	CH_2OH
|	|
$(CHOH)_4$	CO
|	|
CH_2OH	$(CHOH)_3$
	|
	CH_2OH

13.2 CLASSIFICATION

Carbohydrates have been classified in three categories as follows:

(A) *Monosaccharides:* These do not hydrolyze and are white in color, sweet in taste, readily soluble in water. These have been further classified in two categories:

(a) Aldoses: T hese contain aldehyde group. e.g., glucose.

(b) Ketoses: Th ese contain ketonic group. e.g., fructose.

(B) *Oligosaccharides:* These undergo hydrolysis, to yield 2-9 units of monosaccharides. These have been further classified into di, tri-saccharides. Examples of disaccharides are Sucrose, Maltose, and Lactose.

(C) *Polysaccharides:* These upon hydrolysis yield 10 or more units of monosaccharides, e.g., Starch and Cellulose.

13.3 GLUCOSE OR GRAPE SUGAR OR BLOOD SUGAR OR DEXTROSE

Preparation

It is prepared by hydrolysis of sucrose in the presence of H^+

$$\underset{}{C_{12}H_{22}O_{11}} + H_2O \xrightarrow{H^+} \underset{\text{Glucose}}{C_6H_{12}O_6} + \underset{\text{Fructose}}{C_6H_{12}O_6}$$

Properties of monosaccharides (glucose and fructose)

The properties can be discussed under three categories:

(i) Due to —OH group

(ii) Due to —CHO group

(iii) Miscellaneous.

(i) *Due to—OH group:* (1) When glucose is treated with baryta water; it forms calcium glucosate (water-soluble) where as fructose from calcium fructose (water insoluble); hence the reaction has been used to separate glucose and fructose. Moreover, the reaction shows acidic nature of glucose.

$$\underset{\text{Glucose}}{C_6H_{11}O_5OH} + Ca(OH)_2 \rightarrow \underset{\text{Calcium Glucosate}}{C_6H_{11}O_5Ca(OH)}$$

(2) Glucose reacts with acetic anhydrides (consuming 5 moles per glucose mole) forming pentacetyl glucose, thereby confirming presence of 5–OH groups.

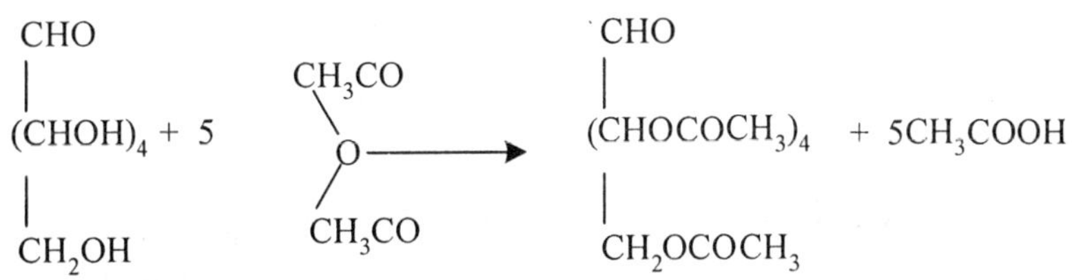

Acetic anhydride Pentacetyl glucose

(ii) *Due to aldehyde groups "—CHO":* They can be discussed as shown below:

Glucose can be shown as:

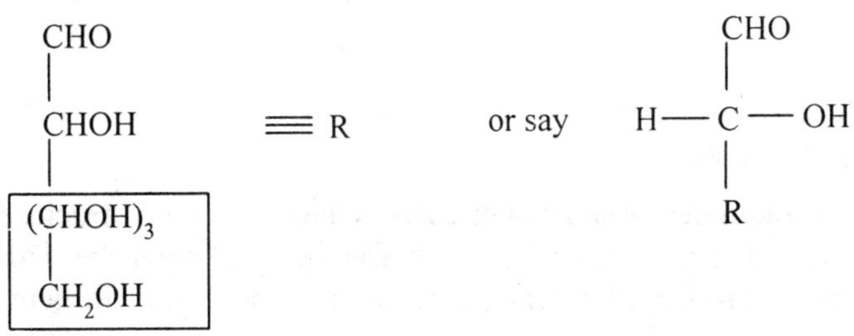

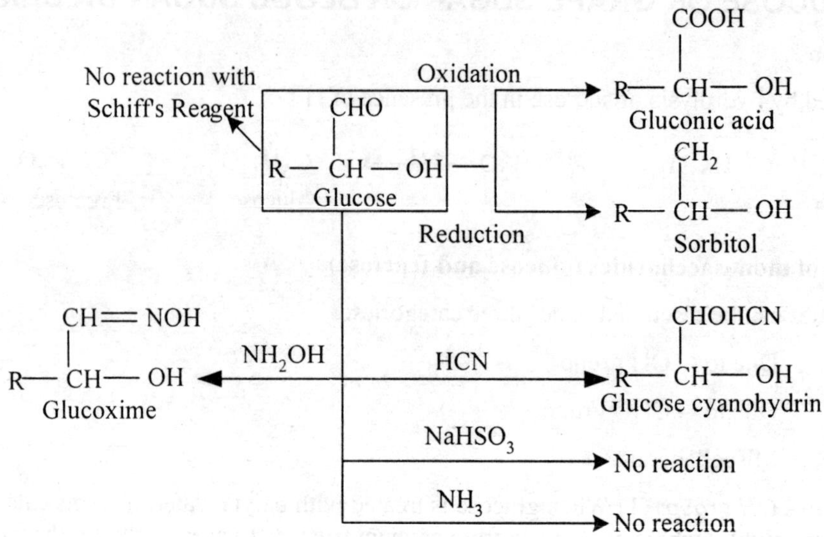

(iii) Miscellaneous: (1) *Reaction with phenyl hydrazine.*

Both glucose and fructose form same osazone hence both have same configuration at all carbon atoms except at C-1 and C-2. The reaction is as follows:

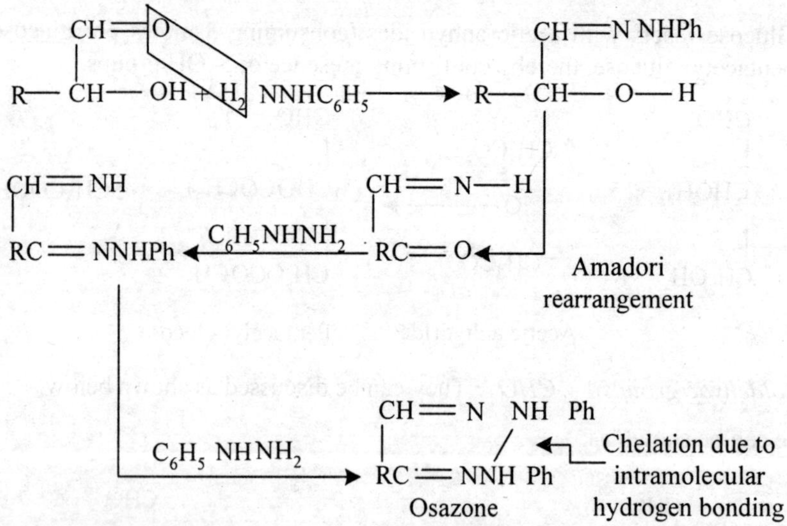

Osazone formation involves only two carbon atoms through formation of chelated ring due to intermolecular hydrogen bonding.

Lobry de Bruyn-Van Ekenstein rearrangement: When glucose or fructose or mannose is treated with alkali, it is rearranged to a mixture of glucose, fructose, and mannose. This explains that fructose, although contains ketonic group, reduces Fehling solution and Tollen's reagent.

Glucose or Fructose or Mannose →Glucose + Fructose + Mannose

Reaction with periodic acid (HIO_4): When glucose reacts with HIO_4, it undergoes oxidative cleavage and one HIO_4 is consumed between 2CH—OH groups and each —CH_2OH group is converted into HCOOH and each CH—OH converted into —CHO group, e.g.,

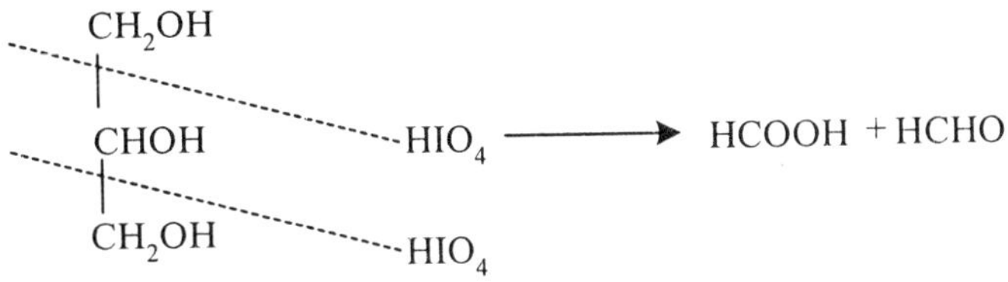

Similarly for glucose

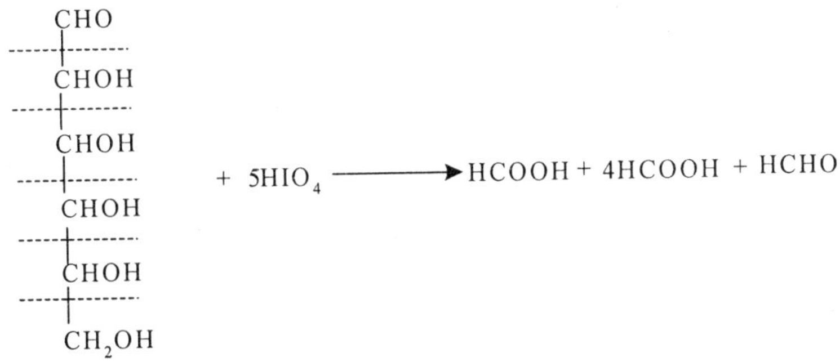

Structure of glucose: Glucose has been assigned following cyclic (6-membered) structure by Haworth because glucose showed the following properties:

(i) No addition reaction with NH_3 and sodium bisulphate.

(ii) Shows mutarotation.

(iii) Does not show pink color with Schiff 's reagent.

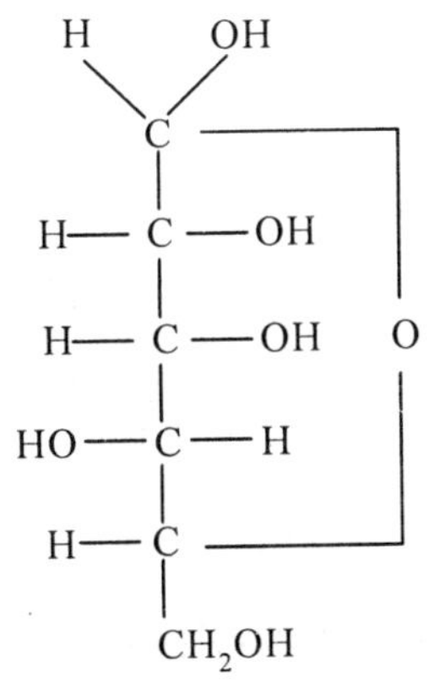

α (+) D – Glucose

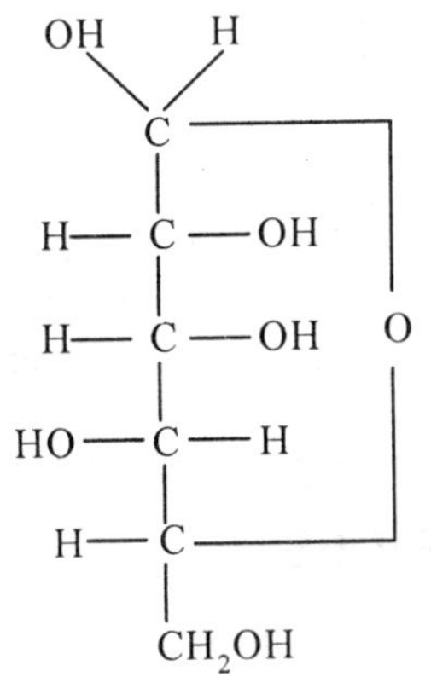

β (+) D – Glucose

The terms used above have following meanings:

α and β-denote anomeric nature (stereoisomeric sugars which have same configuration at all carbon atoms except at first are called *anomers*. Thus, α- and β-D (+) glucose are *anomers*. 'D' indicates configuration and '+' shows that it is dextrorotatory.

Mutarotation: Glucose exists in two anomeric forms α and β anomers, which have specific rotation +112° and +19°, respectively. When either form of glucose is dissolved in amphoteric solvent like water, the values changes to a constant value of +52.5°. This changes in rotation is called *Mutarotation* (Fructose also show mutarotation, α form –133° and β form –21°, the equilibrium values being –92°. It has been suggested that these anomers interchange into each other through open chain structure.

Epimerization: Stereoisomeric sugars which have same configuration at all carbon atoms except at the last but one (carbon number 2) are called *epimers* and their interconversion is called *epimerization*, e.g., glucose and mannose are epimers.

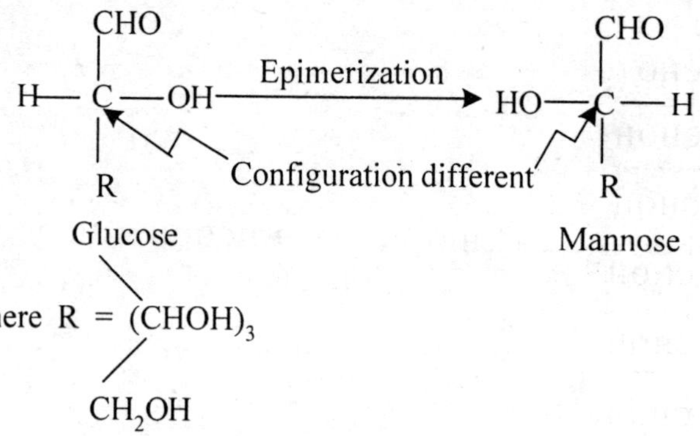

Distinction between glucose and fructose

(i) Selivenoff's Test:

{Fructose + 2 ml of a solution of resorcinol in hydrochloric acid} ⟶ Red color

Glucose gives negative test.

(ii) Pinoff's Test

{Fructose + Freshly prepared solution of ammonium molybdate + glacial acetic acid} ⟶ Greenish blue color

Glucose does not give this test

13.4 DISACCHARIDES

They have general formula $C_{12}H_{22}O_{11}$ and upon hydrolysis yield 2 monosaccharides, e.g.,

***(a)* Sucrose**

$$\underset{\substack{\text{Sucrose of}\\ \text{cane sugar}\\ [\alpha]_D = 66.5^\circ}}{C_{12}H_{22}O_{11}} \xrightarrow{H_3O^+} \underset{\substack{\text{glucose}\\ [\alpha]_D = +52.7^\circ}}{C_6H_{12}O_6} + \underset{\substack{\text{fructose}\\ [\alpha]_D = -92.4^\circ}}{C_6H_{12}O_6}$$

It does not reduce Fehling's solution and Tollen's reagent, hence it is a non-reducing sugar. Moreover, it does not react with reagents like HCN, NH_2OH and phenyl hydrazine and does not show mutarotation.

(b) **Lactose**

$$\underset{\substack{[\alpha]_D = +55.5^\circ\\ \text{Lactose}}}{C_{12}H_{22}O_{11}} \xrightarrow{H_3O^+} \underset{(+)\text{ glucose}}{C_6H_{12}O_6} + \underset{(+)\text{ galactose}}{C_6H_{12}O_6}$$

It reduces Fehling's solution and Tollen's reagent.

(c) **Maltose**

$$\underset{\substack{[\alpha]_D = +136^\circ\\ \text{Maltose}}}{C_{12}H_{22}O_{11}} \xrightarrow{H_3O^+} \underset{\text{glucose}}{2C_6H_{12}O_6}$$

It reduces Fehling's solution and Tollen's reagent.

13.5 POLYSACCHARIDES

Polysaccharides have a general formula $(C_6H_{10}O_5)_n$ (where n is > 10). These are natural polymers, e.g., starch and cellulose.

(a) *Starch* *(i)* When it is hydrolyzed, final product is glucose

$$\underset{\text{Starch}}{2\,(C_6H_{10}O_5)_n} + nH_2O \xrightarrow{\text{Diastage}} \underset{\text{Maltose}}{nC_{12}H_{22}O_{11}}$$

$$\underset{\text{Maltose}}{nC_{12}H_{22}O_{11}} \xrightarrow{H_2O} \underset{\text{Glucose}}{nC_6H_{12}O_6}$$

(ii) It gives blue color with iodine because of formation of a complex in which iodine molecule is trapped.

(iii) Structures of starch can be briefly outlined as below:

Starch →

- *Amylose:* 20% linear, water-soluble fraction, linear part containing α = D=glycopyranose units linked by 1, 4=α=glycosidic linkage
- *Amylopectin:* 80% water=insoluble fraction, branched, containing α-D-glycopyranose units

(b) *Cellulose* *(i)* When boiled with Dil. H_2SO_4, final product is D (+)-glucose

$$\underset{\text{Starch}}{(C_6H_{10}O_5)_n} + nH_2O \longrightarrow \underset{\text{glucose}}{nC_6H_{12}O_6}$$

(ii) On hydrolysis, it yields cellobiose and finally β-d-glucopyranose units linked by 1,4- β-linkages.

PROBLEMS

1. Amongst following answers only one is correct. Write the correct answer:

(i) Consider following reaction:

$A + 4HIO_4 \rightarrow 3HCOOH + HCHO + CHO—COOH$

A is

(a) glucose *(b)* gluconic acid *(c)* sorbitol *(d)* fructose

(ii) When glucose reacts with CH_3OH in the presence of HCl, the number of products is:

(a) 1 *(b)* 2 *(c)* 3 *(d)* 4

(iii) Mutarotation of glucose is observed in:

(a) *P*-cresol *(b)* pyridine *(c)* Mixture of *p*-cresol and pyridine

(d) None of the above

(iv) Final product upon hydrolysis of starch is:

(a) Glucose *(b)* fructose *(c)* amylose *(d)* amylopectin

(v) Which of the following in non-reducing sugar?

(a) Glucose *(b)* fructose *(c)* lactose *(d)* sucrose

(vi) Monosaccharides are sweet in taste due to:

(a) Intermolecular hydrogen bonding

(b) Intramolecular hydrogen bonding

(c) van der Waal's forces

(d) Non-bonded repulsions

(vii) For osazone formation, the effective structural unit that is necessary is

(a)

CH_2OCH_3
|
CO
|

(b) 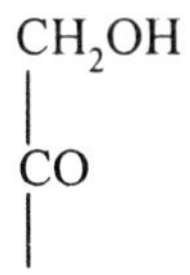

(c) 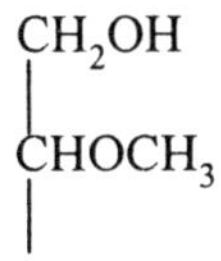

(d) 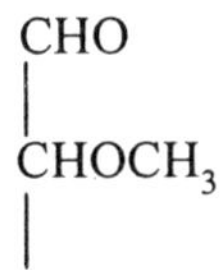

2. State whether true or False:

(i) Bromine water oxidizes both glucose and fructose.

(ii) Fructose has two chiral centers

(iii) Fructose reduces Tollen's reagent **(GATE, 1994)**

3. Identify following glucose derivative

$$A + 3HIO_4 \rightarrow 2HCOOH + 2OHC—COOH.$$

4. Fill in the blanks:

(i) Both glucose and ....on reaction with phenyl hydrazine give the same osazone. **(GATE, 1993)**

(ii) Glucose and mannose are...........

(iii) Glucose does not react with ....

(iv) Maltose and lactose are..........

(v) General formula of polysaccharides is ....

ANSWERS

1. *(i)* (**b**), 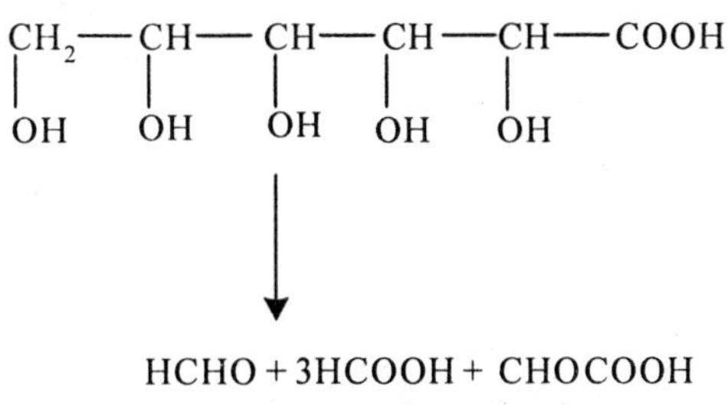

(ii) (**b**) it is because glucose exists in two anomeric forms

(iii) (**c**), it is because solvent should be amphoteric: phenol is acidic and pyridine is basic

(iv) (**a**)

(v) **(d)**
(vi) **(b)**
(vii) **(b)**

2. *(i)* *False.* Bromine water oxidized glucose
(ii) *False.* Fructose has three chiral centers
(iii) *True*

3.

COOH
|
$(CHOH)_4$
|
COOH

Gluconic acid

4. *(i)* Fructose *(ii)* epimers *(iii)* $NaHSO_3$ *(iv)* disaccharides (v) $(C_5H_{10}O_5)_n$

Chapter 14

Biomolecules

The chemistry of biomolecules is largely the chemistry of polyfunctional organic compounds. It includes nucleic acids, amino acids and proteins.

14.1 NUCLEIC ACIDS

Nucleic acids are colorless compounds containing the elements carbon, hydrogen, nitrogen, oxygen and phosphorus. These are natural polymers. The structure of nucleic acids, in brief, can be discussed as below:

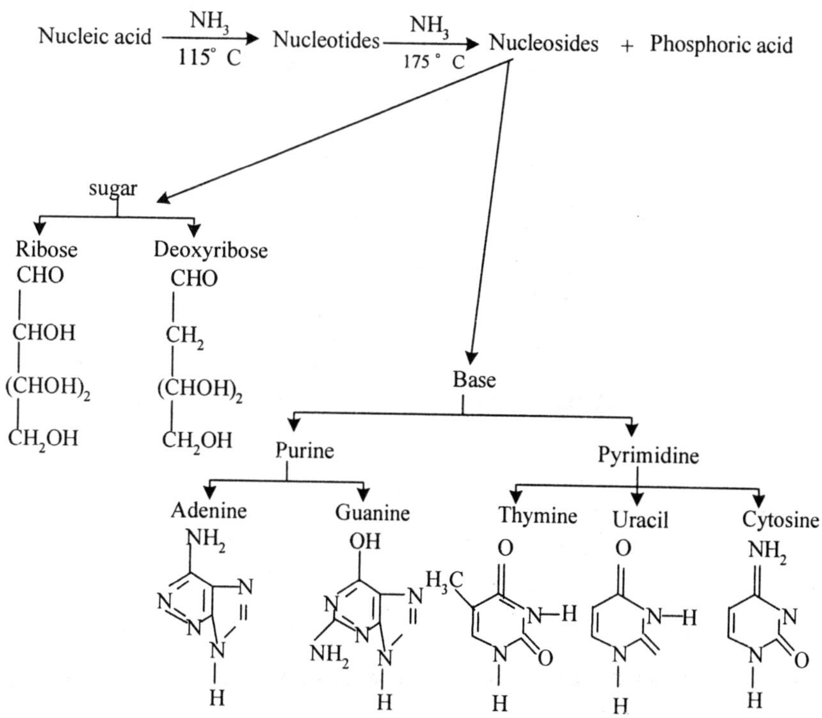

The differences between RNA and DNA are as follows:

RNA (*Ribose nucleic acid*)	*DNA* (*Deoxyribose nucleic acid*)
1. It occurs in cytoplasm	1. It is found inside the nucleus
2. It exists either as single helix or as a double helix	2. It always exist as double helix
3. Sugar is of ribose type	3. Sugar is of deoxyribose type
4. It helps in protein synthesis	4. It carries genetic continuity
5. Bases are uracil, cytosine, adenine, guanine	5. Bases are thymine, cytosine, adenine, guanine

14.2 AMINO ACIDS

Twenty amino acids can be classified on the basis of their R groups:[1]

(i) Amino acids with non-polar 'R' groups: R groups are hydrocarbons in nature and hence hydrophobic. These are eight, e.g., alanine, valine, leucine, isoleucine, proline, phenylanine and tryptophan.

(ii) Amino acids with uncharged polar 'R' groups: The R groups of these amino acids are more hydrophilic than those with non-polar 'R' groups. They contain functional groups that form hydrogen bonds with water hence soluble in water. These are seven, e.g., glycine, serine, cysteine, threonine, asparagines, tytrosine, glutamine.

(iii) Amino acids with negatively-charged (acidic) R groups: The two amino acids namely aspartic acid and glutamic acids, have a net negative charge at pH 7.0, each with a second carboxyl group. The isoelectric point is < 7.

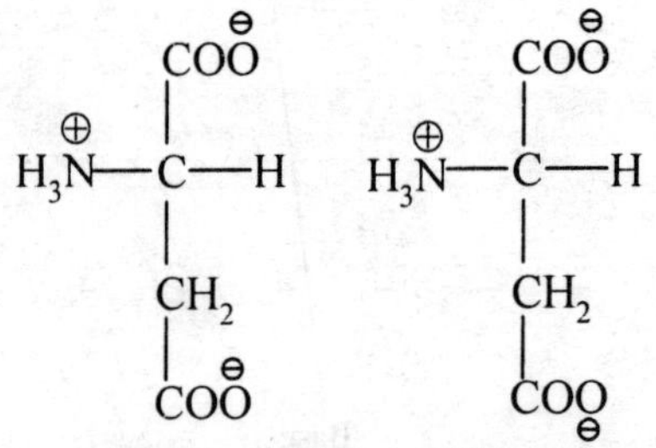

Aspartic acid Glutamic acid

[1] These are the carboxylic acids with an amino group at α-position. These have general formula $(R - \underset{\substack{| \\ NH_2}}{CH}COOH)$.

All except glycine $(R - \underset{\substack{| \\ NH_2}}{CH}COOH)$ have a center of asymmetry at the α-position and belong to the L-series.

(iv) *Amino acids with positively-charged (basic) R groups:* The three amino aoids in which the R groups have a net positive charge at pH 7.0, each with a second amino group. The isoelectric point > 7, e.g., lysine, arginine and histidine.

Preparation of amino acids: α-amino acids can be prepared by following methods:

(i) By reaction of an α-halogen substituted acid with conc.NH_3.

$$\underset{\text{2-chloropropionic acid}}{CH_3-\underset{\displaystyle Cl}{\underset{|}{C}}HCOOH} + 2NH_3 \longrightarrow \underset{\text{Alanine}}{CH_3-\underset{\displaystyle NH_2}{\underset{|}{CH}}-COOH} + NH_4Cl$$

(ii) *Strecker's synthesis:* It consists of reaction of cyanohydrin with conc. NH_3, followed by hydrolysis:

$$RCH_2CH\begin{matrix}CN\\OH\end{matrix} + HNH_2 \longrightarrow RCH_2CH\begin{matrix}CN\\NH_2\end{matrix} + H_2O$$

$$RCH_2CH\begin{matrix}CN\\NH_2\end{matrix} + 2H_2O \longrightarrow RCH_2CH\begin{matrix}COOH\\NH_2\end{matrix} + NH_3$$

(iii) *Gabriel's phthalimide synthesis:* In this method, the α-halogen substituted acids are reacted with potassium phthalimide and followed by hydrolysis of the product:

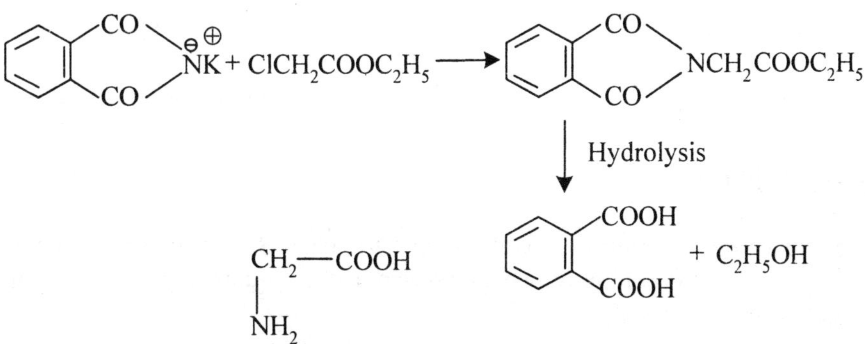

Glycine

Properties

(a) *Physical:* The amino acids are crystalline solids with high melting point and dipole moment. They are soluble in polar solvents like water. They exist as dipolar species called *zwitterions*.

$$R-\underset{\displaystyle \overset{\oplus}{N}H_3}{\underset{|}{CH}}-CO\overset{\ominus}{O} \rightleftharpoons R-\underset{\displaystyle NH_2}{\underset{|}{CH}}-COOH$$

(b) Chemical Properties:

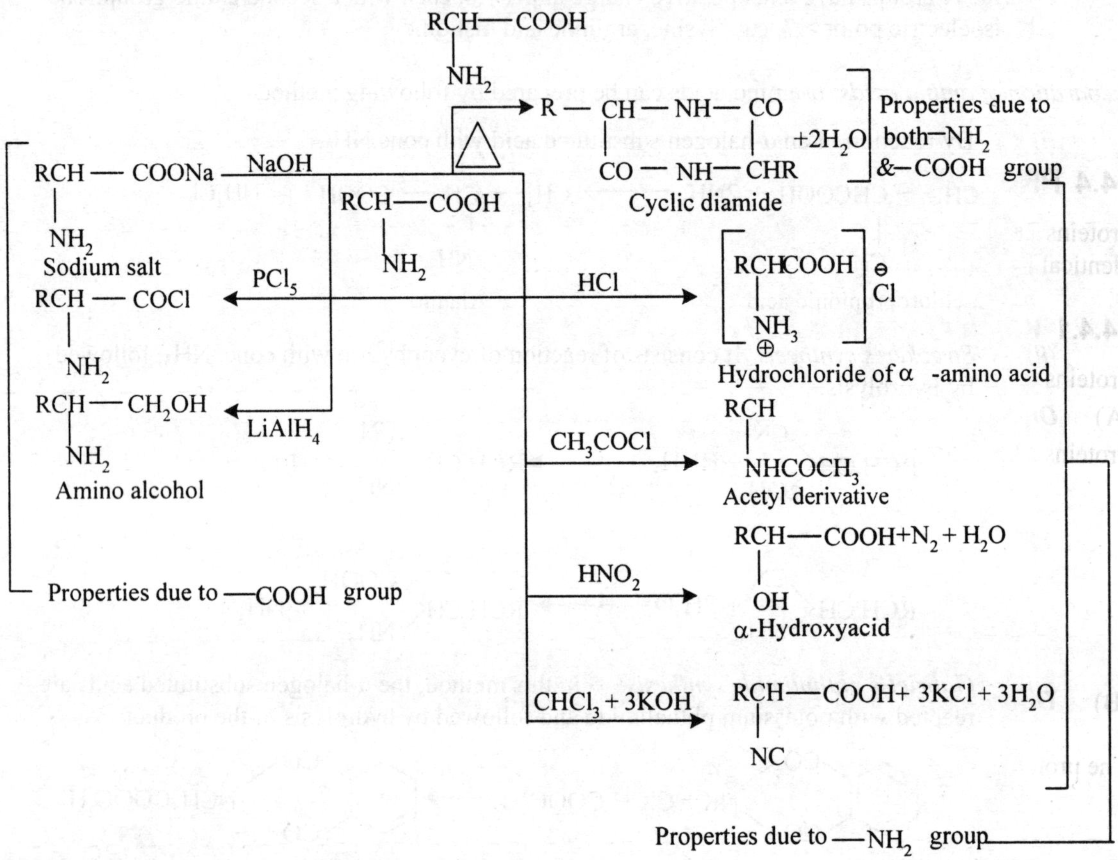

14.3 PEPTIDES

Peptides are the organic molecules with —CONH-linkage. Such a linkage is formed by removal of the water from the carboxyl group of one amino acid and the α-amino group of the other by the action of strong condensing agents, e.g.,

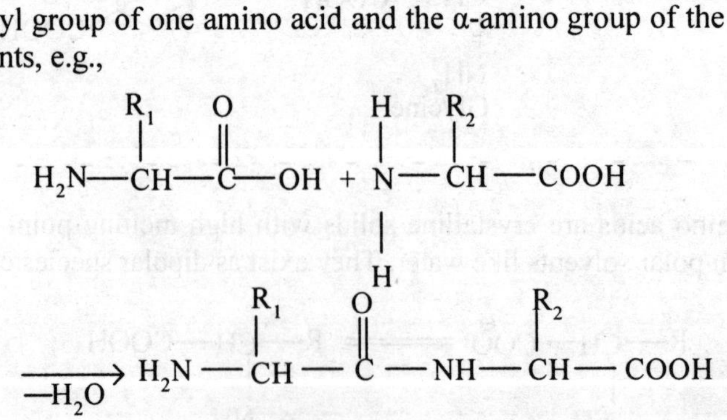

The peptides can be separated on the basis of their ionization behavior. The peptides have characteristic chemical reactions, e.g., peptides can be hydrolyzed by boiling with either strong acid or base to yield the constituent amino acids in free from:

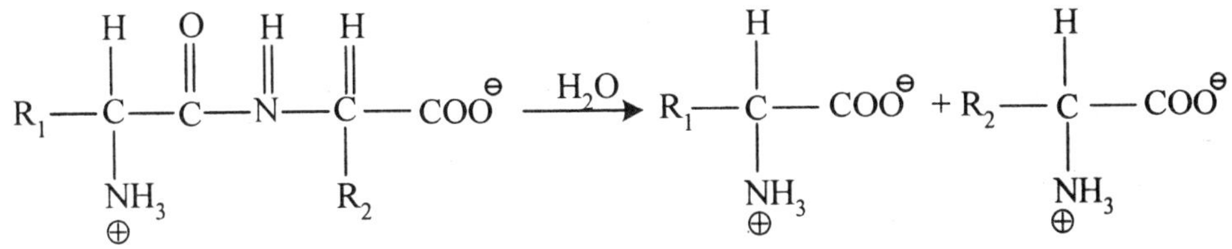

14.4 PROTEINS

Proteins are natural polymers of α-amino acids. All the molecules of any given type of proteins are identical in amino acid composition, sequence, and length of the polypeptide chain.

14.4.1 Kinds of Proteins

Proteins can be classified into following categories:

(A) ***On the basis of solubility and physical properties:***

Proteins are of three types:

- *(a)* *Simple proteins:* These upon hydrolysis yield only amino acids. For example, Albumins.
- *(b)* *Conjugated proteins:* These also contain a non-protein part called *prosthetic group*. For example, Nucleoproteins.
- *(c)* *Derived proteins:* The degradation products obtained at various stages during hydrolysis of proteins are called *derived proteins.* For example, Polypeptides.

(B) ***On the basis of shapes and functions:***

The proteins have been classified in two categories:

- *(a)* *Globular proteins:* These are spherical in shape and water soluble. For example: Enzymes.
- *(b)* *Fibrous proteins:* These are thread like and water insoluble. For example: Collagens.

The bonding in protein may be peptide, hydrogen, disulphide and hydrophobic bonds.

14.4.2 Characteristics of Proteins

- *(i)* Proteins can be separated and purified by dialysis, gel filtration, and electrophoresis.
- *(ii)* Proteins undergo a structural change called *denaturation*.

14.4.3 Tests for Proteins

The various tests are as follows:

- *(i)* *Biuret test:* To 2 ml of test solution and an equal volume of 10% NaOH and one drop of 10% $CuSO_4$ solution → a violet color formation indicates the presence of peptide linkage.
- *(ii)* *Ninhydrin test:* To 1 ml Ninhydrin solution add 1 ml protein solution and heat → formation of violet colour indicates the presence of α-amino acids.

(iii) *Millon's test:* When Millon's reagent is added to a protein solution, a white precipitate, which turns red on heating, may be formed. Millon's reagent is a solution of mercurous and mercuric nitrate in nitric acid.

14.4.4. Structure of Proteins

The arrangement of α-amino acids and disulphide linkages constitute the covalent or primary structure of proteins. The secondary structure of proteins is defined by the local conformation of the polypeptide backbone. The tertiary structure of a protein is its three-dimensional shape that arises from further foldings of its polypeptide chains, foldings superimposed on the coils of the helixes.

14.5 ENZYMES

Enzymes are the functional units of cell metabolism. These are highly specialized proteins, those with catalystic activity, e.g., invertase, amylase. Enzymes have following characteristics:

(i) Enzymes show all the properties of proteins.

(ii) Enzymes enhance the rate of chemical reactions by lowering their activation energy.

(iii) Enzymes can be classified on the basis of the reactions they catalyze.

(iv) There is a quantitative relationship between the substrate concentration and the rate of an enzymatic reaction. The relationship is in the form of Michaelis-Menten equation:

$$V_0 = \frac{V_{max}[S]}{K_m + [S]}$$

Where, V_0 = Initial rate substrate concentration $[S]$

V_{max} = Maximum rate

K_m = Michaelis-Menten constant of enzyme for particular substrate.

(v) Many enzymes catalyze reactions in which there are two substrates.

(vi) Enzymes have an optimum pH.

(vii) Enzymes can be assayed quantitatively.

(viii) Enzymes can be inhibited by specific chemical agents.

(ix) Enzymes show specificity toward their substrates.

Coenzyme

A complexed organic molecule needed for the action of certain enzymes; often contains a vitamin as a component, e.g., thiamine pyrophosphate, flavin, adenine dinucleotide, and some vitamins.

Cofactor

A small molecular weight, heat-stable inorganic or organic substance needed for the action of an enzyme, e.g., Fe^{2+}, Mn^{+2}, Zn^{+2} ions.

Glycolysis

It is a form of fermentation in which glucose is broken down into two molecules of pyruvate. Glycolysis can be classified in two categories:

(i) *Anaerobic glycolysis:* The glycolysis in the absence of O_2 is known as anaerobic glycolysis. It may be summarized as below:

$$\text{Glucose} + 2\text{Pi} + 2\text{ADP} \longrightarrow 2\ \overset{\ominus}{\text{lactate}} + 2\overset{\oplus}{\text{H}} + 2\text{ATP} + 2H_2O$$

(ii) *Aerobic glycosis:* The glycolysis in the presence of O_2 is called aerobic glycolysis. It may be summarized as below:

$$\text{Glucose} + 2\text{Pi} + 2\text{ADP} + 2\overset{\oplus}{\text{NAD}} \longrightarrow 2\text{Pyruvate} + 2\text{ATP} + 2\text{NADH} + 2\overset{\oplus}{\text{H}} + 2H_2O$$

Alcoholic fermentation

It is a method by which carbohydrates break in the presence of enzyme to alcohols, e.g.,

$$\underset{\text{Glucose}}{C_6H_{12}O_6} \xrightarrow[\text{(Yeast)}]{\text{Zymase}} \underset{\text{Ethanol}}{2C_2H_5OH} + 2CO_2$$

$$\underset{\text{Starch}}{(C_6H_{10}O_5)_n} \xrightarrow{\text{B-maceranus}} \underset{\text{Butanol-1}}{CH_3CH_2CH_2CH_2OH} + \underset{\text{Propanone}}{CH_3COCH_3}$$

14.6 GLYCERIDES

The acid may contain 3 to 8 carbon atoms. Hydrolysis yield glycerol and sodium salt of fatty acids called soap.

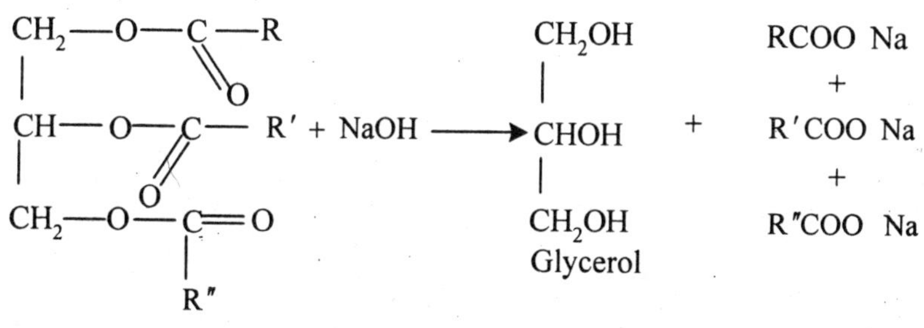

PROBLEMS

1. Tell what class of compounds the following natural polymers belong to and give structure of most likely monomer.
   (i) Protein (ii) Nucleic acid (iii) Starch (iv) Cellulose
2. State whether *True* or False:
   (i) Both fibrous and globular proteins are water-soluble
   (ii) Proteins have only one kind of structure

3. Fill in the blanks:

   (i) Irreversible precipitation of proteins is called ………….. and it is caused by………..

   (ii) Coagulation of egg white by heat is an example of…………

   (iii) Protein mixture can be separated and analyzed by ………………

   (iv) Conjugated protein contain ……………

4. State differences between RNA & DNA

ANSWERS

1. *(i)* Condensation polymers, α amino acids.

   (ii) Condensation polymers, nucleotides.

   (iii) Condensation polymers, α D (+) glucose.

   (iv) Condensation polymers, β (D) (+) glucose.

2. *(i)* *False.* Fibrous proteins are insoluble in water whereas globular proteins are water soluble.

   (ii) *False.* Proteins have two kinds of structure: Primary and Secondary.

3. *(i)* denaturation heat, strong acids or bases

   (ii) denaturation

   (iii) electrophoresis

   (iv) a prosthetic group with a non-peptide protein

Chapter 15

Identification of Organic Compounds

15.1 INTRODUCTION

You have to work like a postman for the identification of an organic compound as shown below:

Name	Specific reaction to identify compound
House no.	Reaction showing possible compound
Street	Class of compound
City	Determination of molecular formula
Postman	Student

15.2 DETERMINATION OF MOLECULAR FORMULA

It is, therefore, suggested that you should examine the molecular formula of the compound to identify the class of compound. Now write all possible structures for the reaction given for the compound. Every question contains one catch point. This has been discussed in the problems.

The steps involved in determination of the molecular formula are given below:

(i) Determination of Percentage Composition

(a) Percentage of carbon =

$$\frac{3 \times \text{Weight of } CO_2 \text{ (in gm)}}{11 \times \text{Weight of organic compound (in gm)}} \times 100$$

(b) Percentage of hydrogen =

$$\frac{1 \times \text{Weight of } H_2O \text{ (in g)}}{9 \times \text{Weight of organic compound (in g)}} \times 100$$

(c) Percentage of nitrogen by Duma's method

$$\frac{7 \times \text{Vol. of } N_2 \text{ (in ml) at NTP}}{5600 \times \text{Weight of organic comound (in gm)}} \times 100$$

(d) Percentage of nitrogen by Kjeldahl's method = $\frac{1.4\,NV}{W}$

Where,
N = Normality of H_2SO_4 in which ammonia was passed;
V = Volume of H_2SO_4 neutralized by NH_3; and
W = Weight of organic compound (in g).

(e) Percentage of sulphur =

$$\frac{32 \times \text{Weight of } BaSO_4 \text{ (in gm)}}{233.4 \times \text{Weight of organic compound (in gm)}} \times 100$$

(f) Percentage of chlorine =

$$\frac{35.5 \times \text{Weight of AgCl (in g)}}{143.5 \times \text{Weight of organic compound (in gm)}} \times 100$$

(g) Percentage of bromine =

$$\frac{80 \times \text{Weight of AgBr (in gm)}}{188 \times \text{Weight of organic compound (mg)}} \times 100$$

(h) Percentage of iodine =

$$\frac{127 \times \text{Weight of AgI (in gm)}}{235 \times \text{Weight of organic compound (in gm)}} \times 100$$

(i) Percentage of oxygen = 100 – (Sum of the percentages of rest of the elements)

(ii) Determination of Empirical Formula

Use the following table:

Percentage of element	*Percentage of element/ Atomic weight of element*	*Each ratio/ Smallest ratio*
.	.	.
.	.	.
.	.	.

(iii) Determination of molecular weight

(a) Molecular weight of acid = Equivalent weight of acid × Basicity.

(b) Molecular weight of base = Equivalent weight of base × Acidity

(c) Molecular weight of a volatile substance = Vapour density × 2

(d) Molecular weight of non-volatile, organic compound

$$= \frac{w \times K}{W \times \Delta T} \times 1000$$

where,

w = Weight of organic compound (in gm)

W = Weight of solvent (in gm)

ΔT = Elevation in boiling point with K as molal elevation constant.

Or

Depression in freezing point with K as molal depression constant.

(e) For molecular weight of non-volatile organic compound:

$$\Delta T = K \times \text{molality}$$

If ΔT is elevation in boiling point K is molal elevation constant and

If ΔT is depression in freezing point then K is molal depression constant.

(f) Equivalent weight of an acid by silver-salt method

$$= \left(\frac{\text{Weight of Ag} - \text{Salt heated} \times 108}{\text{Weight of silver}}\right) - 108 + 1$$

(g) Equivalent weight of a base

$$= \left(\frac{\text{Equivalent weight of choloroplatinate} \quad \text{Molecular weight of } H_2PtCl_6}{2}\right)$$

$$= \frac{\text{Equivalent weight of choloroplatinate} \times 195}{2} - 410$$

$$= \frac{\text{Weight of Platinate obtained}}{2}$$

(iv) Determination of molecular formula of an organic compound

$$= \frac{(\text{Emprical formula} \times \text{Molecular weight})}{\text{Weight of empirical formula}}$$

15.2.1 Detection of Class (functional group of the organic compound)

Examine molecular formula of the compound and proceed in the following manner:

(A) *If the compound contains carbon and hydrogen only:* The compound may be alkane, alkene and alkyne. These can be distinguished with the help of following tests:

S.no.	Tests	Alkane	Alkene	Alkyne
1.	General	C_nH_{2n+2}	C_nH_{2n}	C_nH_{2n-2}
2.	Effect of bromine-water/Baeyer's Reagent	No effect	It decolorized both, bromine water as well as Baeyer's Reagent	It decolorizes both, bromine water as well as Baeyer's Reagent
3.	Ammoniacal/Cu_2Cl_2	No precipitate is obtained	No precipitate is obtained	White precipitate with ammonical $AgNO_3$ and brown precipitate with ammonical Cu_2Cl_2 is obtained in the case of [alkyne-1].
4.	Ozonolysis	No reaction	It undergoes ozonolysis forming to (i) Two or more carbonyl compounds (aldehydes or ketones) in case of (ii) only one carbonyl compounds containing —CHO groups at both ends in case of cycloalkenes like 	It undergoes ozonolysis forming only one carbonyl compound (glyoxal or substituted glyoxal)

(B) *If the compound contains carbon, hydrogen and one halogen atom:* The compound is alkyl halide. Generally, dehydrohalogenation reaction is given in the problem. Use Saytzeff rule, if needed.

(C) *If the compound contains carbon, hydrogen and one oxygen atom:* The compound may be alcohol, ether, aldehyde or ketone. These can be distinguished with the help of following:

S no	Tests	Alcohol	Ether	Aldehyde	Ketone
1.	Reaction with Na	It reacts and forms hydrogen	It does not react	It does not react	It does not react
2.	Reaction with HI	It reacts	It react	It does not react	It does not react
3.	Iodoform test	It gives +ve iodoform test if it has structure $R—\underset{OH}{\overset{H}{C}}—CH_3$ R═H, alkyl or aryl. Iodoform is yellow solid with melting point 119 °C	It does not give test	All Aldehydes except acetaldehyde give negative test	Only methyl ketones give positive test

4.	Reaction with sodium bisulphite	No reaction occurs	No reaction occurs	It forms white precipitate	It forms white precipitate
5.	Fehling solution	No reaction occurs	No reaction occurs	All aldehydes except benzaldehyde form brown precipitate	All ketones except hydroxykenotes do not reduce Fehling Solution.
6.	Tollen's Reagent	No reaction occurs	No reaction occurs	All aldehydes reduce Fehling Solution	Ketones do not reduce
7.	Schiff's Reagent	No reaction occurs	No reaction occurs	It stores pink color	No reaction occurs

(D) *If the compound contains carbon, hydrogen and two oxygen atoms:* The compound may be carboxylic acid or ester. The two compounds can be distinguished with the help of following test:

S.no	*Tests*	*Carboxylic acid*	*Ester*
1.	Effect of sodium bicarbonate	It gives effervescence	No reaction occurs
2.	Reaction with Phosphorus and Iodine	It undergoes Hell-Volhard–Zelinsky reaction if it contains alphahydrogen atom	No reaction occurs

(E) *If the compound contains carbon, hydrogen and nitrogen atom:* The compound may be amine, cyanide or isocyanide. These can be distinguished with the help of following tests:

S.no.	*Tests*	*Amine*	*Cyanide*	*Isocyanide*
1.	Nature of the compound	It is basic	It is neutral	It is neutral
2.	Smell	It has unpleasant odour	It has no characteristic smell	Characteristic smell
3.	Reaction Chloroform & sodium hydroxide (Carbylamine reaction)	Primary amines forms isocyanide having unpleasant odour	No reaction occurs	No reaction occurs
4.	Reaction with acetyl chloride or acetic anhydride in acetic acid	It reacts	No reaction occurs	No reaction occurs
5.	Reaction with nitrous acid	It reacts	No reaction occurs	No reaction occurs
6.	Hydrolysis	No reaction occurs	It undergoes hydrolysis forming carboxylic acid and ammonia	It undergoes hydrolysis forming primary amine and formic acid
7.	Reaction	No reaction occurs	It is reduced to primary amine	It is reduced to secondary amine

(F) *If the compound contains carbon, nitrogen and one oxygen:* It may be an amide ($RCONH_2$). It shows following tests:

(i) It forms ammonia when heated with sodium hydroxide.

(ii) It undergoes Hoffmann bromide reaction, i.e., reacts with bromine and potassium hydroxide to form primary amines.

(G) *If the compound contains carbon, hydrogen and two oxygen atoms:* It may be a nitro compound or nitrite. The compound may be distinguished by reduction. The nitro compound is reduced to primary amine whereas the nitrite compound is not reduced.

15.2.2. The Index of Hydrogen Deficiency (I.H.D.)

The concept of the index of hydrogen deficiency (I.H.D.) has been recently introduced to distinguish compounds which have same general formula. e.g., cycloalkanes and alkenes have same general formula, i.e., C_nH_{2n}. Likewise alkynes & alkdienes have same general formula C_nH_{2n-2}.

The I.H.D. may be defined as the number of pairs of hydrogen atoms that must be subtracted from the molecular formula of the corresponding alkanes to give the molecular formula of the compound under consideration, e.g., two compounds namely cyclohexane or hexane (1 or 2 or 3) have same formula C_6H_{12}. The corresponding alkane is hexane.

formula of corresponding alkane (hexane) = C_6H_{14}

formula of compound (cyclohexane or hexane) under consideration = C_6H_{12} difference of Hydrogen = 2 = 1 pair of hydrogen

Therefore, I.H.D. of the compound is 1.

The I.H.D. as 1 indicates, it may be cyclohexane or hexane. The two can further be distinguished by the fact that hexane consumes 1 molar equivalent of hydrogen whereas cyclohexane does not undergo addition reactions. This is shown in the following table:

[1]I.H.D.	Possible compounds
1	Cycloalkane or alkene
2	Alkyne or cycloalkene or alkdiene
3	Alkyne and cycloalkane
4	Cycloalkane and alkdiene

[1]Kindly see problem 64.

The rules for the calculating I.H.D. for compounds other than hydrocarbons are as follows:

(i) For alkyl halides are simply count the halogen atom as if they were hydrogen atoms, e.g., $C_5H_8Cl_2$. We consider it as C_5H_{10}. The corresponding alkane is C_5H_{12}. Hence, I.H.D. is one. It means the compound contains either one double bond or one ring. Further if compound undergoes, addition reaction, it's alkene otherwise it has a ring.

(ii) In case of compounds containing oxygen. We should ignore oxygen and calculate the I.H.D. from the remaining part of the formula, e.g., $C_5H_{10}O$, the I.H.D. is 1. Therefore, following compounds are possible.

$CH_3CH{=}CH{-}CH_2{-}CH_2OH$; $CH_2{=}CHCH_2CH_2{-}CH_2OH$

$CH_3{-}CH_2{-}CH{=}CH{-}CH_2OH$, 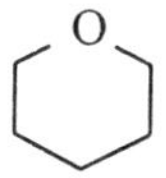 , $CH_3CH_2CH_2{-}\overset{\overset{O}{\|}}{C}{-}CH_3$

(iii) In case compound contains nitrogen, we should substract one hydrogen for each nitrogen atom and than ignore the nitrogen atoms, e.g., for compound C_4H_9N, we treat it as C_4H_8 hence I.H.D. is = 1. The possible compounds are as follows:

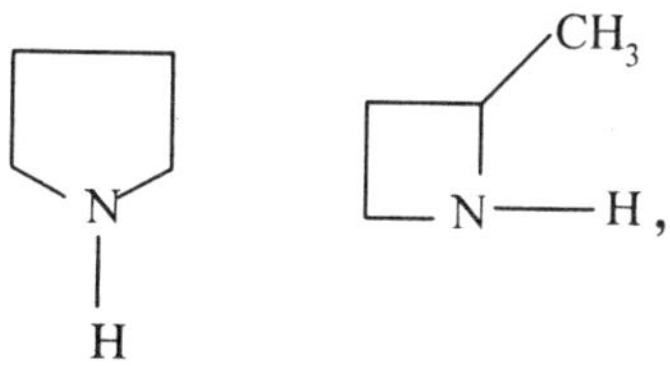 , $CH_3{-}CH_2{-}\underset{\underset{NH}{\|}}{C}{-}CH_3$

PROBLEMS

1. An organic compound 'A' containing 21.18% of carbon, 4.12% hydrogen and 74.7% of iodine on treatment with caustic potash solution gave a compound which responded to the iodoform test. 'A' on heating with methyl iodide and zinc gives as one of the products, a hydrocarbon B containing 82.8% of carbon, 17.2% of hydrogen and with the molecular weight 58. Identify A and B. What product would you get on reduction of A?

2. An organic compounds A containing 85.7% of carbon and 14.3% of hydrogen had vapour density 21. It gave the following reactions:
   (i) With HI, it gave a product B containing 74.7% of iodine.
   (ii) With bromine, it gave a compound C having 79.2% of bromine.
   (iii) A on treatment with alkaline permanganate gave a product D which was also obtained on treating C with dil. aq. caustic potash. Identify A,B,C and D?

3. An aromatic hydrocarbon A containing 93.75% of carbon and rest hydrogen on oxidation gave a dibasic acid which formed an anhydride on heating. What is A?

4. Each of the two compounds A and B contains 88.82% carbon and 11.18% hydrogen. Both the compounds decolorize bromine in CCl_4 and react slowly with conc. H_2SO_4. Oxidation of A gives CO_2 and propionic acid. Compound B fails to give a precipitate with ammonical silver oxide and it yields CO_2 and oxalic acid when oxidized. Identify A and B.

5. Bromination of an organic compound A gives a compound of formula $C_7H_{14}Br_2$. Upon reducing this, 3-methyl hexane is obtained. On ozonolysis it gives a mixture of propanal and 2-butanone. Identify A and write the reactions?

6. 1 g of an organic compound A on burning with oxygen gave 3.197 g of CO_2 and 1.152 g of water. On catalytic hydrogenation it gave a compound B having C = 84.2% and H = 15.8%.

Ozonolysis of A gave acetic acid, acetone and pyruvic acid ($CH_3COCOOH$). If molecular weight of A is 112, identify it.

7. An unknown organic compound A was analyzed and found to contain one atom of bromine per molecule. When the Grignard reagent, prepared from A was treated with water, ethylbenzene was produced. Identify A?

8. Suggest structures compatible with the following data:
   (i) Compound A, C_8H_6 forms a precipitate with ammoniacal cuprous chloride and yield benzoic acid when oxidized.
   (ii) Compound A, C_9H_{12} undergoes electrophilic substitution with Br_2 to yield $C_9H_{11}Br$.

9. An unknown compound A, C_8H_8O gave a negative Tollen's reagent test but gave appositive test with 2,4-dinitrophenyl hydrazine. Iodoform was produced when reacted with I_2 and NaOH. When A was subjected to Clemmensen's reduction, it produced ethyl benzene, Identify A?

10. An unknown A, $C_5H_{12}O$ was oxidized to B $C_5H_{10}O$ by $K_2Cr_2O_7$. B gave a 2,4-dinitrophenyldrazone. Both A and B gave positive iodoform test. A reacted with not H_2SO_4 to produce C, C_5H_{10} as the major product. C rapidly decolorized a solution of Br_2 in CCl_4. When C was dissolved in cold conc. H_2SO_4 and then poured in water and heated, D was produced. D was an isomer of A. D gave a positive iodoform test, and a +ve test with Lucas reagent in less than 1 minute. C was ozonolyzed to produce aldehyde and acetone. Identify A to D and explain reaction.

11. Identify A to D in the following reactions:

$$CH_3CH_2OH + Na \longrightarrow 1/2H_2 + A$$

$$A \xrightarrow{(CH_3)_2CHBr} B$$

$$C_5H_{10}O \xleftarrow{Me_3COOK} C + HBr \xleftarrow{Br_2} B$$

12. Compounds A, B and C all have the molecular formula $C_7H_{15}Br$. Compound A reacts with aq. NaOH to give a C_7H_{16}, an alcohol which gives +ve Iodoform test and compound B reacts with alc. NaOH to give a single C_7H_{14} alkene, which upon ozonolysis yields methanal as a product. C reacts with alc. NaOH to give a mixture containing two alkenes with molecular formula C_7H_{14}. Identify A, B and C?

13. Compound A and B have the molecular formula $C_8H_{10}O$ when compound A is treated with boiling HI acid, a C_6H_6O compound and C_2H_5I is formed. When compound B is heated with HI, a C_7H_7I compound and methyl iodide is formed. Identify A and B?

14. Identify A to E in the following:

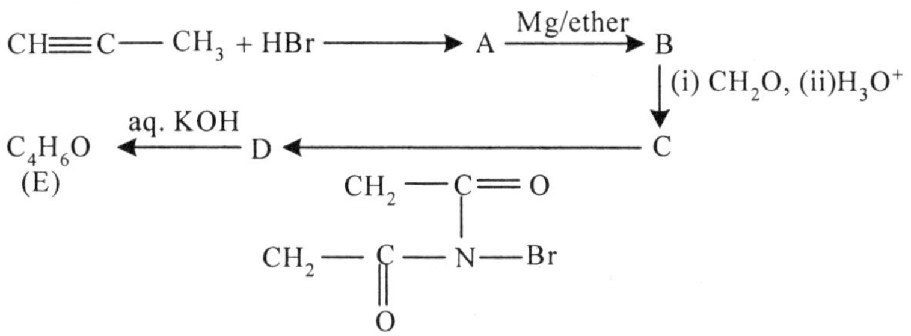

15. An unknown compound A, $C_6H_{14}O$ gave rapid Lucas test. It could not be oxidized easily, but upon treatment with acid, the compound was converted to an alkene B, C_6H_{12}. Ozonolysis of the alkene gave two different carbonyl compounds C and D as products, both of which had the molecular formula C_3H_6O. Identify A to D?

16. An alkene A, C_6H_{12} was found to be optically active. However, when it was hydrogenated the resulting alkane B C_6H_{14} was optically inactive. Identify both?

17. An alkyl bromide A, $C_5H_{11}Br$ reacted rapidly with water to give an alcohol B, $C_5H_{12}O$, but when the A was treated with sodium methoxide in an attempt to convert it to an ether, the only product was an alkene C_5H_{10}. Ozonolysis of this alkene gave two carbonyl compounds, a ketone C_3H_6O and an aldehyde C_2H_4O. Identify all compounds.

18. An aromatic dibromide A, $C_7H_6Br_2$ was found to react readily with aq. NaOH. The product of this reaction however had lost only one bromo group to give the product B, C_7H_7Br. Then the dibromide was converted to a Grignard reagent which was then hydrolyzed to give toluene. What are A and B?

19. An unknown compound A, C_5H_{10}, rapidly decolorized a solution of Br_2 in CCl_4. When A was dissolved in cold conc. H_2SO_4 and heated with water, B resulted. Compound B, $C_5H_{12}O$, reacted with chromic acid to give C, $C_5H_{10}O$. Both B and C gave positive iodoform tests. The iodoform reaction mixture yielded isobutyric acid. What are A, B and C?

20. An alcohol of unknown structure A, $C_4H_{10}O$ gave a positive Lucas test in about 5 minutes. When the alcohol was treated with hot acid, an alkene B, C_4H_8 was obtained. Ozonolysis of B gave only a single, product C, C_2H_4O. What are A, B and C?

21. Identify A to F

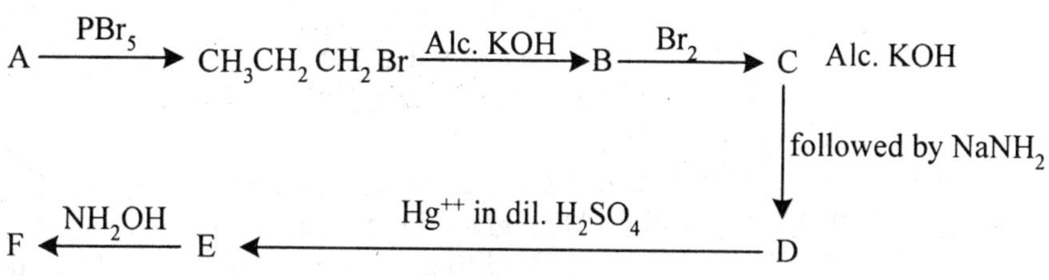

(R, 1990)

22. Identify G, H and I in the following sequence.

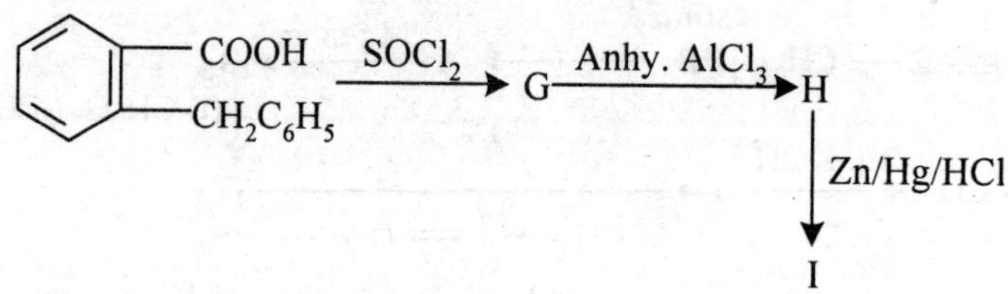

(IIT, 1995)

23. Identify (A) and (B) in the following sequence:

$$CH_3CH_2COOH \xrightarrow{\text{P and } Br_2} A \xrightarrow[\text{(ii) } H^+]{\text{(i) Alc. KOH (excess)}} B$$

(IIT, 1995)

24. Alkenes A and B yield the same alcohol C on hydration. On vigorous oxidation with $KMnO_4$ A gives a carbonyl compound D and an acid E, each containing four carbon atoms. On the other hand, B gives an acid F and a carbonyl compound G. In G, two identical groups are attached to the same carbon atom. Give structures of A and G with proper reasoning? **(R, 1995)**

25. Compounds A, B and C have the same molecular formula C_4H_9NO. All three can be hydrolyzed to bases (NH_3/amines) and carboxylic acid. Compound A gives $C_4H_8O_2$ acid. Compound B gives a $C_3H_6O_2$ acid and compound C gives $C_2H_4O_2$ acid. Identify A, B and C?

26. An aromatic compound A, $C_7H_5NO_2Cl_2$, on reduction with Sn/HCl gives B which on reaction with $NaNO_2$/HCl gives C. Compound B does not form a dye with β-naphthol. However, C gives red color with ceric ammonium nitrate and on oxidation gives an acid D of equivalent weight 191. Decarboxylation of D gives E, which forms a single mononitro derivative F on nitration. Give structures of A to F with proper reasoning? **(R, 1995)**

27. For each of the following descriptions, suggest the amine that satisfies the data:
    (i) An amine dissolved in HNO_2 acid to give only alcohol.
    (ii) An amine reacted with benzene sulphonyl chloride to give a solution of benzene sulphonamide which is insoluble in KOH.
    (iii) It undergoes diazotisation.

28. An acid compound A, $C_4H_8O_3$, loses its optical activity on strong heating yielding B, $C_4H_6O_2$, which reacts readily with $KMnO_4$. B forms a derivative C with $SOCl_2$, which on reaction with $(CH_3)_2NH$ gives D. The compound A on oxidation with dilute chromic acid gives an unstable compound E which decarboxylates readily to give F, C_3H_6O. The compound F gives a hydrocarbon G on treatment with amalgamated Zn and HCl. Gives structures of A to G with proper reasoning. **(R, 1995)**

29. An alkane (A), C_5H_{12}, on chlorination at 300°C gives a mixture of four different monochlorinated derivatives B, C, D and E. Two of these derivatives give the same stable alkene F on dehydrohalogenation. On oxidation with hot alkaline $KMnO_4$, followed by acidification F gives two products G and H. Give structures of A to H with proper reasoning. **(R, 1995)**

30. An organic compound A gives Liebermann reaction and on treatment with $CHCl_3$/KOH followed by hydrolysis gives B and C. Compound B gives color with Schiff's reagent but not C, which is steam volatile. C on treatment with $LiAlH_4$ gives D, $C_7H_8O_2$, which on oxidation gives E. Compound E reacts with $(CH_3CO)_2O/CH_3COOH$ to gives a pain reliever F. Give structures of A to F with proper reasoning? **(R, 1995)**

31. A compound A, $C_9H_7ClO_2$, is insoluble in water but reacts with water to give a compound B, $C_9H_8O_3$. Compound B is stable, is soluble in aqueous $NaHCO_3$ solution and reacts with phenyl hydrazine to form a solid derivative, but fails to react with Fehling solution. Upon strong oxidation B yields compound C, $C_8H_6O_4$ which loses water upon heating and yields and acid anhydride D, $C_8H_4O_3$. Identify A, B and C.

32. An optically-active compound A has the molecular formula C_6H_{10}. The compound gives a precipitate when treated with a solution containing $Ag(NH_3)_2OH$. On catalytic hydrogenation A yields B, (C_6H_{14}), E is optically inactive and cannot be resolved. Identify A and B.

33. Propose structures for compounds G-I.

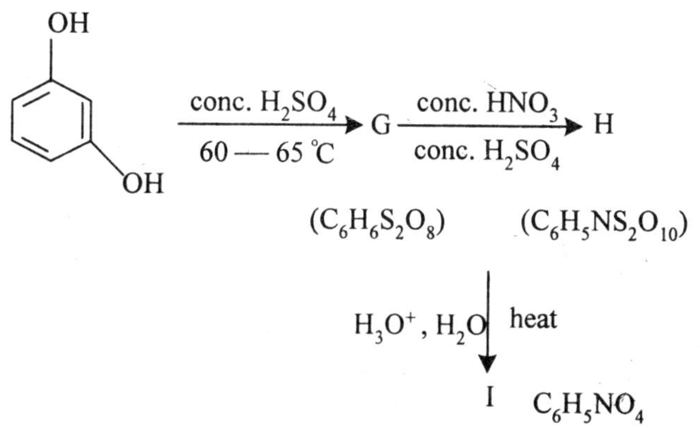

34. Propose structures of compounds A to E.

(i)

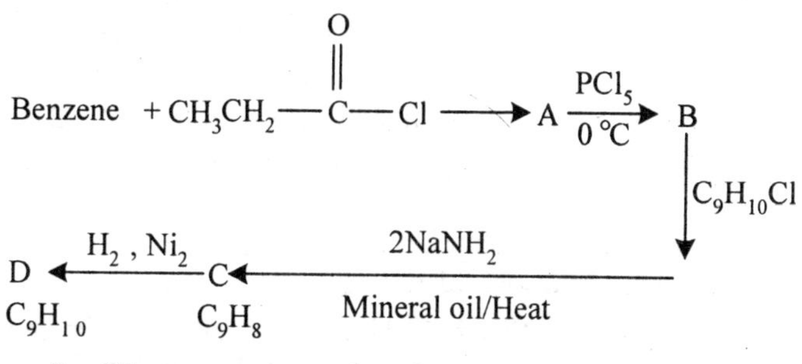

(ii) $\text{D} \xrightarrow[2-5\,^\circ\text{C}]{Br_2,\ CCl_4} \text{E} + \text{enantiomer(major product)}$

35. A colorless oil A has a very characteristic smell. On exposure to air, it slowly forms a colorless solid B which dissolves sparingly in cold water to give an acidic solution. A reduces ammoniacal $AgNO_3$ but not the Fehling's solution and reacts with phenylhydrazine to give a crystalline derivatives. When heated with sodalime, B gives off vapour which burns with a smoky and luminous flame and is 39 times denser than hydrogen under the same condition. Identify A and B?

36. Compound A, C_6H_{10} on successive reduction gave B, C_6H_{12} and C, C_6H_{14}. On ozonolysis followed by hydrolysis A gave two aldehydes D, C_2H_4O and E, $C_2H_2O_2$. The compound B on oxidation with Baeyer's reagent followed by acidification gave and acid F, $C_3H_6O_2$. Identify A to F.

37. Dinitrobenzene A can be reduced with $(NH_4)_2S$ to B, $C_6H_6O_2N_2$, which on reacting with $NaNO_2/HCl$ at 5°C followed by Cu_2Cl_2 gives C, $C_6H_4O_2Cl$. With NaOH, the compound C, in turn, gives D, $C_6H_5O_3N$, which can be steam distilled. Identify the compound A to D?

38. An organic compound A, $C_6H_{12}O$ forms an oxime but does not reduce Tollen's reagent. The compound A on reduction with sodium amalgam forms an alcohol B which on dehydration forms chiefly a single olefin C. The ozonolysis of C produces D and E. The compound D reduces Tollen's reagent but does not respond to iodoform test. The compound E does not reduce Tollen's reagent but does respond to iodoform test. Identify the compounds A to E and explain the reactions.

39. Compound A (C_7H_8) is chlorinated in sunlight to give a product B which on hydrolysis gives C. C shows positive test with Tollen's reagent. C with sodium acetate and acetic anhydride gives D with equivalent weight 148.15. Identify the compounds A to D and explain the reactions?

40. Compound A C_3H_6 reacts with B_2H_6 to give B which reacts with H_2O_2 in presence of OH^- to give C. C reacts with $ZnCl_2$ in HCl to give D (C_3H_7Cl). C has the formula C_3H_8O. B reacts with aq. acid to give E with formula C_3H_8. Identify the compounds A to E.

41. Treatment of C $(CH_2Br)_4$ with Zn yields a hydrocarbon of molecular formula C_5H_8 in which all the hydrogen atoms are equivalent (i.e., they have the same chemical environment). Suggest a structure for C_5H_8 and represent its formation.

42. Two isomeric compounds A and B have molecular formula $C_{11}H_{13}OCl$. Both yield compound C on catalytic hydrogenation and produce 3 ethoxy-4-chlorobenzoic acids on vigorous oxidation. Geometrical isomers D and E, exist in A but not in B. Give structures of the compounds A to E with reasoning.

43. An acid A contains carbon 40.7%, hydrogen 5.1% and its silver salt contains 65.1% silver. The ethyl ester of A has vapour density of 87. What structure A may have? How will you distinguish between the isomers? **(MNR, 1995)**

44. An amine X having molecular formula $C_6H_{15}N$ on treatment with methyl iodide and then caustic potash gives Y, $C_8H_{20}N^+OH^-$. This on heating produces isobutene and an amine Z, $C_4H_{11}N$. Identify X.

45. An organic compound E, C_3H_8 on hydrogenation gives compound F, C_5H_{12}. Compound E on ozonolysis gives HCHO and 2-keto propanal. Identify E. **(IIT, 1995)**

46. An organic compound A contains C=58.53, H=7.31 and N = 34.14%. On reduction it gives a primary amine B which gives ethyl alcohol with nitrous acid. B gives off offensive odour on warming with $CHCl_3$ and KOH and gives compound C which on reduction forms ethylmethyl amine. Assign the structures to A, B and C. Explain the reaction. **(Roorkee, 1980)**

47. An organic compound A contains 69.42% carbon, 5.78% hydrogen and 11.57% nitrogen. Its vapour density is 60.5. It evolves ammonia when boiled with caustic potash. On heating with P_2O_5, it gives a compound B containing 81.5% carbons, 4.85% hydrogen and 13.59% nitrogen. On reduction with sodium and alcohol, B form a base which reacts with HNO_2 giving off nitrogen and yielding an alcohol C. The alcohol can be oxidized to benzoic acid. Explain the above reactions and assign structural formula to A, B, and C. **(Roorkee, 1982)**

48. Hydrocarbon A with molecular formula C_8H_8 gave the following reactions:

(i) On shaking with bromine, a bromoderivative B with the composition $C_8H_8Br_2$ was formed.

(ii) Vigorous oxidation of the hydrocarbon with alkaline $KMnO_4$ gave a monobasic acid C.

(iii) The acid C on distillation with soda lime gave benzene. Deduce structures A, B and C and name the compound A. **(Roorkee, 1983)**

49. An organic compound A contains N = 40% carbon and 6.7% hydrogen. Its vapour density is 16. On reacting with a conc. solution of KOH, it gives two compound B and C. When B is oxidized, the original compound A is obtained. When C is treated with conc. HCl, it gives a compound D which reduces Fehling and ammoniacal $AgNO_3$ solutions and also gives effervescence with sodium bicarbonate solution. Write the structures of A, B, C and D and explain reactions. **(Roorkee, 1983)**

50. An olefin was treated with ozone and the resulting product on hydrolysis gave 2-pentanone and acetaldehyde. What is the structure of the original olefin? **(Roorkee, 1983)**

51. 36.4 g of 1,1,2,2-tetrachloropropane was heated with Zn-dust and the product was bubbled through ammoniacal $AgNO_3$. What is the weight of precipitate obtained. Give equations of the reaction involved. **(Roorkee, 1983)**

52. 0.535 g of mixture of ethanol and acetaldehyde when heated with Fehling solution gave 1.2 g of a red precipitate. What is the percentage of acetaldehyde in the mixture [Cu = 63.5]. **(Roorkee, 1984)**

53. From analysis and molecular weight determination, the molecular formula of a compound A is C_3H_7NO. The compound gives the following information:

(i) On hydrolysis, it gives an amine B and a carboxylic acid C.

(ii) Amine B reacts with benzene sulphonyl chloride and gives product which is insoluble in aq. NaOH.

(iii) An acid C on reaction with Tollen's Reagent gives a silver mirror. What are A, B, and C? Explain the reactions with the help of equations. **(Roorkee, 1984)**

54. An organic compound A, $C_7H_{15}Cl$ on treatment with Al and KOH gives a hydrocarbon B, C_7H_{14}. B on treatment with ozone and subsequent hydrolysis gives acetone and butyraldehyde. What are A and B? Explain the reaction. **(Roorkee, 1984)**

55. 10 g of a mixture of hexane and ethanol reacted with Na to give 200 ml of hydrogen at 27°C and 760 mm pressure. What is the percentage of ethanol in the mixture? **(Roorkee, 1984)**

56. An alkyl halide A, C_4H_9Br reacts with Al/KOH and gives an alkene B which reacts with bromine to give a dibromide C. C is transformed with soda-amide to a gas D which forms a precipitate when passed through an ammoniacal $AgNO_3$ solution. Give the structural formulae of the compounds A to D and explain the reactions involved. **(Roorkee, 1985)**

57. Write the structural formula and IUPAC name for the different alkanes obtained when a mixture of 1-bromopropane and 2-bromopropane is reacted with sodium in the presence of ether. What is the name of the reaction? **(Roorkee, 1985)**

58. Write down the structural formula and IUPAC name of the alkene which gives only pentane-2 on ozonolysis. **(Roorkee, 1985)**

59. Compound A, $C_4H_8Cl_2$ on hydrolysis gives a compound B, C_4H_8O. The compound B reacts with NH_2OH and gives negative test with Tollen's reagent. What are the structural formulae for the compounds A and B? **(R, 1985)**

60. Compound A contains only carbon and hydrogen. It decolorizes bromine in CCl_4 solution and reacts slowly with conc. H_2SO_4. Compound A reacts with HBr to form B. B reacts with NaOH to form C. On oxidation C gives hexanone-3. Identify A to C. **(Roorkee, 1986)**

61. Compound A contains carbon, hydrogen and oxygen and its reacts with sodium with evolution of hydrogen. Controlled oxidation of A gives B which forms a ppt. with $NaHSO_3$ but does not give the silver mirror test. A solution of 7.41 gm of A in 100 g of water freezes at –1.86°C. Identify A and B. **(Roorkee, 1986)**

62. Compound A contains carbon, hydrogen, nitrogen and oxygen. When treated with NaOH and Br_2, it gives compound B. B on reaction with $NaNO_2$ and HCl gives ethanol. What are the structural formulae of A and B?

63. Compound A having empirical formula C_7H_8 is chlorinated in sunlight to give a product B which on hydrolysis gives a compound C. C gives a positive test with Tollen's reagent and with sodium acetate and acetic anhydride gives an acid D with an equivalent weight 148.15. Identify A to D. **(Roorkee, 1986)**

64. A hydrocarbon A, of the formula C_8H_{10}, on ozonolysis gives only the compound B, $C_4H_6O_2$. The compound B can also be obtained from the alkyl bromide C, C_3H_5Br, upon treatment with magnesium in dry ether, followed by treatment with carbon dioxide and acidification. Identify A, B and C and also give equations for the reactions. **(IIT, 1986)**

65. Compound D, $C_8H_{10}O$, upon treatment with alkaline solution of iodine, gives a yellow precipitate. The filtrate on acidification gives a white solid E, $C_7H_6O_2$. Write the structure of D and E, and explain the formation of E. **(IIT, 1996)**

66. An optically-active alcohol A, $C_6H_{10}O$ absorbs two moles of hydrogen per mole of A. Upon hydrogenation, it gives a product B. The compound B is resistant to oxidation by CrO_3 and does not show any optical activity. Deduce the structures of A and B. **(IIT, 1996)**

67. An alkyl halide X, of formula $C_6H_{13}Cl$, on treatment with potassium tertiary butoxide gives two isomeric alkenes Y and Z. C_6H_{12}. Both alkenes on hydrogenation give 2,3-dimethyl butane. Predict the structures of X, Y and Z. **(IIT, 1996)**

68. The percentage of carbon, hydrogen, nitrogen and oxygen in a disubstituted aromatic compound A is 71.11%, 6.67% and 10. 37%, respectively. Its molecule contains a single atom of nitrogen. Compound A evolves NH_3 when heated with NaOH to form a salt B, from which an acid C is obtained on acidification. If an electrophilic substitution is carried out on C, an entering electrophil can occupy ether of the two identical positions available in the ring. Acid C on heating with PCl_5 gives D which reacts with B to give an organic product E. Give structures of A to E with proper reasons.

69. An organic compound A composed of carbon, hydrogen and oxygen and nitrogen characteristic colour with cerric ammonium nitrate, Treatment of A with PCl_5 gives B which reacts with KOH to form C. The reduction of C with warm Na/C_2H_5OH produced D which on heating gives E with evolution of ammonia. Pyridine is obtained on treatment of E with nitrobenzene. Give structures of A to E with proper reasoning. **(Roorkee, 1996)**

70. An organic compound A, C_4H_9Cl on treatment with aq. KOH gives B, and with alcoholic KOH gives C, which is also formed on passing the vapours of B over heated copper. The compound C readily decolorizes bromine water. Ozonolysis of C gives two compounds D and E. Compound D reacts with

NaOH to give an alcohol G and a sodium salt H of an acid, D can also be prepared from propyne on treatment with water in presence of Hg^{2+} and H_2SO_4. Identify A to H with proper reasoning. **(Roorkee, 1996)**

71. Two isomeric mononitro derivatives B and C are obtained by the nitration of an organic compound A, C_7H_8O. Treatment of A with acetyl chloride produces D which on reaction with CrO_2Cl_2, gives E, whose oxidation with neutral $KMnO_4$ followed by acidification gives F. The compound A on heating gives phenol. A on treatment with alkaline $C_6H_5SO_2Cl$ produces G which on oxidation with $KMnO_4$ gives H. Hydrolysis of H also gives F. Give structures of A to H with proper reasoning.

72. An organic compound A, C_5H_9Br which readily decolorizes bromine water and $KMnO_4$ solution, gives B, $C_5H_{11}Br$ on treatment with Sn/HCl. The reaction of A with NaH_2 produces C with evolution of ammonia. C neither reacts with sodium nor forms any metal acetylide but reacts with Lindlar catalyst to give D and on reaction with Na/liq. NH_3 produces E, Both the compounds D and E are isomeric. Give structures of A to E with proper reasoning. **(Roorkee, 1996)**

73. The hydrocarbon A, adds one mole of hydrogen in the presence of a plantinum catalyst to form *n*-hexane when A is oxidized vigorously with $KMnO_4$, a single carboxylic acid containing three carbon atoms, is isolated. Give the structure of A and explain. **(IIT, 1997)**

74. Two isomeric forms of an organic compound (A), $C_{11}H_{13}OCl$ readily decolorizes Br_2/H_2O and give same compound (B) on catalytic hydrogenation. Both the isomeric forms on vigorous oxidation give (C) which on treatment with sodalime gives 2-chloroethoxybenzene. However (C), on treatment with Ni/Al alloy in alkaline medium gives 3-ethoxybenzoic acid. Only one of the isomers of (A) gives geometrical isomers (D) and (E). Identify (A) to (E) with proper reasoning. **(Roorkee, 1997)**

75. Compound (A), $C_{10}H_{12}O$ gives off H_2 on treatment with sodium metal and also decolorizes Br_2 in CCl_4 to give (B), $C_{10}H_{12}O\ Br_2$. (A) on treatment with I_2 in NaOH gives iodoform and an acid (C) after acidification. Give structures of (A) to (C) and also of all the geometrical optical isomers of (A). **(Roorkee, 1997)**

76. An organic acid (A) $C_5H_{10}O_2$ reacts with Br_2 in the presence of phosphorous to give (B). Compound (B) contains an asymmetric carbon atom and yields (C) on dehydrobromination. Compound (C) does not show geometric isomerism and on decarboxylation gives an alkene (D) which on ozonolysis gives (E) and (F). Compound (E) gives a +ve Schiff test but (F) does not. Identify A to F. **(Roorkee, 1997)**

77. A hydrocarbon (A) of molecular weight 54 reacts with an excess of Br_2 in CCl_4 to give a compound (B) whose molecular weight is 593%, more than that of (A). However, on catalytic hydrogenation with excess of hydrogen, (A) forms (C) whose molecular weight is only 7.4% more than that of (A). (A) reacts with CH_3CH_2Br in the presence of $NaNH_2$ to give another hydrocarbon (D) which upon ozonolysis yields diketone (E). (E) on oxidation gives propionic acid. Identify A to E with reasons. **(Roorke, 1997)**

78. Dehydrobromination of compounds (A) and (B) yields the same alkene (C). Alkene (C) can regenerate (A) and (B) by the addition of HBr in the presence and absence of peroxide respectively. 1,1-diphenyl ethane is obtained on reaction of (C) with benzene in presence of $H^{\oplus}$. Identify (A) to (E) with reasons. **(Roorkee, 1997)**

79. A phenolic compound (A), $C_7H_8O_2$ on mild oxidation gives a highly volatile oil (B). (A) forms (C) on reaction with dimethyl sulphate in alkali. Oxidation of (C) with hot $KMnO_4$ gives (D) which then reacts with Br_2/H_2O to give (E) containing about 72% bromine. Give structure of A to E with proper reasoning. **(Roorkee, 1997)**

80. Cyclobutyl bromide on treatment with Mg in dry ether forms an organometallic A. The organometallic reacts with ethanal to give an alcohol B after mild acidification. Prolonged treatment of alcohol B with an equivalent amount of HBr gives 1-bromo-1-methylcyclopentane (C). Write the structure of A, B and explain how C is obtained from B. **(IIT, 2001)**

81. An alkene A (C_6H_{16}) on ozonolysis gives only one product B (C_8H_8O) compound. B on reaction with NaOH/ I_2 yields sodium benzoate. Compound B reacts with KOH/NH_2NH_2 yielding a hydrocarbon C (C_8H_{10}). Write the structure of compounds B and C. Based on this information the isomeric structures can be proposed for alkene A. Write their structures and identify the isomer which on catalytic hydrogenation (H_2/Pd—C) gives a racemic mixture. **(IIT, 2001)**

82. Identify A, B, C, D, and E in the following schemes and write their structures

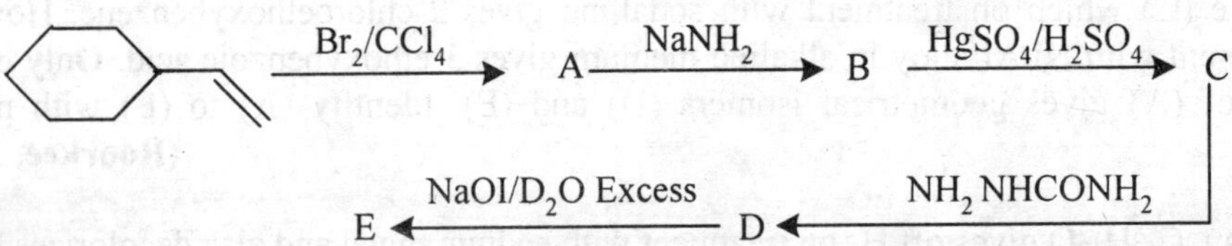

(IIT, 2001)

83. Identify X, Y, and Z in the following synthetic scheme and write their structures. Explain the formation of labeled (CH_2O) as one of the products when compound Z is treated with HBr and subsequently ozonolysed. Mark the C^* carbon in the entire scheme. **(IIT, 2001)**

C^* denotes C^{14}

$$BaC^*O_3 + H_2SO_4 \longrightarrow X \text{ gas}$$

$$CH_2{=}CH_2Br \xrightarrow[\text{(ii) X} \quad \text{(iii) } H_3O^{\oplus}]{\text{(i) Mg/ether}} Y \xrightarrow{LiAlH_4} Z$$

84. Aspartane, an artificial sweetener, is a peptide and has the following structure:

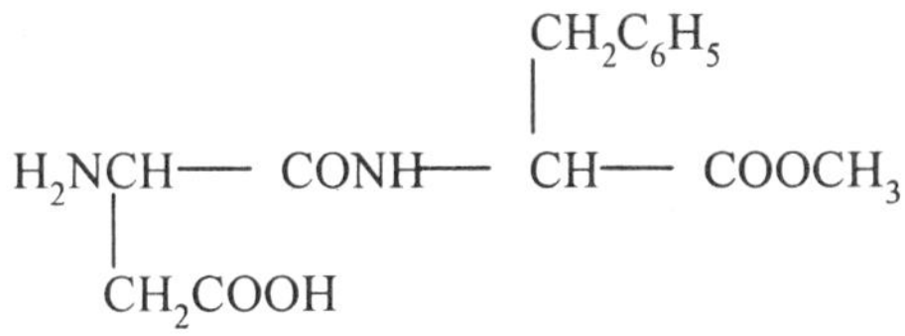

(i) Identify the four functional groups.

(ii) Write the Zwitterionic structure.

(iii) Write the structures of the amino acids obtained from the hydrolysis of aspartame.

(iv) Which of the two amino acids is more hydrophobic? **(IIT, 2001)**

ANSWERS

1. For empirical formula of compound A:

I Percentage	II Percentage	III $\frac{\text{Each ratio}}{\text{Smallest ratio}}$
C = 21.18	$\frac{21.18}{12} = 1.765$	$\frac{1.765}{0.588} = 3$
H = 4.12	$\frac{4.12}{1} = 4.12$	$\frac{4.12}{0.588} = 7$
I = 74.70	$\frac{74.70}{127} = 0.588$	$\frac{0.588}{0.588} = 1$

Hence, empirical formula or molecular formula of A is C_3H_7I. A may be

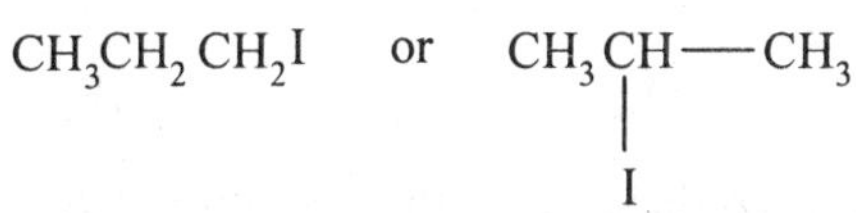

(I) (II)

A on hydrolysis with KOH gives compound which gives +ve iodoform test, i.e., the latter compound should have structure

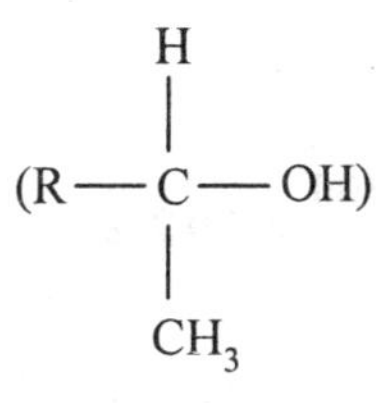

i.e., it should alkanol-2 (see Chapter 10) hence A has structures —II, i.e., 2-iodopropane. A on reaction with CH_3I and Zn would undergo Wurtz reaction yielding 2-methylpropane (isobutene).

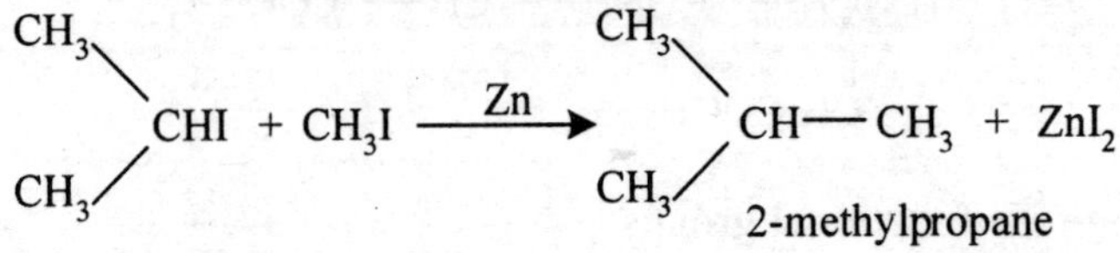

(A) (B)

has C = 82.8%, H = 17.2% and has molecular weight 58. A on reduction gives B.
This is in accordance with the data hence C_3H_7I is the molecular formula of A and A is $CH_3CH—CH_3$
Hence, A is 2-I-Iodopropane B is 2-methylpropane

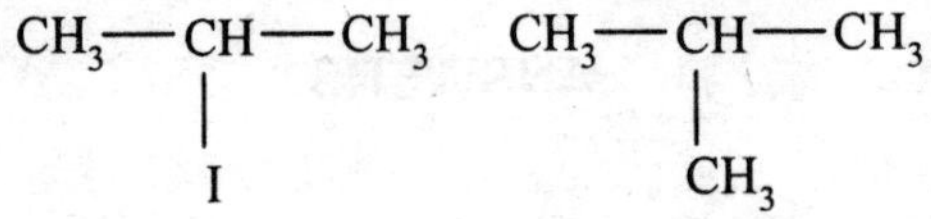

2.

Percentage	$\frac{\text{Percentage}}{\text{Atomic weight}}$	$\frac{\text{Each ratio}}{\text{Smallest ratio}}$
For A, C = 85.7%	$\frac{85.7}{12} = 7.14$	1
H = 14.3%	$\frac{14.3}{1} = 14.30$	2

Empirical formula = CH_2, since vapour density = 21, mol. wt = 42. Molecular formula, C_3H_6? The formula shows that it is an alkene hence it can only be propene, i.e.,$CH_3—CH = CH_2$. When HI reacts, addition takes place according to Markovnikov's rule. So, B is 2-Iodo-propene and the reactions.

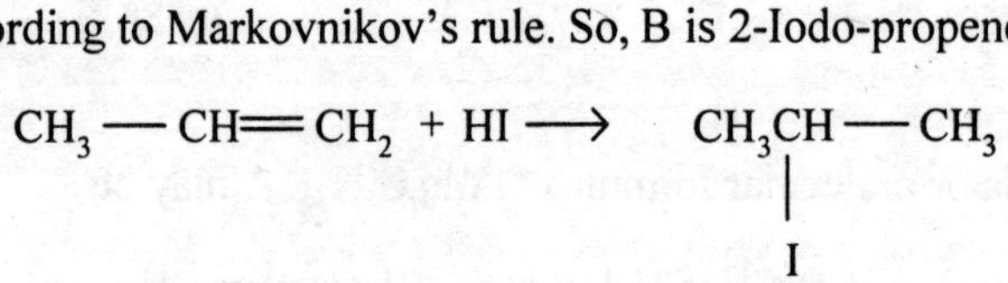

(A) (B)

'B' has C = 74.7%, which is in accordance with the data provided. When A reacts with bromine, we get 'C'. The reaction is as follows:

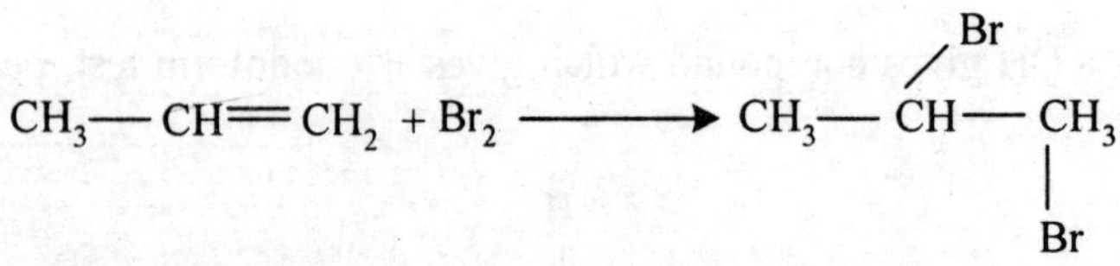

The percentage of bromine in C $= \frac{160 \times 100}{202} = 79.2\%$.

This also matches with the given data. The oxidation of propene gives propane 1,2-diol (See chapter 9). Oxidation of propene and reaction of (C) with aq. KOH are shown below:

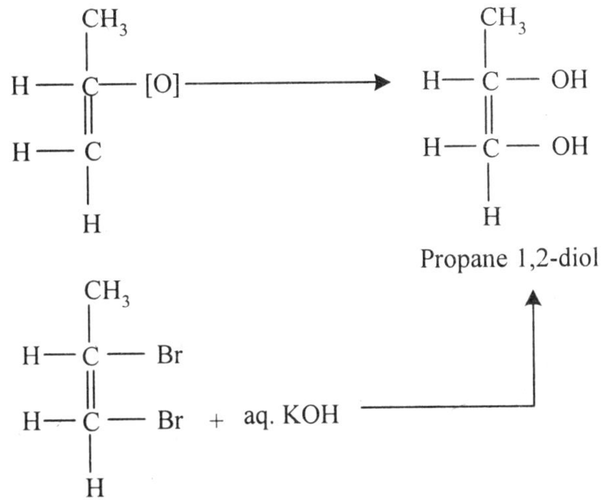

All this confirms that (A) is propene.

3.

Percentage	$\frac{\text{Percentage}}{\text{Atomic weight}}$ ratio	$\frac{\text{Each ratio}}{\text{Smallest ratio}}$
C = 93.75%	$\frac{93.75}{12} = 7.81$	$\frac{7.81}{6.25}\ \frac{1.25}{1} = 5$
H = 6.25%	$\frac{6.25}{1} = 6.25$	$\frac{6.25}{6.25}\ \frac{1.0}{1} = 4$

Since it is aromatic compound hence it should be $C_{10}H_8$. Therefore, it is naphthalene which on oxidation gives phthalic acid,

Hence, A is naphthalene, 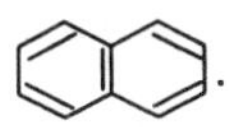.

4.

Percentage	$\frac{\text{Percentage}}{\text{Atomic weight}}$	$\frac{\text{Each ratio}}{\text{Smallest ratio}}$
C = 88.82 %	$\frac{88.82}{12} = 7.40$	$\frac{7.40}{7.40} = 1$
H = 11.18 %	$\frac{11.18}{1} = 11.18$	$\frac{11.18}{7.40} = 1.5$

This shows that the ratio of carbon to hydrogen is 1:1.5. It should be multiplied by 4 because alkene and alkyne have general formula C_nH_{2n} and C_nH_{2n-2}. Therefore, molecular formula of the compound is C_4H_6.

Molecular formula is C_4H_6. Two compounds are possible

$$\underset{(A)}{CH_3CH_2—C\equiv CH} \xrightarrow{[O]} \underset{\text{(See Chapter 9)}}{CH_3CH_2COOH + CO_2}$$

$$\underset{(B)}{CH_2=CH—CH=CH_2} \xrightarrow{[O]} CO_2 + \begin{array}{c}COOH \\ | \\ COOH\end{array}$$

Note: A is alkyne-1, since it reacts with ammoniacal silver oxide.

Hence, (A) $CH_3CHC\equiv CH$
Butyne—1

(B) $CH_2\equiv CH—CH=CH_2$
But-1,3-diene

5. Short summary

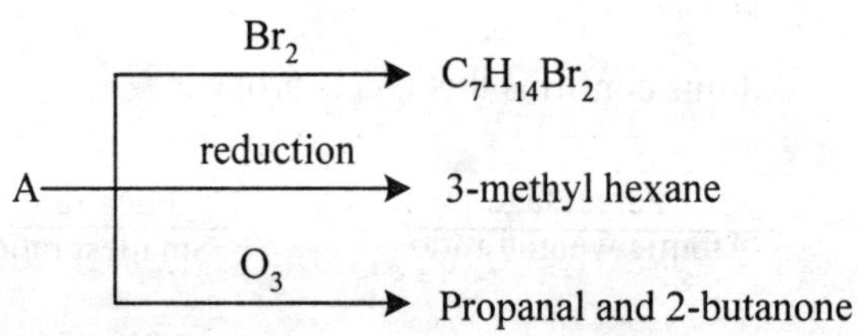

Use ozonolysis step for identification of A. Write structure of alkene which would yield propanal and 2-butanone. Such alkene is 3-methylhexene-3. Use Reduction and bromination product to confirm the alkene. Hence

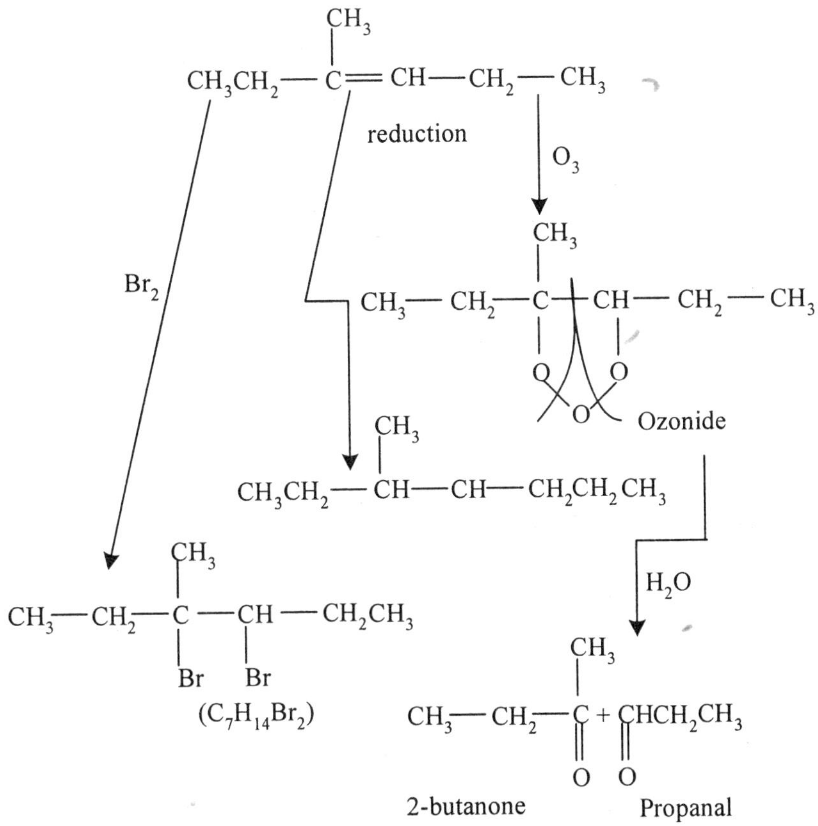

Hence, A = 3 methylhexene— 3($H_3C—CH_2—\overset{\overset{\large CH_3}{|}}{C}=CH—CH_2—CH_3$)

6. Short summary: Determined molecular formula of compounds A and B

$$\text{Compound A, } \%C = \frac{3.197 \times 3 \times 100}{11 \times 1} = 87.2\%$$

$$\%H = \frac{1 \times 1.152 \times 100}{9} = 12.8\%$$

Percentage	$\frac{\text{Percentage}}{\text{Atomic weight}}$	$\frac{\text{Each ratio}}{\text{Smallest ratio}}$
C = 87.2%	$\frac{87.2}{12} = 7.25$	$\frac{7.25}{7.25} = 1$
H =12.8%	$\frac{12.8}{1} = 12.8$	$\frac{12.8}{7.25} = 1.76$
(B), %C = 84.2%	$\frac{84.2}{12} = 7.01$	$\frac{7.01}{7.25} = 1$
%H =15.8	$\frac{15.8}{1} = 15.8$	$\frac{15.8}{7.01} = 2.25$

Molecular formula of B = C_8H_{18}

The compound A is 2,4-dimethyl hex-2,4-diene.

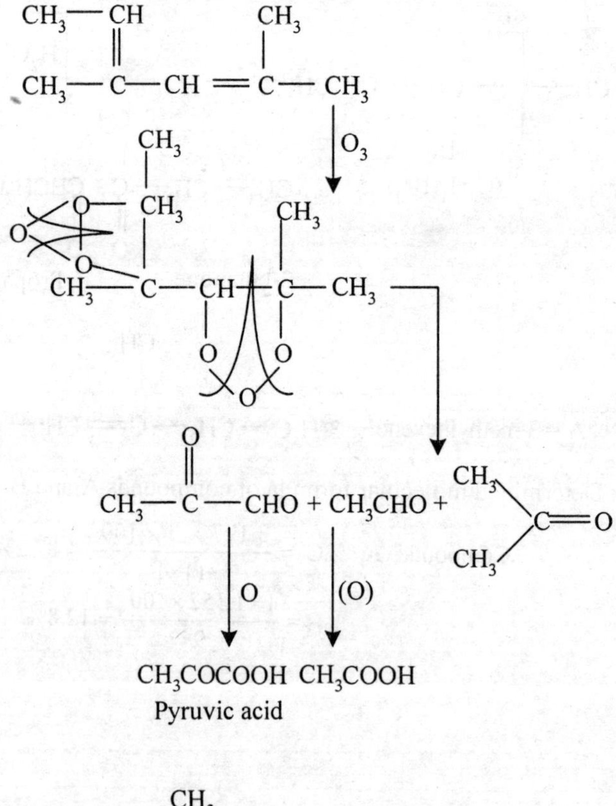

Hence,

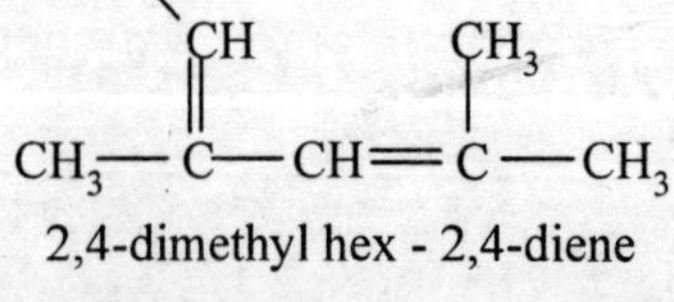

2,4-dimethyl hex - 2,4-diene

7. Short summary

$$(A) \xrightarrow{Mg} RMgX \xrightarrow{H_2O} C_6H_5—C_2H_5$$

Obviously R is $C_6H_5CH_2CH_2$, hence (A) is $C_6H_5CH_2CH_2Br$. The reaction is as follows:

$$\underset{\text{Phenyl ethyl bromide}}{C_6H_5CH_2CH_2Br} \xrightarrow{Mg} C_6H_5CH_2CH_2MgBr$$

$$C_6H_5CH_2CH_2MgBr \xrightarrow{HOH} C_6H_5CH_2CH_3 + Mg(OH)Br$$

Hence, A is phenyl ethyl bromide ($C_6H_5CH_2CH_2Br$)

8. *(i)* Short summary

$$(A) \xrightarrow{\text{Amm. CuCl}} \text{ppt}$$

$$C_8H_6 \longrightarrow \underset{\text{Benzoic acid}}{C_6H_5COOH}$$

Since it forms ppt. with ammoniacal CuCl hence it should be an alkyne with acidic H-atom. Since it yields benzoic acid on oxidation hence it must have a phenyl group. So, the compound is $C_6H_5—C \equiv CH$ (Phenyl acetylene).

The reactions are as follows:

$$C_6H_5—C \equiv CH \xrightarrow{CuCl} \underset{\text{Acetylide (ppt.)}}{C_6H_5C \equiv C—Cu}$$

$$C_6H_5—C \equiv CH \xrightarrow{[O]}$$

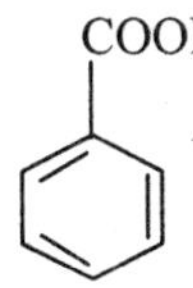

Note that side chain is oxidized to —COOH group

Hence, A is Phenylacetylene ($C_6H_5C \equiv CH$)

(ii) $C_9H_{12} \xrightarrow{Br_2} C_9H_{11}Br$

Since it undergoes electrophilic substitution reaction, it is aromatic compound. Since only bromine atom is substituted, hence rest of the positions should be occupied by alkyl groups. So the compound is:

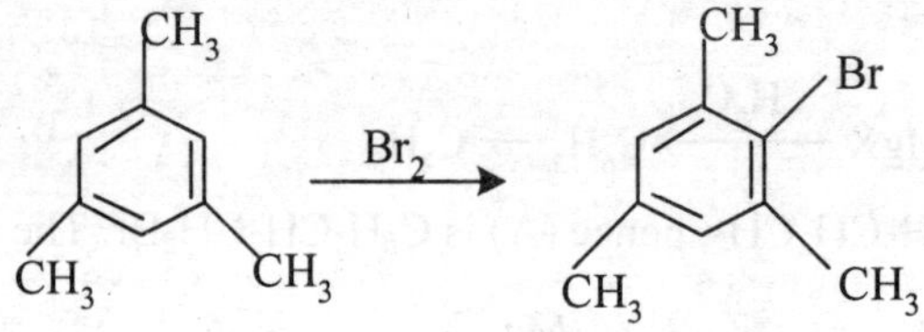

A is 1,3,5-Trimethyl benzene

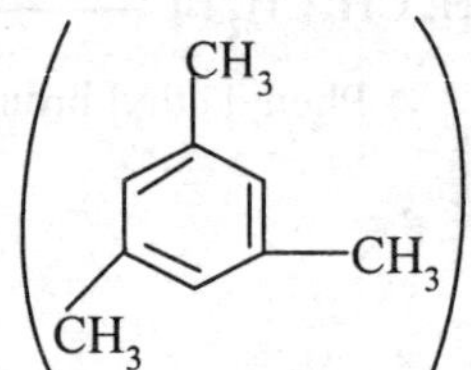

9. *(i)* Brief summary

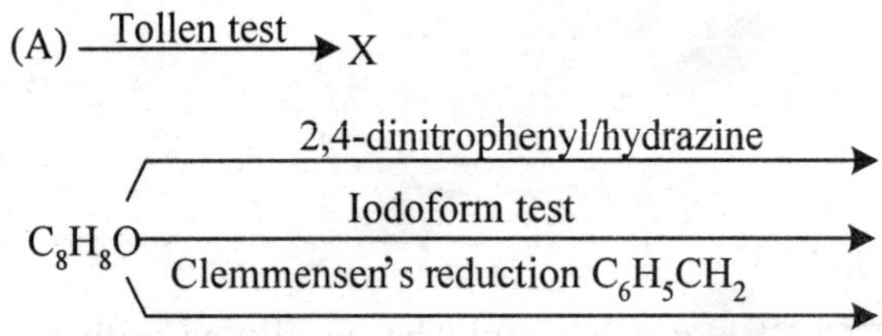

(ii) It contains more than 6 C atoms hence it is likely to be aromatic.

(iii) Since it contains C, H, O hence it may be phenol, aldehyde, ketone or ether.

(iv) Since it does not give positive test with Tollen's reagent aldehyde is excluded. But it reacts with 2,4-dinitrophenyl hydrazine, hence it is a ketone.

(v) Since it gives +ve iodoform test hence it must contain —$COCH_3$ group.

(vi) This is further confirmed by the fact that on reduction it yields ethyl benzene.

Hence, A is Acetophenone $(C_6H_5—\overset{\overset{O}{\|}}{C}—CH_3)$

10. *(i)* Brief summary

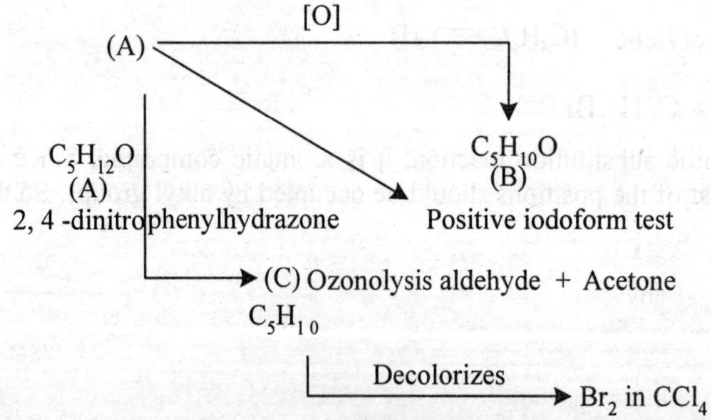

Dissolved in cold H_2SO_4 and poured in water and heated.

(i) It is isomer of A

(ii) Gives +ve Iodoform test

(iii) Gives +ve with Lucas reagent in less than 1 minute.

(ii) It contains C, H, O so it may be alcohol, aldehyde, ketone or ether.

(iii) Since B forms 2,4-dinitrophenyldrazone, it must be a ketone. It is, therefore, conclude that A is a secondary alcohol.

(iv) Since D gives +ve iodoform test and is an isomer of A, and B should have 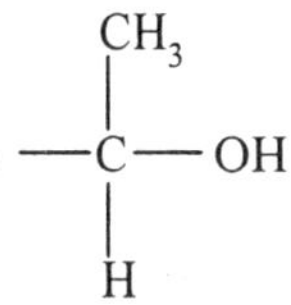

and 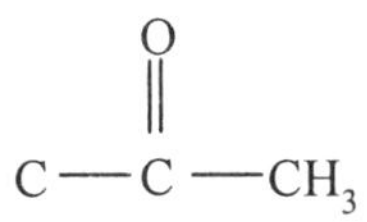 groups, respectively.

(v) Since D gives +ve iodoform test and is an isomer of A, it must contain 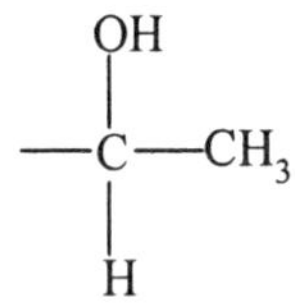

group. Moreover, it gives +ve response to Lucas reagent within 1 min, hence it must be a secondary alcohol.

(vi) Based on above discussion two structures may be assigned to A.

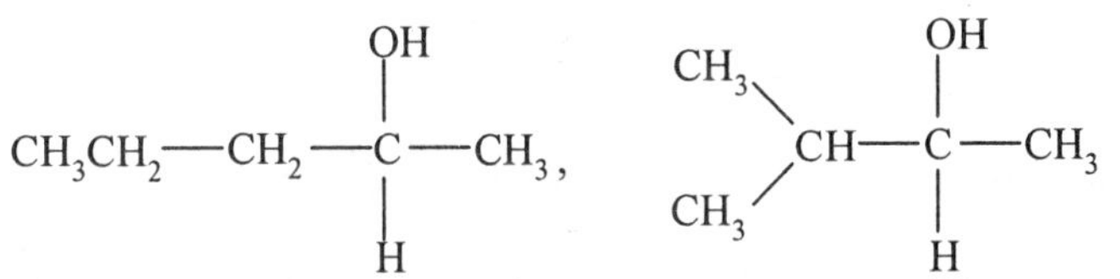

(I) (II)

Alkene C, upon ozonolysis yields aldehyde and acetone. This excludes structure I, because it will yield alkene $CH_3CH_2CH{=}CH{-}CH_3$ which, upon ozonolysis will yield only aldehyde. Therefore, structure II may be assigned to A. The reactions are as follows:

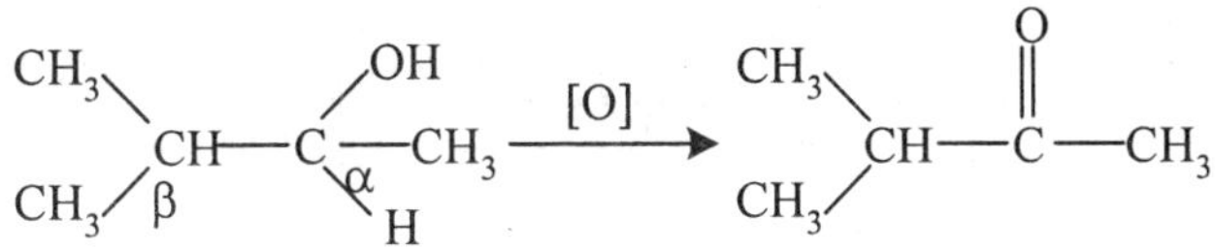

3-Methyl butanol—2 (A)
(See Saytzeff rule)

(B)

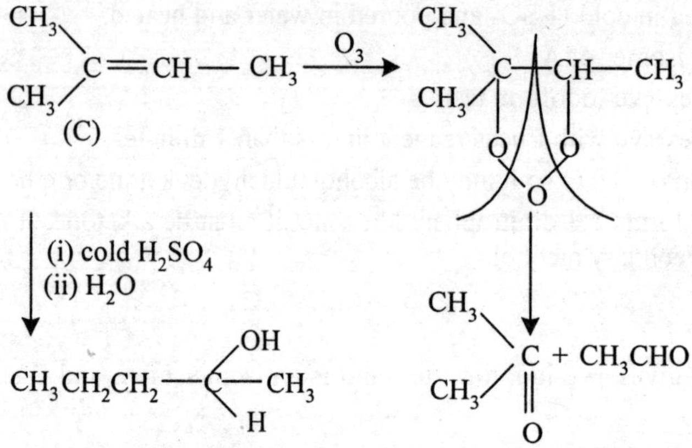

(D) Pentanol—2

Hence,

A is 3 methyl butanol-2 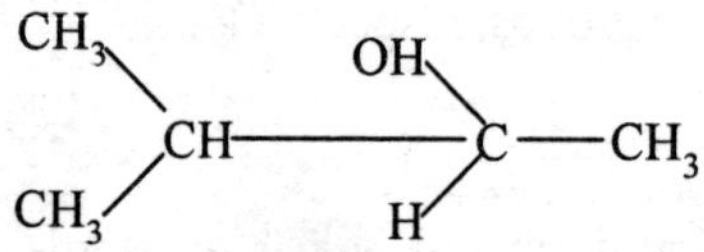

B is 3-methyl butanol-2 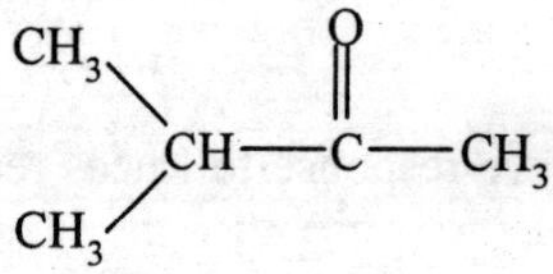

C is 2-methyl butanol-2 $(CH_3)_2C{=}CH{-}CH_3$

D is pantanol-2 $CH_3CH_2CH_2{-}CH(OH){-}CH_3$

11. 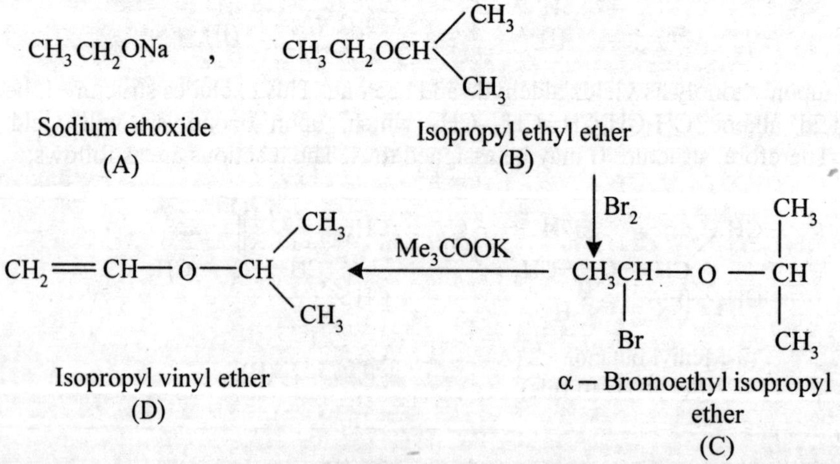

Sodium ethoxide (A), Isopropyl ethyl ether (B), α – Bromoethyl isopropyl ether (C), Isopropyl vinyl ether (D)

Hence,

A is Sodium ethoxide CH_3CH_2ONa

B is Isopropyl ethyl ether 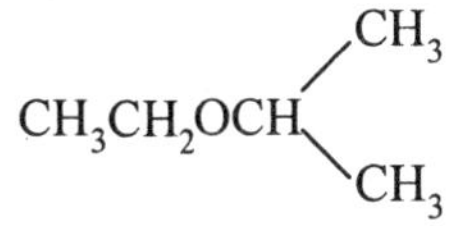

C is Bromoethyl isopropyl ether 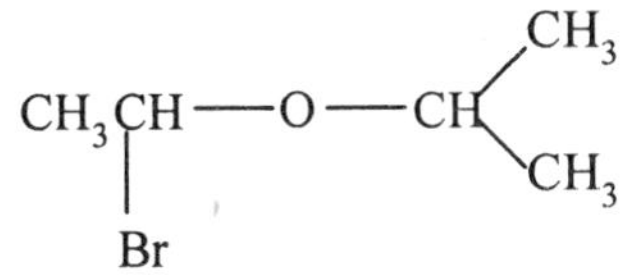

D is Isopropyl vinyl ether 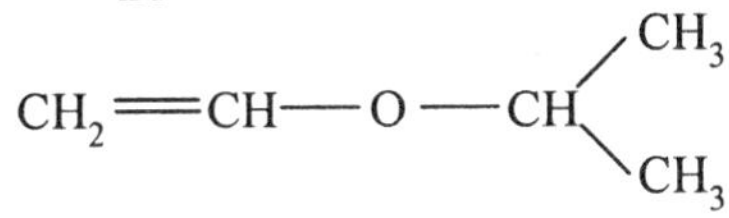

12. *(i)* Since upon reaction with aq. NaOH A gives an alcohol which gives +ve iodoform test it contains 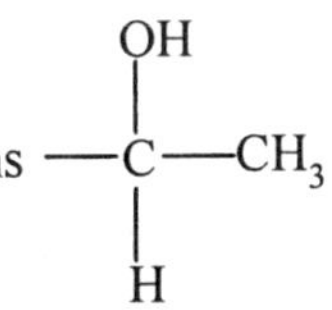 group. So, A is

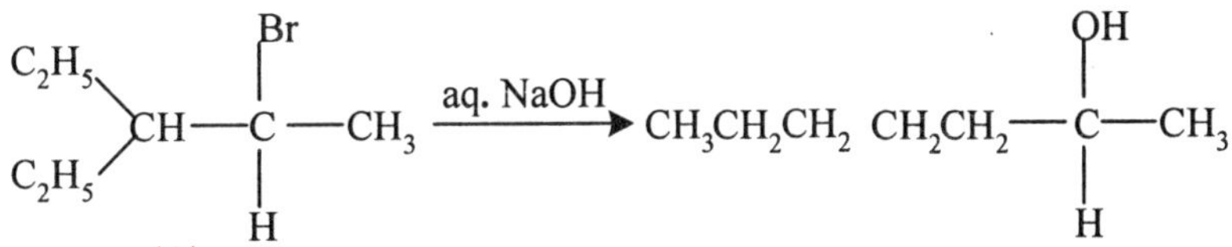

(A) Heptanol–2

(ii) Since compound B with alc. NaOH gives only one alkene C_7H_{14}, it should be 1-bromoheptane, otherwise in case of 2-bromoheptane two alkenes would be obtained according to Saytzeff's Rule $CH_3CH_2CH_2CH_2CH_2CH_2CH_2Br$

$$\xrightarrow{\text{alc. NaOH}} CH_3CH_2CH_2CH_2CH_2CH{=}CH_2$$

alkene upon ozonolysis yields methanol as one of the products.

(iii) Since (C) on reaction with alc. NaOH yields two alkenes, it must be 2-bromoheptane. Rest of bromoalkanes would yield only one alkene.

$$\underset{\beta}{}\ CH_3CH_2CH_2CH_2CH_2\underset{\alpha}{\overset{Br}{\overset{|}{CH}}}-\underset{\beta}{CH_3}$$

$$\downarrow \text{Saytzeff rule}$$

$$\underset{\text{Heptene—2}}{CH_3CH_2CH_2CH_2\,CH{=}CH{-}CH_3} +$$

$$+\underset{\text{Heptene— 1}}{CH_3CH_2CH_2CH_2CH_2{-}CH{=}CH_2}$$

Hence, A is 3-Ethyl-2-bromopentane $CH_3{-}CH_2{-}\overset{C_2H_5}{\overset{|}{CH}}{-}\overset{Br}{\overset{|}{CH}}{-}CH_3$

∴ B is 1-Bromopeptane $CH_3CH_2CH_2CH_2CH_2CH_2CH_2Br$

C is 2-Bromoheptane $CH_3\,CH_2CH_2CH_2CH_2{-}\underset{Br}{\underset{|}{CH}}{-}CH_3$

13. 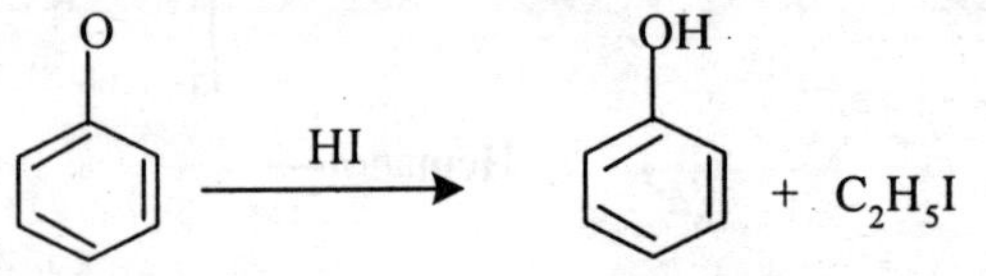

Phenyl ethyl ehter (A) Phenol Ethyl iodide

$$\underset{(B)}{C_6H_5CH_2OCH_3} \xrightarrow{2HI} C_6H_5CH_2I + CH_3I$$

Hence, A is Phenyl ether $C_6H_5OC_2H_5$

B is Benzyl methyl ether $C_6H_5CH_2OCH_3$

14.

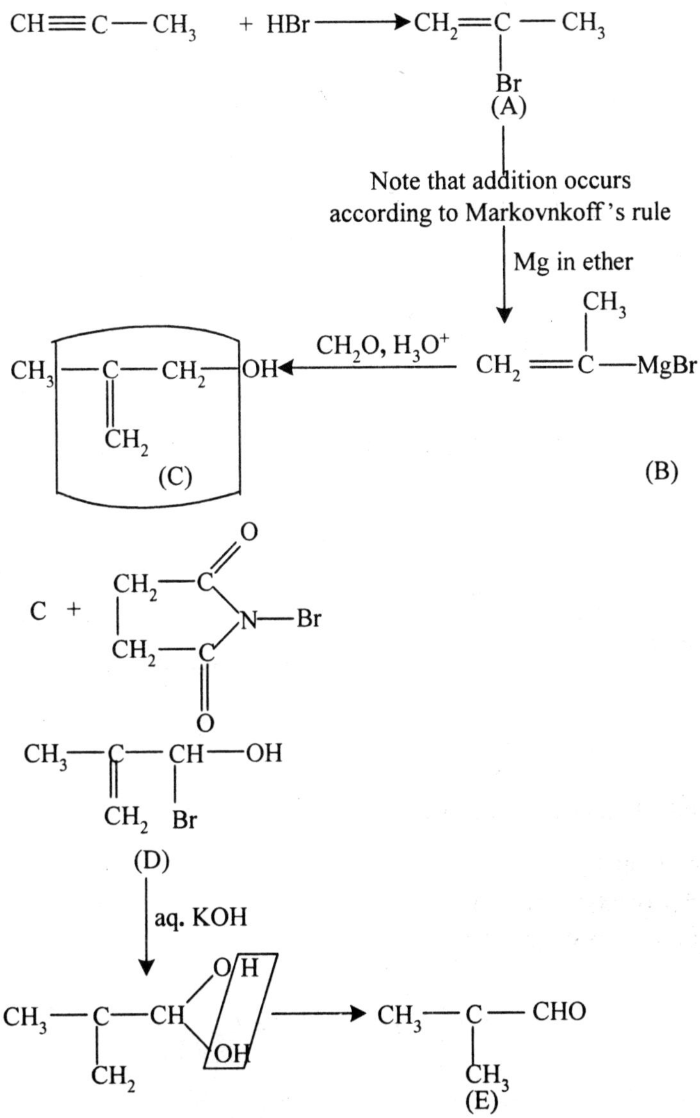
CH≡C—CH3 + HBr → CH2=C—CH3
Br
(A)
Note that addition occurs
according to Markovnkoff's rule
Mg in ether
CH3
CH2=C—MgBr
(B)
CH2O, H3O+
CH3—C—CH2—OH
CH2
(C)
C +
CH2—C
CH2—C
O
O
N—Br
CH3—C—CH—OH
CH2 Br
(D)
aq. KOH
O H
CH3—C—CH
CH2
OH
CH3—C—CHO
CH3
(E)

A is 2 - bromopropene 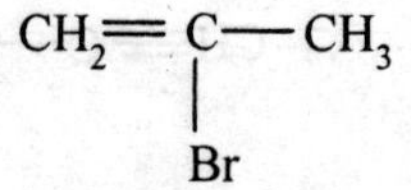

B is Propenyl magnesium bromide 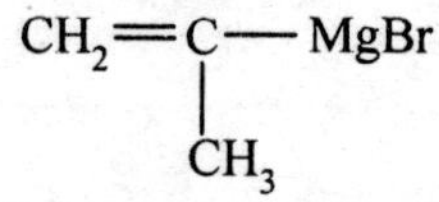

∴ C is 2 - methyl prop-2-en-1-ol 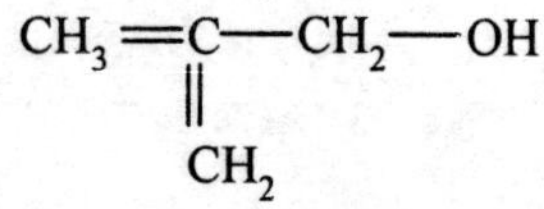

D is 2 - methyl prop-2-ene-1-bromo-1-ol 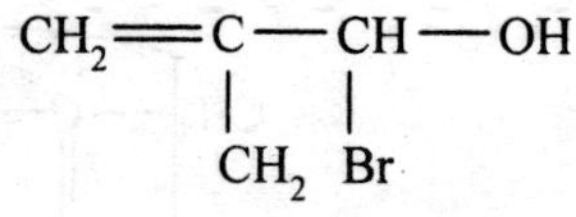

E is 2 - methyl propanal 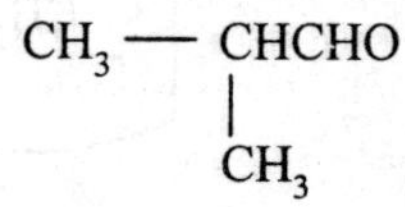

Hence,

* Note that 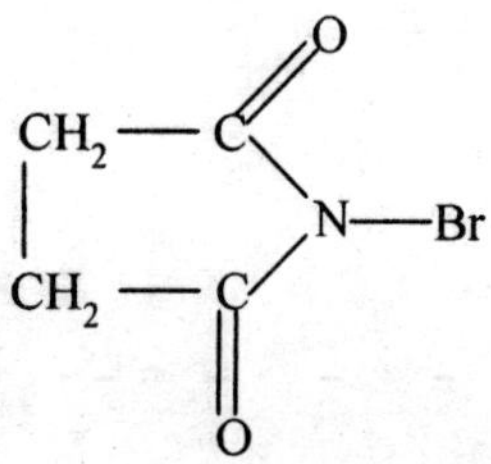 is N-bromo-succinicimide which substitutes H atom at *allyl position*

15. *(i)* Brief summary

A → gives rapid Lucas test

$C_6H_{14}O$ → oxidation difficult

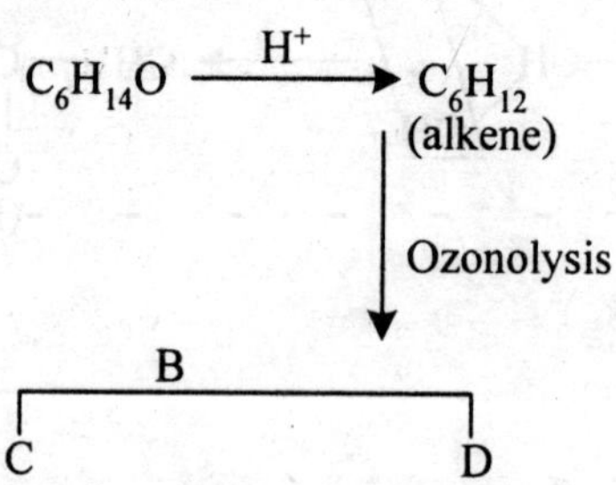

Molecular formula C_3H_6O

Both contain carbonyl group

(ii) Since A gives rapid Lucas test, it must be a tertiary alcohol. The possible compounds are:

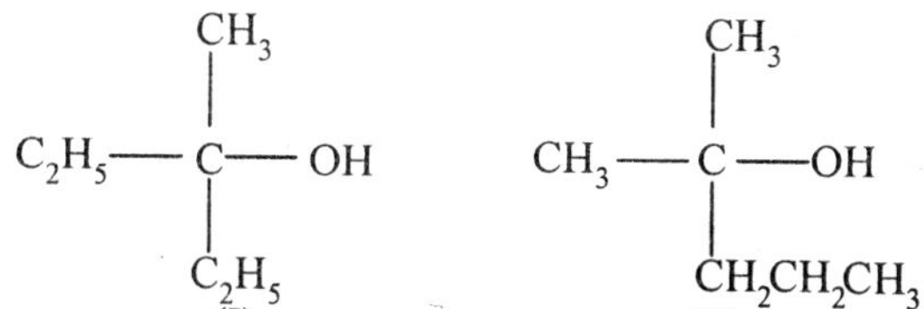

(I) (II)

The alkene 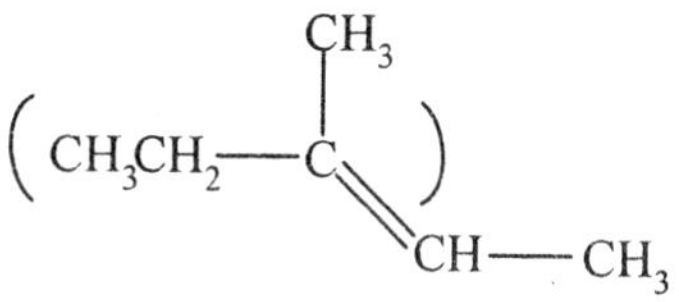 obtained from I, upon ozonolysis yields carbonyl compound having different molecular formula hence it is rejected. So, the compound is **II**.

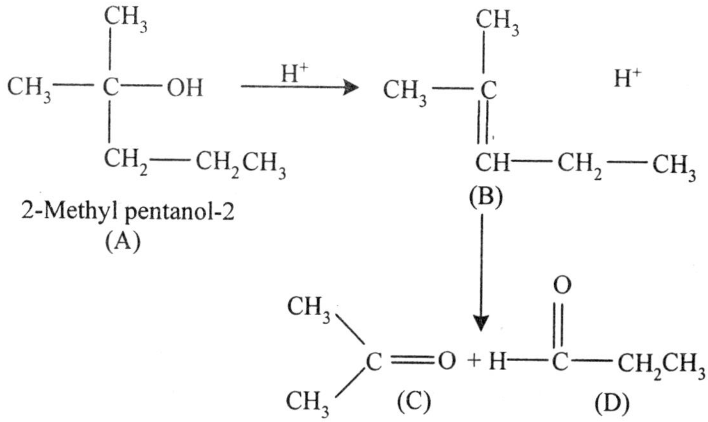

Both have mol. formula C_3H_6O

So A is:

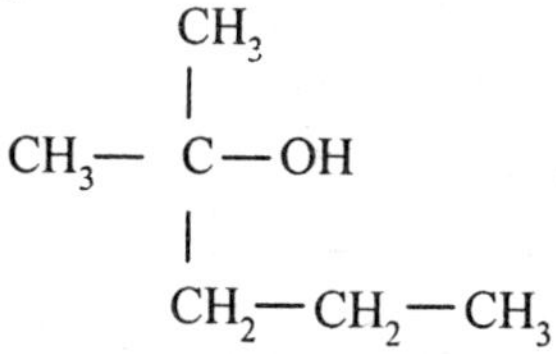

Hence,

A is 2 ethyl pentanol-2 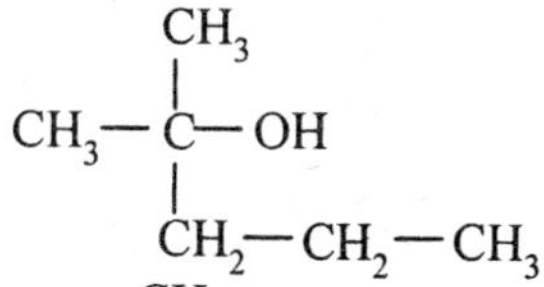

B is 2-methyl pentene-2 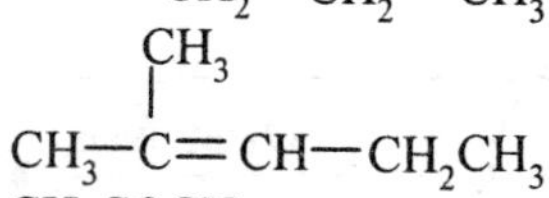

C is propanone CH_3COCH_3

D is propanal CH_3CH_2CHO

16. *(i)* Brief summary

$$C_6H_{12} \xrightarrow{H_2} C_6H_{14}$$

Alkene (optically active) Optically inactive alkene.

(ii) Since alkene is optically active, it must contain one chiral carbon atom, so

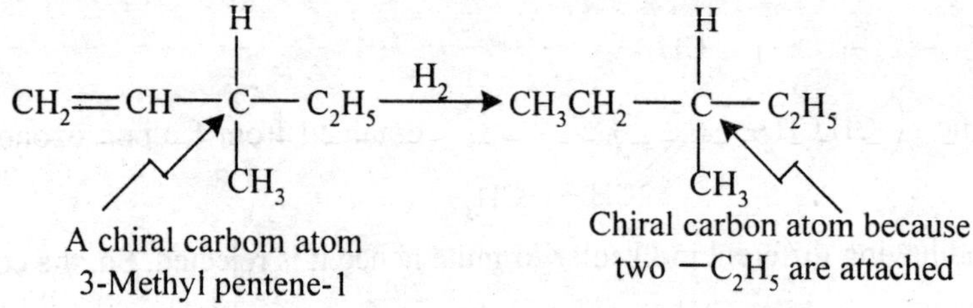

Hence, A is 3-methyl pentene-1

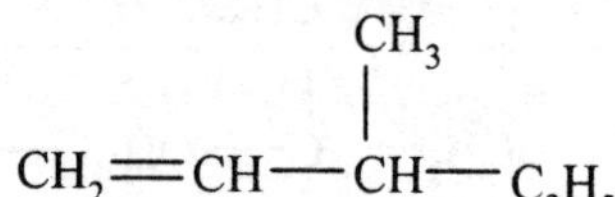

B is 3-methyl pentane

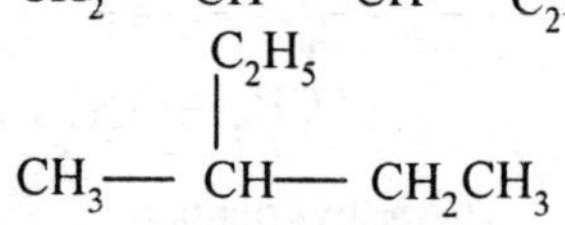

17. Brief summary

$$\underset{A}{C_5H_{11}Br} \xrightarrow{\text{Hydrolysis}} \underset{B}{C_5H_{12}O}$$

$$\downarrow CH_3ONa$$

$$\underset{\text{Alkene}}{C_5H_{10}} \xrightarrow{O_3} \underset{\text{Ketone}}{C_3H_6O} + \underset{\text{Aldehyde}}{C_2H_4O}$$

(i) Since A does not form an ether with sodium methoxide, it must be a tertiary alkyl halide. Therefore, A is 3-methyl-2-bromobutane.

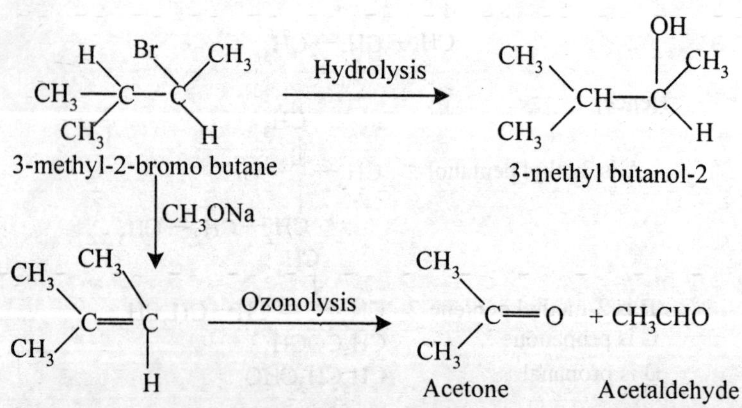

Hence,

A is 3-methyl-2-bromo butane $(CH_3)_2CH—\underset{\underset{Br}{|}}{CH}\ CH_3$

B is 3-methyl butanol-2 $(CH_3)_2CH—\underset{\underset{CH}{|}}{OH}—CH_3$

18. *(i)* Brief summary

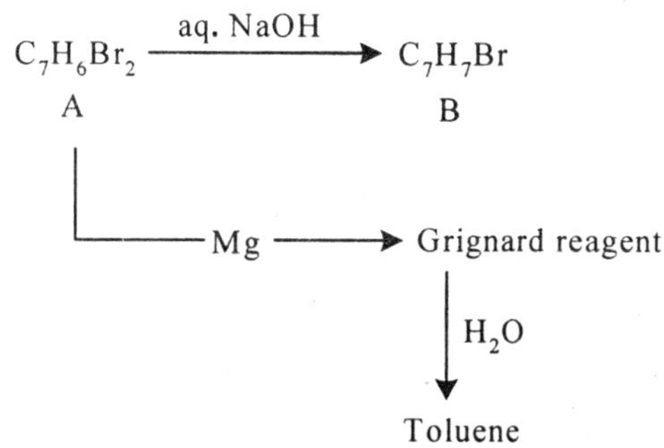

(ii) It is likely to be aryl bromide since it contains 7 carbon atoms.

(iii) Since only one bromine atom is readily lost upon hydrolysis, it is present in the side chain. It has already been mentioned (See Chapter 3 'Resonance') that the halogen atom is firmly held to benzene ring hence it cannot be easily replaced.

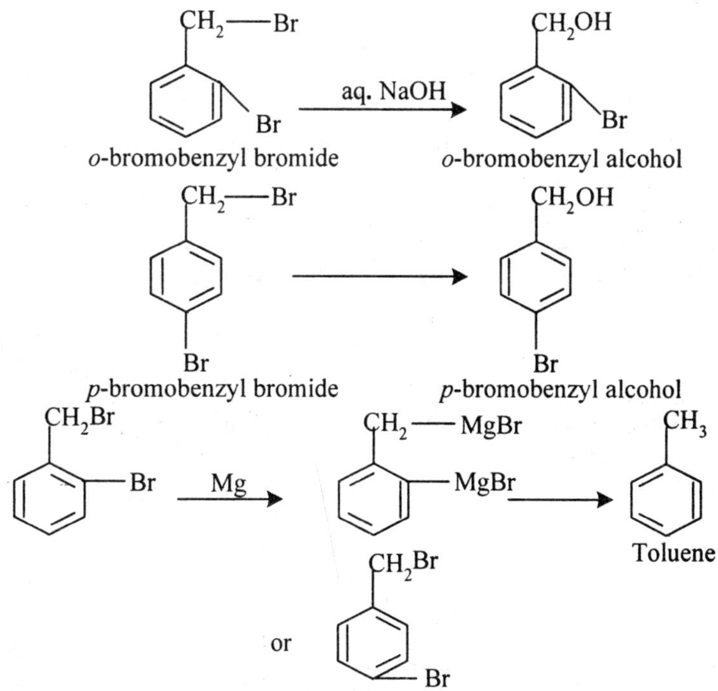

Hence, A may be *o*-bromobenzyl bromide

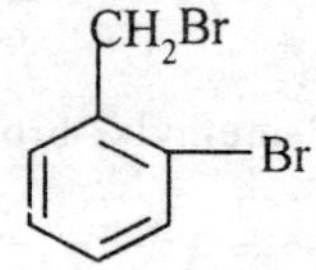

or

p-bromobenzyl bromide

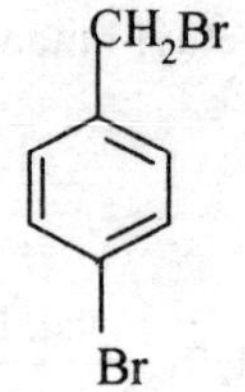

19. *(i)*

$$C_5H_{10} \xrightarrow{Br_2 \text{ in } CCl_4} \text{Decolorizes}$$

$$A \xrightarrow{\text{(i) } H_2SO_4 \text{ (ii) } H_2O} \underset{B}{C_5H_{12}O} \xrightarrow{\text{Chromic acid}} \underset{C}{C_5H_{10}O}$$

Both B and C gave positive iodoform test.

(ii) A is an alkene because it decolorized Br_2 in CCl_4. Since B and C both give positive iodoform test. Two alkenes are possible:

$$CH_3CH_2—CH_2—CH{=}CH_2$$

I

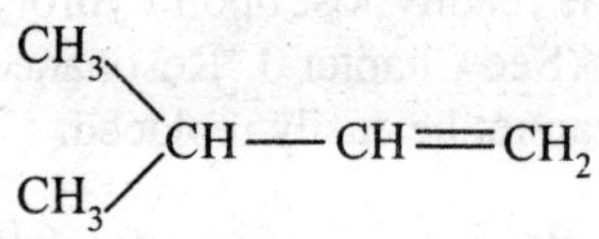

II

Alkene I will not produce isobutyric acid as one of the products of iodoform reaction. Hence, alkene II is the organic compound (A).

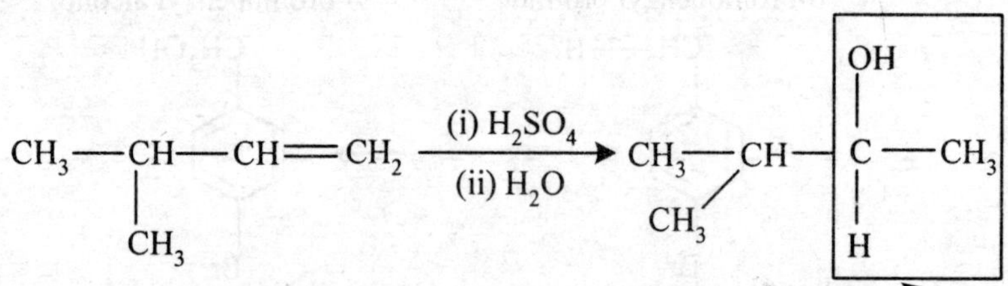

3-methyl butene-1 (A)

group giving iodoform test

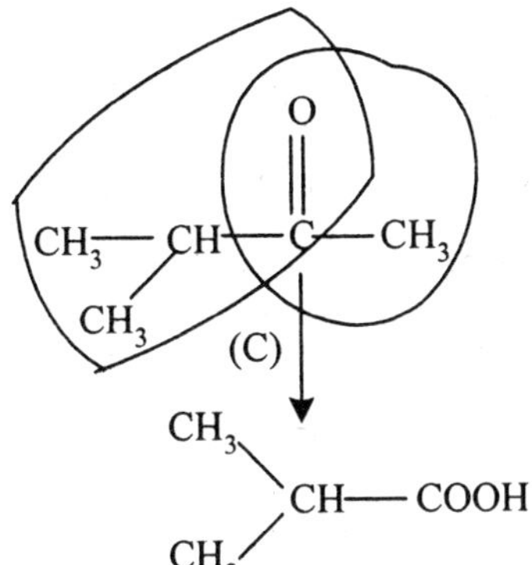

Isobutyric acid

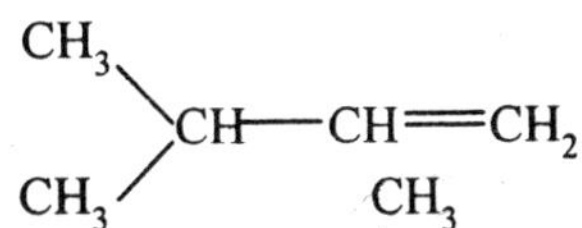

Hence, A is 3-Methylbutene-1

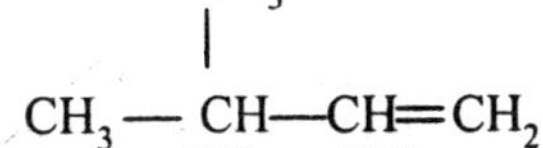

B is 3-Methylbutanol-2

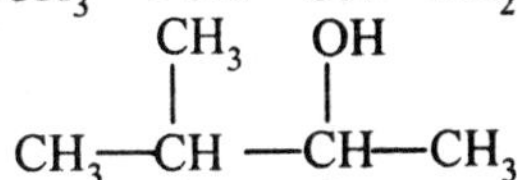

C is 3-methyl butanone-2

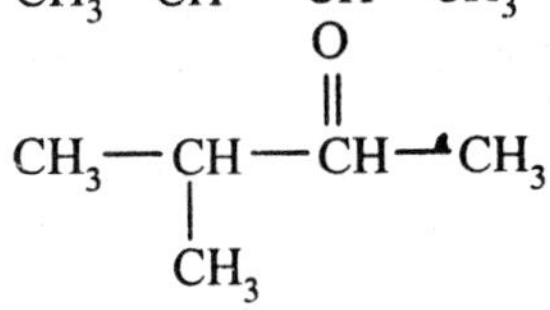

20. Brief summary

(i)

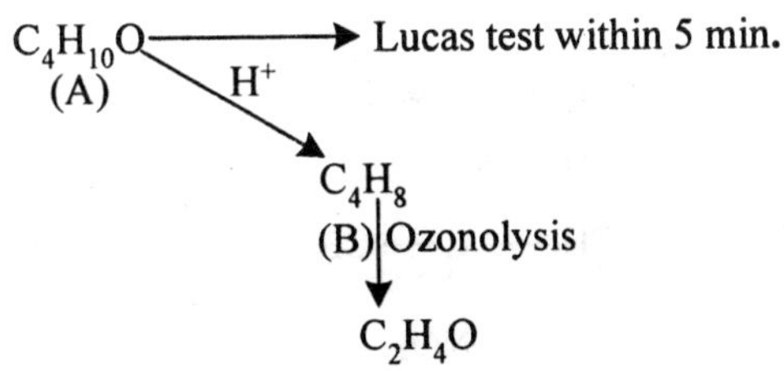

(ii) Since it gives +ve Lucas test within 5 minutes, it must be a secondary alcohol, i.e., Butanol-2

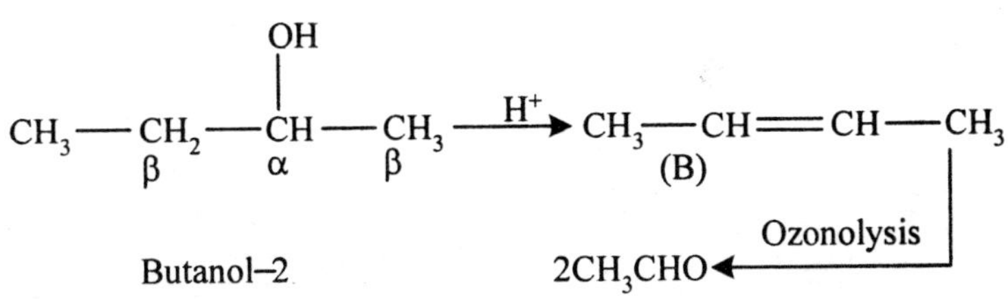

Note: Elimination according to Saytzeff rule

Hence, A is Butanol-2 $(CH_3-CH_2-\underset{\large |}{\overset{OH}{CH}}-CH_3)$

B is Butene-2 $(CH_3-CH=CH-CH_3)$

C is Ethanal (CH_3CHO)

21. $\underset{\text{A}}{\underset{\text{Propanol-1}}{CH_3CH_2CH_2OH}}$ $\underset{\text{B}}{\underset{\text{Propene}}{CH_3CH=CH_2}}$

$$B \xrightarrow{Br_2} \underset{\text{1, 2-Dibromopropane}}{CH_3\ \underset{\displaystyle Br}{\underset{|}{CH}}-CH_2\,Br} \xrightarrow[NaNH_2]{alc.KOH} \underset{\text{(Propyne)}}{CH_3-C\equiv CH}$$

$$\underset{}{(D)} \xrightarrow[\text{dil } H_2SO_4]{Hg^{++}\text{ in}} \underset{\text{(E)}}{\underset{\text{acetone}}{CH_3COCH_3}} \xrightarrow{NH_2OH} \underset{\text{(F) Acetoxime}}{CH_3-\underset{\displaystyle NOH}{\underset{\|}{C}}-CH_3}$$

Hence, A: Propanol-1 $CH_3CH_2CH_2OH$

B: Propene $CH_3CH=CH_2$

C: 1,2-Dibromo propane $CH_3-\overset{Br}{\overset{|}{CH}}-\underset{\displaystyle Br}{\underset{|}{CH_2}}$

D: Propyne $CH_3-C\equiv CH$

E: Acetone CH_3COCH_3

F: Acetoxime $CH_3-\underset{\displaystyle NOH}{\underset{\|}{C}}-CH_3$

22. 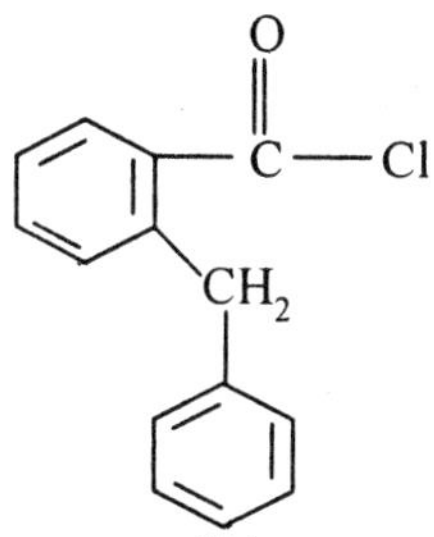

(A)

o-benzyl benzoyl chloride

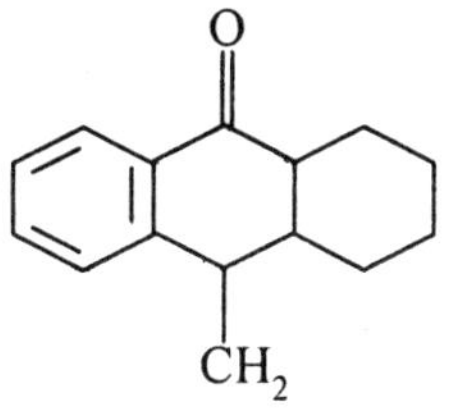 , 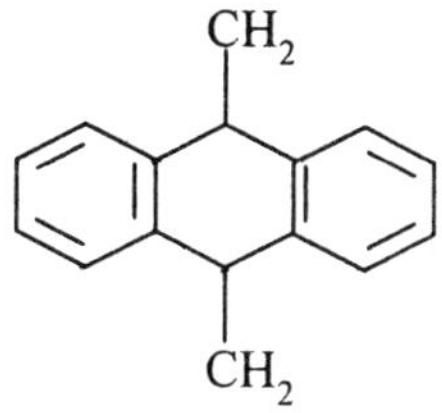

23. $CH_3-\underset{\displaystyle Br}{\underset{|}{CH}}-COOH$, $CH_2=CH-COOH$

α-Bromopropionic acid (A) Acrylic acid (B)

Note: See Hell-Volhard-Zelinsky reaction.

24. (*i*) Since D and E both contain 4 carbon atoms each, A is an alkene and it must have molecular formula C_8H_{16}. Likewise B must be an alkene with the molecular formula C_8H_{16}. The possible structure of A is

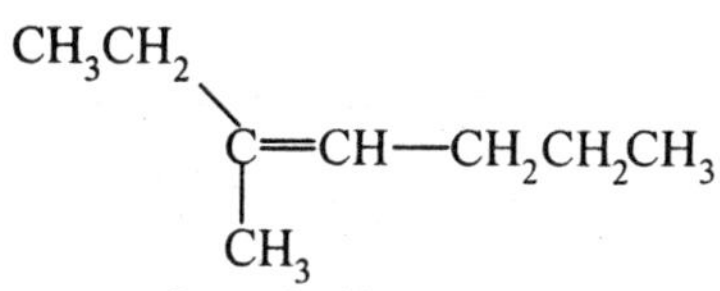

3-methyl heptene-3

The reaction and various compounds are as follows:

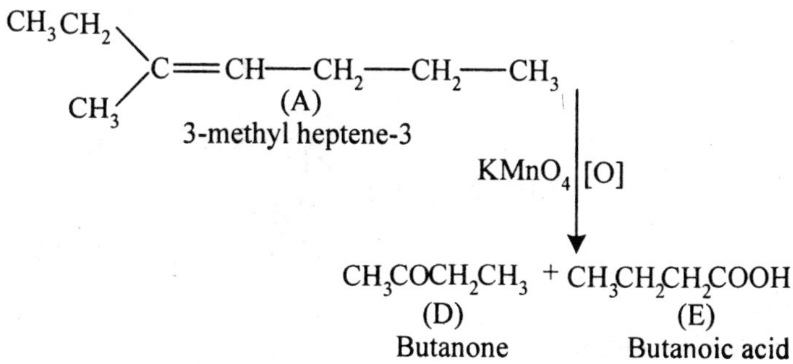

(ii)

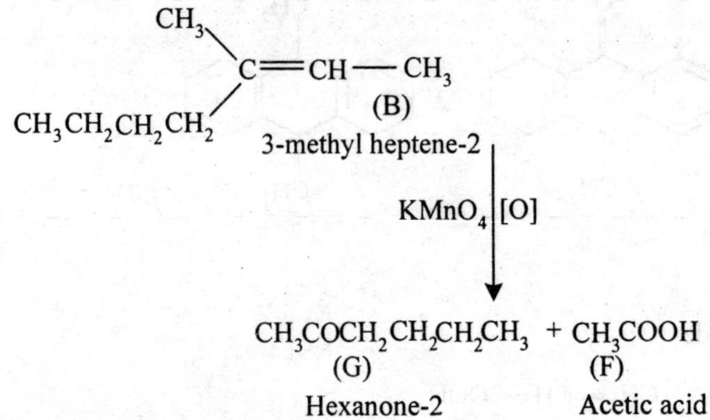

A and B, both alkenes give some alcohol on hydration:

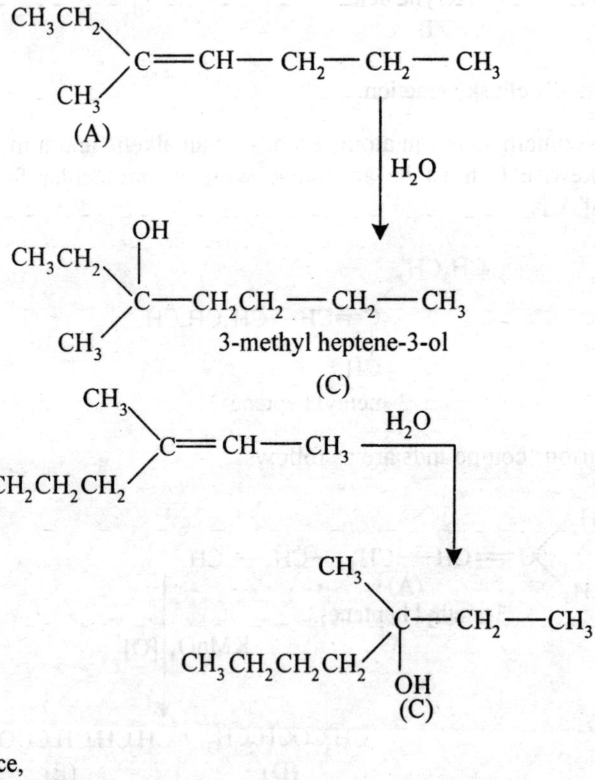

Hence,

A: 3-methyl heptene-3

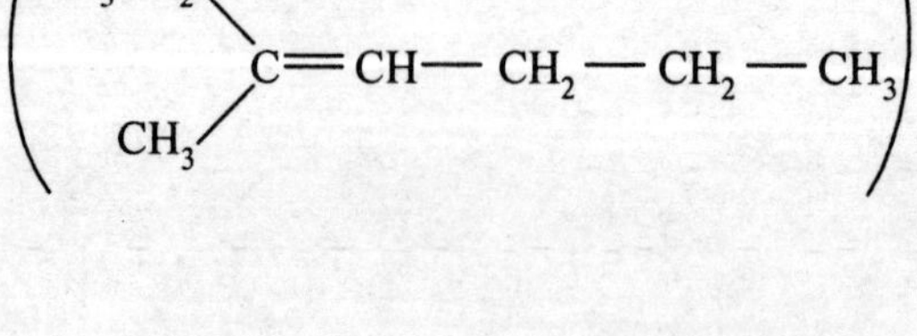

B: 3-methyl heptene-2

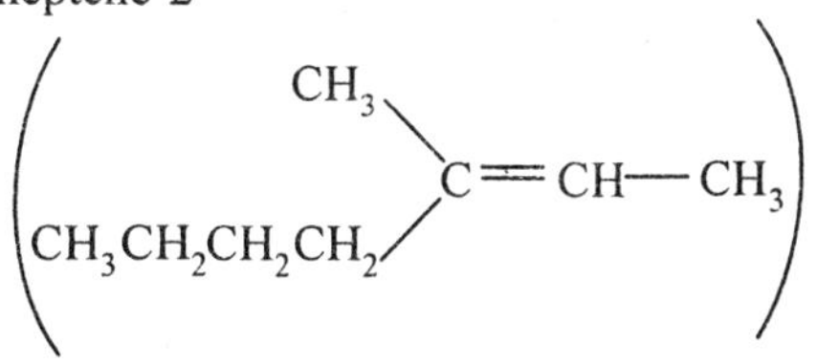

C: 3-methyl heptene-3-ol

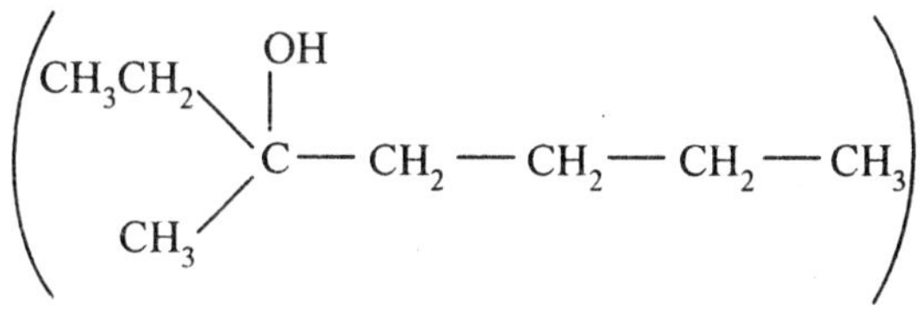

D: Butanone $(CH_3COCH_2CH_3)$
E: Butanoic acid $(CH_3CH_2CH_2COOH)$
F: Acetic acid (CH_3COOH)
G: Hexanone-2

$$(CH_3CO - CH_2 - CH_2 - CH_2 - CH_3)$$

25. *(i)* $CH_3CH_2\,CO\,NH_2 \xrightarrow{\text{Hydrolysis}} CH_3CH_2\,COOH + NH_3$
OH H
(A)
Butanamide

(ii) $CH_3CH_2CONHCH_3 \xrightarrow{\text{Hydrolysis}} CH_3NH_2 + CH_3CH_2COOH$
OH H
N-methyl propionamide

(iii) $CH_3CONHCH_2CH_3 \xrightarrow{\text{Hydrolysis}} CH_3COOH + CH_3CH_2NH_2$
OH H
N-methyl acetamide

Hence, A is Butanamide $(CH_3CH_2CH_2CONH_2)$
B is N-Methyl propionamide $(CH_3CH_2CONHCH_3)$
C is N-Ethyl acetamide $(CH_3CONHCH_2CH_3)$

26.

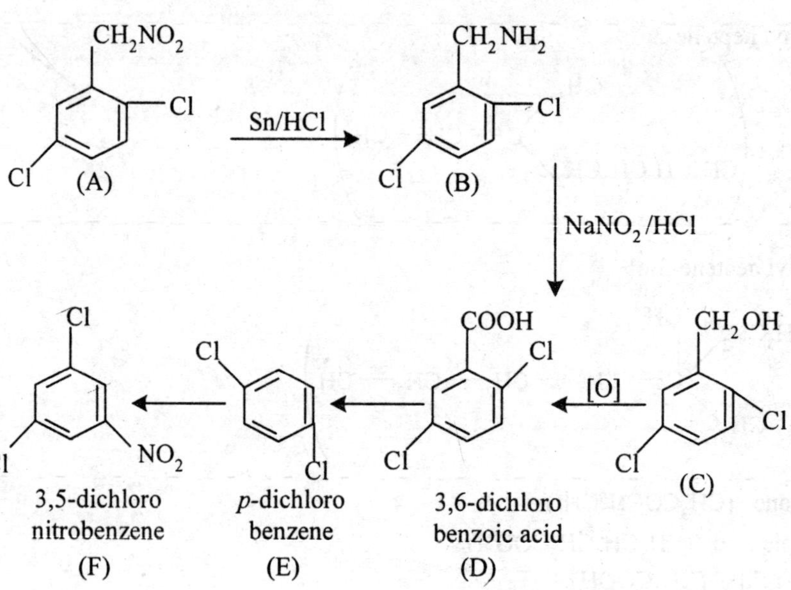

Hence,

A is 2-nitromethyl *p*-dichlorobenzene

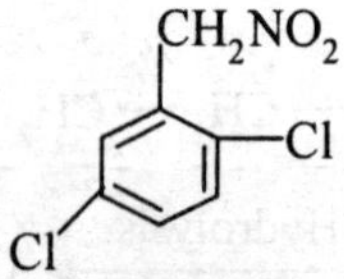

B is 2, 5 - Dichlorobenzylamine

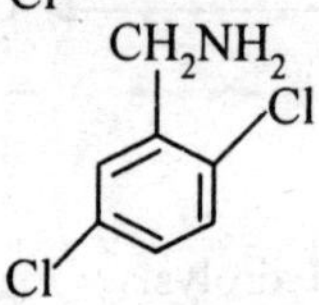

C is 2, 5 - Dichloro benzylalcohol

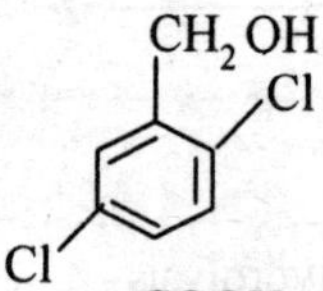

D is 3, 6 - Dichlorobenzoic acid

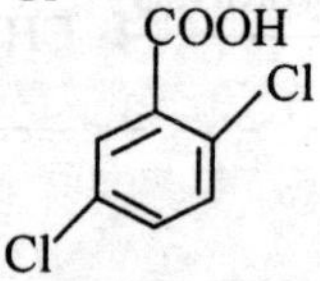

E is *p*-Dichlorobenzene

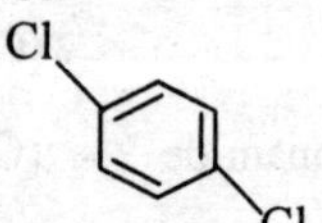

F is 3, 5-Dichloronitorbenzene

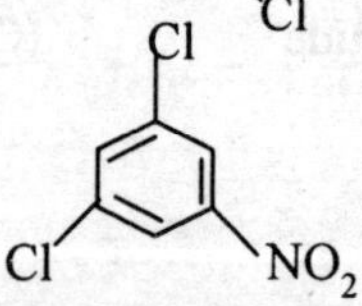

(i) Since compound (C) does not form a dye with naphthol, it means —NH_2 group and hence—NO_2 is present in the side chain and not to the benzene ring.

(ii) Since (C) gives red colour with Cerric ammonium nitrate, it must contain alcoholic group.

(iii) Since (E) forms only a single mono-nitroderivative (F) on nitration, both—Cl atoms must be present in 'p' position, both —Cl atoms must be present in 'p' position to each other. This is known as **ORIENTATION** (Convert disubstituted benzene derivative into trisubstituted derivative and count number of isomers. Remember formula 231, i.e., if two groups are present in ortho position, we will get 2 trisubstituted derivative, if two groups are present in meta position we will get 3 trisubstituted derivatives. Likewise, if two groups are at para-position, we will get only one trisubstituted derivative).

27. (i) Any primary amine.

(ii) Any secondary amine since product is insoluble in KOH.

(iii) It is aromatic amine in which —NH_2 group is directly attached to the ring.

28.

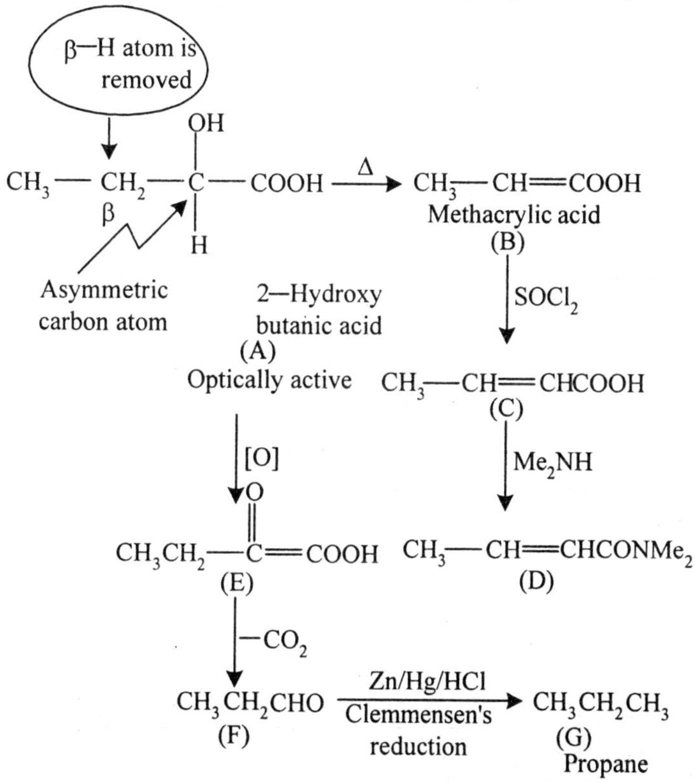

Hence,

A is 2 - Hydroxybutanoic acid $(CH_3—CH_2—C(OH)(H)—COOH)$

B is Methacrylic acid $(CH_3CH{=}CH—COOH)$

C is Butanoic acid $(CH_3CH{=}COOH)$

D is N,
N-Dimethyl-but-2-enamide $(CH_3CH{=}CHCONMe_2)$

E is 2-Keto butanoic acid $(CH_3CH_2—C({=}O)—COOH)$

F is Propanal $(CH_3\,CH_2\,CHO)$

G is Propane $(CH_3\,CH_2CH_3)$

29. Since it gives four different mono-chloroderivatives it is:

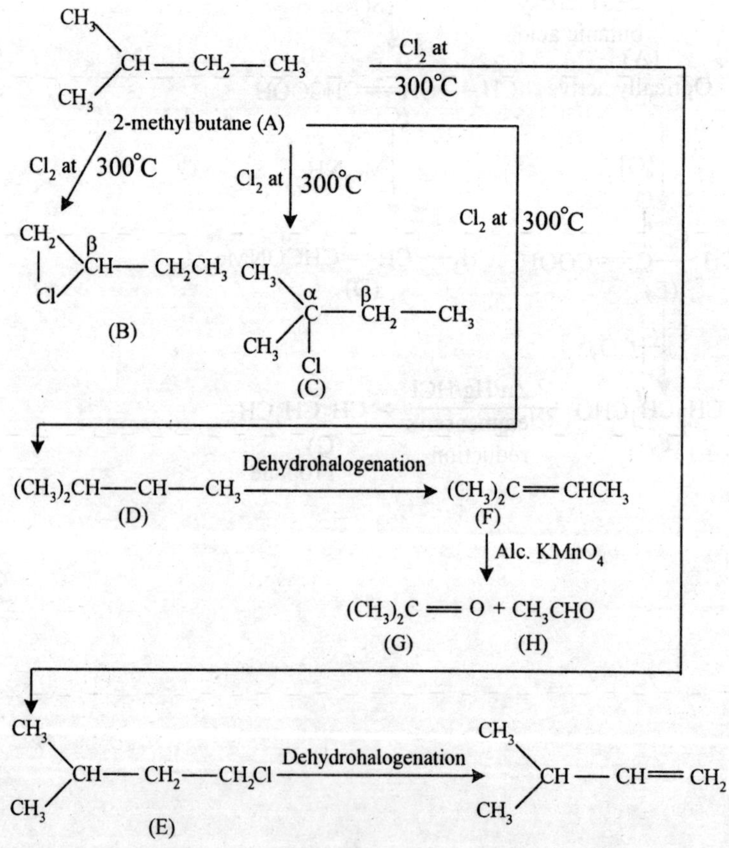

30. *(i)* Brief summary

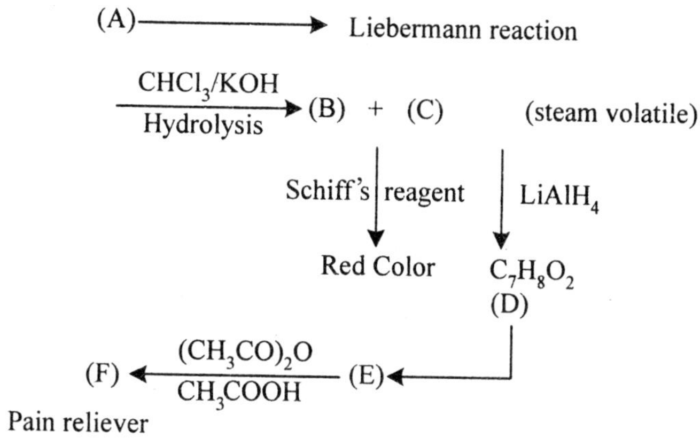

(ii) A is phenol because it undergoes Liebermann reaction. It undergoes Reimer-Tiemann reaction.

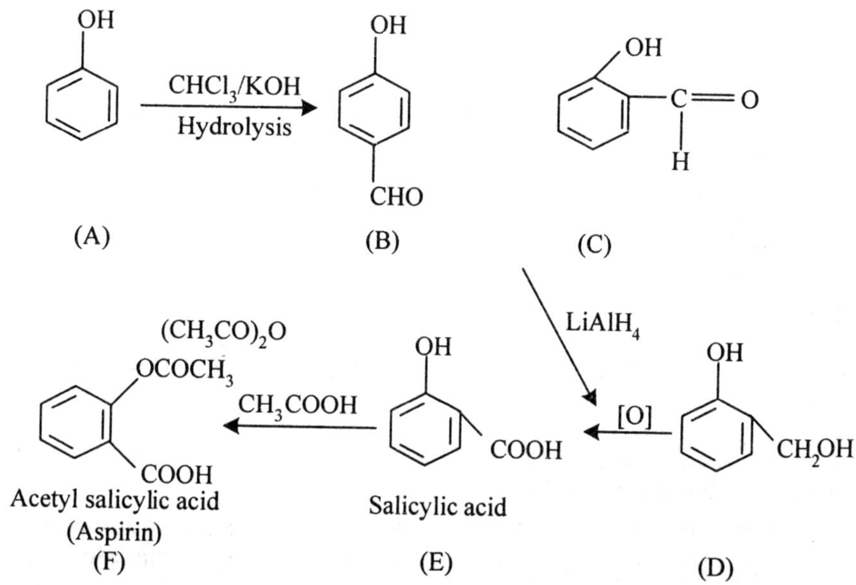

Hence,

A is Phenol

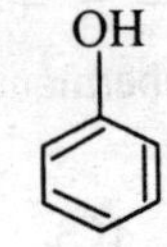

B is *p*-Hydroxybenzaldehyde

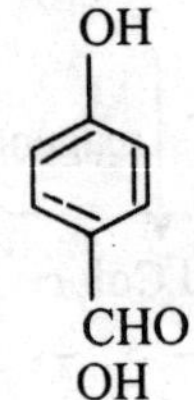

C is Salicylic aldehyde

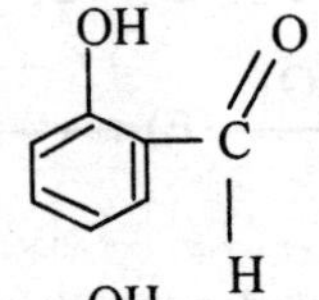

D is *o*-Methyloyl phenol

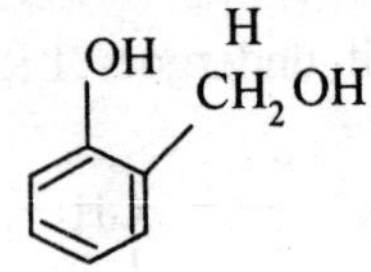

E is Salicylic acid

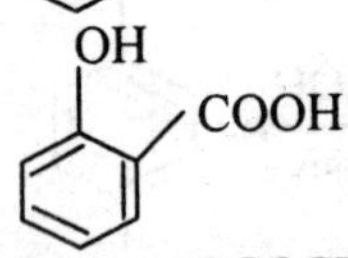

F is Acetyl salicylic acid Aspirin

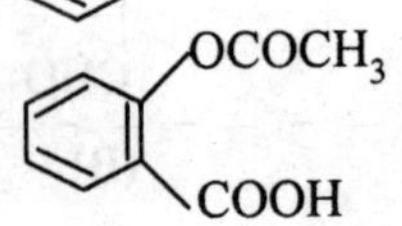

31. (i) Brief summary:

$$\underset{A}{C_9H_7ClO_2} \xrightarrow{H_2O} \underset{B}{C_9H_8O_3} \xrightarrow{[O]} \underset{C}{C_8H_6O_4} \xrightarrow{Heat} \underset{D}{C_8H_4O_3}$$

Insoluble in water, stable, soluble in acid $NaHCO_3$, reacts with anhydride phenyl hydrazine. No reaction with Fehling solution.

(ii) Since A is insoluble in water, hence –Cl atom should be attached to benzene ring. Since compound B is soluble in $NaHCO_3$ it must contain —COOH group. Moreover, B contains ketonic group, as it evidenced from it does not reduce Fehling solution. Compound C is an acid since D is an acid anhydride. Therefore, compounds A to D are identified as below:

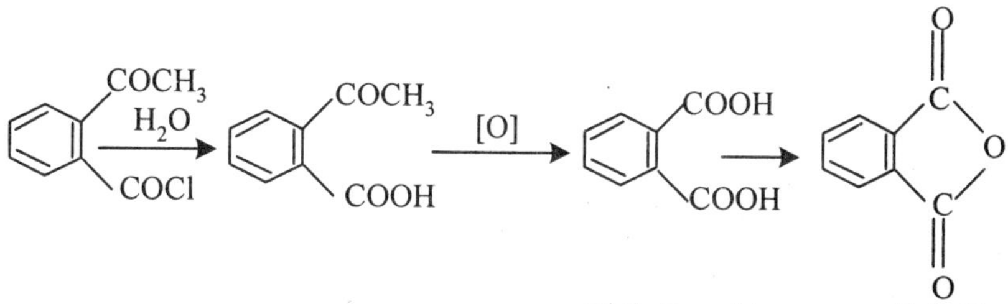

Phthalic acid Phthalic anhydride

A B C D

Hence, A is *o*-methyl ketobenzoyl chloride

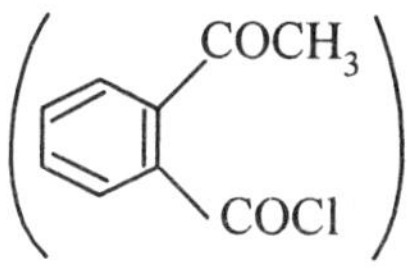

B is *o*-methyl ketobenzoic acid

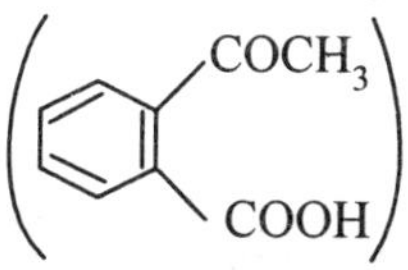

C is phthalic acid

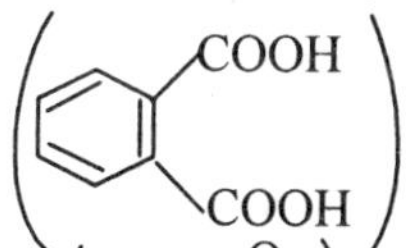

D is phthalic anhydride

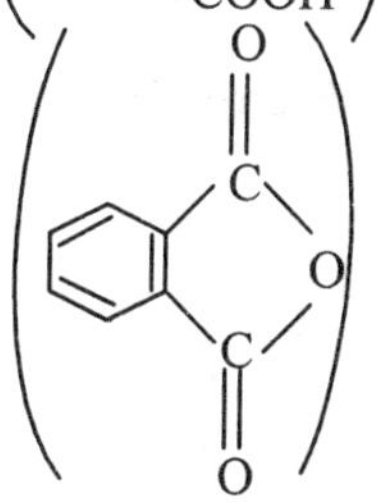

32. *(i)* Brief summary

$$\underset{\text{Optically active}}{C_6H_{10}} \xrightarrow{H_2} \underset{\text{Optically inactive}}{C_6H_{14}}$$

(ii) A must contain one chiral carbon atom since it is optically active. It must also contain —C≡CH group because it forms ppt. with $Ag(NH_3)_2OH$. Therefore,

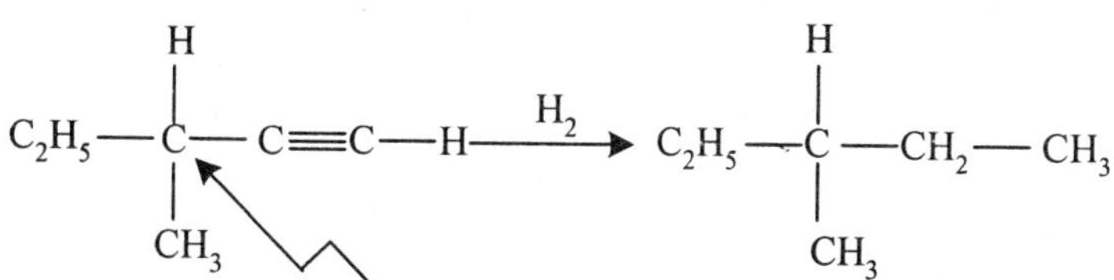

3-methylpentene -1 (A) Chiral carbon atom 3-methylpentane (B)

Hence,

A is 3-methylpentyne-1 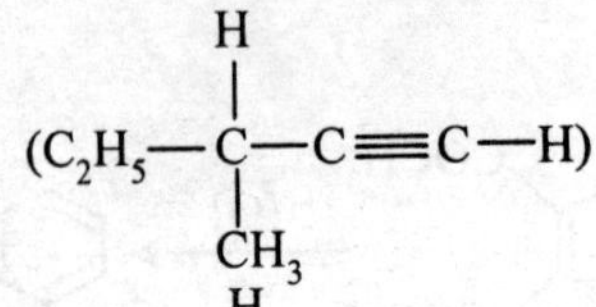

B is 3-methyl pentane 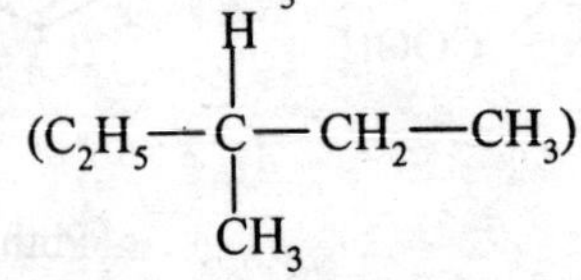

33. 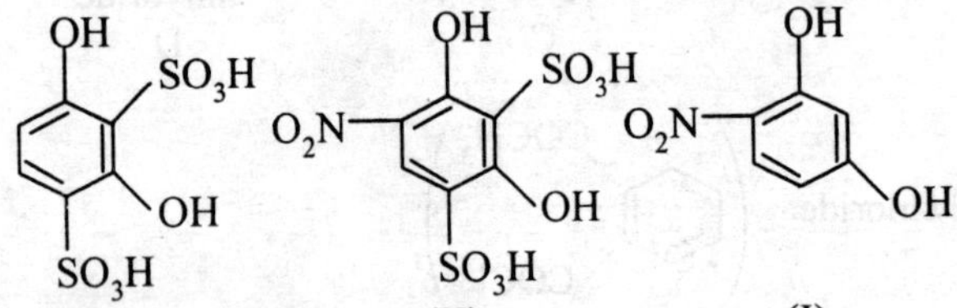

(G) (H) (I)

Hence,

G is 2,6- Dihydroxy 1,3 benzene disulphonic acid 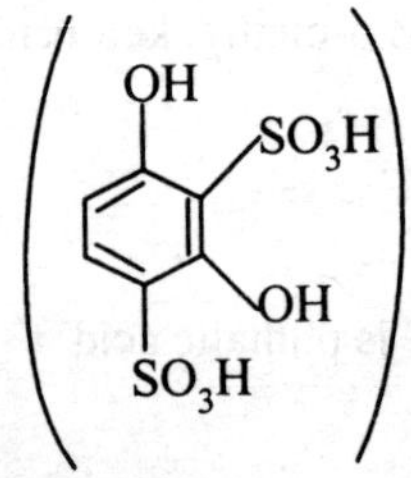

H is 2,6- Dihydroxy-5 nitro 1,3 benzene disulphonic acid

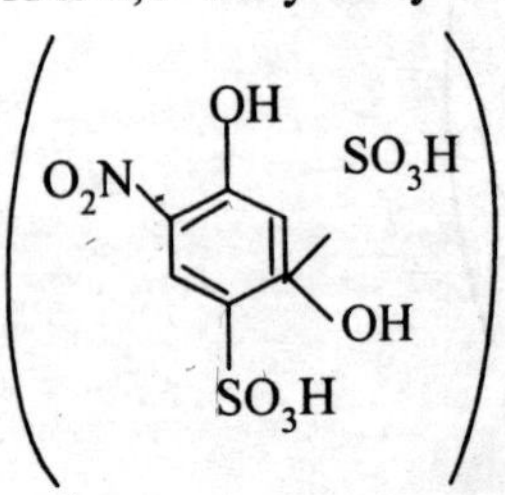

I is 5 - Nitro 1,3-dihydroxy benzene 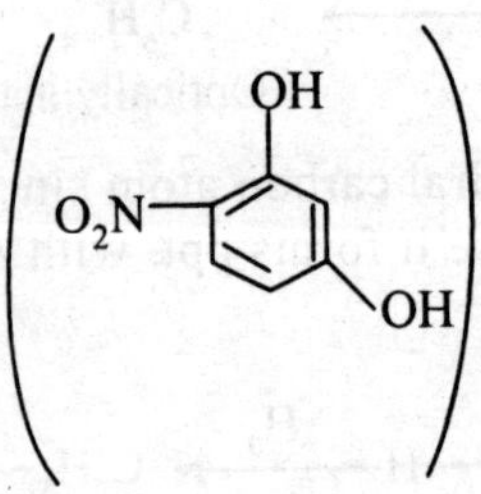

34. 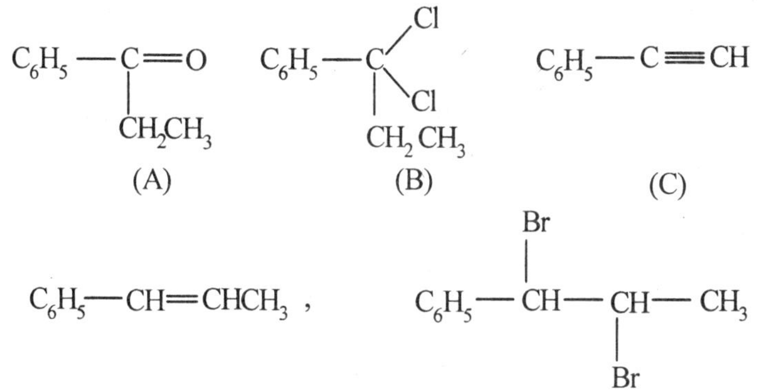

(A) (B) (C) (D) (E)

Hence,

A is Phenyl ethyl ketone 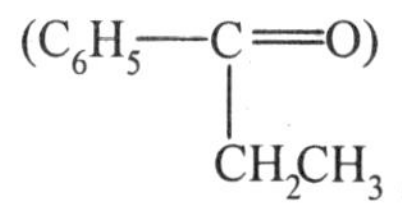

B is Ethyl dichlorotoluene 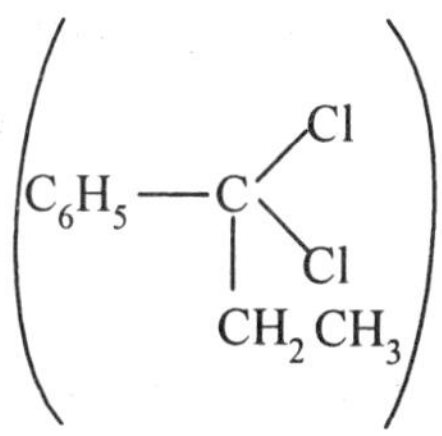

C is Phenyl acetylene (C_6H_5—C≡CH)

D is Phenyl methyl styrene (C_6H_5—CH=$CHCH_3$)

E is 3-phenyl, 1,2-Dibromopropane 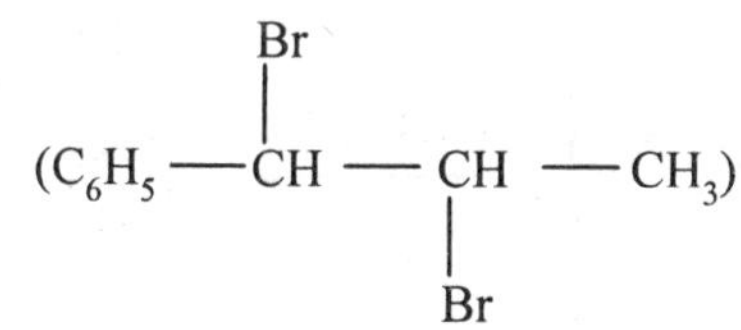

35. A reduces ammoniacal $AgNO_3$ but not the Fehling solution. Hence A is likely to be an aromatic aldehyde getting oxidized to acid B on exposure to air. B gives a vapour which burns with a smoky and luminous flame, on heating with sodalime. The vapour is that of a hydrocarbon of molecular weight 78.

Hence A is likely to be an aromatic aldehyde getting oxidized to acid B on exposure to air. B gives a vapour which burns with a smoky and luminous flame, on heating with sodalime. The vapour is that of a hydrocarbon of molecular weight 78. The vapour is C_6H_6. Hence acid B is C_6H_5COOH (Benzoic acid) and A is C_6H_5CHO (Benzaldehyde).

Hence, A is Benzaldehyde (C_6H_5CHO)

B is Benzoic acid (C_6H_5COOH)

36. $$\underset{A}{C_6H_{10}} \xrightarrow{[H]} \underset{B}{C_6H_{12}} \xrightarrow{[H]} \underset{C}{C_6H_{14}}$$

This shows that A has two double bonds. The product on ozonolysis followed by hydrolysis indicates the structure of

$$\underset{A}{CH_3CH{=}CHCH{=}CHCH_3} \longrightarrow \underset{D}{CH_3CHO} + \underset{E}{OHCCHO} + \underset{D}{OCHCH_3}$$

The product of oxidation of B indicates the structure of B as

$$\underset{B}{CH_3CH_2CH{=}CHCH_2CH_3} \xrightarrow{[O]} \underset{F}{2CH_3CH_2COOH}$$

Thus, $A = CH_3CH{=}CHCH{=}CHCH_3$ Hex 2,4-diene

$B = CH_3CH_2CH{=}CHCH_2CH_3$ hexene-3

$C = CH_3CH_2CH_2CH_2CH_2CH_3$ *n*-hexane

$D = CH_3CHO$ Acetaldehyde

$E = OHC{-}CHO$ Glyoxal

$F = CH_3CH_2COOH$ Propanoic acid

37. In the reaction A $\xrightarrow{(NH_4)_2S}$ B only one of the nitro groups is reduced to NH_2 group. The reaction B $\xrightarrow[(ii)\ CH_2Cl_2]{(i)\ NaNO_2/HCl}$ C involves converting B to a diazonium by replacing compound followed replacing diazonium group chlorine atom. In the reaction C →NaOH → D chlorine atom is replaced by hydroxyl group. Hence D is nitrophenol and it can be steam distilled.

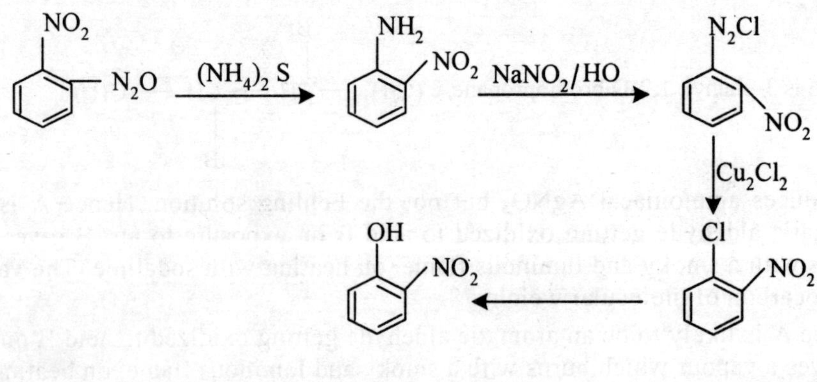

Hence,

A = *o*-Dinitrobenzene

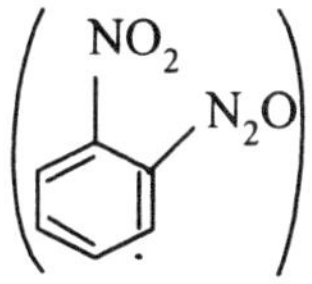

B = *o*-Nitroaniline

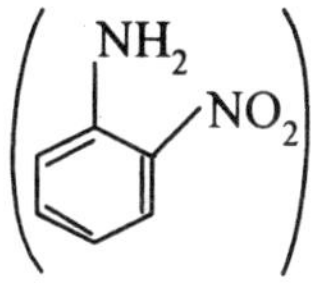

C = *o*-Chloronitro benzene

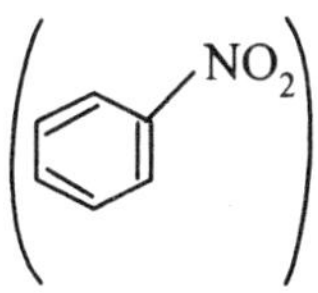

D = *o*-Nitrophenol

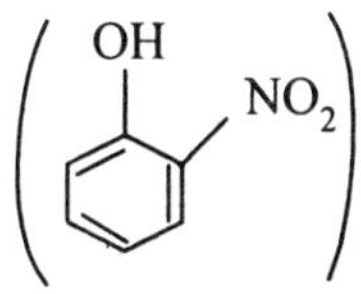

38. A must be a ketone RCOR′. On reduction it gives (B) RCHOR′ which on dehydration gives an alkene (C). The alkene on ozonolysis gives an aldehyde (D) other that acetaldehyde since CH_3CHO answers iodoform test. Therefore, A is 2-methyl pentanone-3. Therefore, reactions are as follows:

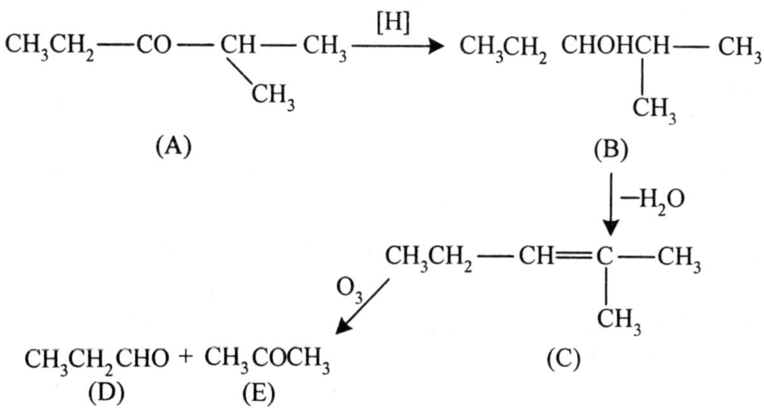

Hence,

A is 2-methyl pentanone-3

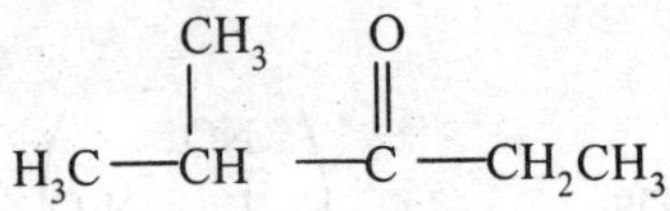

B is 2-methyl pentanol-3

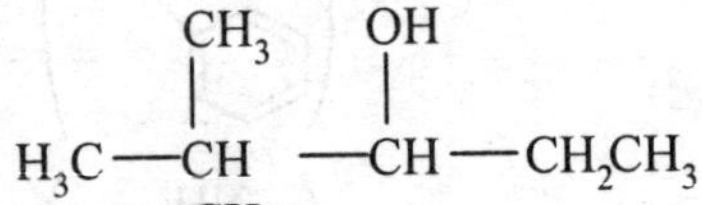

C is 2-methyl pentene-2

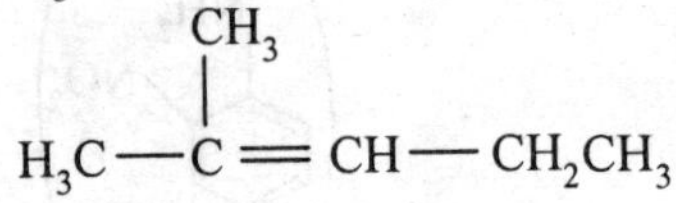

D is Propanal CH_3CH_2CHO

E is propanone CH_3COCH_3

39. $\underset{(A)}{C_7H_8} \xrightarrow{Cl_2} B \xrightarrow{H_2O} C \xrightarrow[CH_3COONa]{CH_3(CO)_2O} D$

C shows +ve reaction with Tollen's reagent. Hence, it is an aldehyde. The conversion of C into D is Perkin reaction. Therefore, C is an aromatic aldehyde. B should be a dichloride.

$$—CHCl_2 \xrightarrow{H_2O} CH(OH)_2 \longrightarrow —CHO$$

A contains phenyl ring

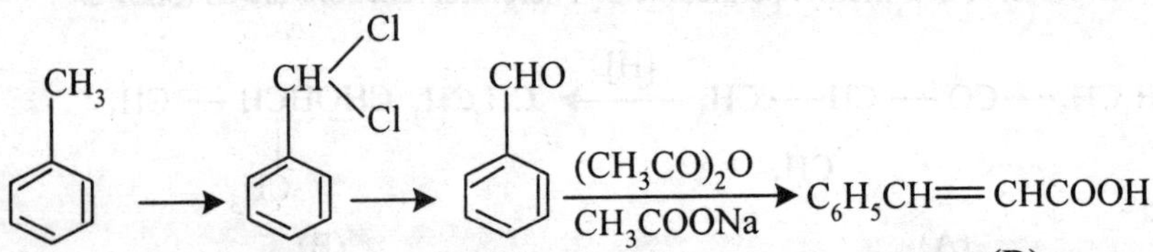

(A) (B) (C) (D) Cinnamic acid

Molecular weight of D = 72 + 6 +12 +12 +12 2 +32 = 148

This problem involves hydroboration reaction of alkene (See Chapter 9).

Hence,

A is Toluene 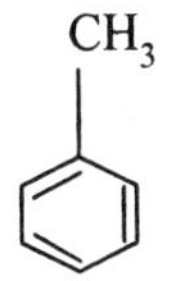

B is Benzal chloride 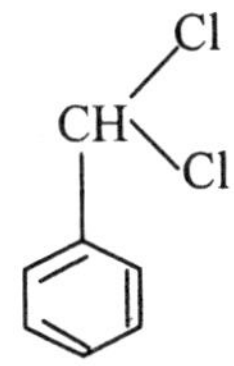

C is Benzaldehyde 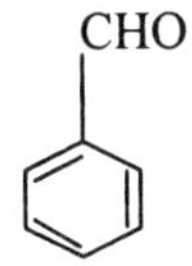

40. $$\underset{A}{6CH_3CH{=}CH_2} + B_2H_6 \longrightarrow \underset{\substack{B \\ \text{Trialkyl borane}}}{2(CH_3CH_2CH_2)_3B}$$

$$\underset{B}{(CH_3CH_2CH_2)_3B} + 3H_2O_2 \xrightarrow{OH^-/25^\circ C} \underset{B}{3CH_3CH_2CH_2OH} + B(OH)_3$$

$$CH_3CH_2CH_2OH + ZnCl_2 \xrightarrow{HCl} \underset{D}{CH_3CH_2CH_2Cl} + H_2O$$

$$(CH_3CH_2CH_2)_3B + 3H_2O \xrightarrow{H^+} \underset{E}{3CH_3CH_2CH_3} + B(OH)_3$$

Hence,

A : $CH_3CH{=}CH_2$
Propene

B : $(CH_3CH_2CH_2)_3B$
Tripropyl borane

C : $CH_3CH_2CH_2OH$
Propanol-1

D : $CH_3CH_2CH_2Cl$
1-chloropropane

E : $CH_3CH_2CH_3$
Propane

41. 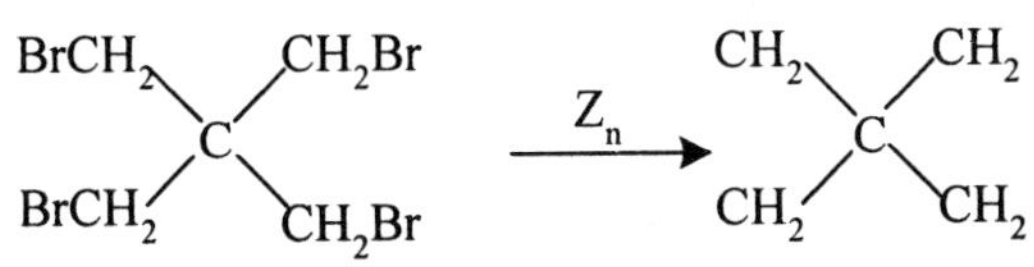

(C) Spiropentane

42. $$(A), (B) \xrightarrow{\text{Hydrogen}} (C)$$
$$C_{11}H_{13}OCl \xrightarrow{[O]}$$

Same isomeric compounds

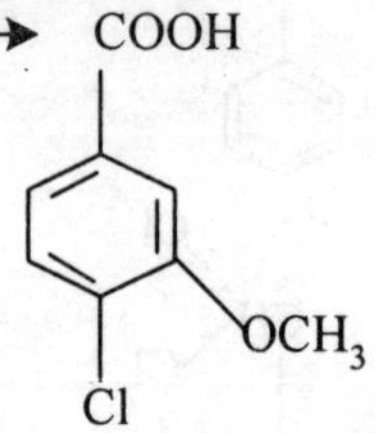

Thus, the possible structures of A and B may be written as

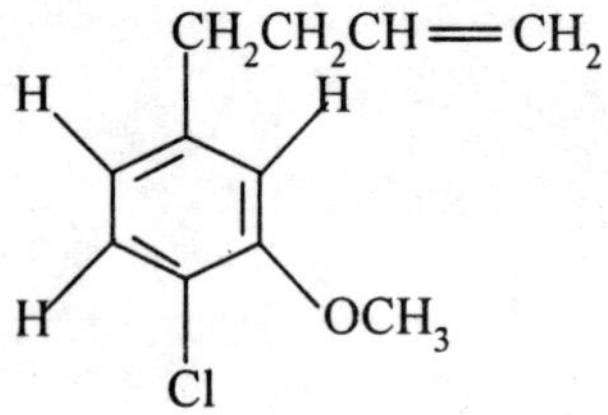

(1)

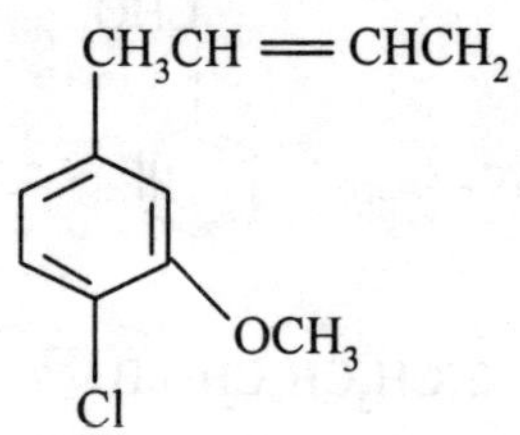

(2)

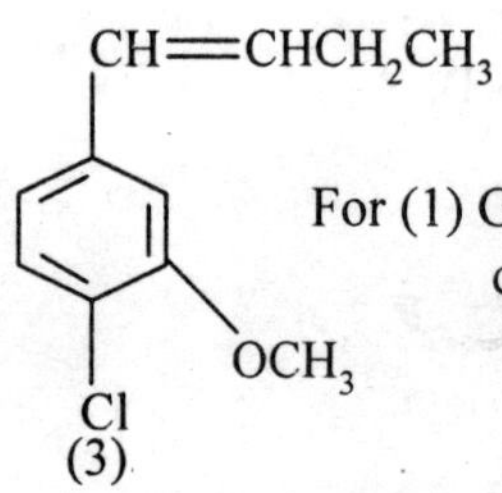

(3)

For (1) Geometrical isomers do not exist

(A) is (D) + (E)

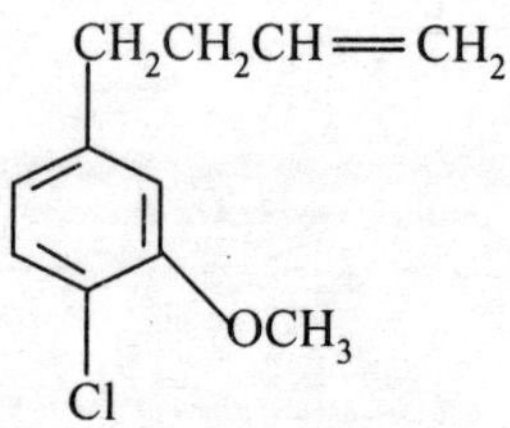

(1)

p-(*p*-chloro-m-methoxy butene-1)

Hence,

A: D + E

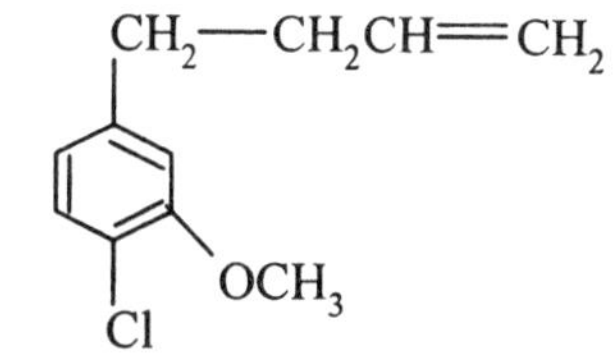

B:

C: *p* (*p*-chloro-m-methoxy butene-1)

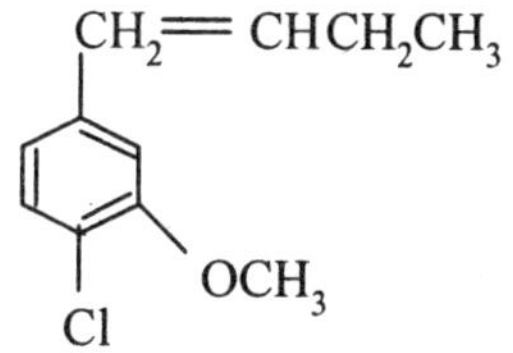

D:

1-(*p*-chloro-m-methoxy phenyl butene-1)

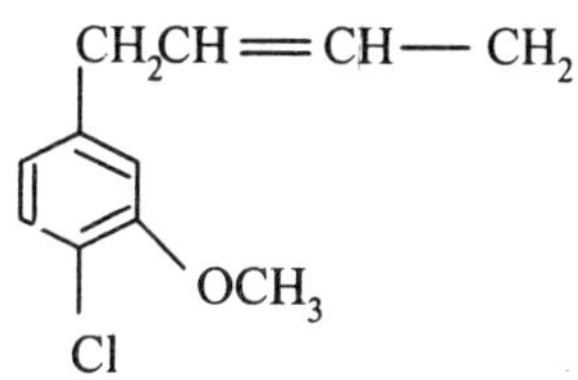

E:

1-(*p*-chloro-m-methoxy phenyl butene-1)

43. Percentage of O = 100 45.8 = 54.2%. Thus, the possible acid is $\begin{matrix} CH_2COOH \\ \diagdown \\ CH_2COOH \end{matrix}$

It's ethyl ester will be $\begin{matrix} CH_2COOC_2H_5 \\ | \\ CH_2COOC_2H_5 \end{matrix}$. M. Wt. = 87 × 2, which agress with the given data. Hence, A is $\begin{matrix} CH_2COOH \\ | \\ CH_2COOH \end{matrix}$. Its silver salts will be

$\begin{matrix} CH_2COOAg \\ | \\ CH_2COOAg \end{matrix}$. Percentage Ag in the silver = $\dfrac{2(108) \times 100}{332.2} = 65.1\%$

Possible isomers are:

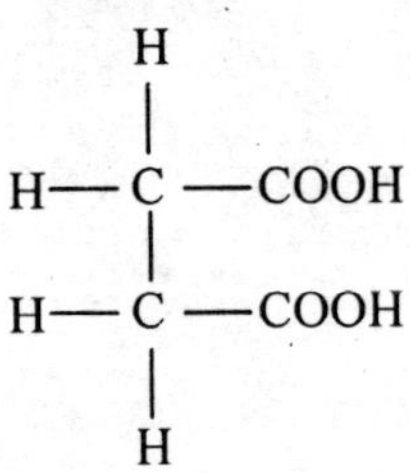

(i)

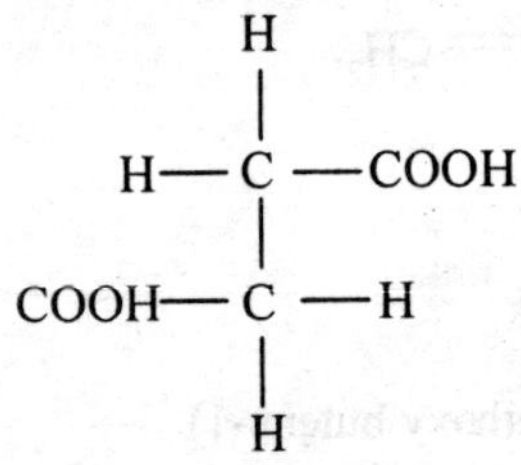

(ii)

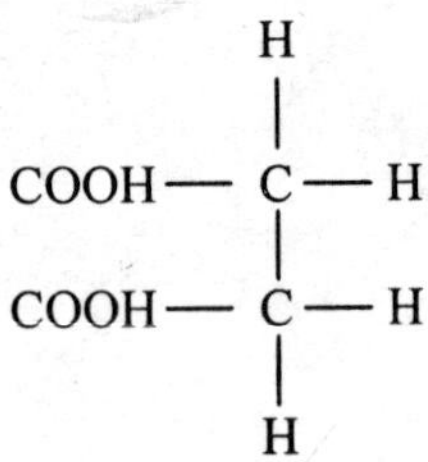

(iii)

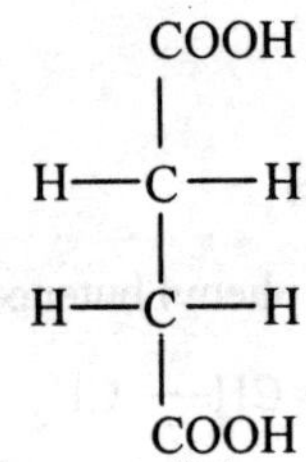

(iv)

Hence, A is succinic acid $\begin{array}{l} CH_2COOH \\ | \\ CH_2COOH \end{array}$

44. 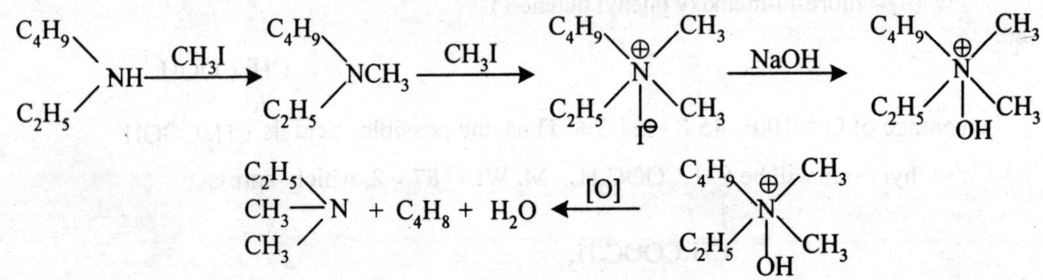

Isobutene may be produced either of the following chains in the alkyl group

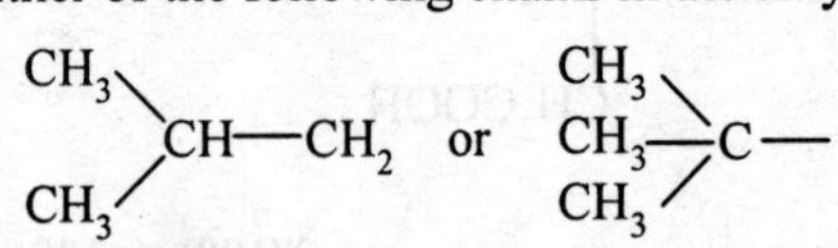

Thus, x is $(CH_3)_2CHCH_2C_2H_5NH$ (ethyl isobutyl amine)

or 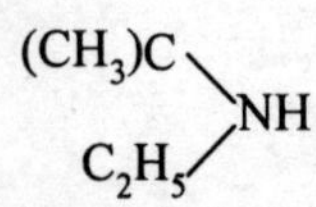

Ethyl tertiary butyl amine

Hence, May be ethyl isobutyl amine or ethyl tertiary butyl amine.

45. Since compound E (C_5H_8) absorbs 2 moles of hydrogen to form compound F (C_5H_{12}) therefore, it may be alkyne or alkadiene. However, the formation of two ozonolysis products exclude alkyne. Hence, E must be an alkadiene.

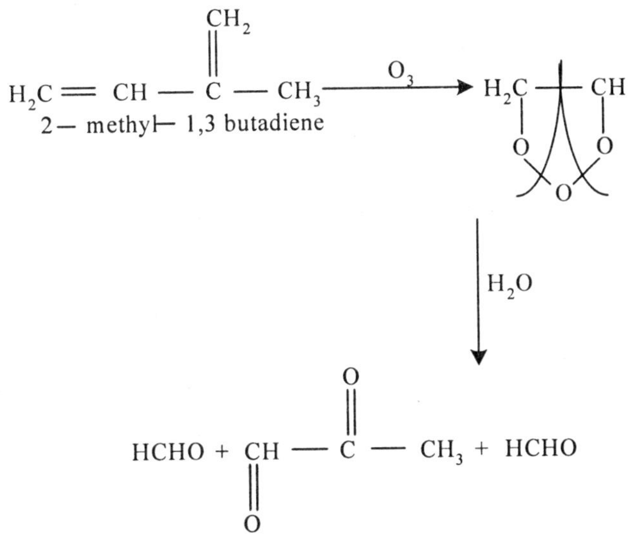

Hence, E is 2-methyl -1,3 - butadiene ($CH_2=CH-C(=CH_2)-CH_3$)

46. *(i)* Brief summary

(A) $\xrightarrow{\text{Reduction}}$ (B) Pri. amine $\xrightarrow{HNO_2}$ Ethyl alcohol

(B) $\xrightarrow{CHCl_3 \text{ and } KOH}$ (C) with the offensive odour $\xrightarrow{\text{Reduction}}$ Ethylmethyl amine

(ii) B is ethylamine because of reaction with HNO_2 it gives ethyl alcohol. Hence A is methyl cyanide.

$$\underset{\text{(A) Methyl cyanide}}{CH_3CN} \xrightarrow{4H} \underset{\text{(B) Ethyl amine}}{CH_3CH_2NH_2} \xrightarrow{HNO_2} \underset{\text{Ethyl alcohol}}{CH_3CH_2OH}$$

$$CH_3CH_2NH_2 \xrightarrow{CHCl_3/KOH \text{ carbylamine reaction}} \underset{\text{Ethyl isocyanide}}{CH_3CH_2NC} \xrightarrow{H} \underset{\text{Ethyl methyl amine}}{C_2H_5NH\,CH_3}$$

Hence,

A - methylcyanide (CH_3CN)

B - Ethyl amine $(CH_3CH_2NH_2)$

C - Ethyl isocyanide (CH_3CH_2NC)

47. *(i)* Brief summary

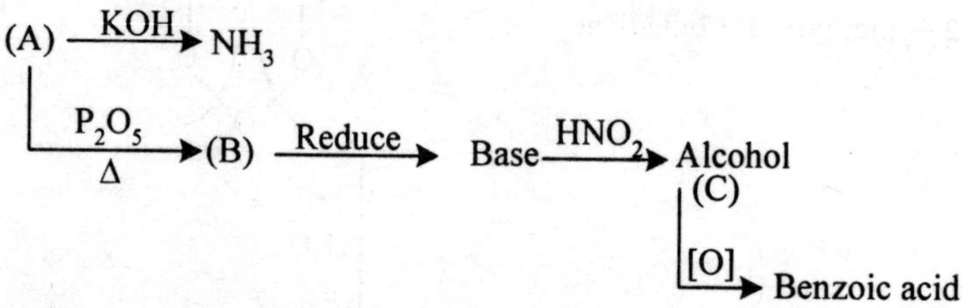

(ii) Calculate molecular formula of A and B.

(iii) Alcohol C is benzyl alcohol since, on oxidation, it yields benzoic acid.

(iv) Base should be Benzylamine and therefore B should be phenyl cyanide. Obviously A is benzamide because when heated with KOH, it evolves NH_3.

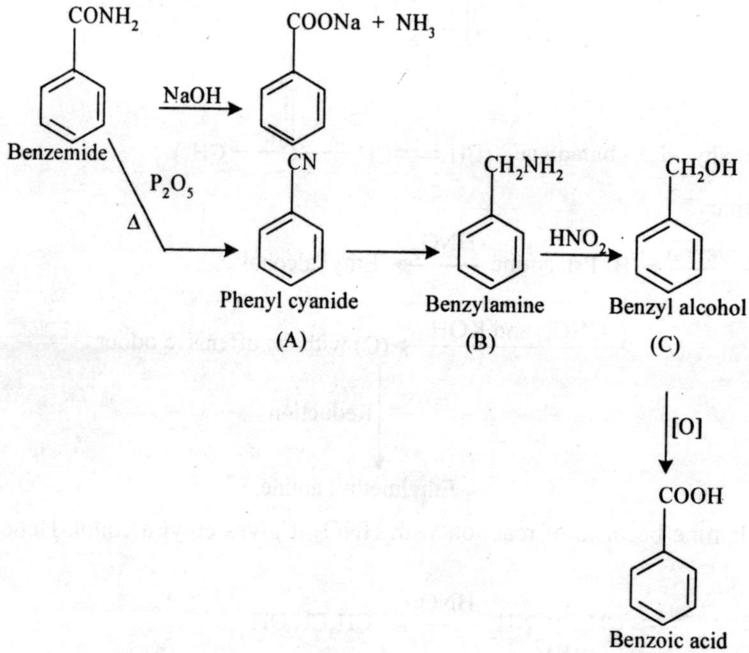

Hence, A - Benzamide

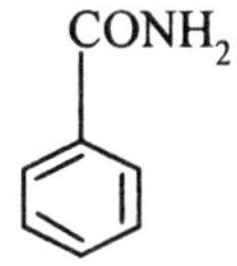

B - Phenyl cyanide

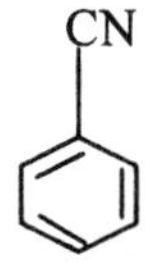

C - Benzyl alcohol

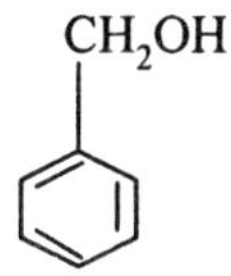

48. *(i)* Brief summary

$$\underset{(A)}{C_8H_8} \xrightarrow{Br_2} \underset{(B)}{C_8H_8\,Br_2}$$

(A) $\xrightarrow{[O]}$ (C) Monobasic acid

(C) $\xrightarrow{\text{Distillation with sodalime}}$ Benzene

(ii) Since Bromine is added A must contain a double bond. Moreover, final product is benzene. Hence A must be a aromatic compound, i.e., it is styrene

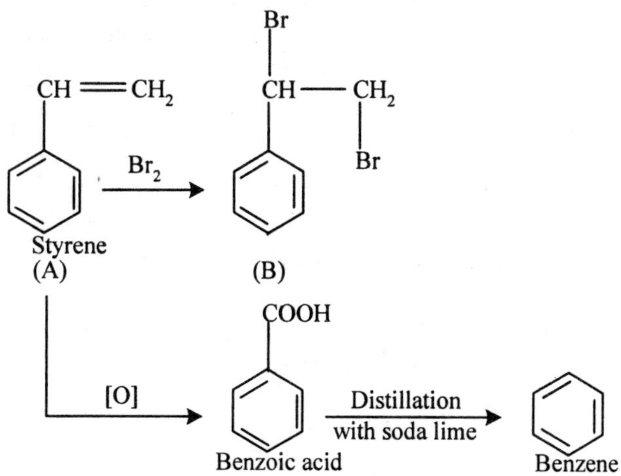

A - Styrene

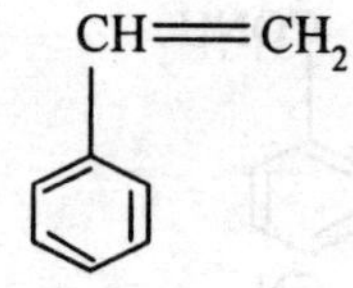

B - *p*-phenyldibromo styrene

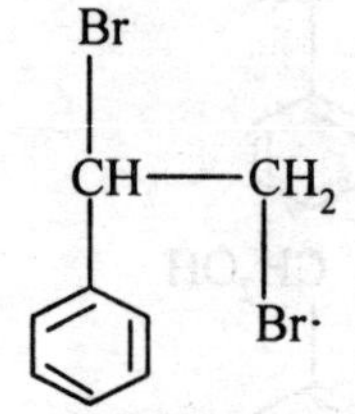

C - benzoic acid

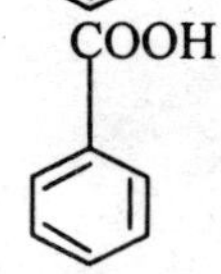

49. *(i)* Brief summary

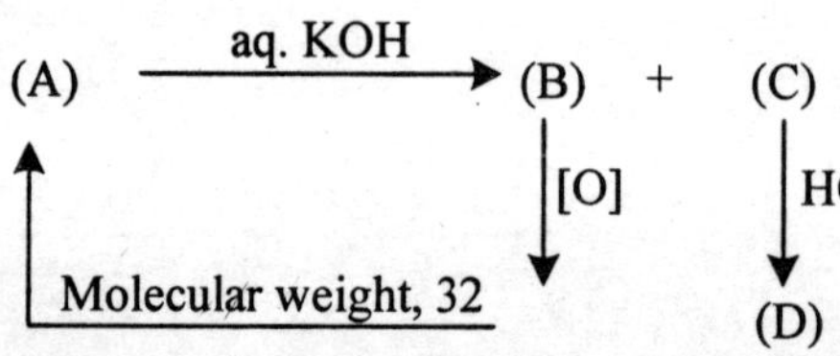

It reduces Fehling solution and ammoniacal $AgNO_3$. It gives effervescence with $NaHCO_3$.

(ii) D is formic acid (HCOOH) because it reacts with $NaHCO_3$ due to —COOH group. Moreover, it is the only acid which reduces Fehling solution as well as ammoniacal $AgNO_3$. So, A is formaldehyde, which undergoes Cannizzaro reaction with aq. KOH (Molecular weight confirms this). Therefore,

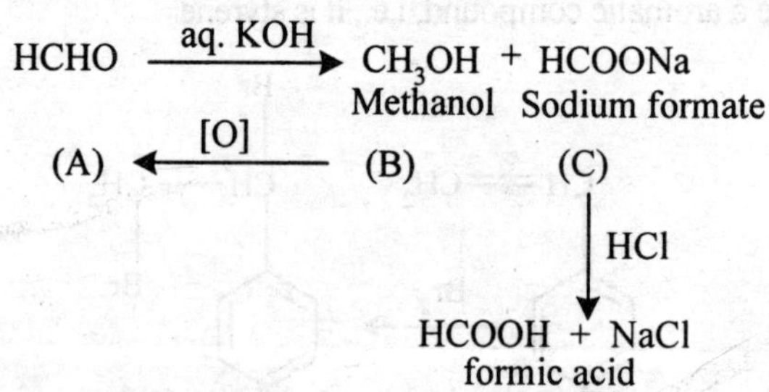

Hence, A - Formaldehyde (HCHO)

B - Methanol (CH_3OH)

C - Sodium formate (HCOONa)

D - Formic acid (HCOOH).

50.

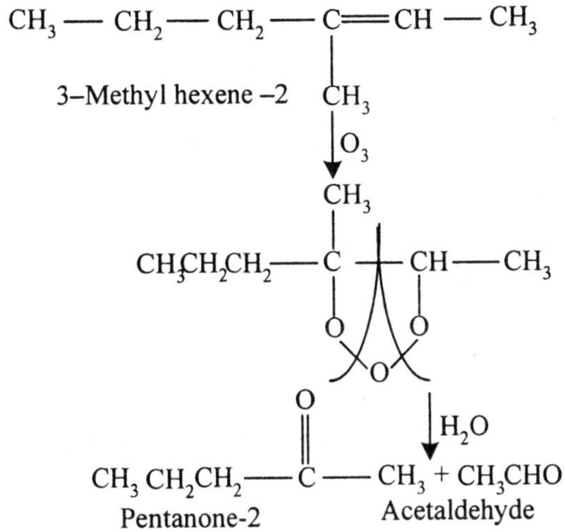

51. The olefin is 3-methyl hexane-2

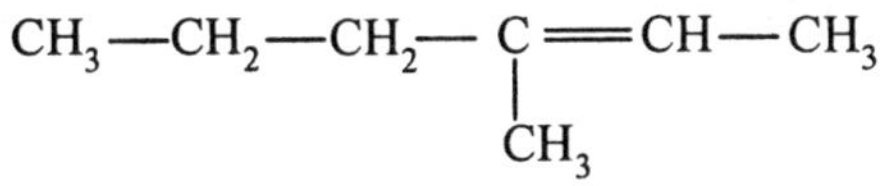

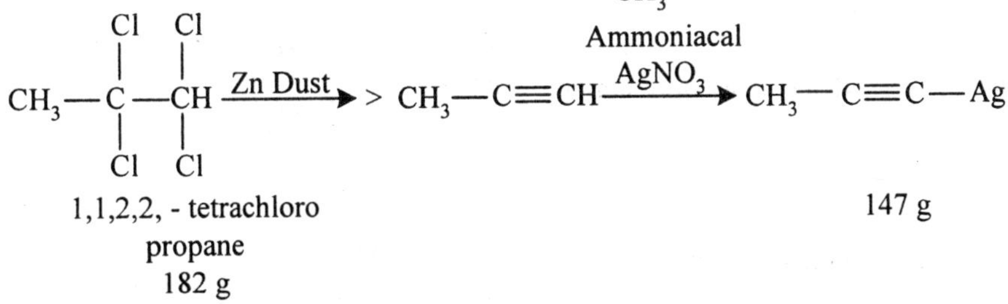

Therefore, the weight of precipitate = 147 × 36.4/182 = 29.4g

52. Ethanol does not react with Fehlilng solution; hence the red ppt. is obtained from acetaldehyde.

$$\underset{44\text{ g}}{CH_3CHO} + 2CuO \longrightarrow CH_3COOH + \underset{\text{Red ppt., 143 g}}{Cu_2O}$$

Therefore, the weight of acetaldehyde = 44 × 1.2/143g = 0.369g.

Percentage of acetaldehyde = 0.369 × 100/0.535 = 68.9%

53. *(i)* Brief summary

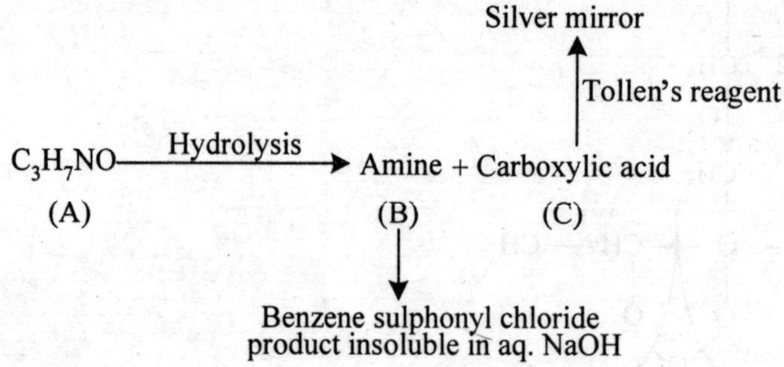

(ii) C is formic acid because it reduces Tollen's reagent and B is a secondary amine because product of the reaction with the benzene sulphonyl chloride is insoluble in aq. NaOH. Therefore,

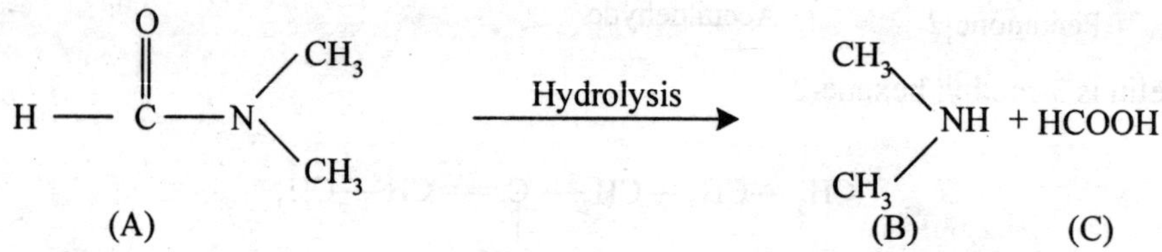

Hence, A - NN Dimethyl formamide

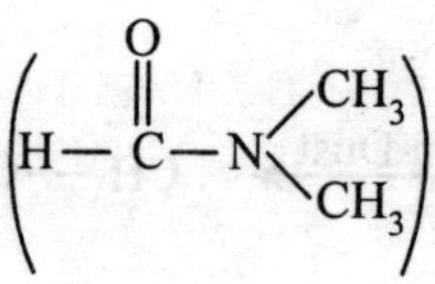

B - Dimethyl amine

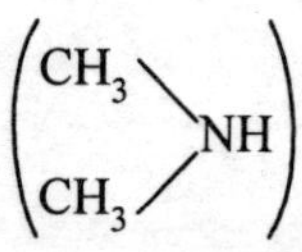

C - Formic acid (HCOOH)

54. *(i)* Brief summary

$$\underset{(A)}{C_7H_{15}Cl} \xrightarrow{\text{Alc. KOH}} \underset{(B)}{C_7H_{14}}$$

$$A \longrightarrow B \xrightarrow{O_3} \text{Acetone + butyraldehyde}$$

(ii) Find out alkene from ozonolysis product and then alkyl halide using Saytzeff 's rule.

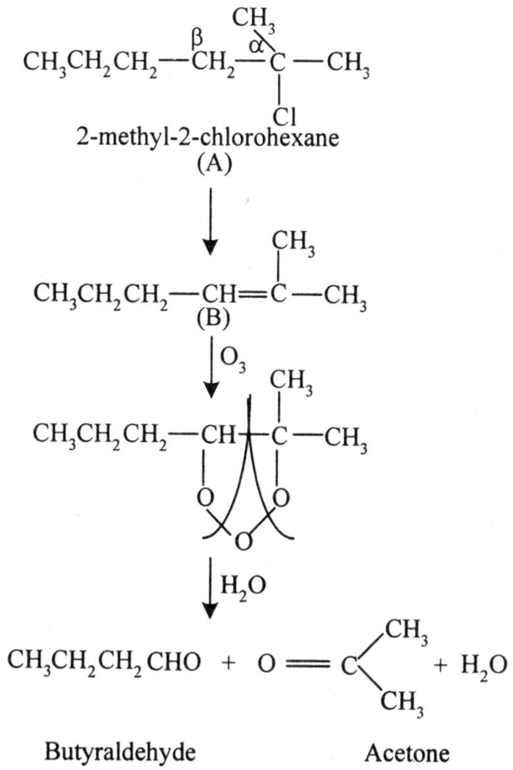

Hence, (A) 2-methyl-2-chlorohexane

(B) 2-methyl hexene-2.

55. *(i)* Brief summary: 10 g of a mixture $\xrightarrow{\text{Na}}$ 200 ml H_2 at 27 °C and 760 mm of hexane and ethanol.

(ii) Let us find out moles of H_2 using Avogadro's hypothesis volume of H_2 at NTP = 273 × 200/300 = 182 ml moles of H_2 = 182/22400

(iii) The reaction is as follows:

$$\underset{46\text{ g}}{C_2H_5OH + Na} \longrightarrow 1/2\ H_2 + \underset{0.5\text{ moles}}{C_2H_5ONa}$$

Remember that hexane will not react with sodium. Therefore, the hydrogen is obtained from ethanol

$$= \frac{46 \times 182}{.5 \times 22400}\ \text{g}$$

Percentage ethanol in mixture

$$= \frac{46 \times 182 \times 100}{10 \times .5 \times 22400} = 7.4\%$$

56. *(i)* Brief summary:

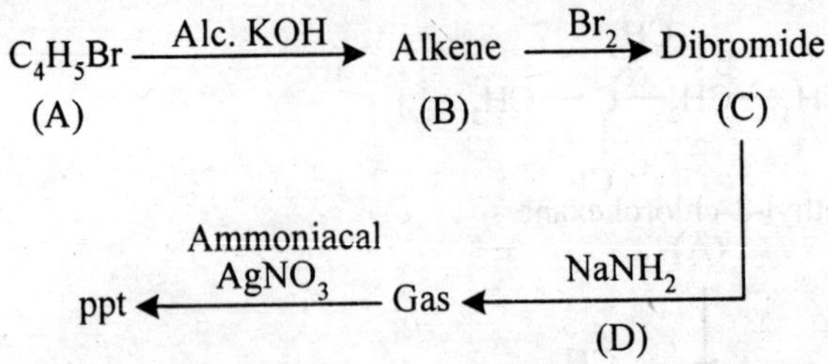

(ii) Gas D is ethyne or alkyne-1, i.e., butyne-1 because these give white ppt. with ammoniacal $AgNO_3$.

(iii) Since only one alkene is formed hence bromine atom should be attached to carbon-1, because in rest of the cases, two alkenes would have been obtained according to Saytzeff's rule.

Therefore,

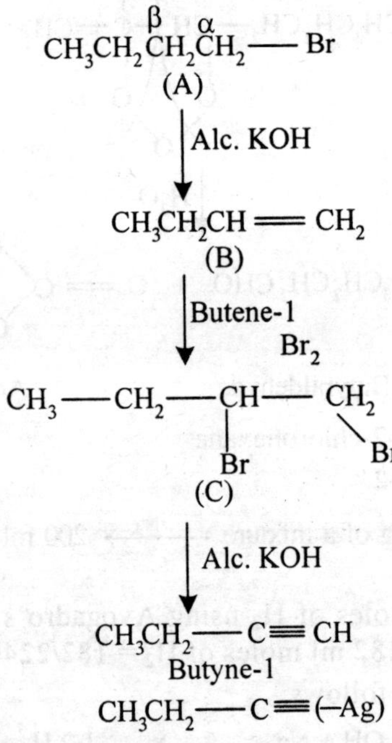

Hence,

A is 1 - Bromobutane $(CH_3CH_2CH_2—CH_2Br)$

B is Butene-1 $(CH_3CH_2CH{=}CH_2)$

C is 1, 2 - Dibromobutane $(CH_3—CH_2—\underset{|}{\overset{Br}{C}}H—CHBr)$

D is Butyne-1 $(CH_3CH_2C{\equiv}CH)$

57. Alkyl halides undergo Wurtz reaction. Three alkanes are obtained.

(i)

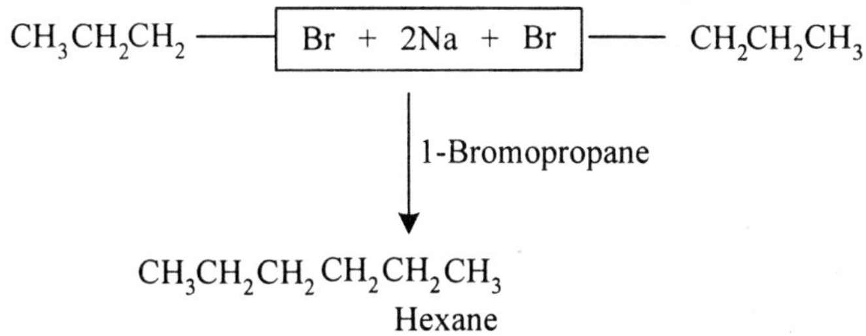

(ii)

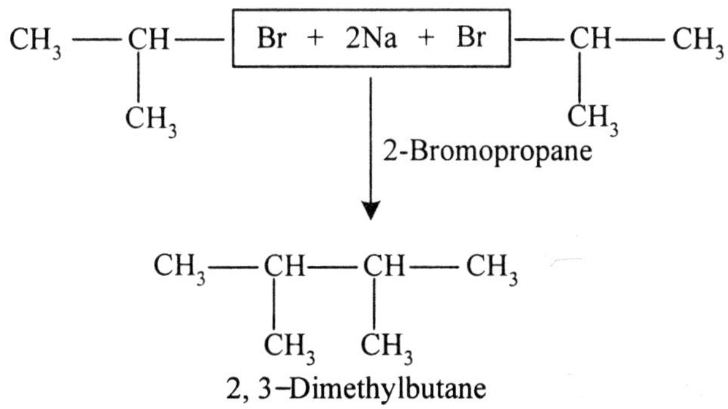

(iii)

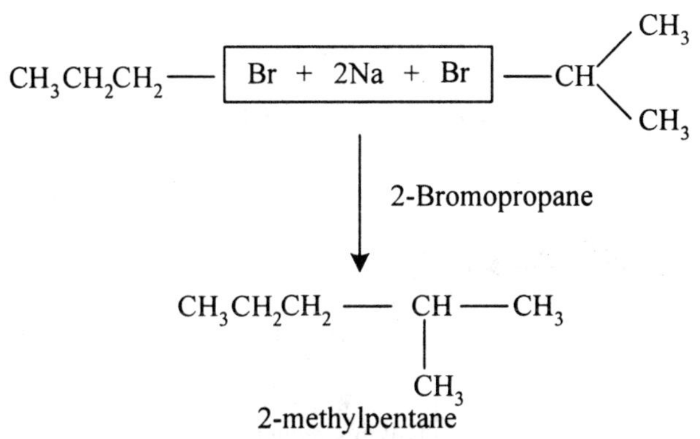

58.

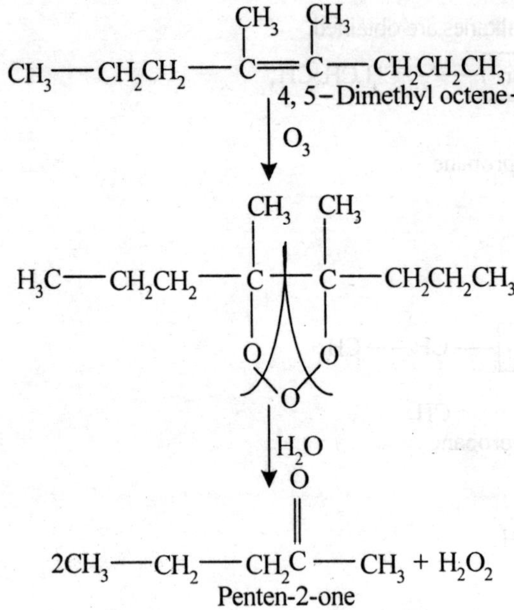

Hence, A is 4,5 – Dimethyl octene-4

$$(CH_3—CH_2CH_2—\underset{}{\overset{CH_3}{\overset{|}{C}}}=\overset{CH_3}{\overset{|}{C}}—CH_2CH_2CH_3)$$

59. *(i)* Brief summary:

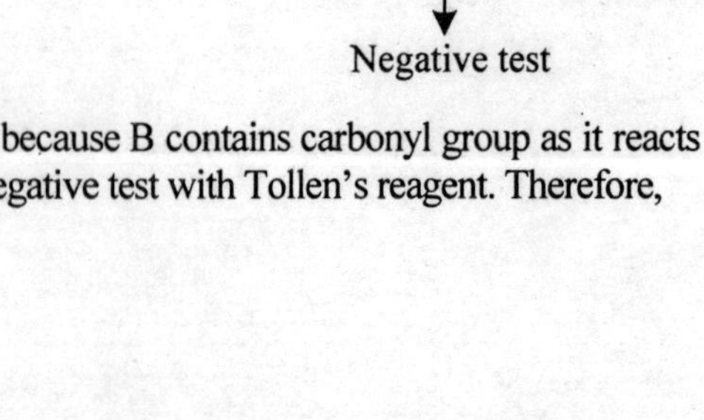

(ii) A is a gem dihalide because B contains carbonyl group as it reacts with NH_2OH. Moreover B is a ketone because of negative test with Tollen's reagent. Therefore,

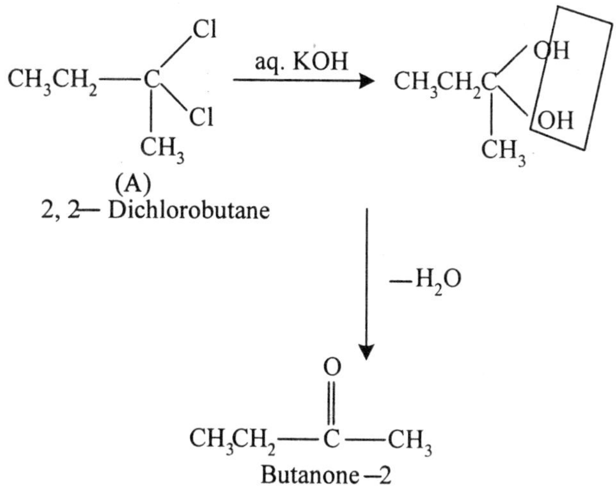

Hence, A - 2,2 - Dichlorobutane 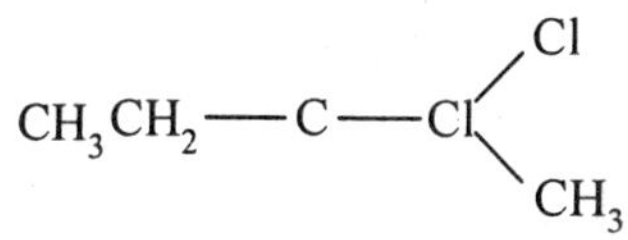

B - Butanone-2 $(CH_3CH_2-\overset{\overset{O}{\|}}{C}-CH_3)$

60. *(i)* Brief summary

A reacts with conc. H_2SO_4 and decolorizes Br_2 in CCl_4

$\xrightarrow{HBr} B \xrightarrow{NaOH} C \xrightarrow{[O]}$ Hexanone–3

(ii) Start from hexanone-3 and go back to identify A.

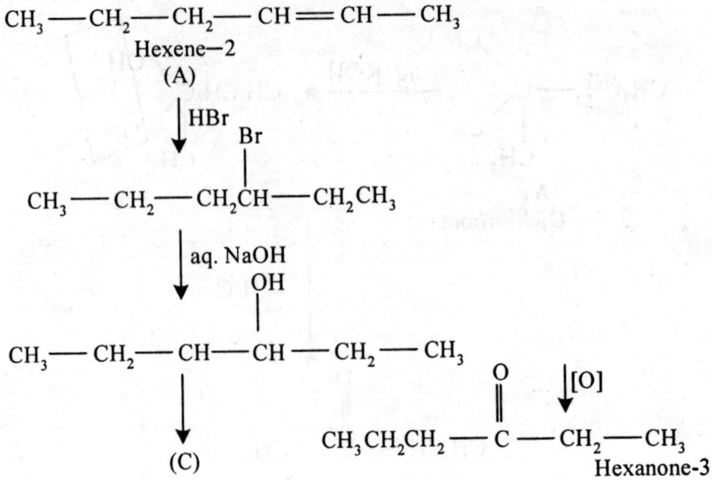

Hence,

(A) Hexene -2 (CH_3—CH_2CH_2—CH—CH_3)

(B) 3 - bromohexane (CH_3—CH_2—$CH_2CH(Br)$—CH_2CH_3)

(B) - Hexanol - 3 (CH_3—CH_2—CH_2—$CH(OH)$—CH_2CH_3)

61. *(i)* Brief summary

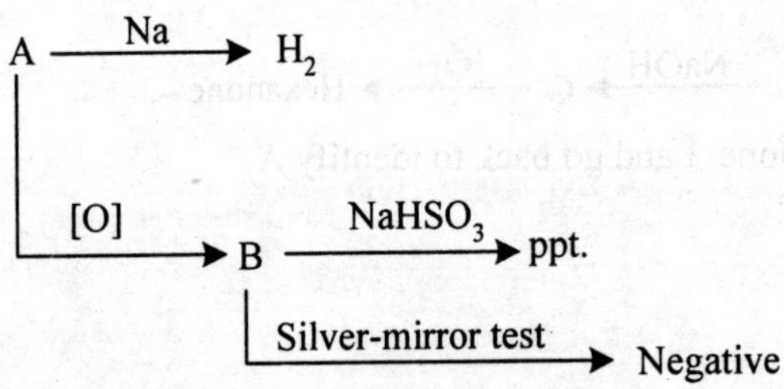

(ii) Calculate molecular weight of A using colligative properties. If molecular weight of A is M

Molality	=	7.41/M × 1000/100 = 74.1/M
ΔT	=	KB × Molality
Where ΔT	=	Elevation in boiling point
KB	=	motal elevation constant
1.86	=	1.86 × 74.1/M
M	=	74.1

(iii) Since it evolves hydrogen with sodium hence it is alcohol. Compound B is a ketone because it reacts with $NaHSO_3$ and does not give silver-mirror test. Therefore, A is a secondary alcohol with molecular weight 74.

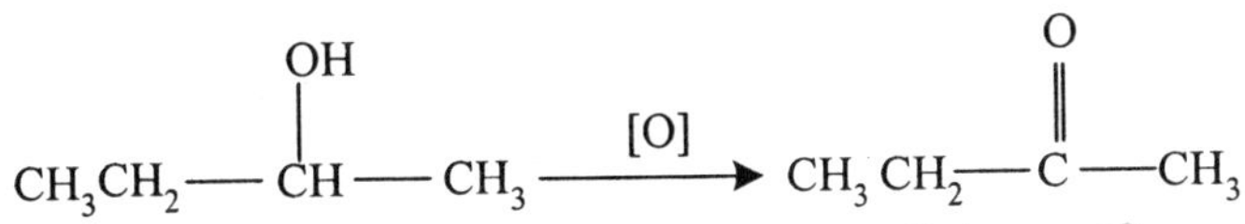

Butanol -2 (A) Butanone -2 (B)

62. *(i)* Brief summary

$$A \xrightarrow{\text{NaOH and } Br_2} B \xrightarrow{NaNO_2/HCl} \text{Ethanol}$$

(ii) Since A contains C, H, N, O, it may be an amide. This is further confirmed because it reacts with NaOH and Br_2 (Hoffmann bromide reaction). Since ethanol is obtained from B, B is ethylamine and A is propanamide

$$\underset{(A)}{CH_3CH_2CONH_2} \xrightarrow{\text{NaOH and } Br_2} \underset{(B)}{CH_3CH_2NH_2} \xrightarrow{HNO_2} \underset{\text{Ethanol}}{CH_3CH_2OH}$$

63. *(i)* Brief summary

$$C_7H_8 \xrightarrow{Cl_2} B \xrightarrow{\text{Hydrolysis}} C \xrightarrow{\text{Tollen's reagent}} \text{+ve test}$$

$$\downarrow$$

CHCOONa and acetic anhydride acid

D

molecular weight,148

(ii) C gives +ve test with Tollen's reagent hence it is an aldehyde. It undergoes Perkin reaction, hence C is benzaldehyde. Therefore, A is toluene

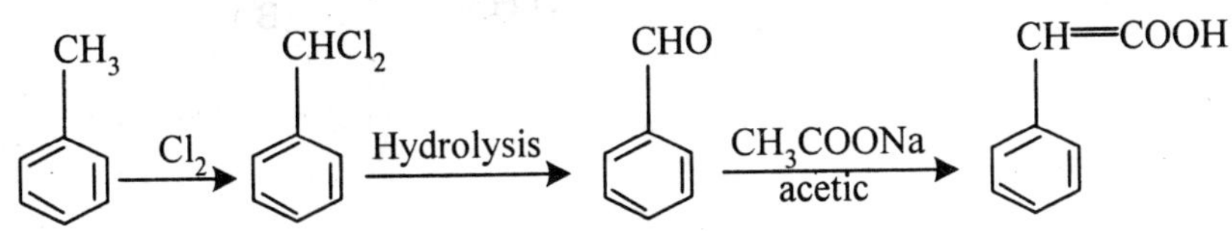

Toluene	Benzyl chloride	Anhydride benzaldehyde	Cinnamic acid
(A)	(B)	(C)	(D)

mol. wt. of cinnamic acid (148) confirms the compound A to D.

Hence, A — Toluene 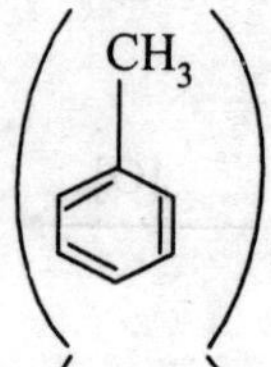

B — Benzyl chloride 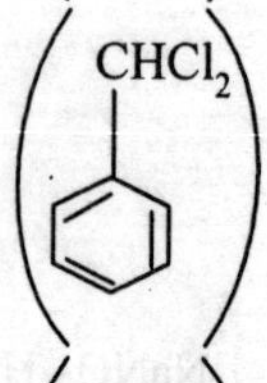

C — Benzaldehyde 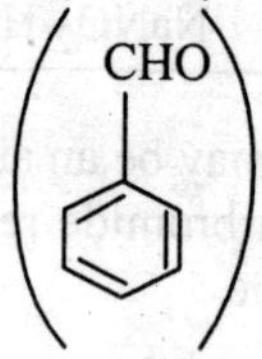

D — Cinnamic acid 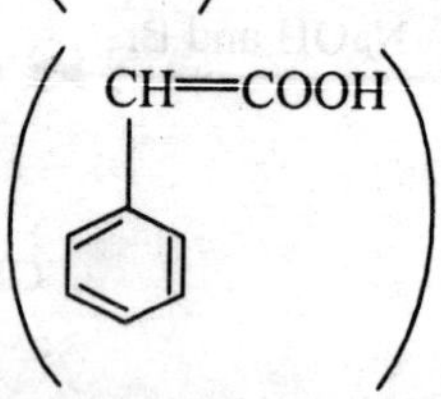

64. The I.H.D of the compound is 4 (C_8H_{18}–C_8H_{10}), hence it is likely to contain triple bond (I. H. D. = 2) and two rings (I. H. D. = 2),

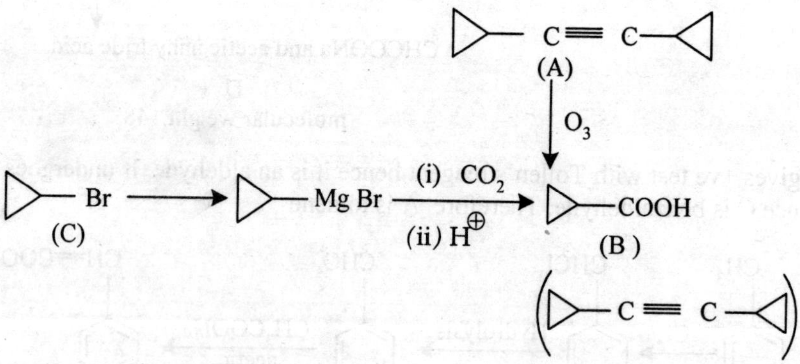

Hence, A — Dicyclopropyl ethyne
B — Cyclopropyl methanoic acid
C — Cyclopropyl bromide

65. Since the compound gives yellow precipitate with I_2 and alkali, i.e., it undergoes iodoform test,

therefore, it must contain 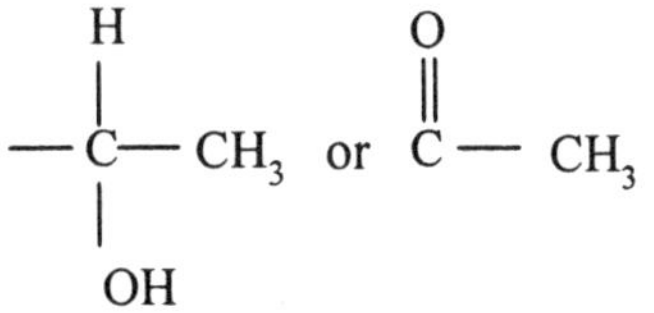 group. Hence D is $C_6H_5CHOHCH_3$

$$\underset{(D)}{C_6H_5\,CHOH\,CH_3} \xrightarrow{I_2/\text{Alkali}} \underset{\text{Yellow ppt.}}{CHI_3} + C_6H_5COO^{\ominus}$$

$$\downarrow H^{\oplus}$$

$$C_6H_5\,COOH$$

Hence, D - Phenyl ethanol 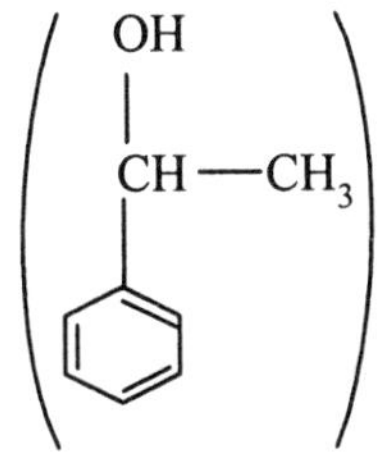

E - Benzoic acid 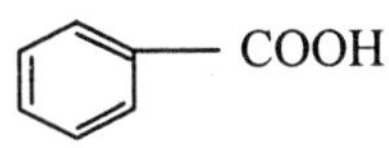

66. 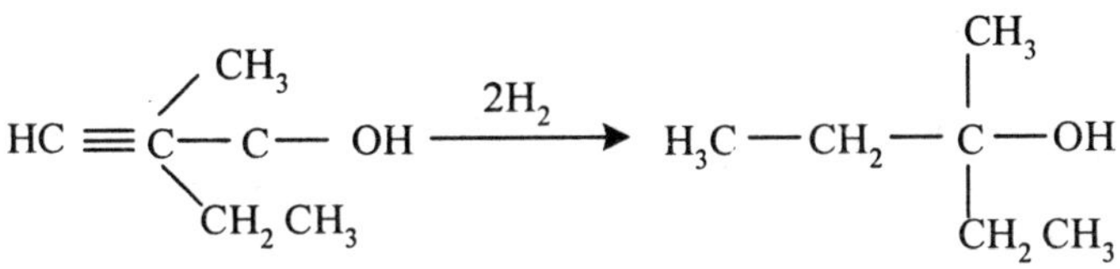

It is optically active because it contains one asymmetric carbon atom.

B is optically inactive because the carbon atom is symmetric. It has also resistance to oxidation because is tertiary alcohol.

A : 2 methyl-pent-1-yne-2-ol 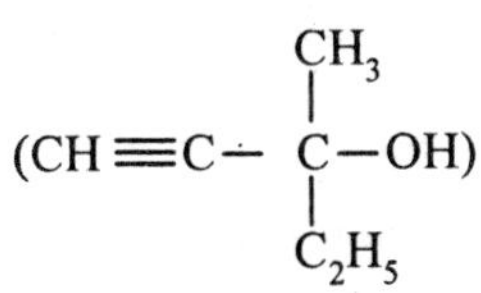

B : 2-methyl-2-pentanol 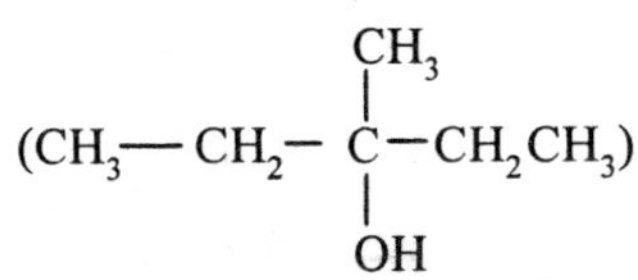

67. Formation of two alkene upon treatment of alkyl halide (x) suggests that there are two B carbon atoms. Hence, the halogen should be present on second carbon atom. Further the formation of 2,3-dimethyl butane suggests that 'X' is

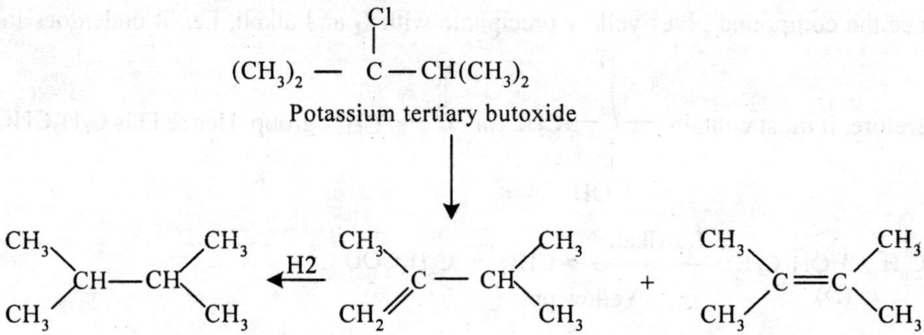

Hence, X : 2,3-Dimethyl-2-Chlorobutane
Y : 2,3-Dimethyl butene-2
Z : 2,3-Dimethyl butane.

68. Brief summary

$$A\ \ NaOH \longrightarrow B\ \ NH_3 \xrightarrow{H^+\ acid} C \xrightarrow{PCl_5} D \xrightarrow{B} E$$

S.no.	*Element*	*Percentage atomic weight*	*Ratio*
1.	C	71.11/12 = 5.926	5.926/.740 = 8
2.	H	6.670/1 = 6.670	6.670/0.740 = 9
3.	N	10.37/14 = 0.740	.740/.740 = 1
4.	O	11.85/16 = 0.740	.740/.750 = 1

Empirical formula C_8H_9NO
Molecular formula C_8H_9NO
It is a disubstituted aromatic compound it should have following structure:

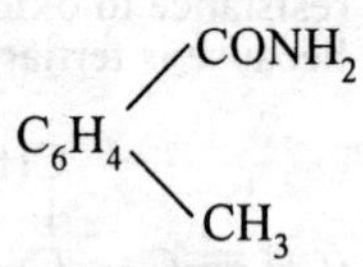

Since amide group is meta-directing hence it should be

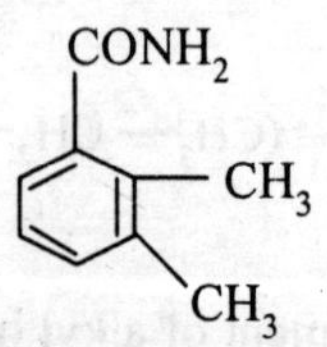

m-methyl benzamide

The reactions are as follows:

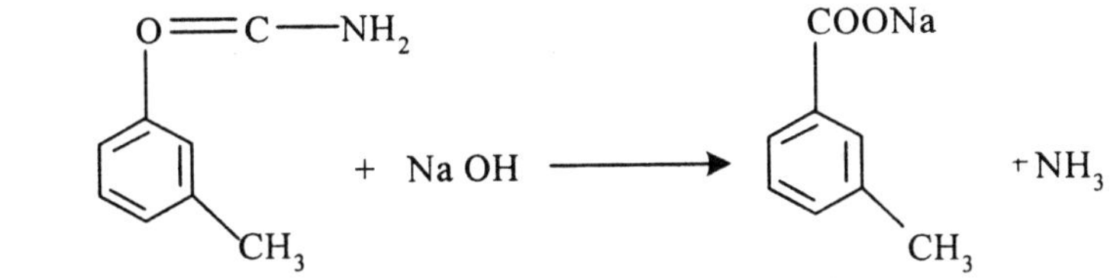

m-methyl benzamide *m*-methyl sodium benzoate

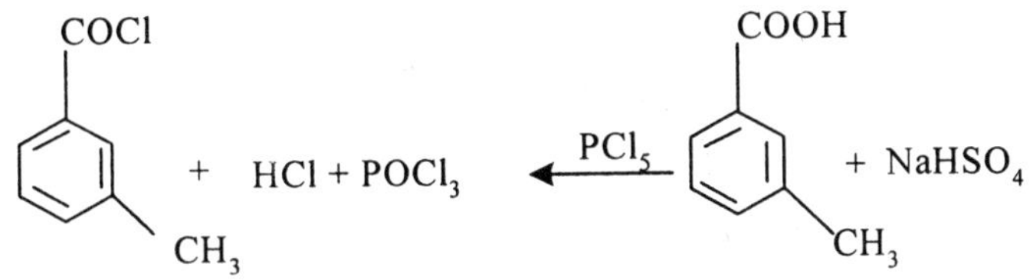

Benzoylchloride *m*-methyl benzoic acid

(D)

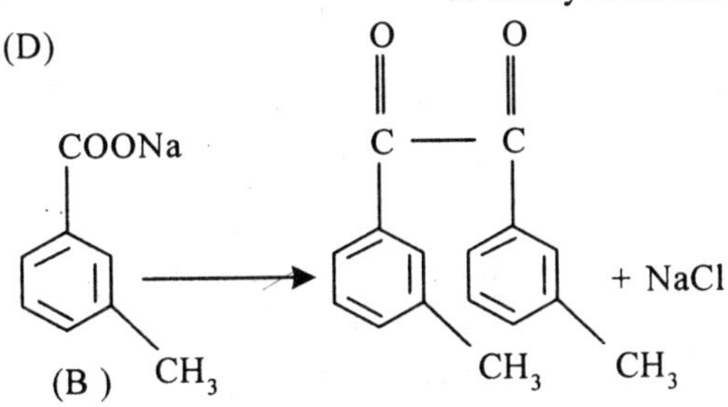

m,m-Dimethyl benzoic anhydride

Hence,

A = *m*-methyl benzamide

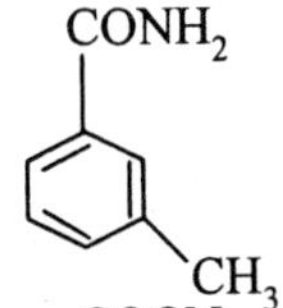

B = *m*-methyl sodium benzoate

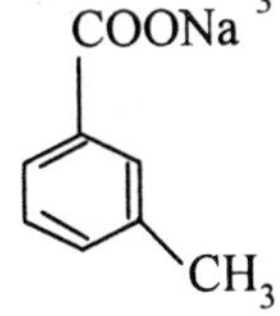

C =*m*-methyl benzoic acid

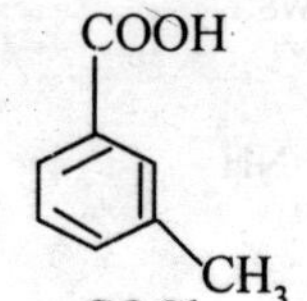

D = *m*-methyl benzoyl chloride

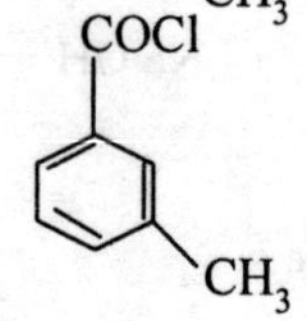

E = *m*,*m*′-Dimethyl benzoic anhydride

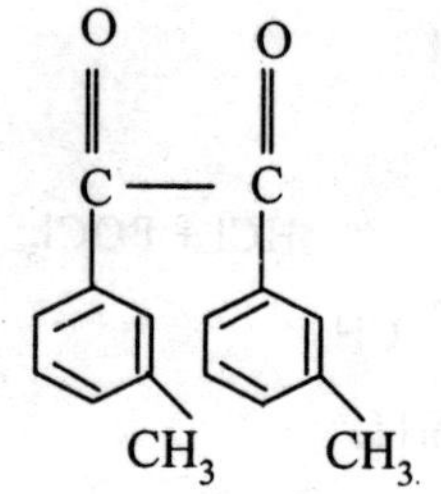

69. Brief summary

(i) $A \xrightarrow{\text{Cerric ammonium nitrate}} \text{Characteristic color} \xrightarrow{PCl_5} B \xrightarrow{KOH}$

$C \xrightarrow{Na/C_2H_5OH} D \xrightarrow{-NH_3} E + NH_3 \xrightarrow{\text{Nitrobenzene}} \text{Pyridine}$

(ii) Since A contains C, H, and O and it gives characteristic color with cerric ammonium nitrate, hence A must be an alcohol, i.e., ethyl alcohol. Therefore, the chemical reactions may be written as below:

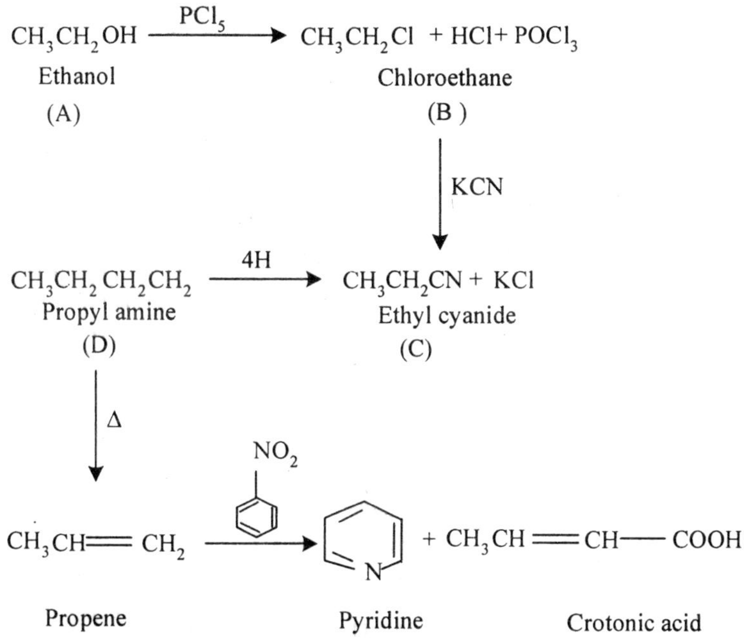

Hence,	A = Ethanol	C_2H_5OH
	B = Chloroethane	C_2H_5Cl
	C = Ethyl cyanide	C_2H_5CN
	D = Propyl amine	$CH_3CH_2CH_2NH_2$
	E = Propene	$CH_3CH{=}CH_2$

70. Compound A undergoes hydrolysis with aq. KOH, whereas it undergoes dehydrohalogenation with alc. KOH to yield alkene. The reaction of propyne with Hg^{++} in the presence of H_2SO_4 yield propanone (see Chapter 9), therefore 'D' is propanone.

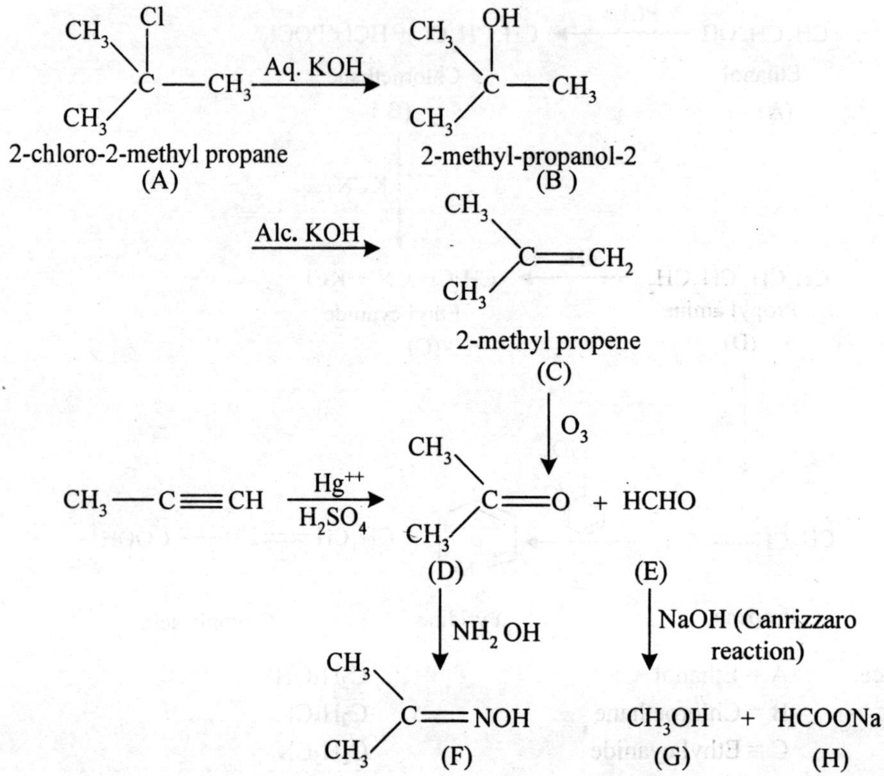

Hence,		
A —	2-chloro-2-methyl propane	$(CH_3)_2—C(Cl)—CH_3$
B —	2-methyl propanol-2	$(CH_3)_2—C(OH)—CH_3$
C —	2-methyl propene	$(CH_3)_2\ C{=}CH_2$
D	Propanone	$(CH_3)_2\ C{=}O$
E —	Methanal	HCHO
F —	Acetoxime	$(CH_3)_2C{=}NOH$
G —	Methanol	CH_3OH
H —	Sodium formate	HCOONa

71. The oxygen atom of the compound A is present in the form of —OH group because A reacts with acetyl chloride. Since compound D reacts with CrO_2Cl_2 (Etard reaction), therefore, A must contain one methyl group attached to benzene ring. A is therefore, a cresol. Further the formation of two isomeric mononitroderivative suggests that it is *ortho cresol.*

Hence,

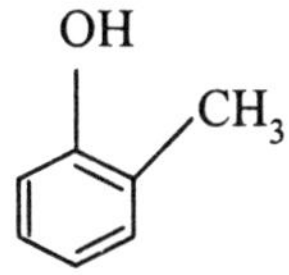

A : *o*-cresol

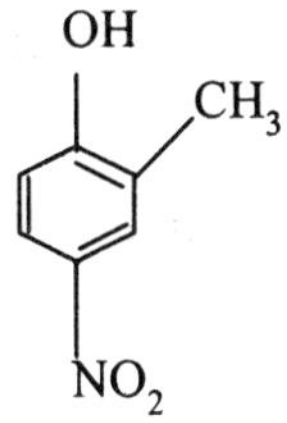

B : *p*-nitro-*o*-cresol

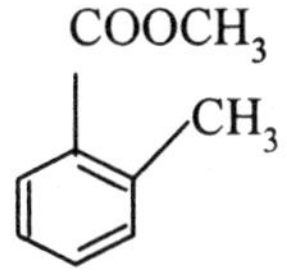

C : 3-nitro-*o*-cresol

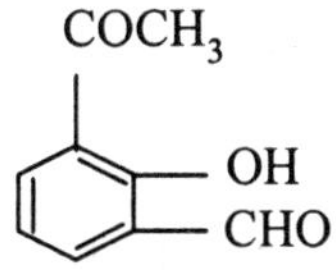

D : *o*-methyl phenyl acetate

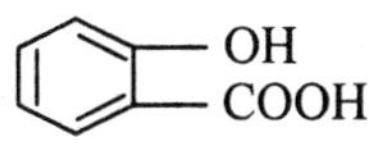

E : *o*-acetyle salisyaldehyde

F : Salicylic acid

G : *o*-methyl diphenyl sulphonate

H : *o*-carboxyl diphenyl sulphonate

72. 'A' is an alkene containing bromine since it decolorizes bromine water and its molecular formula is C_5H_9Br. Sn/HCl reduces 'A' to bromopentane. Since C does not form any acetylide, it does not contain acidic hydrogen atom. Further Lindlar's catalyst reduces alkyne to alkene (see Chapter 9).

$$\underset{\text{(A)}}{CH_3-CH_2-CH=\underset{\displaystyle Br}{\underset{|}{C}}-CH_3} \longrightarrow \underset{\text{(B)}}{CH_3CH_2-CH_2-\underset{\displaystyle Br}{\underset{|}{CH}}-CH_3}$$

$$\downarrow NaNH_2$$

$$\underset{\text{(C)}}{CH_3CH_2-C\equiv C-CH_3} \xrightarrow[\text{catalyst}]{\text{Lindlar's}} \underset{\text{(D)}}{CH_3CH_2-CH=CH-CH_3}$$

$$\downarrow$$

$$\underset{\text{(E)}}{CH_2=CH-CH_2-CH_2-CH_3}$$

Hence, A : 2-Bromopentene-2 $(CH_3CH_2—CH{=}\underset{\displaystyle Br}{\underset{|}{C}}—CH_3$

B : 2-Bromopentane $(CH_3\ CH_2\ CH_2—\ CH—CH_3)$

C : Butyne-2 $(CH_3CH_2\ —C{\equiv}C—CH_3)$

D : Pentene-2 $(CH_2—CH_2—\ CH{=}CH—CH_3)$

E : Pentene-1 $CH_2{=}CH—CH_2—CH_2—CH_3$

73. Since the 'A' adds 1 mole of hydrogen, it indicates it contains one double bond. The formation of *n*-hexane indicates that it is hexane. The position of double bond at position '3' is confirmed by the fact that on oxidation we set a carboxylic acid with three carbon atoms.

Therefore 'A' is $CH_3CH_2CH{=}CH\ CH_2CH_3$

The reactions are as follows:

$$CH_3CH_2\,CH = CHCH_2CH_3 \xrightarrow{H_2,\ Pt} CH_3CH_2CH_2\,CH_2\,CH_2\,CH_3$$

$$CH_3CH_2\,CH = CHCH_2CH_3 \xrightarrow{KMnO_4} 2CH_3\ CH_2\,COOH$$

74. The decolorization of Br_2/H_2O indicates unsaturation. Therefore, the isomer of A is:

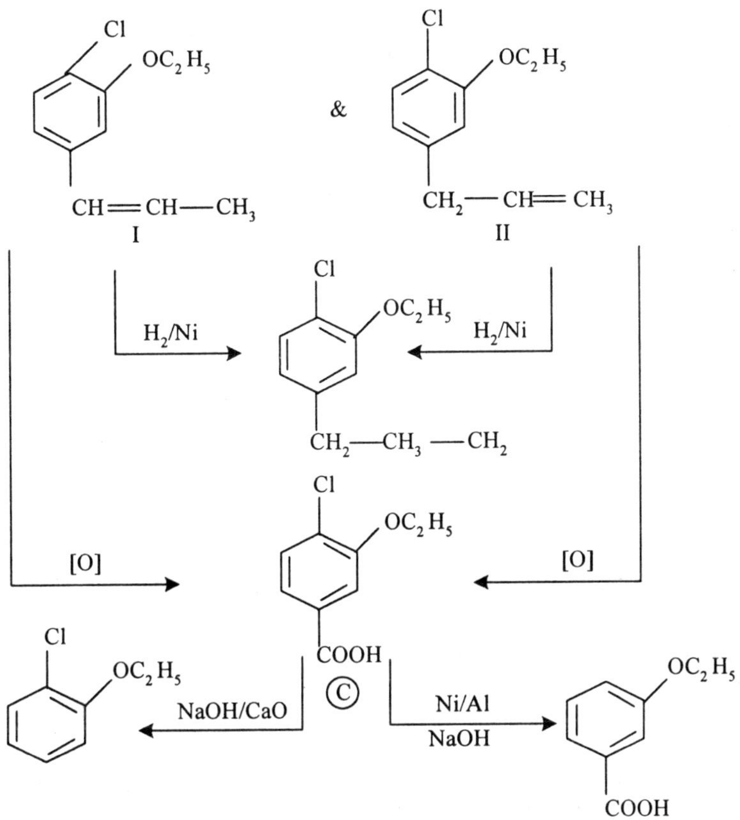

Geometrical isomers of (I) are

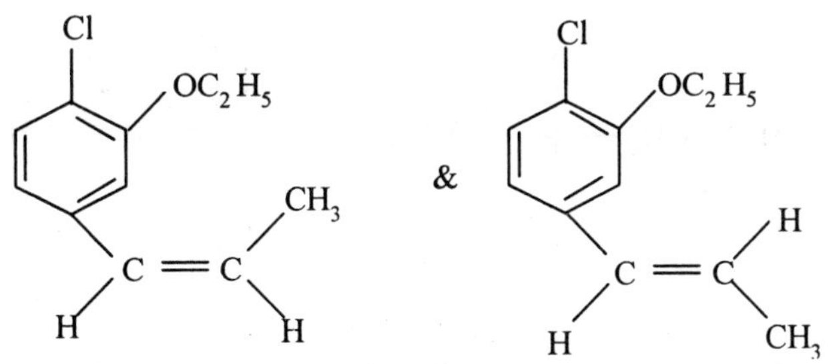

75. It liberates H_2 with Na, hence it is an alcohol. Further the decolorization of Br_2/CCl_4 indicates unsaturation. Therefore,

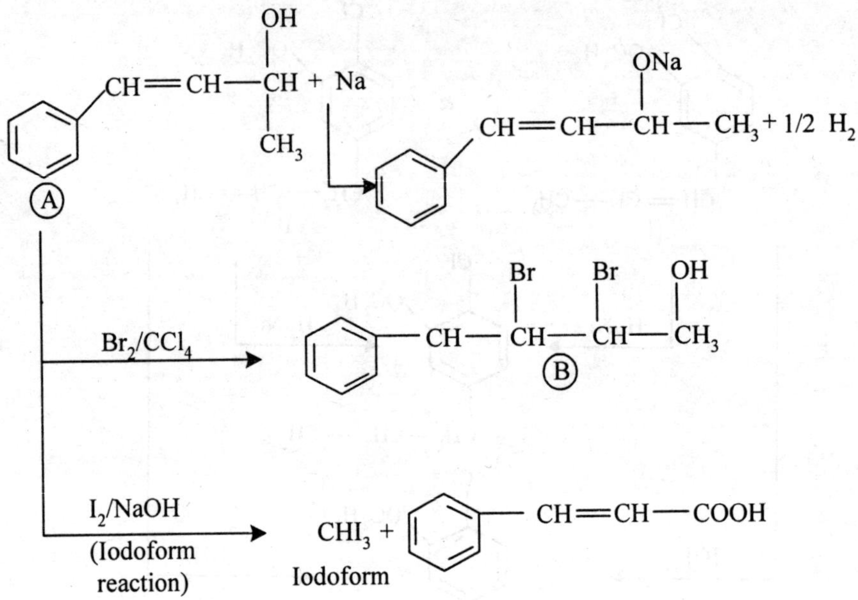

Geometrical isomers of (A) are :

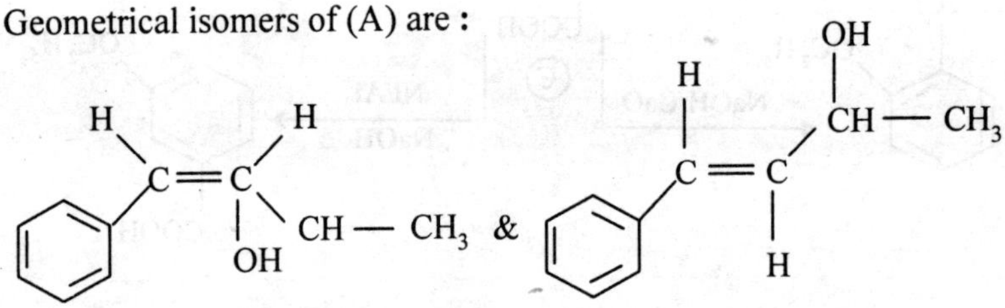

Optical isomers of A

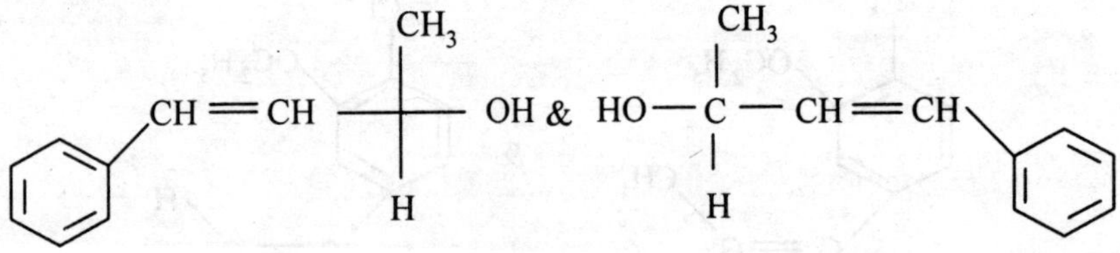

76. An examination of the formula of (A) shows that (A) may be an acid or an ester. Since (A) reacts with Br_2 and phosphorus, it is an acid and undergoes Hell-Volhard-Zelinsky reaction, Therefore,

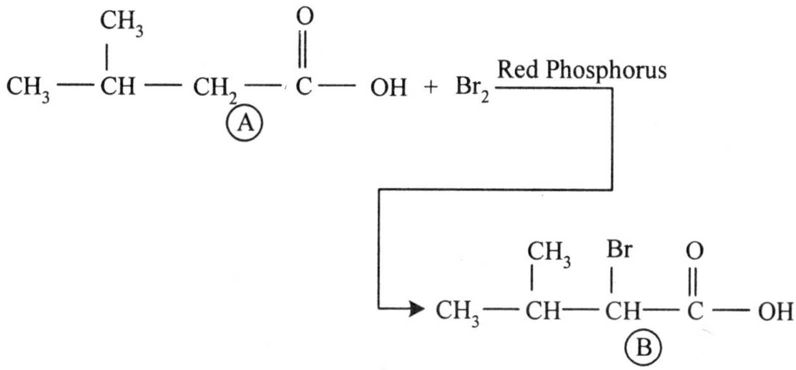

(B) Contains chiral carbon atoms and on dehydrobromination gives (C).

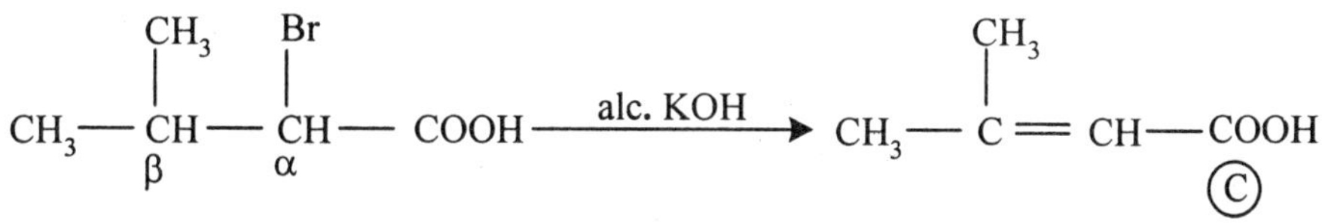

The compound (C) does not show geometrical isomerism and gives an alkene (D) on decarboxylation. Thus:

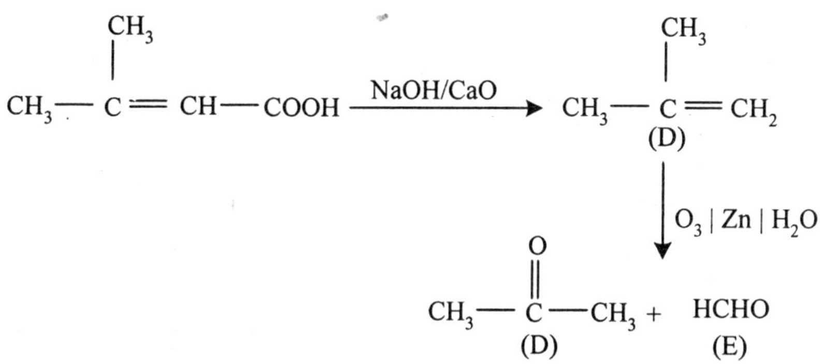

Compound (E), being aldehyde, gives a positive Schiff test, but (F), being ketons, does not give a positive test.

77. Reaction of hydrocarbon (A) with Br_2/CCl_4 indicates unsaturation. Therefore,

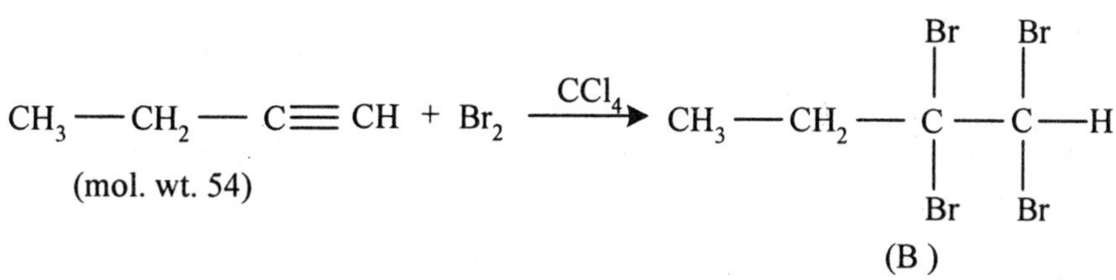

The molecular weight of (B) is 374, which is 59.3% more than the molecular weight of compound (A), which is 54. Compound (A) on catalytic hydrogenation with excess of H_2 gives compound (C), whose molecular weight is 7.4% more than that of (A). Thus,

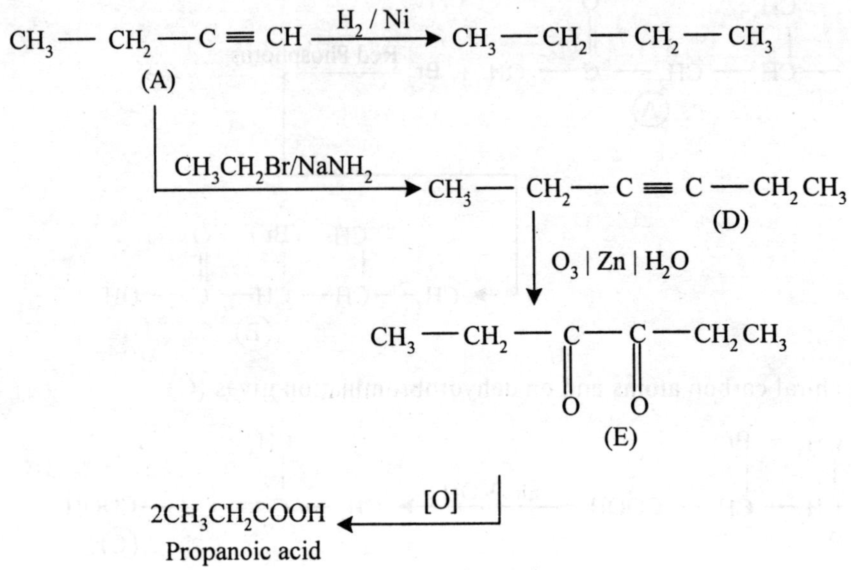

78.

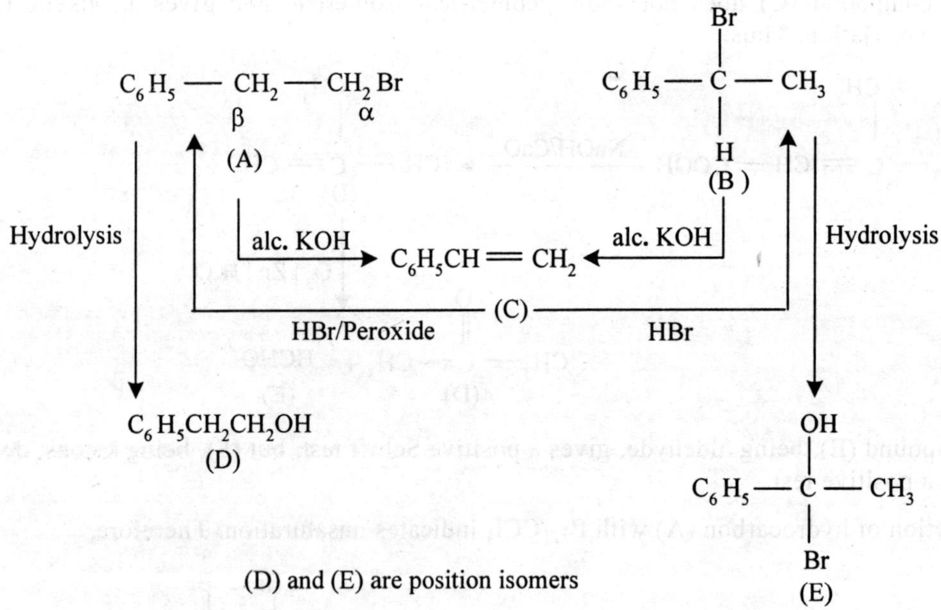

(D) and (E) are position isomers

Compound (C) reacts with benzene in the presence of $H^{\oplus}$ and gives 1,1-diphenyl ethene.

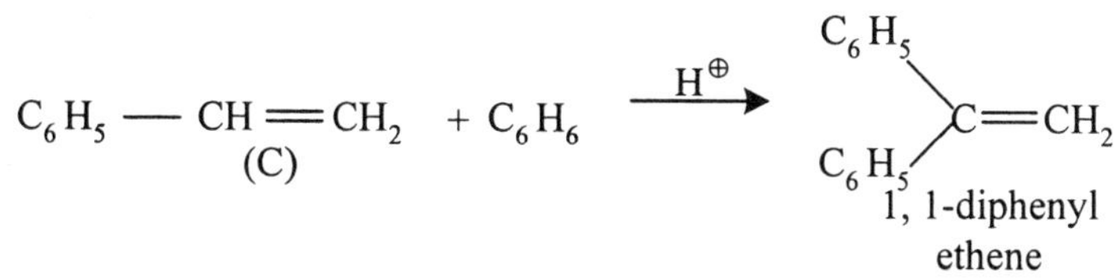

79.

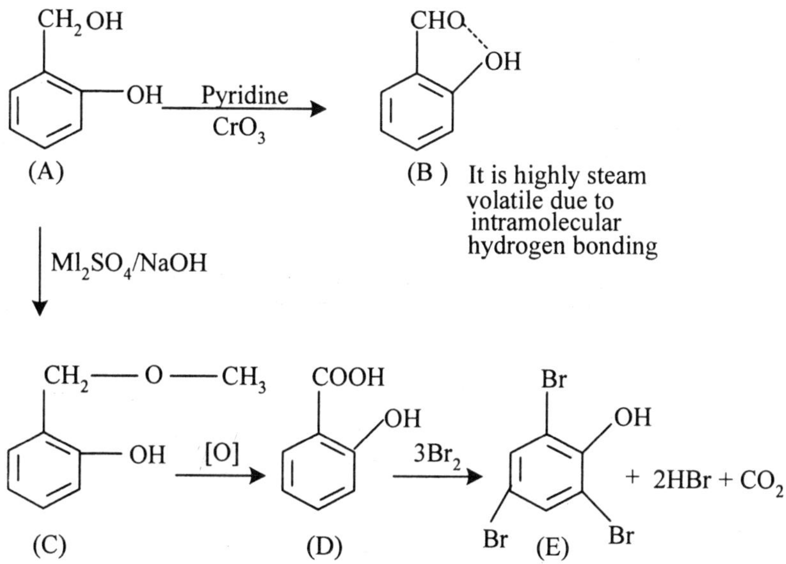

$A = C_6H_5COCH_3$; $B \& C = C_6H_5C(CH_3){=}NOH$ (Syn & Ants)

$D = CH_3CONHC_6H_5$

$E = C_6H_5CONHCH_3$, $F = C_6H_5NH_2$

$G = C_6H_5COOH$

80.

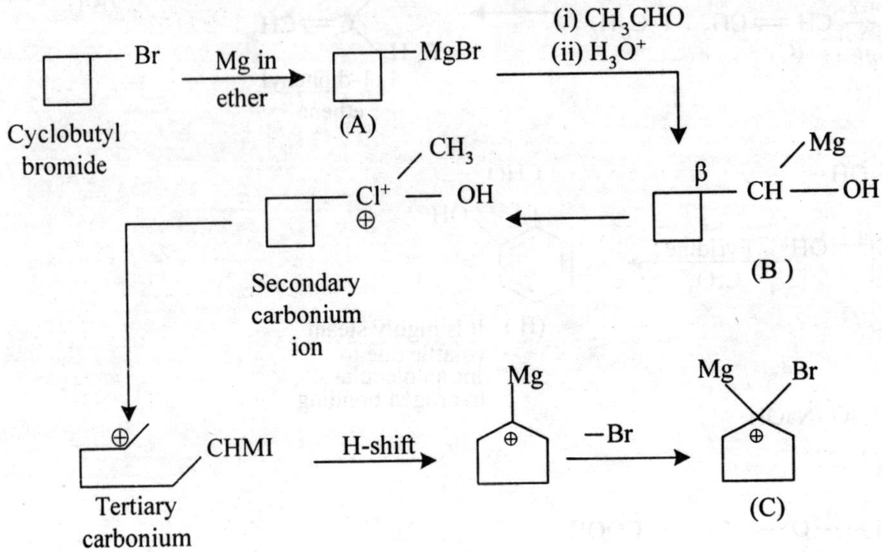

81.

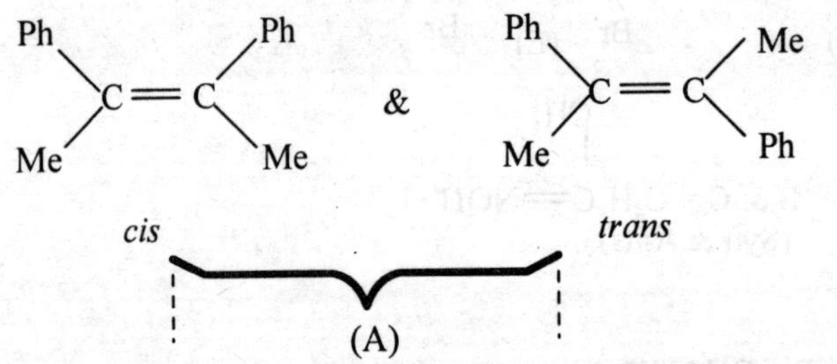

PhCOCH$_3$ (Acetophenone) (B)

$C_6H_5CH_2CH_3$ Ethyl benzene (C)

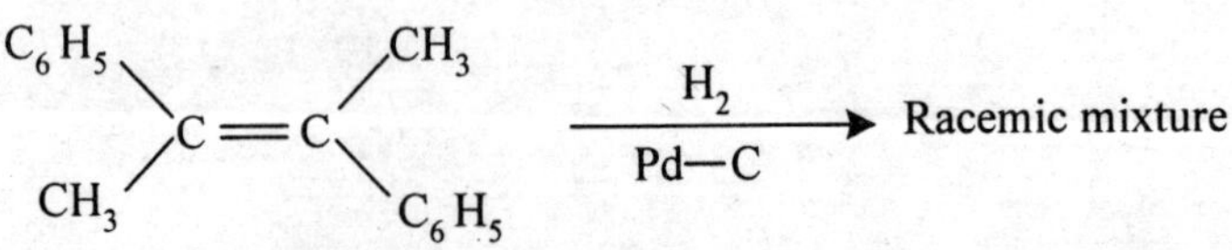

82.

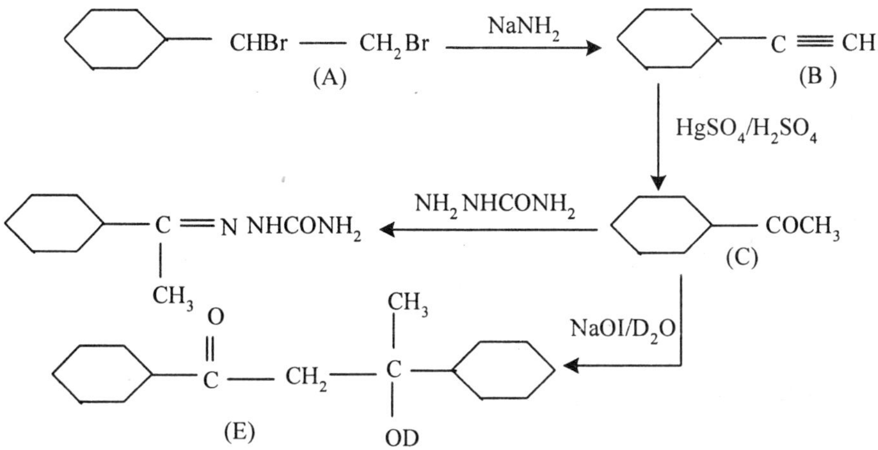

83.

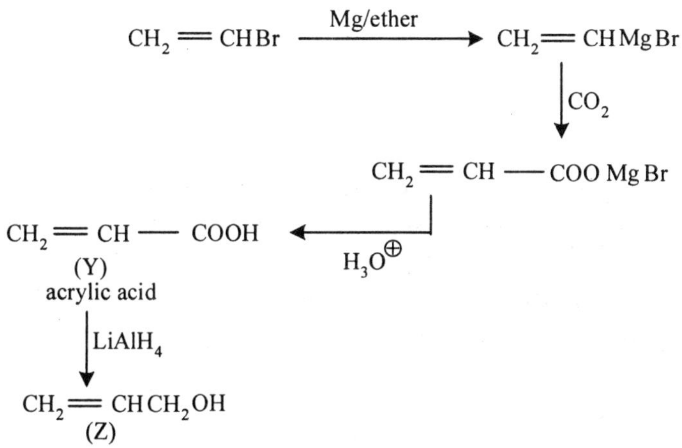

84. (i) Ester (—COOCH3); amino (—NH2)
acid (—COOH); substituted amide (—CONH)

(ii)

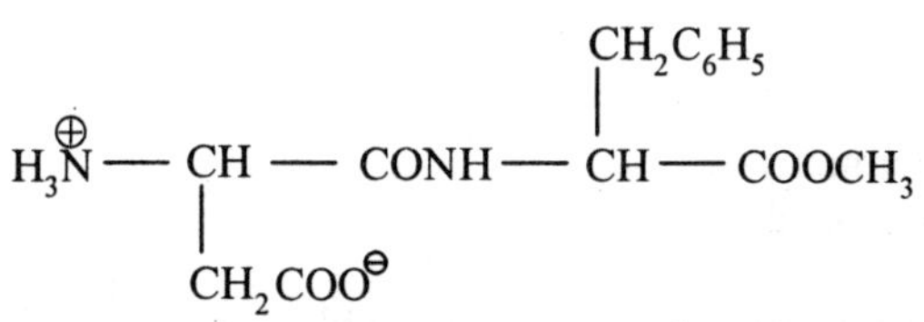

(iii)

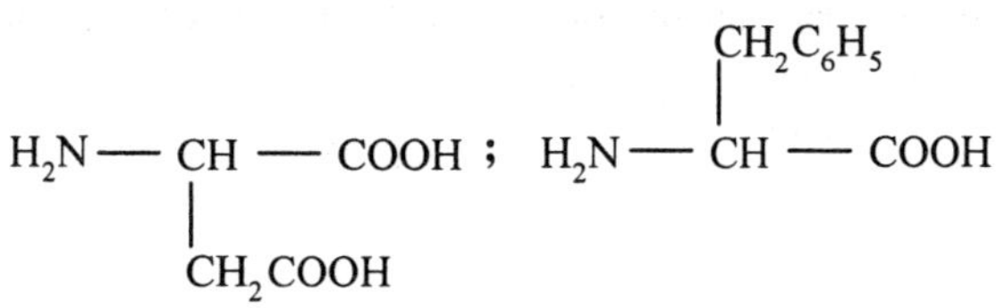

(iv)

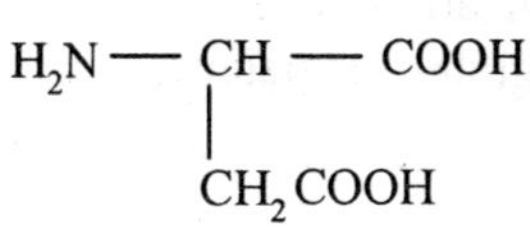

is more hydrophobic

Chapter 16

Molecular Rearrangements

Molecular Rearrangements: During the Rns, often the basic skeleton of molecule rearranges to form new structures with same number of atoms and double bonds. Such Rn are known as Molecular Rearrangements.

In a rearrangement Rn, a group moves from one atom to another in the same molecule. Mostly are migration from an atom to an adjacent one (called 1, 2 — shift).

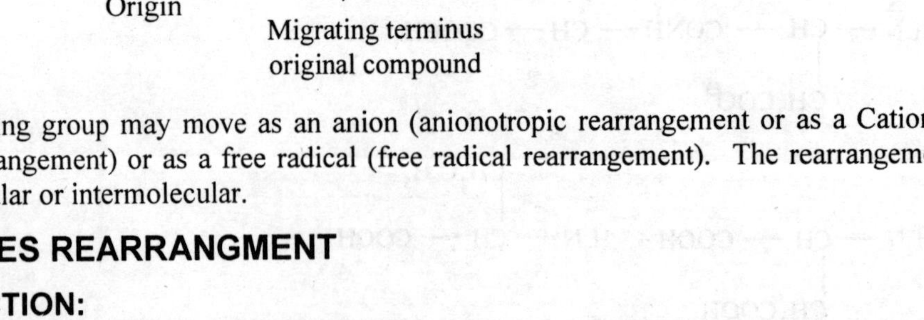

The migrating group may move as an anion (anionotropic rearrangement or as a Cation (Cationo - tropic rearrangement) or as a free radical (free radical rearrangement). The rearrangements may be intramolecular or intermolecular.

16.1 FRIES REARRANGMENT

INTROUCTION:

It consists of conversion of *Phenyl ester* into an o or p hydroxyl aromatic ketone, or both, in the presence of Anhy $AlCl_3$. The acyl or aryl group from side chain migrates to nucleus.

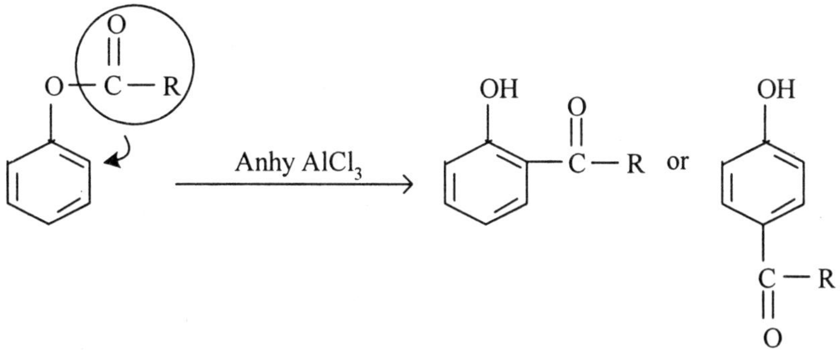

e.g.,

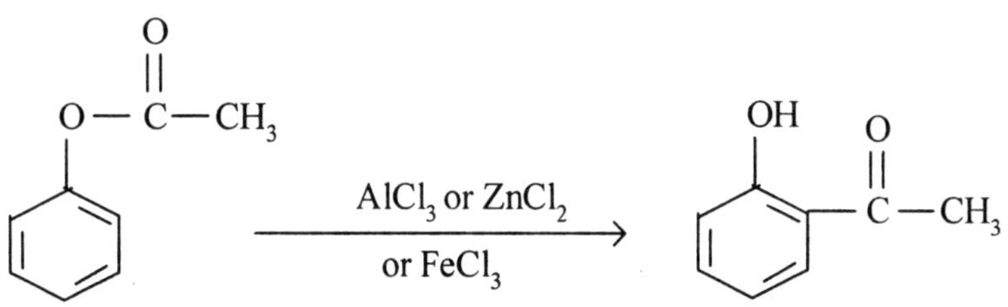

The function of Anhy $AlCl_3$ is the withdrawal of electron density from the acyl oxygen bond by Coordination, which facilitates the initial heterolysis.

Features:

(i) High temperature ∝ the ortho isomers

Low temperature ∝ the para isomers.

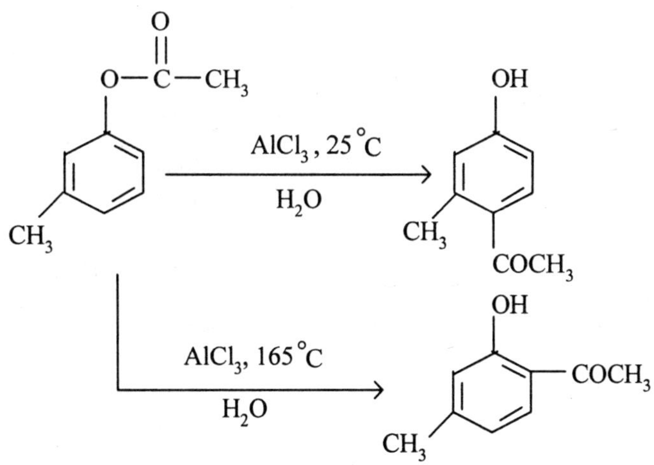

(ii) The rearrangement generally does not need any solvent. However, in some cases the solvents like nitrobenzene reduce the temperature at which the Rn proceeds at a useful rate.

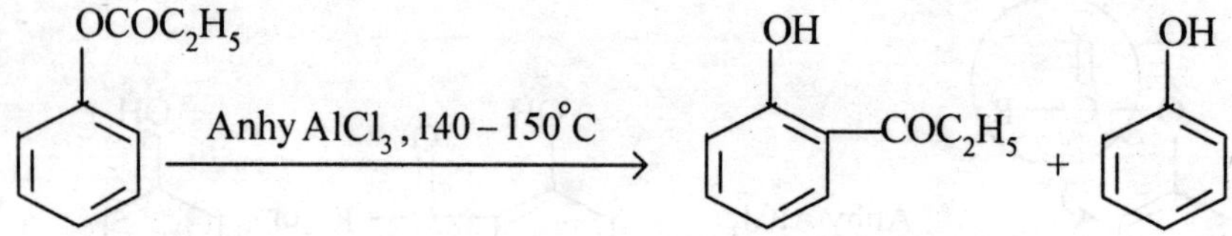

(iii) Usually ester and the reagent are employed in equimolar ratio. However, the ratio of ester to reagent may influence the ratio of p to ortho product.

$AlCl_3$ /Phenylacete	p to o isomer
0.5	2.93
1.0	2.41
2.0	10.1

(iv) The nature and position of the substituent in the ester is also of significance. The presence of m directing groups interferes with the rearrangement. The influence of methyl group is as follows:

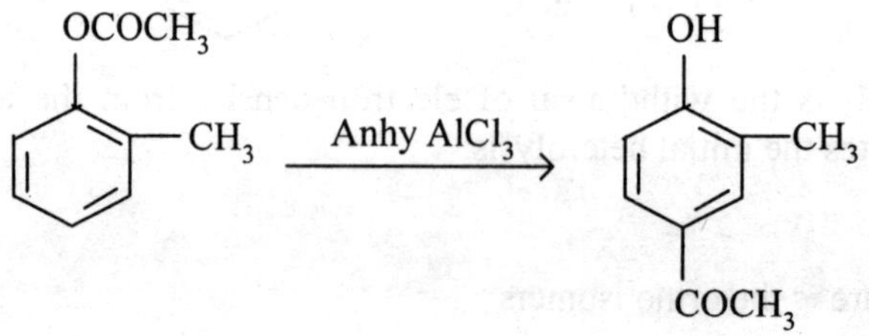

Major

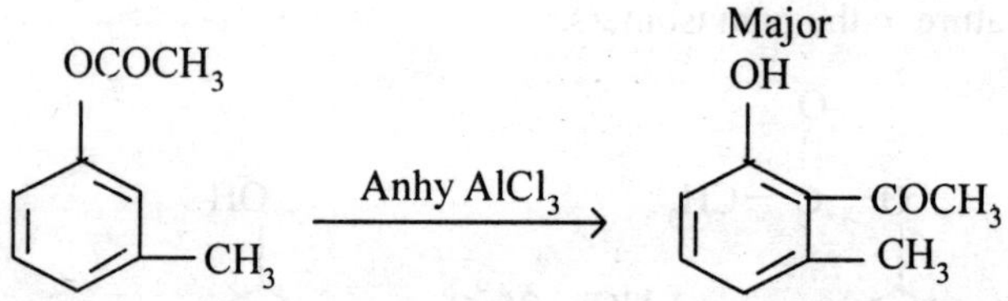

Major

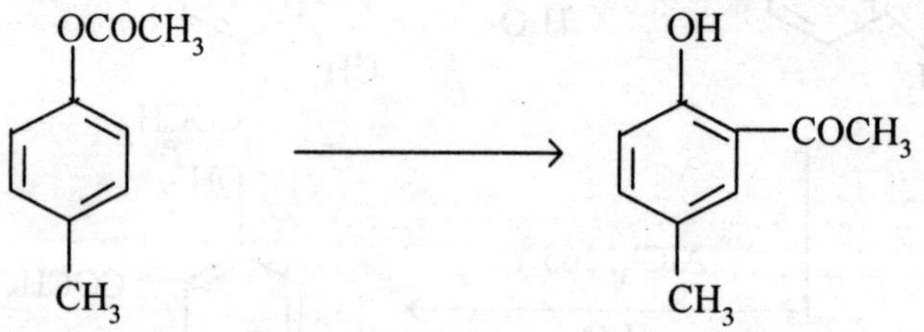

98%

Mechanism:

(a) It involves intramolecular change. *Intramolecular mechanism.*

(i)

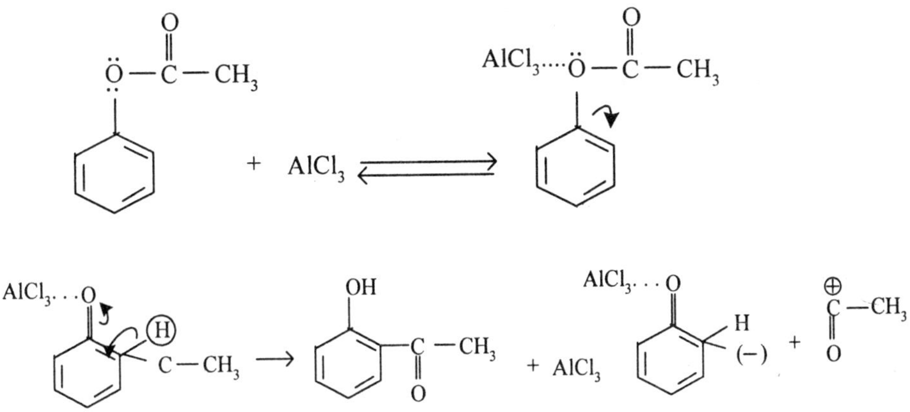

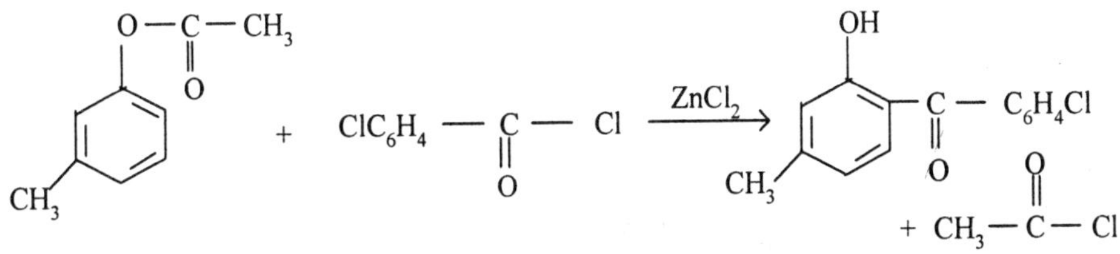

(b) Whitmore mechanism

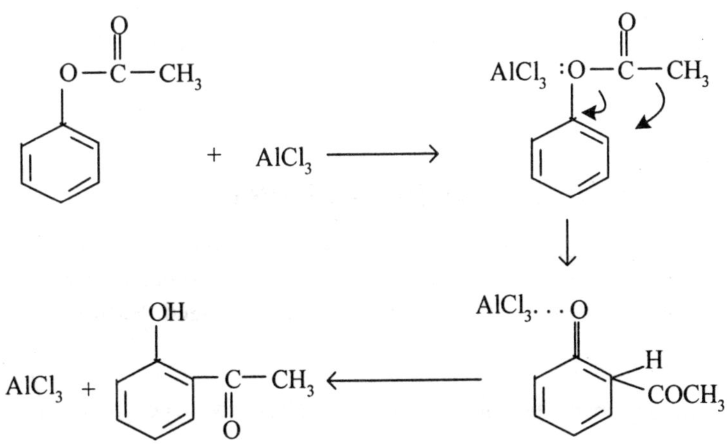

(c) Baltzly in 1956 also supported Whitmore mechanism:

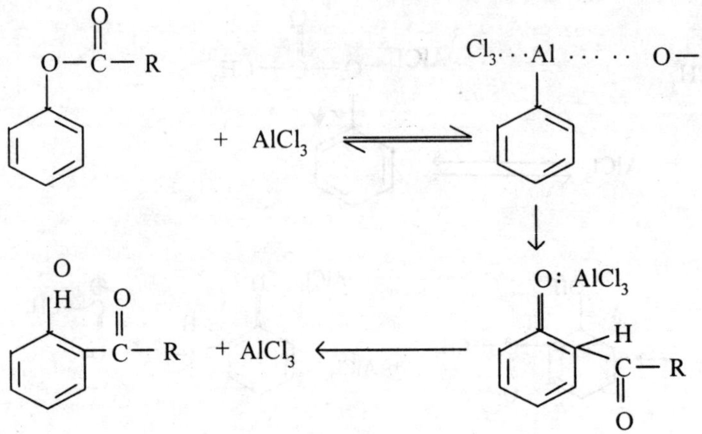

Applications:

(1) For the preparation of ortho & para hydroxyl ketones

(2) o and p – isomers can be obtained by varying the experimental conditions. These can be separated by steam distillation.

(3) A series of o – hydroxyl ketones of following structures have been prepared:

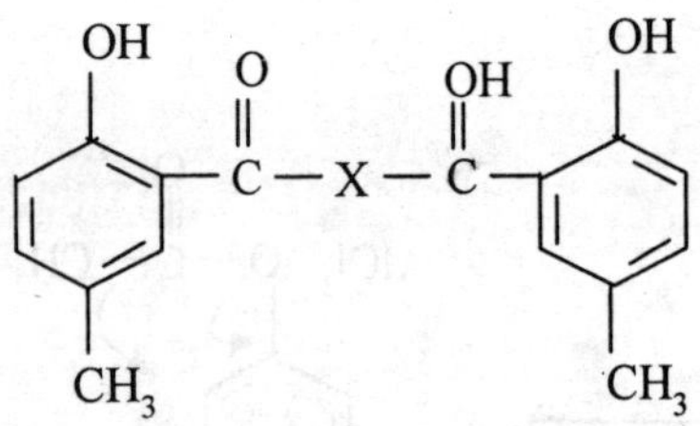

$X = (CH_2)_2$ to $(CH_2)_4$

16.2 WAGNER MEERWEIN REARRANGEMENT

INTRODUCTION: When alcohols are treated with acids, simple substitution or elimination usually accounts for most or all of the products. But in many cases, especially two or three alkyl or aryl groups are on the β carbon, some or all of the product is rearranged. Such rearrangements are called Wagner-Meerwein Rearrangement. In this rearrangement Neopentyl type group, 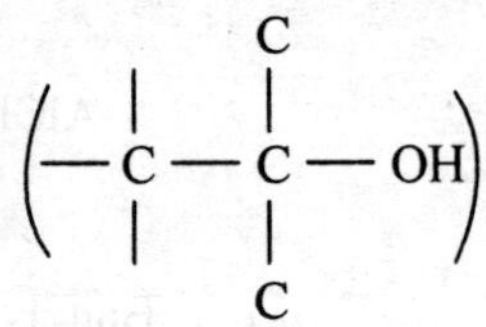

under influence of catalyst forms 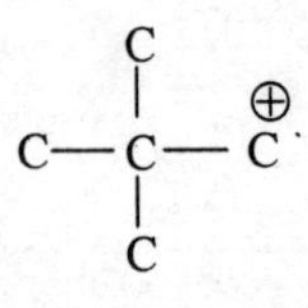

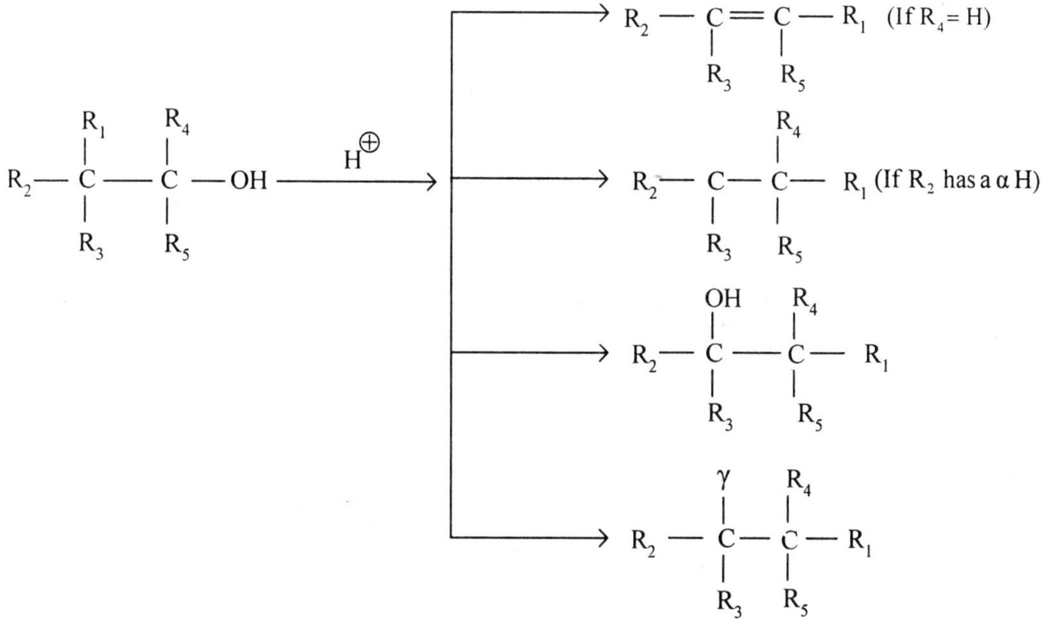

R = alkyl, aryl or H

(1) The Carbonium ion stabilizes itself by loss of a H, β to it; so that the rearrange products is usually stable.

Some examples are:

(i)

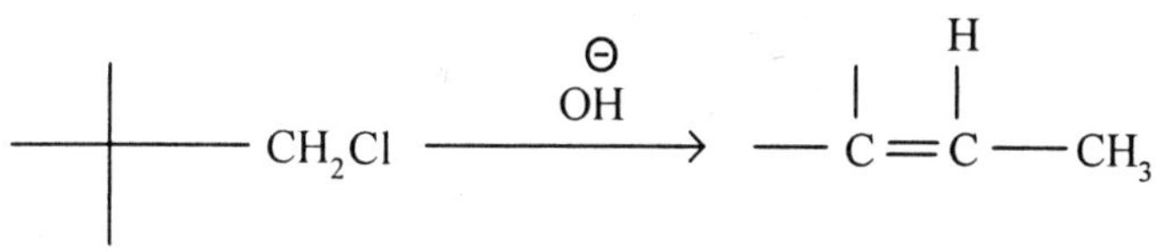

(ii)

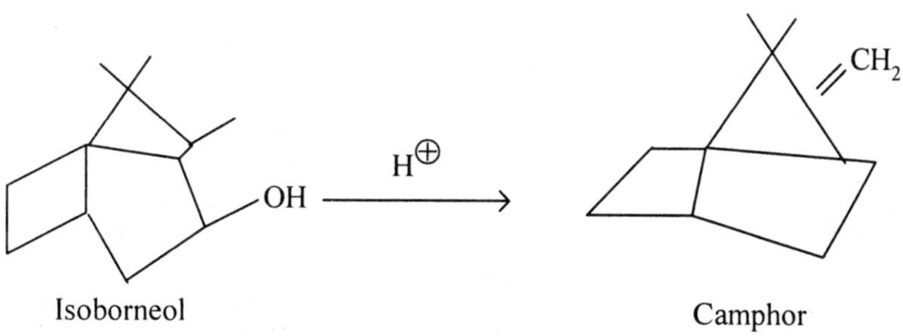

Features: *(i)* $H^{\ominus}$ may migrate e.g.,

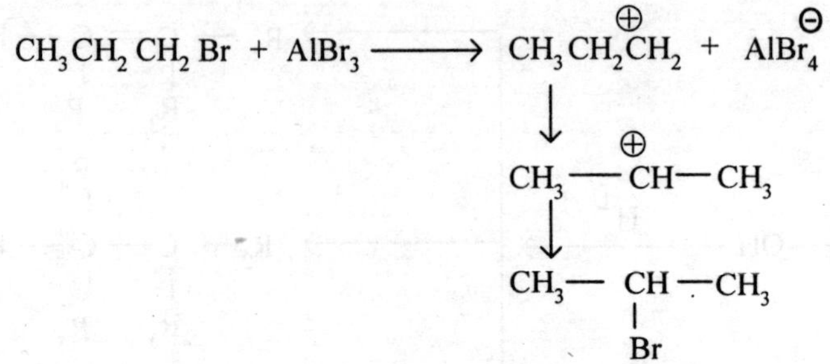

(2) The leaving group does not have to be water, but it may be any departing species whose loss creates a Carbonium ion, including N_2 from diazonium ions.

(3) The direction of the rearrangement is usually in the order tert > sec > Pri. Sometimes products is a mixture corresponding to the mixture of possible Carbonium ion.

(4) The groups which are most suspectible to electrophile attack will migrate most readily.

(5) The relative migratory aptitudes is as follows.

Ar > R > H
and among Aryl groups

$$p-CH_3OC_6H_5 > p-CH_3C_6H_4 > C_6H_5 > p-ClC_6H_5$$

Mechanism:

It involves Carbonium ions involving whitmore mechanism.

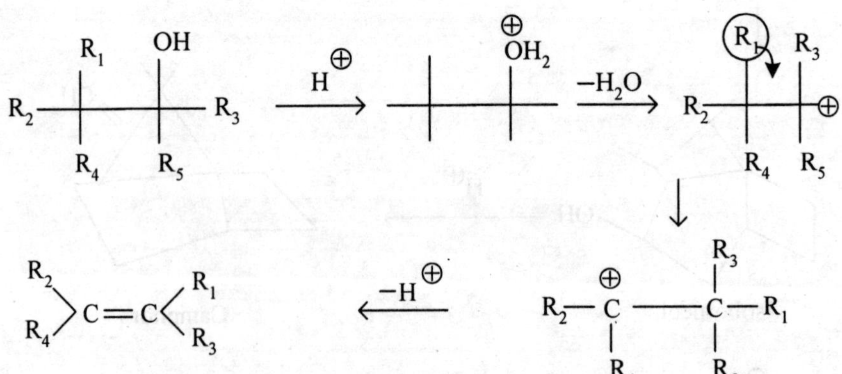

(If $R_5 = H$)

e.g.,

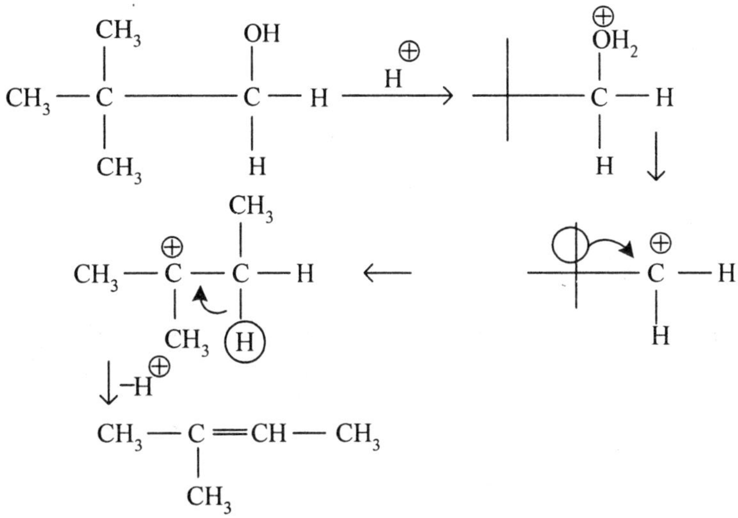

Applications:

(1)

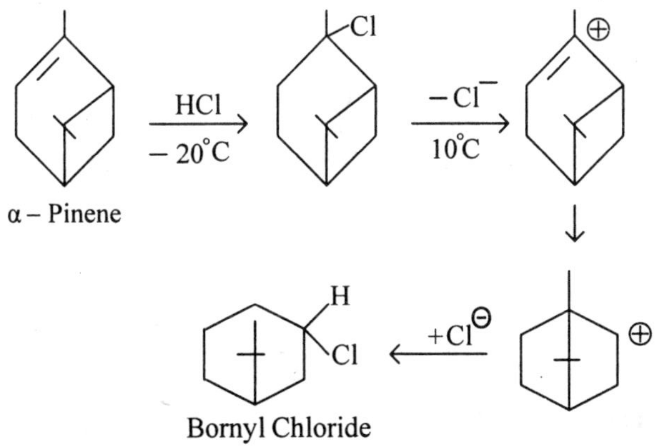

(2)

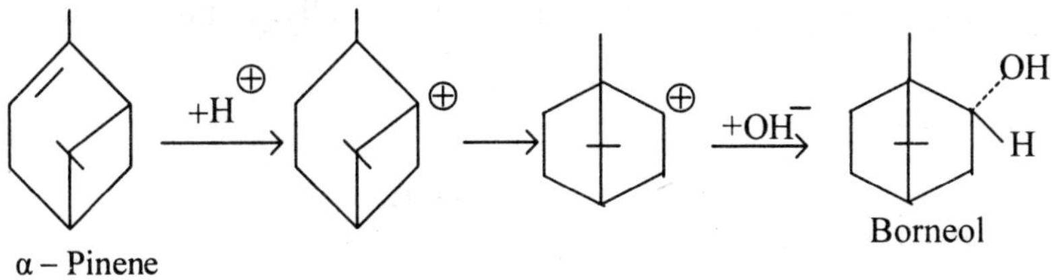

(3) Bornyl chloride → Camphene.

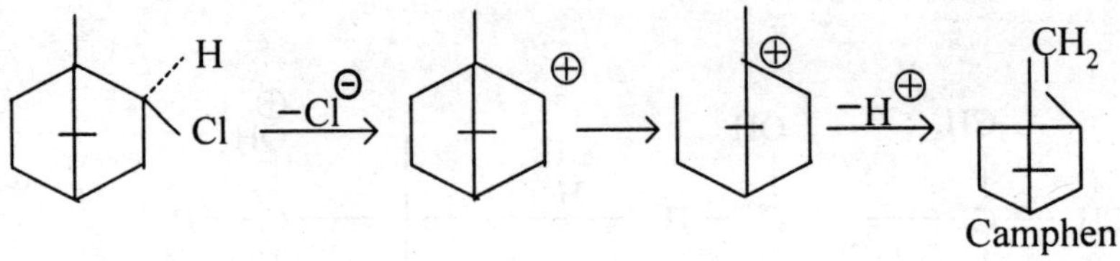

(4) It is also used for ring expansion and contraction

16.3 CLAISEN REARRANGEMENT

INTRODUCTION: Allyl aryl ethers, when heated rearranges to ortho-allyl phenol is in a reaction which is known as Claisen rearrangement. In this rearrangement allyl group migrate to o-position but if both positions are filled then allyl group migrates to para position but never to meta position. The Rn sometimes need either heat or catalyst like HCl, H_2SO_4, $ZnCl_2$, BF_3, $AlCl_3$.

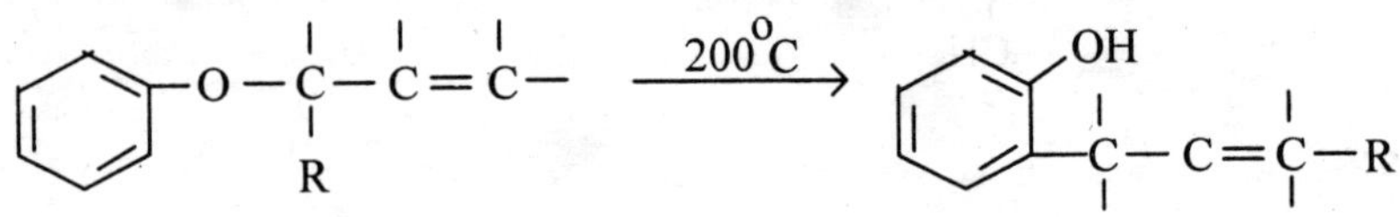

e.g.,

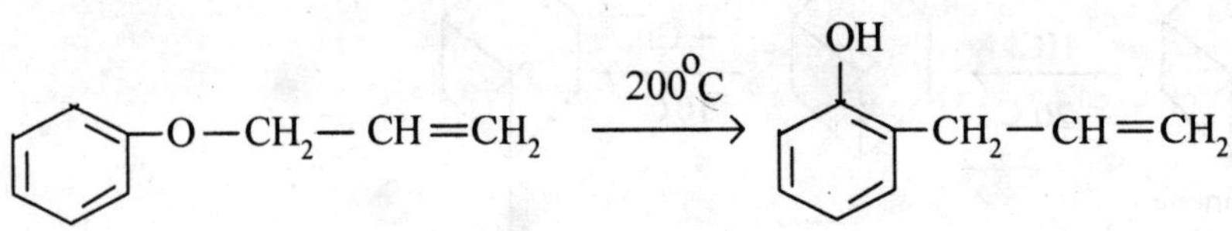

Features

(1) Sometimes some para product is obtained even if one or both ortho-positions are vaccant.

(2) If both ortho positions are occupied the product is para.

(3) No Rn when both ortho and para position is blocked.

(4) In the ortho migration the allyl group always undergoes an allylic shift i.e., α γ bonding is reversed.

Schmidt in 1953 proved it.

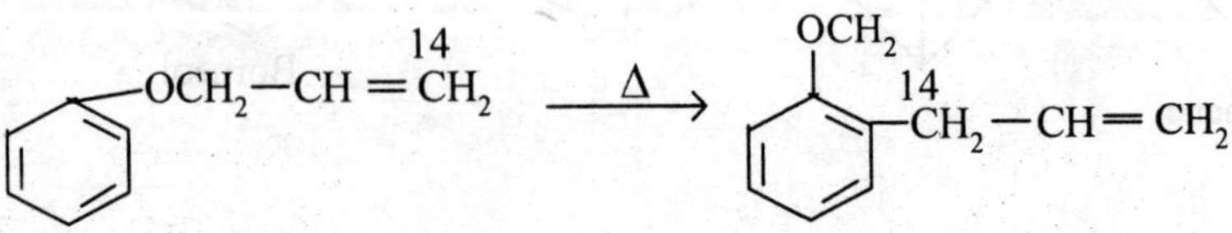

(5) In case of para migration there is never an allylic shift.

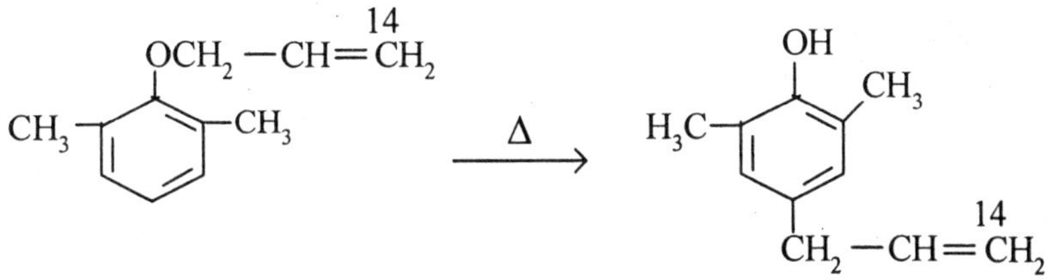

It is explained on the basis of two allylic shift.

(6) During migration to p position allyl group does not react with solvent.

(7) Since the rearrangement mechanism does not involve ions it is not greatly dependent on the presence or absence of substituent group on the ring. Generally, electron donating groups increase the rate, and the withdrawing groups decrease it, but the effect is small.

(8) Allyl ethers of enols (allyl vinyl ethers) also give the Claisen rearrangement.

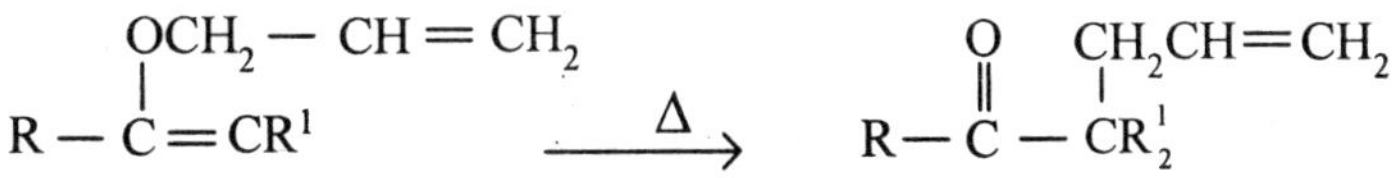

Mechanism:

O – rearrangement is a intramolecular process since no cross product is obtained when a mixture of Cinnamyl phenyl and allyl – β – naphthyl ether is heated

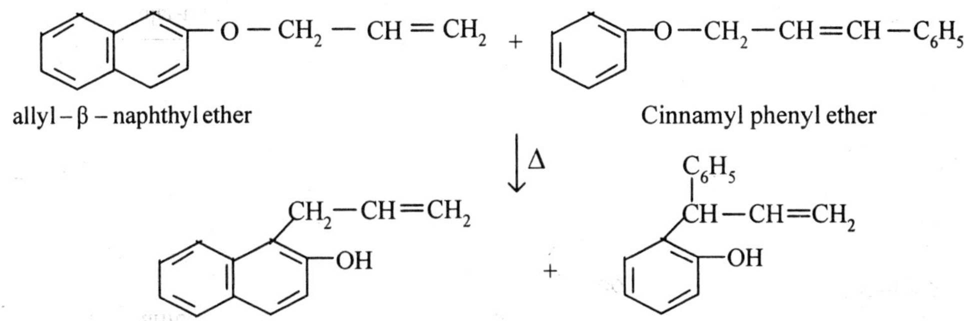

Therefore a cyclic mechanism has been proposed:

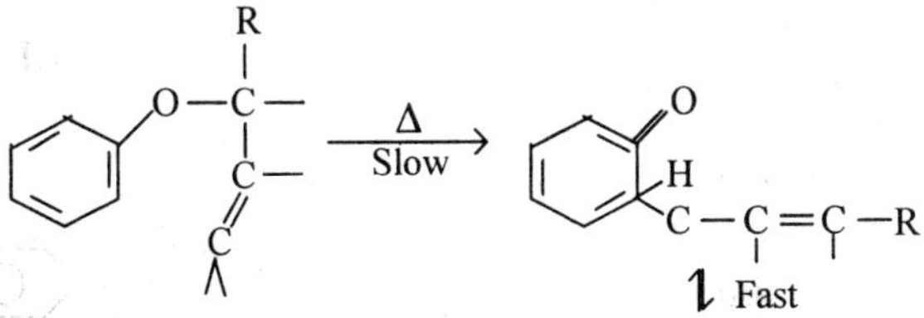

It should be noted that the breaking of a C – O bond and the attachment of γ carbon atom to O – position must be simultaneous and it is rate determining step.

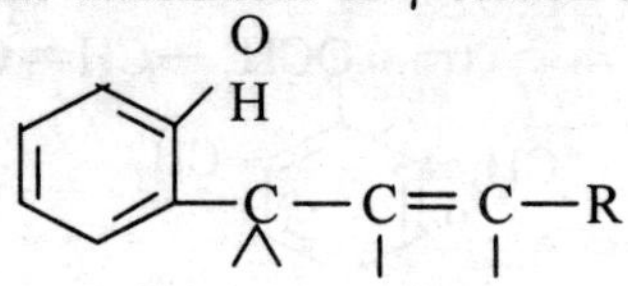

The above mechanism is supported by the following facts.

(i) The lace of a catalyst

(ii) The Rn is first order in the ether

(iii) Absence of cross over products when mixtures are heated.

(iv) Presence of allylic shift.

The mechanism for allylic shift to para position is as follows.

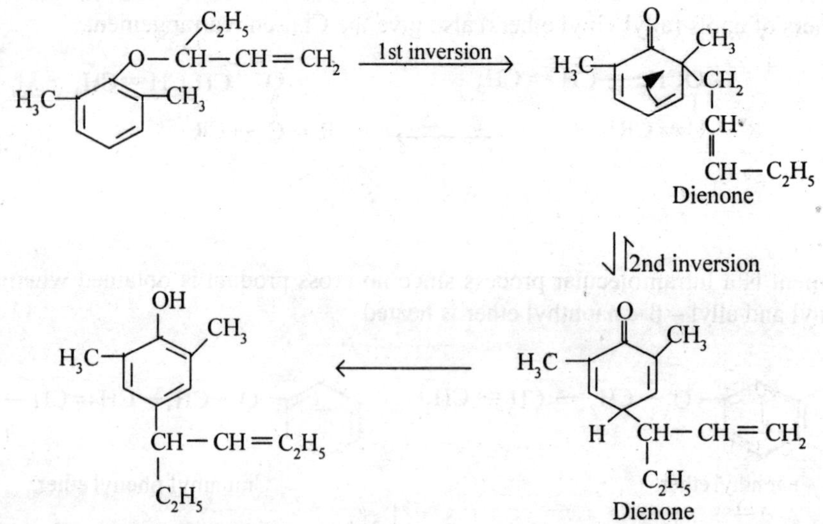

The formation of dienone has been confirmed by Curtin 1956 who prepared dienone and heated above to yield mixture of II and III.

Stereochemistry:

The rearrangement occurs with retention of configuration

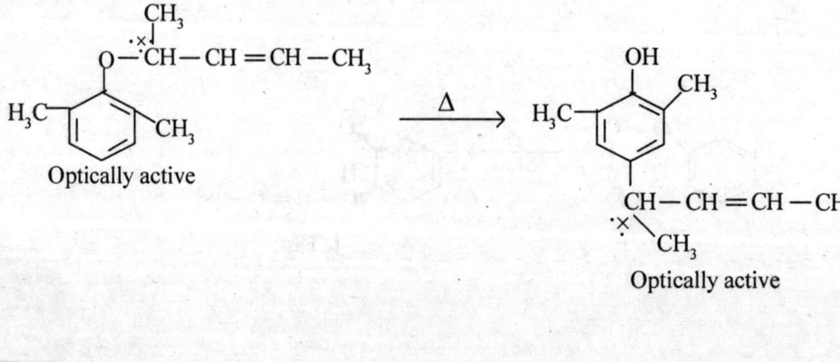

Applications:

1. For introducing allyl groups into a wide variety of phenolic compounds

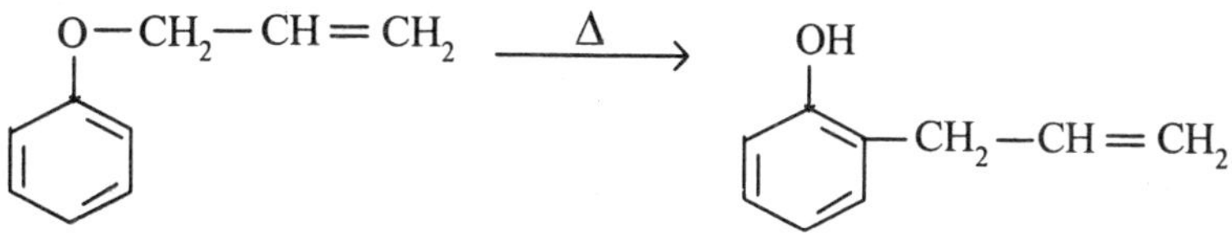

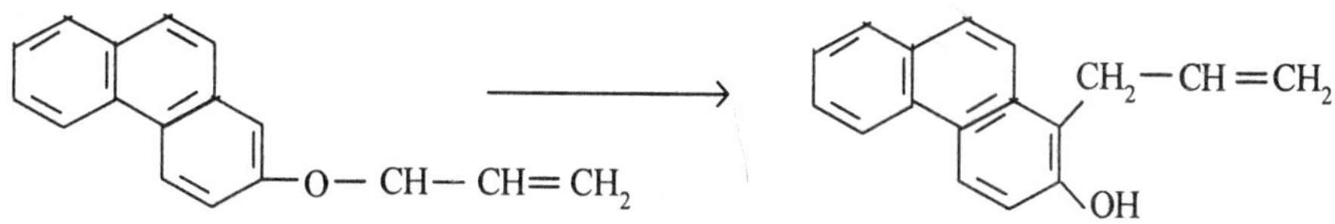

2. For the rearrangement of 1:5 — heptadienes

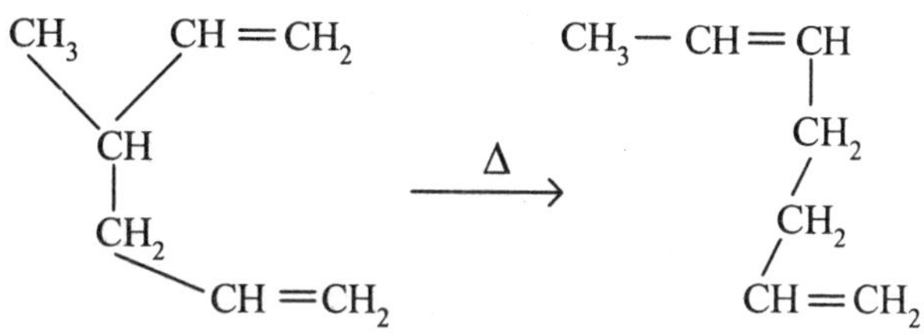

16.4 BENZIDINE REARRANGEMENT

INTRODUCTION: The conversion of hydrazobenzene into 4, 4' diamine diphenyl (benzidine) by means of dil aq acids like H_2SO_4 or HCl is called Benzidine rearrangement.

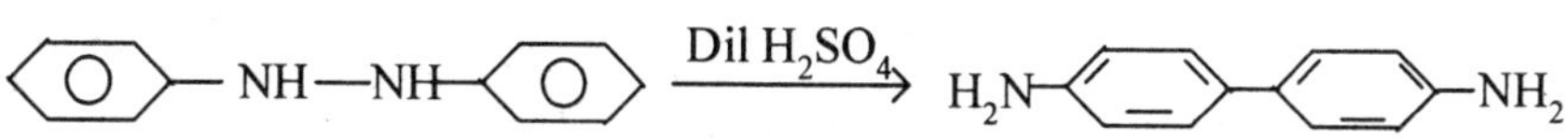

Benzidine
70%

Features: 1 **Effect of Substituents:**

(a) For rearrangement to occur both para positions should be free. If however, one of these is occupied then o: p[1] — coupling occurs.

(b) When a methoxyl or an ethoxyl group is in p-position, the main product is an O-semidine accompanied by Small amount of p-semidine.

---

[1] It is called "Polar-transition state" mechanism & was proposed by Banthorpe, Hughes & In gold

e.g.,

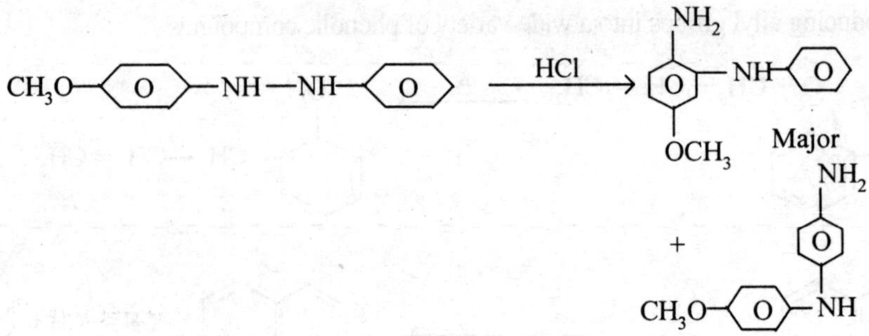

(c) When Cl, Br, I or NMe_2 is in p position then main product is diphenylamine derivative along with small amount of o – & p semidines e.g.

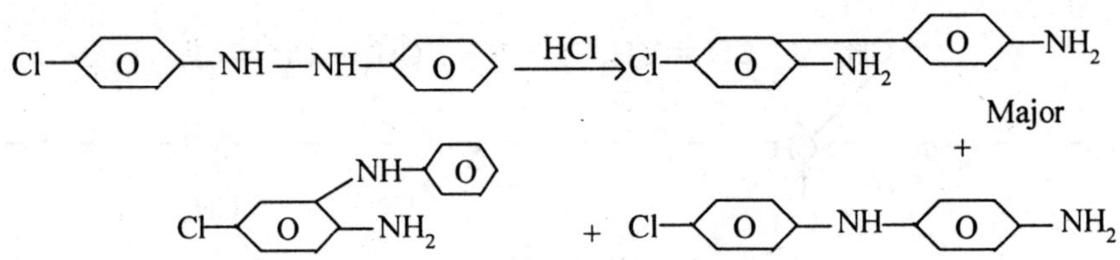

(d) If p position is occupied by NH_2 or NHAC, the main product is p-semidine.

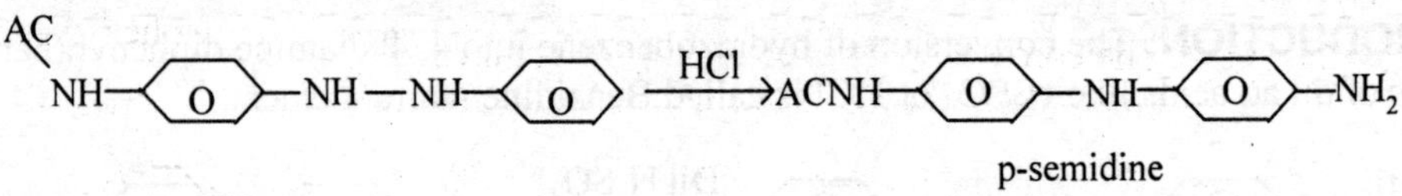

(e) When both p positions are occupied only o-semidine is obtained e.g.,

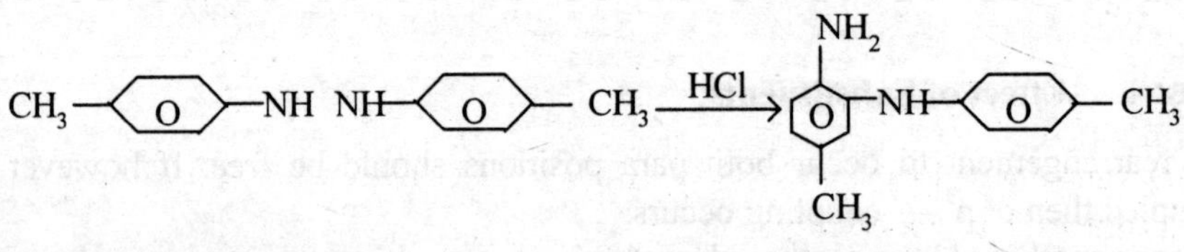

Mechanism: It is purely an intramolecular rearrangement. The cleave into two independent species i.e. ions or radicals have been ruled out by many cross over experiments. No cross over products was obtained. Two mechanisms.

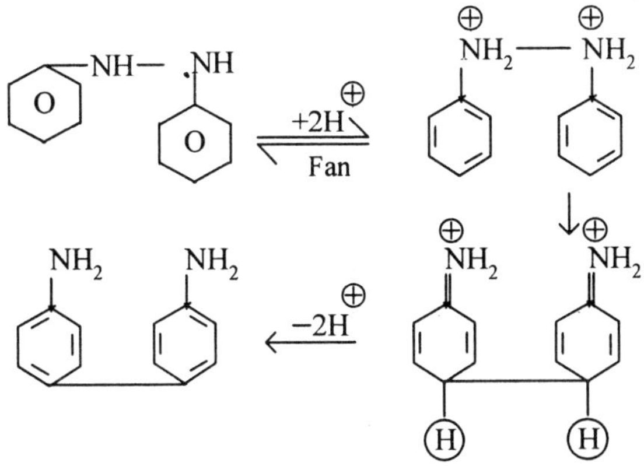

According to another mechanism called the π complex mechanism, proposed by Dewan includes cleavage of the monoprotonated substance into two parts which are held together by π orbital overlap between a filled orbitals of the benzene type ring and an empty orbital of the other.

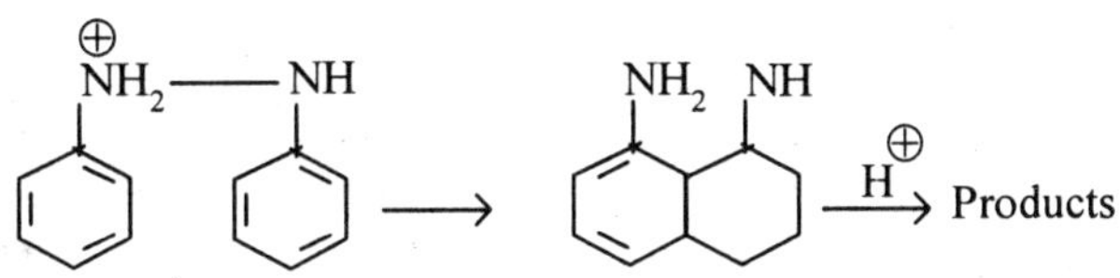

The rings can easily rotate to for various products.

PROBLEMS

1. Define molecular rearrangements. Give some examples.
2. What is Fries rearrangement? State features and mechanism of the rearrangement.
3. Discuss mechanism of Wagner-Meerwein rearrangement.
4. What is Claisen rearrangement? Discuss mechanism of the reaction.
5. What is Benzidine rearrangement? Discuss mechanism.
6. Carry out following conversions:
   (i) α — Pinene → Bornyl chloride
   (ii) α — Pinene → Borneol
   (iii) Bornyl chloride → Camphene

Appendix I

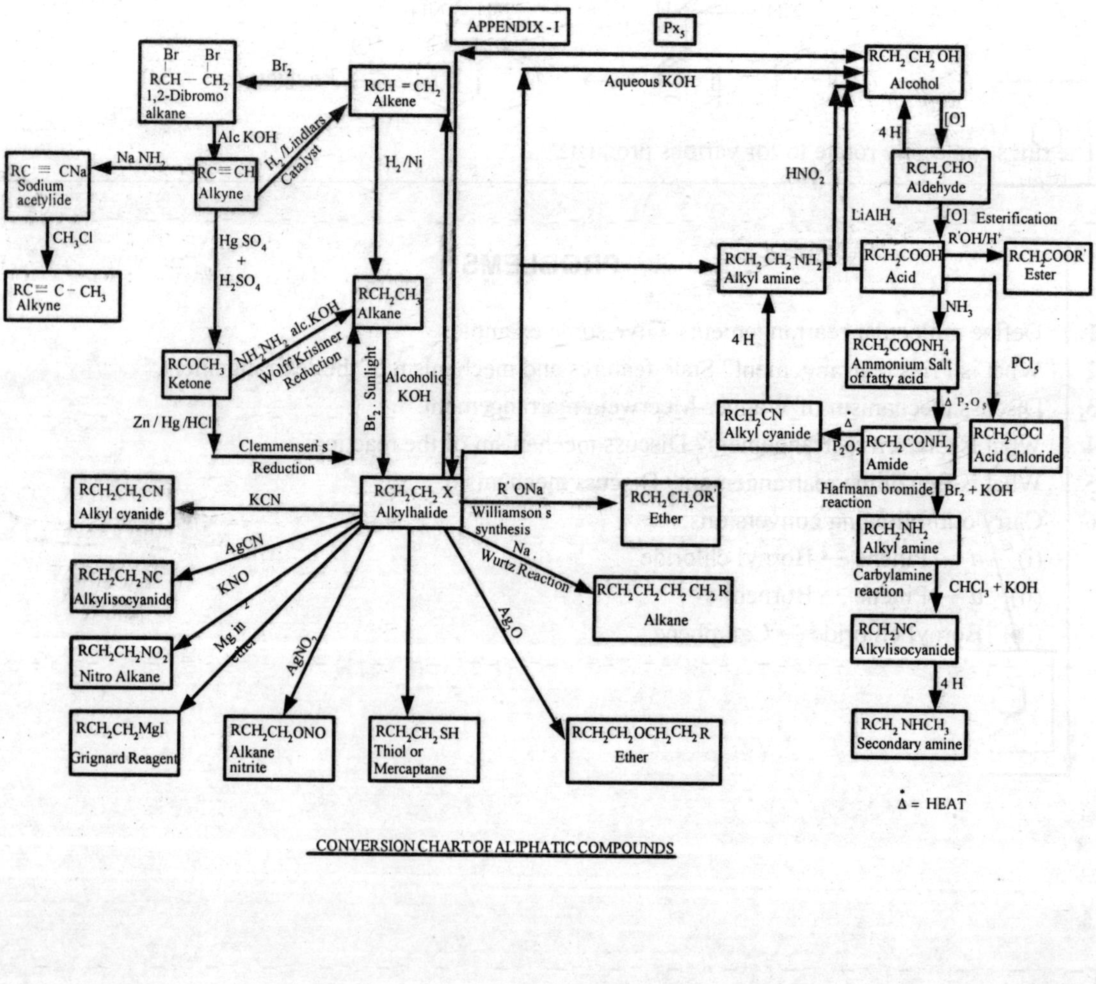

CONVERSION CHART OF ALIPHATIC COMPOUNDS

Appendix II

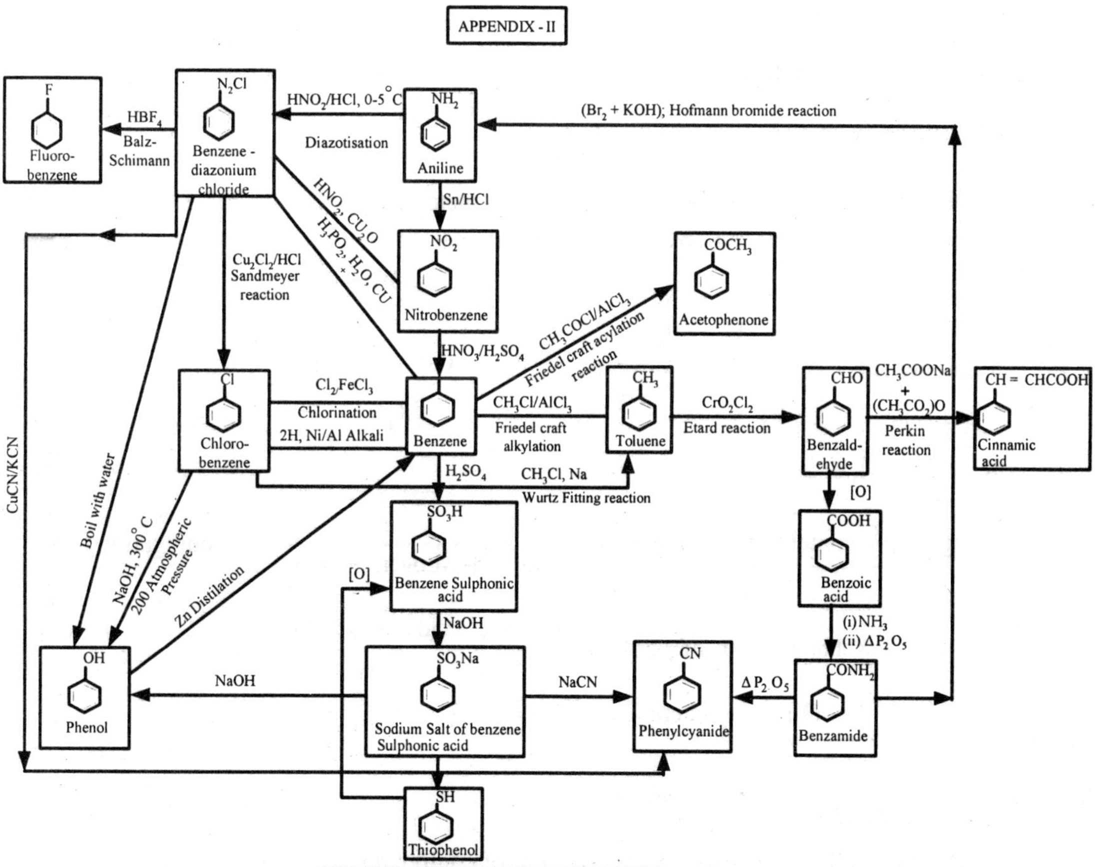

CONVERSION CHART OF AROMATIC COMPOUND

Appendix III

SOME IMPORTANT CONCLUSIONS

1. Percentages *s* orbital character $\propto \begin{cases} \text{Acidity of hydrocarbons} \\ 1/\text{C}—\text{C and C}—\text{H bond} \\ \text{lengths in Hydrocarbons} \end{cases}$
2. Heterolytic bond dissociation energy > Homolytic bond dissociation energy
3. Bond dissociation energy $\propto \begin{cases} \text{Electronegativity difference} \\ \text{of two atoms} \end{cases}$
4. Electrophile: +ve charge or central atom with incomplete octet.
5. Nuclephile: –ve charge, e (CO_2, SO_3) exception lone pair of *e*.
6. Tautomerism 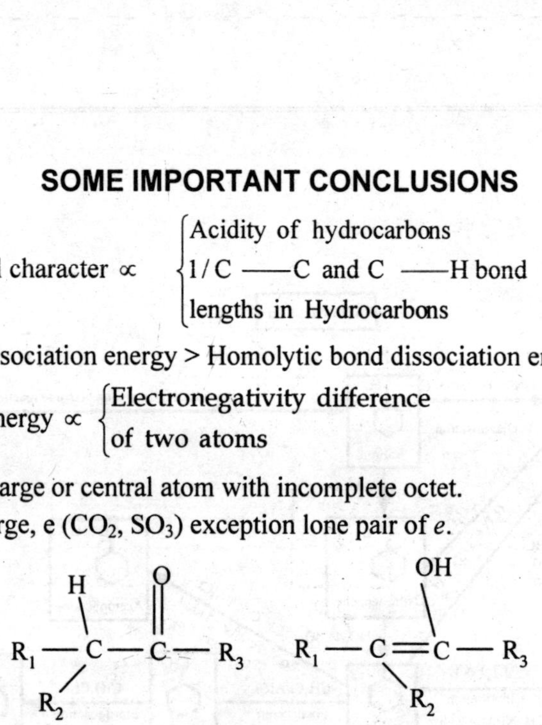

   R_3 = H, alkyl, or aryl group.
7. Percentage enoic content of a tautomer $\propto \dfrac{1}{\text{Temp}} \propto \dfrac{1}{\text{Hydrogen bonding}} \propto$ Conjugation in R_3.
8. $-\text{IE} \propto \dfrac{1}{>\text{C}—\text{X bond length of alkyl halide}} \propto$ Acid strength

   $\propto \dfrac{1}{\text{Basic strength}} \propto$ Dipole moment (exception $CH_3Cl > CH_3F$).
9. $+\text{IE} \propto \dfrac{1}{\text{Acid strength}} \propto$ Base strength.
10. No. of π bonds $\propto$ Resonance energy $\propto$ Stability of compound
11. Stability of contributing structure $\propto$ Maximum number of covalent bonds

i.e.,

Stability of contributing structure $\propto$
- $\dfrac{1}{\text{Formal charge maximum}}$ distance between like charges. minimum distance between unlike charges.
- −ve formal charge on more electonegative element. +ve formal charge on less electronegative element.

12. Structure for hyperconjugation

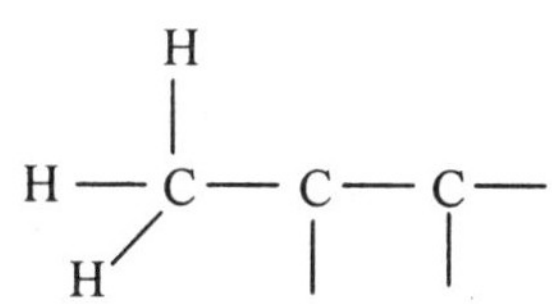

13. No. of hyperconjugation structures: Group

Group	No. of hyperconjugation structures
$—CH_3$	3
$—C_2H_5$	2

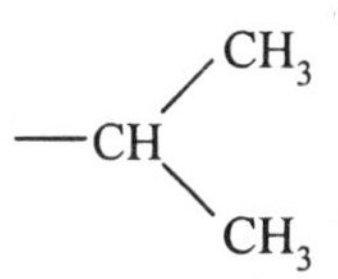

1

14. Stability of carbonium ion: (order) 3 > 2 > 1
tert sec pri

15. Stability $\propto$ No. of hyperconjugation structure

16. Stability of carbanion: 1 > 2 > 3
pri sec tert

$$\text{Stability} \propto \frac{1}{\text{No. of alkyl groups}}$$

1

17. Stability of free radicals $\propto$ 3 > 2 > 1
tert sec pri

18. $RMgX + R'OH \longrightarrow RH + Mg\begin{smallmatrix} X \\ OR \end{smallmatrix}$

19. Number of geometrical isomers (N):

(i) If both ends are different, $N = 2^{n-1} + p - 1$ where $p = n/2$, for even and p for odd number and n is the number of double bond.

(ii) If both ends are identical, $N = 2^n$ where n is the number of double bonds.

20. Number of optical isomer = 2^n

Where n is the number of asymmetric carbon atom.

21. $$\text{Boiling point} \propto \begin{cases} \text{No. of carbon atoms} \\ \dfrac{1}{\text{Branching of number of carbon atom is same}} \end{cases}$$

Glossary

Acetal (Chapter 10): A function group, having general formula $RCH(OR_1)_2$ derived by adding two molar equivalents of an alcohol to an aldehyde or ketone.

Achiral molecule (Chapter 7): A molecule that is superimposable on its mirror image. It has two or more identical atoms or groups.

Activating group (Chapter 11): A group that when presented on a benzene ring makes the ring more reactive in electrophilic substitution than benzene itself.

Acyl group (Chapter 10): The group with the structure RCO— or ArCO—.

Acyl halide (Chapter 10): A general name for compounds with structure RCOX or ArCOX.

Acylation (Chapter 10): The introduction of an acyl group into a molecule.

Addition polymer (Chapter 12): A polymer that results from addition polymerization, e.g., Polystyrene.

Addition reaction (Chapter 5): An organic reaction in which 1σ and 1π bond is converted into 2σ bond.

Aliphatic compound (Chapter 9): An open chain, non-aromatic compound like alkane, alkene, or alkyne, alcohols, etc.

Alkylation (Chapter 11): The introduction of an alkyl group into a molecule.

Allyl group (Chapter 8): $CH_2{=}CH{-}CH_2$.

Anomers (Chapter 13): A term used in carbohydrates. Monosaccharides that have some configuration on all carbon atoms except at carbon no.1, e.g., α & β D (+) glucose.

Aprotic solvent (Chapter 5): A solvent whose molecules do not have a hydrogen atom attached to a strongly electronegative element like oxygen., e.g.,

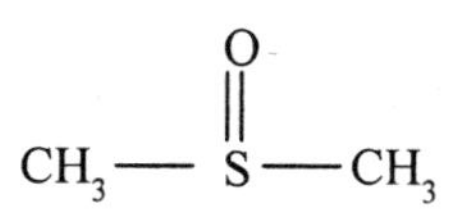

Dimethyl Sulfoxide

Arene (Chapter 11): A general name for an aromatic hydrocarbon.

Aromatic compound (Chapter 11): A conjugated, planar, cyclic polyene having delocalized π electrons equal to $(4n + 2)$ (where n is integer) and that resists oxidation and undergoes substitution rather than addition reactions, e.g., Benzene.

Aryl group (Chapter 11): A group obtained by the removal of a hydrogen from a ring position of an aromatic hydrocarbon.

Atomic orbital (Chapter 1): The volume of space about the nucleus of an atom where there is maximum probability of finding an electron.

Axial bond (Chapter 9): The bond present in the plane of the carbon atom.

Benzyl group (Chapter 11): $C_6H_5CH_2$—.

Boat conformation (Chapter 11): A conformation of cyclohexane that resembles a boat.

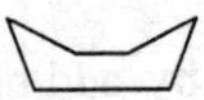

Bond angle (Chapter 1): The angle between two bonds originating at the same atom.

Bond length (Chapter 1): The equilibrium distance between two bonded atoms or groups.

Bromination (Chapter 11): An organic reaction in which bromine atoms are introduced into a molecule.

Carbanion (Chapter 4): A reaction intermediate in which carbon carries negative charge

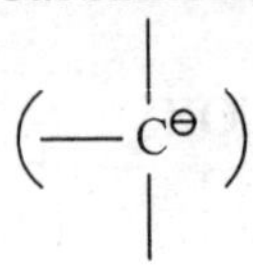

Carbonium ion (Chapter 4): A reaction intermediate in which carbon carries positive charge

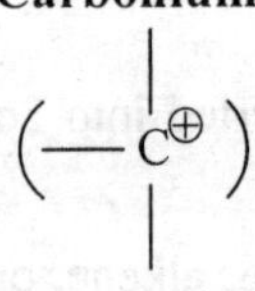

Carbohydrate (Chapter 13): Organic compounds which are chemically polyhydroxyl aldehyde or ketone or the substance which yield them upon hydrolysis, e.g., Glucose, Fructose or Sucrose.

Carbonyl group (Chapter 10): A functional group consisting of a carbon atom doubly bonded to an oxygen atom. It occurs in aldehydes, ketones, amides, carboxylic acides, ester, etc.

Chain reaction (Chapter 4): A reaction that proceeds by a sequential stepwise mechanism, in which each step generates the reactive intermediate that causes the next step to occur.

Chair conformation (Chapter 11): The conformation of cyclohexane. 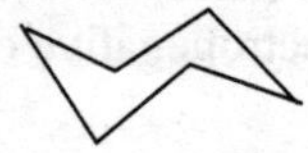

Chiral molecule (Chapter 7): A molecule containing chiral carbon atom, i.e., a carbon atom with four different groups.

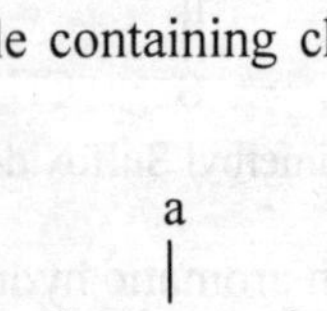

Condensation polymerization (Chapter 12): A polymerization in which something like water, ammonia may or may not be eliminated.

Configuration (Chapter 11): The particular arrangement of atoms or groups in space that is possible by making and breaking covalent bonds.

Conformer (Chapter 9): Various possible structures obtained due to rotation of C—C in space.

Covalent bond (Chapter 1): The type of bond that results when atoms share electrons.

Crown ether (Chapter 10): Cyclic polyethers that have the ability to form complexes with metal ions.

D & L designations (Chapter 13): A way of expressing configuration of organic compounds. The reference compounds is (+) or (–) glyceraldehydes.

Deactivating group (Chapter 11): A group that when present on a benzene ring makes the ring less reactive in electrophilic substitution than benzene itself.

Decarboxylation (Chapter 9): An organic reaction where a carboxylic group (—COOH) is removed in the form of CO_2.

Dehydration reaction (Chapter 5): An organic reaction in which water is lost from the substrate.

Dehydrohalogenation (Chapter 5): An elimination reaction in which one proton and one halide ion is lost and unsaturated molecule is obtained.

Delocalization (Chapter 11): The dispersal of electrons.

Dextrorotatory (Chapter 13): An optical isomer that rotates the plane of polarized light clockwise.

Diastereomers (Chapter 7): Pair of optical isomers that are nonsuperimposable upon each other.

Dipolar ion (Chapter 14): A species where atoms carry two charges, e.g., amino acids.

Dipole moment (Chapter 1): A physical property used to express polarity of a molecule.

Disaccharide (Chapter 13): A carbohydrate which upon hydrolysis yield two units of monosaccharide.

E-Z system (Chapter 8): A system for designating the steroid chemistry of tri- or tetra-substituted alkenes.

Elimination reaction (Chapter 5): An organic reaction where two groups or atoms are removed simultaneously, one is the form of an electrophile and other in the form of a nucleophile organic compound.

E-1 reaction (Chapter 5): An elimination reaction where the rate of reaction depends only upon the concentration of substrate. It involves carbonium ion as reaction intermediate.

E-2 reaction (Chapter 5): An elimination reaction where the rate of reaction depends upon the concentration of both substrate and reagent. A transition state is formed.

Electrophile (Chapter 2): A species that accepts electron pair.

Empirical formula (Chapter 15): A formula that expresses the relative proportions of atoms in a molecule as smaller whole number.

Enantiomers (Chapter 7): Optical isomers that are mirror image of each other, e.g., lactic acid.

Epoxide (Chapter 8): A three-membered ring containing one oxygen atom.

Equatorial bond (Chapter 9): A bond that lies in the plane of carbon.

Fatty acid (Chapter 10): A long-chained carboxylic acid obtained by the hydrolysis of a fat.

Fischer projection (Chapter 7): A two-dimensional formula used to represent configuration of a chiral molecule.

Fluorination (Chapter 11): A reaction in which fluorine atoms are introduced into a molecule.

Formal charge (Chapter 3): The difference of core charge and that of electron ownership is called formal charge. It is calculated by the formula.:

Formal charge = Core charge – electron ownership = No. of valence electrons –1/2 shared electrons + unshared electron.

Freon (Chapter 10): CF_2Cl_2.

Functional group (Chapter 10): A group of atoms in a molecule that is responsible for the properties of the molecule.

Furanose (Chapter 13): Five-membered cyclic acetal or hemiacetal.

Geminal substituents (Chapter 10): Substitutents present on the same carbon atom.

Grignard reagent (Chapter 6): An organometallic halide having general formula RMgX.

Hemiacetal (Chapter 10): A functional group, having general formula

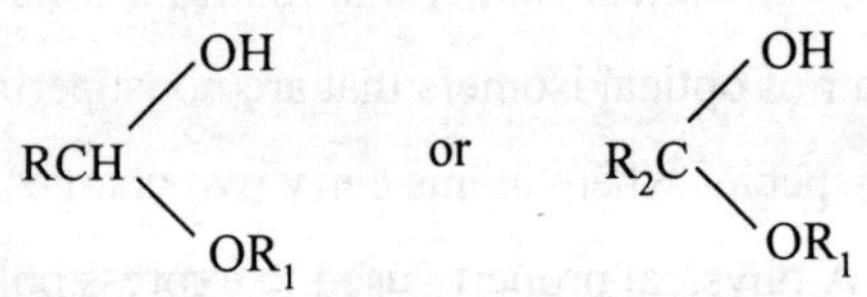

Hemiketal (Chapter 10): A functional group, having general formula

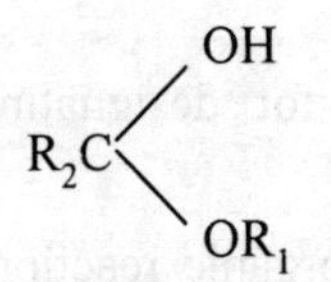

Heterocyclic compounds (Chapter 8): A cyclic compound which contains at least one atom like N, S, O others than C.

Heterolytic fission (Chapter 2): A method of fission of covalent bond in which one species is deprived of its own electron while other species gets both electrons. The products are cation and anion.

Hofmann rule (Chapter 5): According to this rule, in the elimination reactions of changed organic compounds, hydrogen is eliminated from that β carbon atom that carries maximum number of hydrogen atoms.

Homolytic fission (Chapter 2): A method of fission of covalent bond in which both species gets one electron each.

Huckel's Rule (Chapter 11): According to this rule, if a conjugated, planer, cyclic polyene contains $(4n + 2)$ delocalised π electrons it would be aromatic in nature.

Hybridization (Chapter 1): Phenomenon of mixing of atomic orbitals of nearly equal energy to form new orbitals of equal energy.

Hybrid orbitals (Chapter 1): Orbitals obtained after Hybridization. They have equal energy.

Hydroboration (Chapter 9): Reaction of alkene with diborane to produce alcohol.

Hydrogen bond (Chapter 3): A weak bond which results in association of molecule containing hydrogen attached to more electronegative element like O, N and F.

Hydrogenation (Chapter 9): Addition of hydrogen to an unsaturated molecule in the presence of a catalyst like Ni, Pt, Pd, Ru.

Inductive Effect (Chapter 3): Induction of polarity in an otherwise non-polar bond because of incomplete shifting of electron pair due to difference in electronegativity of two atoms.

Ion (Chapter 4): A chemical species that bears an electrical charge.

Isoelectric point (Chapter 14): The pH at which migration of amino acid stops.

Isomers (Chapter 7): Compound having same molecular formula but differ in properties.

Isoprene unit (Chapter 12): A structural unit found in all terpenes and natural rubber.

Kekule structure (Chapter 11): A structure in which lines are used to represent bonds in benzene.

Levorotatory (Chapter 7): An optically-active compound that rotates the plane of polarized light in anticlockwise direction.

Lewis dot formula (Chapter 3): A representation of a molecule showing electron pairs as a pair of dots.

Macromolecule (Chapter 12): A compound of high molecular weight formed from repetitive units.

Markovnikov's rule (Chapter 9): According to this rule, when an unsymmetrical molecule like Hx adds to an unsymmetrical alkene or alkyne, the negative part of addendum attaches to the carbon atom bearing least number of carbon atoms.

Meso compound (Chapter 7): An optical isomer containing plane of symmetry and is optically inactive due to internal compensation.

Molecular formula (Chapter 15): A formula that gives the total number of each kind of atom in a molecule.

Monomer (Chapter 12): Starting material of a polymer.

Monosaccharide (Chapter 13): The simplest carbohydrates that do not undergo hydrolysis.

Mutarotation (Chapter 13): The spontaneous change in optical rotation that occurs when α or β *anomers* of a monosaccharide or disaccharide is dissolved in amphoteric solvent like water.

Nucleic acids (Chapter 14): Natural polymer containing C, H, N, O and P.

Nucleophile (Chapter 14): Species, which donate electrons, e.g., NH_3.

Olefin (Chapter 9): Alkene.

Optically-active compound (Chapter 7): A compound that rotates the plane of polarized light.

Orbital (Chapter 1): The volume of space around nucleus of an atom where probability of finding an electron is maximum.

Organometallic compound (Chapter 6): A compound in which metal is directly attached to carbon atom of an alkyl group.

Oxymercuration (Chapter 9): Reaction of mercuric acetate with alkene to produce an alcohol.

Ozonolysis (Chapter 9): Reaction of O_3 with an unsaturated hydrocarbon to form an ozonide, followed by hydrolysis to yield carbonyl compounds.

Paraffin (Chapter 9): Alkane.

Pi (π) bond (Chapter 1): A bond formed by parallel overlapping of two p orbital.

Plane of symmetry (Chapter 7): An imaginary plane that divides a molecule in such way that both are related to each other as an object and mirror image.

Polarimeter (Chapter 7): An instrument used to measure optical activity of molecule.

Polymer (Chapter 12): A compound of high molecular weight containing repetitive units.

Polysaccharide (Chapter 13): A carbohydrate which upon hydrolysis yield 10 or more units of monosaccharides.

Primary carbon (Chapter 8): A carbon which is attached to only one carbon atom.

Primary structure (Chapter 14): The covalent structure of a polypeptide or protein.

Protein (Chapter 14): Natural polymer of α-amino acids linked by amide linkage.

Protic solvent (Chapter 5): A solvent where hydrogen is directly attached to more electronegative element like oxygen, e.g., alcohol, water.

Pyranose (Chapter 13): A sugar in which the cyclic acetal or hemiacetal ring is six membered.

(R-S) System (Chapter 8): A method for designing the configuration of tetrahedral stereocentres.

Racemic form (Chapter 7): Equimolar mixture of enantiomers. It is optically inactive due to external compensation.

Racemization (Chapter 7): Conversion of an optical-active compound into racemic mixture.

Reaction mechanism (Chapter 5): Path by which a reactant is converted into product.

Rearrangement (Chapter 5): A reaction that results in a product with a different carbon skeleton from the reactant.

Reducing sugar (Chapter 13): A sugar that reduce Tollen's or Benedict's reagents.

Resolution (Chapter 7): Separation of racemic mixture into dextro and laevo form.

Resonance (Chapter 3): An effect by which a substituent exerts either an withdrawing or electron releasing effect with π electrons in the molecule.

Resonance energy (Chapter 3): The energy content of resonance hybrid is always less than that of most stable contributing structure by an amount called resonance energy.

Secondary carbon (Chapter 8): A carbon atom that is attached to two carbon atoms.

Secondary structure (Chapter 14): The local conformation of a polypeptide backbone.

Sigma (σ) bond (Chapter 1): A bond formed by coaxial overlapping of atomic orbitals.

SN^1 reaction (Chapter 5): An organic reaction in which a group or an atom of an organic compound like tertiary alkyl halide is substituted by a nucleophile and the rate of reaction is independent of concentration of nucleophile.

SN^2 reaction (Chapter 5): An organic reaction in which a group or atom of an organic compound like primary or secondary alkyl halide is substituted by a nucleophile along with transfer or electron pair from nucleophile to the leaving group through reaction center. The rate of reaction depends upon concentration of both substrate as well as nucleophile.

Sp^3 hybrid orbital (Chapter 1): An orbital obtained by mixing of one 2*s* and three 2*p* orbitals the bond angle is 109.5°.

Sp^2 hybrid orbital (Chapter 1): An orbital obtained by mixing of one 2*s* and 2*p* orbital. The bond angle is 120°.

Sp hybrid orbital (Chapter 1): An orbital obtained by mixing of one 2*s* and one 2*p* orbital. The bond angle is 180°.

Steric hindrance (Chapter 3): An effect because of which the reactivity of an organic molecule is hampered.

Structural formula (Chapter 15): A formula that shows how the atoms of a molecule are attached to each other.

Substitution reaction (Chapter 5): A reaction in which one group replaces another in a molecule.

Tautomerism (Chapter 7): Two forms of a compound which are interconvertible into each other by migration of an atom from one part to other part of the molecule.

Tertiary carbon (Chapter 8): A carbon atom linked to three carbon atoms.

Vicinal (vic) substituents (Chapter 10): Substituents that are on adjacent atoms.

Vinyl group (Chapter 8): $CH_2{=}CH—$.

Witting reaction (Chapter 9): A reaction used to convert aldehydes or ketones into alkene.

Wilkinson's catalyst (Chapter 9): A catalyst used to convert alkynes into cis-alkene only. $[Ph_3P]_3$, i.e., tris [triphenyl phosphine] Chloro rhodium II.

Wurtz reaction (Chapter 9): A reaction in which two molecules of alkyl halide react with sodium to form alkane.

Ylide (Chapter 12): A dipolar species in which a carbanion is attached to a hetero atom like N, P, As, S, Sb, Bi carrying positive charge.

Zwitterion (Chapter 14): A dipolar species in which the terminal atoms carry different charge.

Subject Index